Organic Chemistry

Butterworths Intermediate Chemistry is a new series of three books—Physical Chemistry, Organic Chemistry and Inorganic Chemistry—giving a comprehensive coverage of all modern A-level syllabuses. The treatment is designed to consolidate the student's knowledge after each new topic has been dealt with, by means of concise summaries and pertinent examination questions at the end of each chapter. The authors avoid unnecessary confusion by adopting throughout the latest recommendations on nomenclature of the Association for Science Education. In addition to their particular suitability for students preparing for A-level, the rigorous, modern and concise treatment of the subject that the books provide makes them an ideal introduction for first-year university students and those reading chemistry as a subsidiary subject.

Butterworths Intermediate Chemistry

Organic Chemistry

C. B. Hunt, B.Sc., Ph.D., C.Chem., M.R.S.C.
Assistant chemistry master, Birkenhead School

A. K. Holliday, Ph.D., D.Sc., C.Chem., F.R.S.C.
Grant Professor of Inorganic Chemistry,
The University of Liverpool

Butterworths
London Boston Sydney Wellington Durban Toronto

First published 1981

© **Butterworth & Co (Publishers) Ltd 1981**

British Library Cataloguing in Publication Data

Hunt, C B
        Organic Chemistry. (Butterworths intermediate chemistry).
        1. Chemistry, Organic
        I. Title   II. Holliday, Arthur Kenneth
        547   QD251.2   80-41939

        ISBN 0-408-70915-4

Printed by Page Bros (Norwich) Ltd.

Preface

The aim of this book is to present, as clearly and as accurately as possible, the theoretical and descriptive organic chemistry required by the GCE Advanced Level Chemistry Syllabuses of the major Examining Boards. The book should also be of value to students in Further and Higher Education. We have not adopted a 'teach yourself' approach to this presentation but, rather, have assumed that the reader is having some formal teaching in the subject. Accordingly, we have tried to provide, in a conventional but 'tried and trusted' format, a clear explanation of essential concepts and a detailed description of those organic compounds which the reader is likely to encounter. We hope that students will thus be encouraged to concentrate on the exposition of topics in their lessons or lectures, in the knowledge that they may refer subsequently to the book in compiling notes. Experience suggests to us that this is the major function which students expect their textbook to serve.

Following an introductory chapter, the basic organic chemistry is divided into that of **aliphatic** (open-chain) compounds and **aromatic** (benzenoid, see p. 172) compounds (Chapters 2–11 and 12–14 respectively). This division may be unfashionable at the present time but we feel that it has advantages from the teaching point of view. The separate treatment of aromatic compounds can serve as a valuable exercise in the revision of functional group chemistry (particularly if it is presented in the second year of a traditional two-year sixth-form course).

In dealing with functional group chemistry, we have emphasized *reaction types* and essential reaction conditions, before introducing any mechanistic detail. We feel that it is important to stress that the unambiguous identification of reaction products and, where possible, the establishment of the stoichiometric equation, come before the (usually) more speculative mechanistic description. Whereas a mechanistic approach to organic reactions is a *sine qua non* of advanced work, it can prove a mixed blessing at an elementary level. The inevitable oversimplifications which are introduced rob the approach of much of its predictive power so that it becomes essentially a means of rationalizing molecular behaviour, of being 'wise after the event'. The more able student can certainly benefit from a mechanistic approach; the weaker student is apt to find it confusing and an additional burden to carry. Discussions of reactivities of organic compounds really require consideration of thermodynamic and kinetic stabilities (and of what they refer to) and of the nature of the transition state leading to the product, considerations which are obviously too complex to be presented in the early stages of study. Yet if explanations offered are too simple, the seeds of much later confusion can unwittingly be sown unless care is taken. Nevertheless, it is only right that students should be given some glimpse of the remarkable sophistication which the organic chemist brings to his investigations of organic reactions and most examination syllabuses require a knowledge of fundamental reaction mechanisms. We have been able to do no more than touch on the ways in which mechanisms can be investigated (for example, in Chapter 16) but students should be given some indication of the experimental work which underpins them.

In recent years, more emphasis in school chemistry has been placed on principles and concepts, sometimes perhaps to the detriment of the more descriptive parts of the subject. In organic chemistry, in particular, there is a danger that students may come to view organic reactions as essentially paper exercises involving the manipulation of curly arrows and fractional charges. To help counteract this, we have provided, where appropriate, physical data such as melting point and boiling point, in addition to structural formulae, in the hope that they may impress upon the reader the real, rather than abstract, nature of the compounds. However, we have omitted full practical details, or recipes, for the preparation of organic compounds. Most teachers like to plan their own courses of practical work and the inclusion of such details would have lengthened the book considerably. Methods of preparing the various classes of compounds are not dealt with in individual chapters but are gathered together in Chapter 15 and this should be useful for revision purposes.

Chapters 17–20 describe the applications of basic organic reactions and principles to such important fields as biological molecules, petrochemicals, fats and synthetic polymers. Here, the coverage is extended beyond the confines of the usual syllabuses, both to stimulate interest and to help those studying syllabuses with an 'applied' bias, such as the JMB *Chemistry Syllabus A* with its distinctive emphasis on chemistry-based industries and technologies.

Chapters dealing with the separation and purification of organic compounds and with structural determination have been deliberately placed last as we feel that it is only in the light of what has preceded them that they can be appreciated properly. The introduction of spectroscopic methods into school chemistry has provoked some controversy and disagreement; whether the introduction is desirable or not, spectroscopic methods are touched upon in some syllabuses (for example, London and Nuffield) and hence brief outlines of mass spectrometry and infra-red spectroscopy have been included.

Summaries are provided at the ends of Chapters 1–14 and we hope that these will be found useful by at least some readers during revision. Questions taken from GCE and other examination papers have also been provided and we strongly urge our readers to test their comprehension and assimilation of chapter material by tackling these. Effective revision depends very much on this sort of self-testing; repeated readings of the text on their own are not sufficient and excessive rote-learning should, of course, be discouraged. The questions are classified in a loose and subjective way. The answers to unstarred questions are generally to be found directly in the text and hence these are the easiest ones. Single-starred questions require the application and extension of principles and reactions already described to slightly less familiar situations and examples. Double-starred questions are the most difficult, requiring more perception and imagination.

We are grateful to Universities and Examining Boards for permission to reproduce questions set in examinations in recent years, viz.

Joint Matriculation Board (JMB), GCE

Oxford and Cambridge Schools Examination Board (Oxford & Cambridge), GCE

University of London University Entrance and School Examinations Council (London, Nuffield), GCE

University of Cambridge, Cambridge Colleges Examination (Cambridge)

Colleges of Oxford University, Joint Scholarship and Entrance Examination (Oxford)

Imperial College of Science and Technology, Royal and Entrance Scholarship Examination (Imperial College)

University of St Andrews, Entrance Scholarships and Bursaries Examination (St Andrews)

The methods of melting point and boiling point determination described in Chapter 21 are due to R. L. Parry Jones and A. M. Stumbles and first appeared in *School Science Review* **58**, 473 (1977). The Schöniger oxygen flask technique, described in Chapter 22, is due to W. Schöniger and is thoroughly described by H.-J. Schmidt in *School Science Review* **54**, 718 (1973).

Finally, our sincere thanks go to Dr R. J. S. Beer, Reader in Organic Chemistry, University of Liverpool for reading the manuscript and for his helpful comments and constructive criticism. Needless to say, what is in the text is the sole responsibility of the authors.

C.B.H.
A.K.H.

Contents

Nomenclature, Symbols and Units

In general, the naming of organic compounds used in the text follows the recommendations in the Second Edition (1979) of the Association for Science Education report *Chemical Nomenclature, Symbols and Terminology for Use in School Science*. The systematic nomenclature is designed to be unambiguous and clear and replaces many older, well-established but non-systematic, or trivial, names by more informative ones. For example, *chloroform* becomes trichloromethane and *cumene* (1-methylethyl)benzene. It must be admitted that systematic names can be cumbersome (particularly so in speech; it is much easier to order *Rochelle Salt*, or even *potassium sodium tartrate*, from the stores, than it is potassium sodium 2,3-dihydroxybutanedioate). Occasionally, some are rather unsatisfactory in that they derive from a parent compound which does not, indeed cannot, exist; for example, *acetophenone* is now named phenylethanone, although there can be no ketone ethanone (see p. 109). Nevertheless, on balance the advantages are deemed to outweigh any disadvantages.

However, one cannot ignore the fact that many simple, trivial names are in widespread use throughout the chemical and allied industries, for example *acetic acid* for ethanoic acid, and are likely to remain so (chemical road tankers are unlikely to carry the label benzene-1,2-dicarboxylic anhydride in place of the current *phthalic anhydride*, for example). Nor can one ignore the vast chemical literature in which many trivial names are still retained.

A balance has therefore to be struck between these usages and students are recommended to familiarize themselves with the dual names of important, widely-used organic chemicals. Since this text is for educational use, it is the systematic names which are emphasized with common trivial names italicized in parentheses.

The units employed in the text are generally SI or SI-derived units:

mass	gram g (SI unit kilogram kg)
volume	cm^3 or litre l (SI unit cubic metre m^3)
	1 litre is 10^{-3} m^3 or 1 dm^3
temperature	degree Celsius °C or, occasionally, kelvin (SI unit of thermodynamic temperature kelvin K)

wavelength	metre m
time	second s
energy	joule J (i.e. $kg\ m^2\ s^{-2}$)
force	newton N (i.e. $kg\ m\ s^{-2}$)
pressure	$N\ m^{-2}$, sometimes called the pascal Pa (i.e. $kg\ m^{-1}\ s^{-2}$). 1 atmosphere (atm) $= 101\ 325$ Pa $(\approx 10^5\ N\ m^{-2})$
amount of substance	mole mol

The following prefixes are used to indicate multiples and submultiples of SI base units and SI-derived units:

nano n 10^{-9}
micro μ 10^{-6}
milli m 10^{-3}
centi c 10^{-2}
deci d 10^{-1}
kilo k 10^3
mega M 10^6

Additionally, the following symbols are used in the text:

A_r relative atomic mass
M_r relative molecular mass
ΔH enthalpy change
$\Delta H^{\oplus}$ standard enthalpy change (temperature 298 K)
R ideal gas constant (i.e. $8.314\ J\ K^{-1}\ mol^{-1}$)

The abbreviations m.p. and b.p. stand for melting point and boiling point respectively.
The following information may be required in calculations:

Relative atomic masses: H = 1.0   C = 12.0   N = 14.0
O = 16.0   Cl = 35.5
The molar volume of an ideal gas at standard temperature and pressure (273 K and $101.3\ kN\ m^{-2}$) = 22.4 litres

Chapter 1
Structure and reactivity of organic compounds

1.1 Introduction

Historically, organic chemistry was associated with the isolation and study of chemical substances derived from living sources—hence the adjective 'organic'. At one time it was widely believed that such substances, which invariably contained the element carbon, were of a fundamentally different nature from inorganic substances which were obtained from inanimate matter like rocks and minerals. It was assumed that organic substances originated from some undefined 'vital force' in living matter and that they were not susceptible to synthesis by ordinary laboratory methods.

This belief was disproved by the German chemist Friedrich Wöhler who, in 1828, evaporated a solution of ammonium cyanate (essentially an inorganic substance) in his laboratory and produced urea, previously only obtained by isolation from urine. From that date, chemists gradually came to appreciate that the properties and reactions of organic compounds were subject to exactly the same chemical principles or laws as those of other compounds. This being so, organic chemistry may be defined, as Gmelin wrote in 1848, as **the chemistry of the compounds of carbon**. As such, it embraces both naturally occurring substances like proteins, carbohydrates, fats and 'essential oils', and synthetic ones like plastics, dyestuffs, pharmaceuticals and detergents. The number of known carbon compounds exceeds that of known compounds of all the other elements put together. Therefore, it is simply a matter of convenience that the distinction between organic and inorganic chemistry is preserved. In fact, these two areas of chemistry are linked together by the more recent, and very important, one of **organometallic chemistry**, in which compounds containing carbon bonded to metals are studied.

The existence of so many compounds of carbon is a result of several characteristics of the element. First, the carbon–carbon bond is particularly strong, having a bond energy of about $350 \, \text{kJ} \, \text{mol}^{-1}$.* This means that carbon atoms can link with other carbon atoms apparently indefinitely

---

* This may be taken as the amount of energy by which two carbon atoms, linked by a single covalent bond, are more stable than when isolated as separate atoms, the value being expressed per *mole* (mol) of bonds, i.e. $2\text{C(g)} \rightarrow \text{C—C(g)}; \Delta H = -350 \, \text{kJ}$.

to form molecular structures which may consist of chains, straight and branched, and closed rings of atoms, e.g.

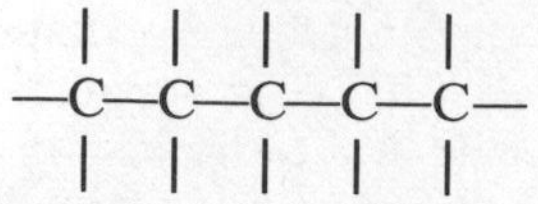

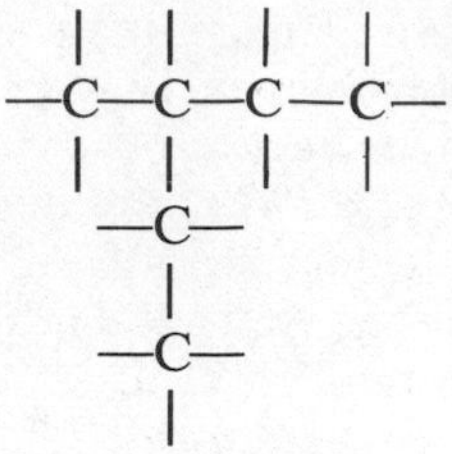

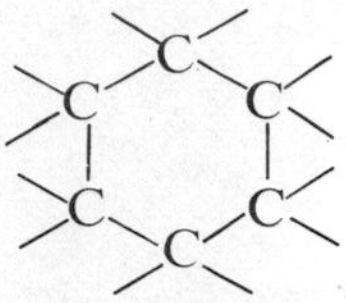

Other elements, such as nitrogen and oxygen, have much weaker bond energies (about 160 kJ mol^{-1} for N—N and 146 kJ mol^{-1} for O—O), so that such chain or ring structures would be too unstable and would simply break down into smaller fragments (in fact, no more than three N atoms and two O atoms are to be found bonded together in the majority of compounds).*

Secondly, the carbon atom can exhibit its covalency of four in many ways: (1) by forming **single** bonds with other carbon atoms (as in the structures shown above), or with hydrogen, oxygen, halogens or other elements; (2) by forming **multiple** (double or triple) bonds (e.g. $>C=C<$, $—C\equiv C—$, $—C\equiv N$, $>C=O$, etc.). These features are responsible for the impressive diversity of compounds based on carbon. All terrestrial life-forms are based on carbon compounds and their transformations. In this sense, carbon is uniquely important among the elements.

1.2 Formulae, structures and isomerism

Quantitative analysis of a pure sample of a compound enables the atomic ratio of the elements present to be found. For example, analytical figures of 92.4% carbon and 7.6% hydrogen for a compound indicate that in 100 g there are 92.4/12.0 mol of carbon atoms and 7.6/1.0 mol of hydrogen atoms combined together (taking the relative atomic masses of carbon and hydrogen to be 12.0 and 1.0, respectively). The ratio of moles, and therefore atoms, of carbon to hydrogen is thus 7.7 to 7.6, i.e. within experimental error the ratio is 1:1. The compound is said to have an **empirical formula** CH, since there are no other atoms present. However,

---

* Silicon, the element immediately below carbon in Group 4 of the Periodic Table, also has a covalency of four but the Si—Si bond energy of 226 kJ mol^{-1} is much weaker than that for carbon and the stronger Si—O bond dominates the chemistry of silicon. Nevertheless, **silanes** Si_nH_{2n+2} are known with n up to 10–12.

an empirical formula is insufficient to characterize a compound since there may be more than one compound with the same such formula; for example, both benzene and ethyne have the formula just derived. If the relative molecular mass of the compound can be determined, then a **molecular formula** may be arrived at, i.e. one which indicates the actual number of each type of atom present in a molecule of the compound. The relative molecular mass of benzene is found by experiment to be 78. The molecular formula of benzene must therefore be $(CH)_6$, written more conventionally as C_6H_6 (since $(6 \times 12) + (6 \times 1) = 78$). Similarly, the relative molecular mass of ethyne is found to be 26 and its molecular formula is therefore $(CH)_2$ or C_2H_2 (since $(2 \times 12) + (2 \times 1) = 26$).

A molecular formula does *not* show how the constituent atoms of a molecule are linked together, i.e. its structure. This is the purpose of a **structural formula**, as illustrated in *Table 1.1*.

Table 1.1

Name of compound	Empirical formula	Molecular formula	Structural formula
methane	CH_4	CH_4	 $\begin{array}{c} H \\ \vert \\ H-C-H \\ \vert \\ H \end{array}$
ethane	CH_3	C_2H_6	$\begin{array}{cc} H & H \\ \vert & \vert \\ H-C-C-H \\ \vert & \vert \\ H & H \end{array}$
propane	C_3H_8	C_3H_8	$\begin{array}{ccc} H & H & H \\ \vert & \vert & \vert \\ H-C-C-C-H \\ \vert & \vert & \vert \\ H & H & H \end{array}$

In a structural formula, a single line — represents a single covalent, or electron-pair, bond. It is not always necessary to write out such formulae in full, as characteristic arrangements of atoms soon become clearly recognizable and permit a more concise representation to be used. In methane, one hydrogen atom is linked to a CH_3 group, which is therefore called the **methyl group**. In ethane, a hydrogen atom is linked to a CH_3CH_2 group which is called the **ethyl group**, and so on. The general name for such groups is **alkyl group** since, when a hydrogen atom is linked to one, the resulting structure is that of an **alkane** (*see* Chapter 2).

Thus, an ethane molecule consists of two methyl groups bonded together and a more concise structural formula can be written as CH_3—CH_3 or just CH_3CH_3. Similarly, propane may be written as CH_3—CH_2—CH_3, or simply $CH_3CH_2CH_3$.

In a molecule of ethane, all six hydrogen atoms are identically situated in the molecule but the same is not true of a molecule of propane. Here, there are two types of hydrogen atom—those in the methyl groups and those bonded to the central carbon atom. If a methyl group replaces one of the hydrogen atoms in propane to produce C_4H_{10}, two different structures are now possible, depending on which type of hydrogen atom has been replaced:

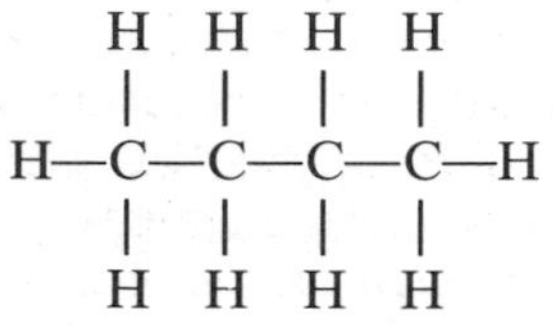

$CH_3CH_2CH_2CH_3$

b.p. $= -0.5°C$

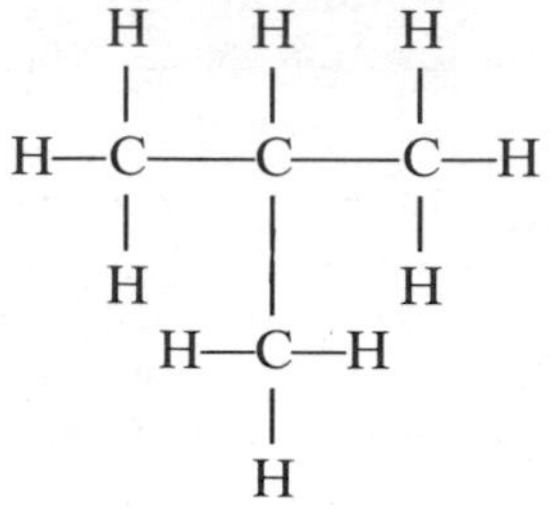

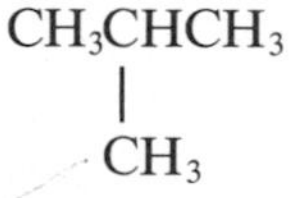

CH_3CHCH_3
$\quad\ \ |$
$\quad\ \ CH_3$

b.p. $= -12°C$

In the straight-chain structure above, the two carbon atoms at the ends of the chain are each bonded to *one* other carbon atom and are described as **primary** carbon atoms. The middle two carbon atoms are each bonded to *two* other carbon atoms and are described as **secondary** carbon atoms. In the branched-chain structure shown above, the central carbon atom is bonded to *three* other carbon atoms and is accordingly described as a **tertiary** carbon atom. These two structures show that there can exist two or more compounds having the same molecular formula but possessing different chemical and physical properties. Such compounds are called **isomers** and the general phenomenon is referred to as **isomerism**. When the difference in properties is due, as in the above example, to a difference in structure the type of isomerism involved is **structural isomerism**. As a molecular formula becomes more complex, so the number of possible structural isomers increases (although the structures may not all correspond to known compounds).

In drawing structural formulae, it is necessary to appreciate that two-dimensional structural formulae actually represent three-dimensional molecular structures. The four covalent bonds formed by a carbon atom are *directional* and experimental evidence shows that, when carbon forms single covalent bonds with four other atoms or groups, the bonds are disposed in a regular tetrahedral fashion around the central carbon atom:

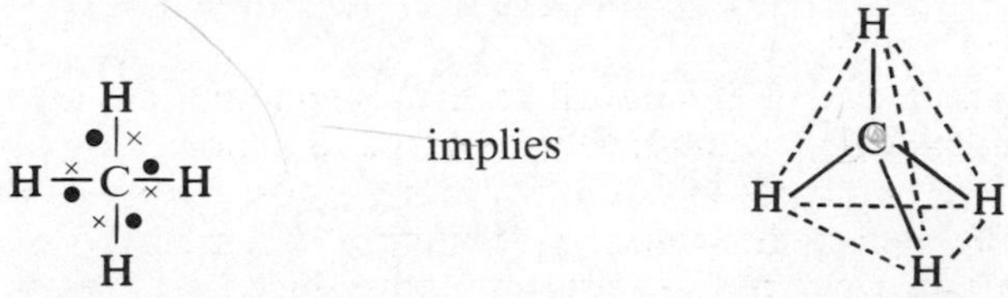

This is in accord with the Valence Shell Electron Pair Repulsion (VSEPR) theory which considers that four electron-pair bonds will be arranged about a central atom in such a manner that electrostatic repulsions between them will be minimized. In a regular tetrahedral arrangement, the outer electron-pairs are all to be found at equal distances from one another. The shape, or **stereochemistry**, of a molecule like methane is not, therefore, square-planar, as implied by the 'flat' structural formula, but tetrahedral. It is seldom necessary, at an elementary level, to depict the shapes of organic molecules (although the three-dimensional nature of these molecules should be constantly borne in mind*), but when it is necessary to do so, use must be made of a **stereochemical formula**. By convention, two of a carbon atom's four bonds, which may be considered to lie in the plane of the paper, are represented by single solid lines; the bonds which must project out of, and recede behind, the plane of the paper are represented by a heavy solid line and a broken, or dotted, line respectively. Thus

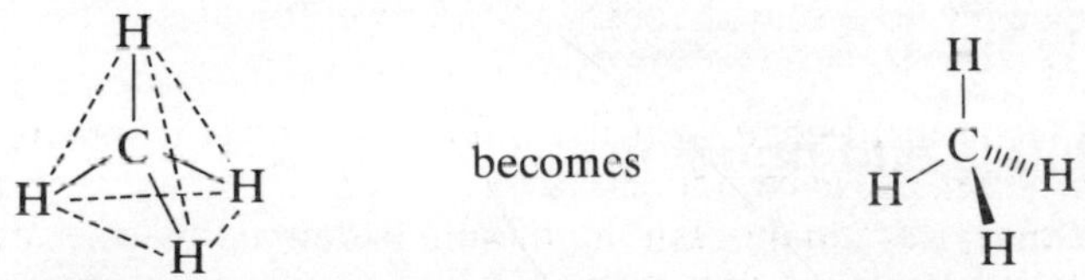

Familiarity with these shapes and formulae will soon lead to the recognition that the following structural formulae are equivalent:

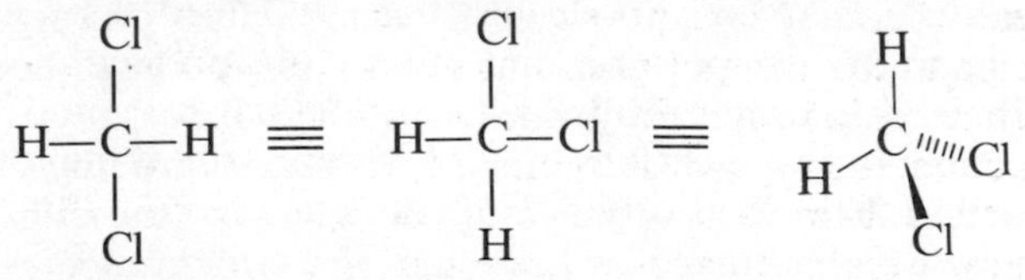

* Familiarity with molecular shapes is best gained with the aid of molecular models, either of the 'ball-and-stick' variety which emphasizes molecular geometry or else of the 'space-filling' type which emphasizes the general molecular shape and volume.

and

$$H-\underset{\underset{H}{|}}{\overset{\overset{H}{|}}{C}}-\underset{\underset{H}{|}}{\overset{\overset{H}{|}}{C}}-\underset{\underset{H}{|}}{\overset{\overset{H}{|}}{C}}-H \equiv H-\underset{\underset{H}{|}}{\overset{\overset{H}{|}}{C}}-\underset{\underset{\underset{H}{\underset{|}{C}-H}}{|}}{\overset{\overset{H}{|}}{C}}-H$$

It can be seen from the last pair that the description *straight* chain means unbranched, since the actual shape of such a carbon chain involves a 'zig-zag', rather than geometrically straight, sequence of carbon atoms.

The spatial arrangements of atoms in comparatively simple organic molecules are well established from experimental evidence, supplied by various (X-ray, electron and neutron) diffraction studies. The problem of determining the shapes of much larger, complex molecules like the proteins is more formidable. In 1964, Professor Dorothy Hodgkin became the third woman to win the Nobel Prize for Chemistry, in recognition of her determination of the stereochemistry of Vitamin B_{12} ($C_{63}H_{90}O_{14}N_{14}PCo$, *see also* Section 15.12 Organic synthesis). In 1969, she succeeded in unravelling the complete stereochemical structure of the hormone insulin $C_{254}H_{377}N_{65}O_{75}S_6$, after many years of intensive work. Molecular shapes can be of the utmost significance in contributing to certain properties. For example, the remarkable catalytic properties of the enzymes arise from specific three-dimensional arrangements of atoms, and the pharmacological effects of some drugs arise from the interaction of specific molecular shapes with the receptor sites in the nervous system.

1.3   Orbitals and bonding

The energy of an electron in an atom is defined by a set of four **quantum numbers**. Restrictions on their values limit the number of electrons which may possess a given energy in an atom. The situation is perhaps more readily comprehended by saying that electrons may only occupy certain **energy levels**, and only in certain numbers. Since electrons possess both wave and particulate properties, the older view of an electron as simply a negatively charged particle has been superseded by a more abstract view, in which the position and energy of an electron in an atom are described by an **atomic orbital**. Each orbital can accommodate, or contain, a maximum of two electrons only. A vacant orbital therefore contains no electrons, a half-filled orbital contains one electron, and a filled orbital contains two electrons.

When the *energy* of an electron in an atom is known precisely, there is necessarily some uncertainty as to its *precise position* (the Heisenberg Uncertainty Principle) and it is only possible to refer to the **probability** of finding an electron at any point. The probability distribution of an orbital can be calculated and it is 'visualized' by depicting it as a charged

electron cloud, whose density at any point is directly proportional to the probability of finding an electron there. In the case of the carbon atom, with its six electrons, only orbitals of the s and p type are involved. Electron clouds associated with s orbitals are spherically symmetrical, while those associated with p orbitals are 'dumb-bell' shaped, and are disposed along a set of orthogonal axes, arbitrarily designated x, y and z:

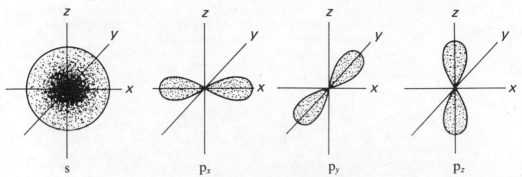

According to bonding theory, when a covalent bond is formed, two atomic orbitals overlap to form a **molecular orbital** and the strength of the bond is dependent on the degree of overlap achieved.

A single, discrete carbon atom in its lowest electronic energy state— the **ground state**—has an electronic configuration of $1s^2 2s^2 2p^2$. There are three *degenerate* (i.e. of equal energy) 2p orbitals and, according to Hund's rule, the electrons are distributed in such orbitals in such a way that there is the maximum number of unpaired electrons. Thus, the configuration may be written more precisely as $1s^2 2s^2 2p_x^1 2p_y^1$ (the choice of x and y rather than y and z, etc., is quite arbitrary):

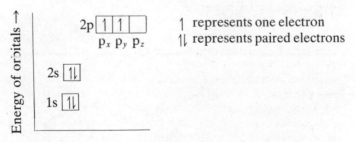

From such considerations it might be expected that carbon would therefore display a covalency of two, the two unpaired electrons being shared with one electron from each of two other atoms. Nevertheless, the quadrivalency (i.e. four) of carbon has long been a well-established experimental fact. However, when a carbon atom bonds to four other atoms, as, for example, in methane CH_4, it clearly interacts electronically with these atoms. The carbon atom is then no longer in its ground state, but is perturbed electronically into a higher energy **excited state**. It may, therefore, be reasonable to consider the promotion of a 2s electron into the vacant $2p_z$ orbital, so producing *four* unpaired electrons. The energy required to effect this promotion may be more than compensated for by the energy released when four covalent bonds are formed. Since three of the unpaired electrons are now in 2p orbitals and one in a 2s orbital, one of the bonds should be different (in strength and in spatial position) from the other three; yet all the pertinent physical and chemical evidence shows that the four covalent bonds in methane are identical in every respect and, moreover, are arranged in a tetrahedral fashion. If the three 2p orbitals each overlapped with the 1s orbitals of three hydrogen atoms, these bonds would be expected to be at right-angles.

In order to reconcile the experimental evidence with bonding theory, the concept of orbital **hybridization** has been introduced. Some, or all, of the 2s and 2p orbitals of carbon may become blended together, or **hybridized**, to form new **hybrid** orbitals, each of which is associated with a 'pear-shaped' electron cloud, well suited to overlap with orbitals of other atoms. When carbon forms single bonds with four other atoms or groups, the single 2s orbital and the three 2p orbitals are considered to become blended together to form *four* **sp³** hybrid orbitals (so called because they are formed from one s and three p orbitals):

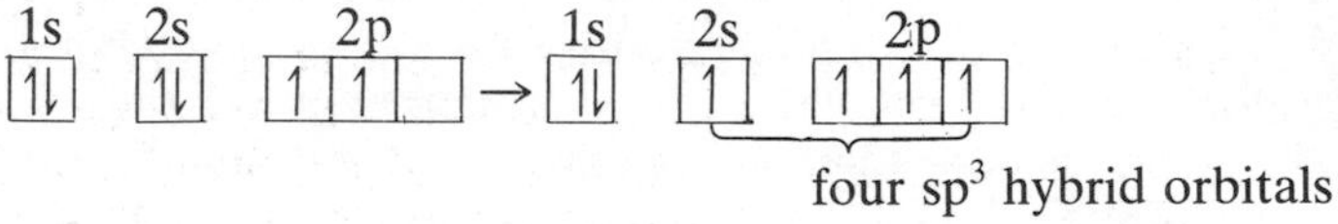

four sp³ hybrid orbitals

These hybrid orbitals have a greater degree of p character than s character, as is only to be expected, and are tetrahedrally disposed as four pear-shaped electron clouds. Thus, in methane CH_4 the four valency electrons of carbon occupy, singly, four sp³ hybrid orbitals, which each overlap with the spherically symmetrical 1s orbital of hydrogen to form four tetrahedrally arranged covalent bonds of identical strengths, with a bond angle for HCH of 109° 28':

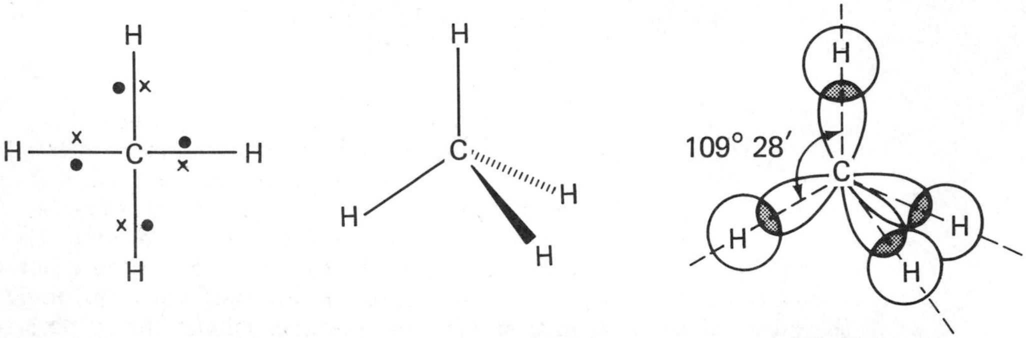

Such bonds, formed by the simple, *linear* overlap of atomic orbitals, are called **sigma (σ) bonds**. It should be emphasized, at this point, that the concept of hybridization provides a simple description, rather than explanation, of the bonding which is in agreement with experimental facts. Whether carbon is *really* hybridized or not is immaterial. It may seem that, since the simple electron-pair view of covalent bonds, along

with VSEPR theory, satisfactorily predicts the known facts, the introduction of hybridization has added little to our understanding of the
bonding and is indeed unnecessary.

The usefulness of the hybridization concept becomes apparent only
when multiple bonds between carbon and either carbon, nitrogen or
oxygen are considered. In the molecule of ethene C_2H_4, the six atoms are
coplanar (i.e. all lie in the same plane) and the carbon atoms are linked
by a carbon–carbon **double bond** C=C. An alternative mode of hybridization has to be devised. If one 2s orbital and *two* 2p orbitals are
hybridized, then *three* **sp²** hybrid orbitals will be formed:

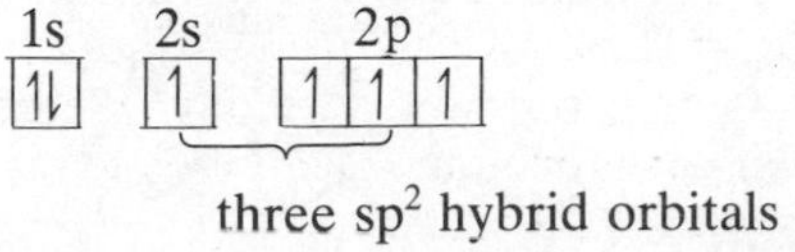

1s   2s   2p

three sp² hybrid orbitals

These are arranged as three pear-shaped electron clouds in a **trigonal
planar**, rather than tetrahedral, fashion with a HCH bond angle of 120°.
On each carbon atom, two of the three sp² hybrid orbitals overlap with
the 1s orbitals of two hydrogen atoms to form C—H sigma bonds. The
third hybrid orbital overlaps with a similar orbital of the other carbon
atom to form a C—C sigma bond. Each of the carbon atoms still possesses
a half-filled p orbital, perpendicular to the plane of the molecule; two
such orbitals overlap 'sideways-on' to form a C—C pi (π) **bond**, whose
molecular orbital is associated with two electron clouds, lying between
the carbon atoms and above and below the plane of the molecule:

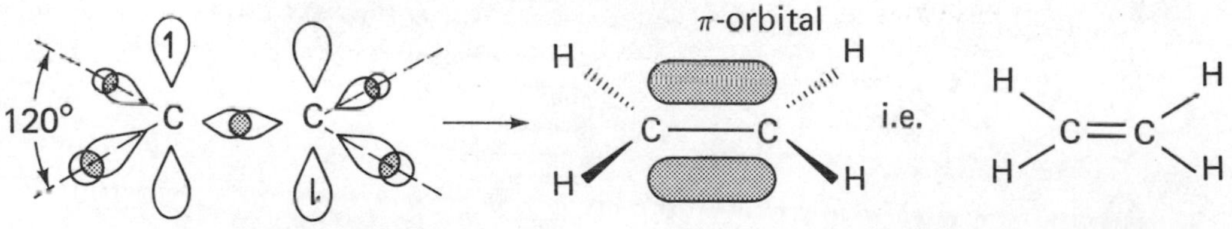

Thus, the two bonds linking the carbon atoms together are not identical
according to this treatment, one being a σ-bond and the other a π-bond.
This is consistent with the chemical behaviour of ethene (*see* Section 3.3),
and with thermochemical calculations. The latter indicate that one bond
(the σ-bond) has a bond energy of 346 kJ mol⁻¹, while the other (the
π-bond) has a value of 252 kJ mol⁻¹. The carbon–carbon double bond
C=C is thus *less* than twice as strong as a single one. Not surprisingly,
too, the distance between the carbon atoms in ethene (0.134 nm) is
shorter than in ethane C_2H_6 (0.154 nm). A simpler treatment of the
bonding in ethene may fit in with the number of bonds formed, and the
shape of the molecule, but would hardly anticipate the unequal nature
of the two carbon–carbon bonds.

In a molecule like ethane C_2H_6, the two carbon atoms may be rotated
about an axis joining them without destroying, or indeed affecting, the

overlap of the orbitals involved in the σ-bond. The C—C bond is said to permit relatively free rotation of the groups about it, and at most temperatures sufficient thermal energy is available to lead to this. In ethene, however, the double bond between the carbon atoms prevents such rotation at normal temperatures, since the overlap of the p_z orbitals which produces the π-orbital would be destroyed and the π-bond therefore broken. If sufficient energy *is* supplied, then this bond *will* eventually break and rotation may take place. The lack of free rotation under ordinary conditions can have important consequences (*see* Section 3.1):

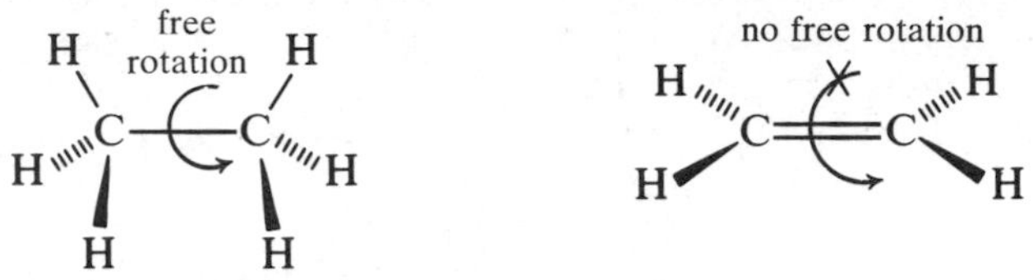

When carbon forms double bonds with oxygen, or nitrogen, sp^2 hybridization of both carbon and oxygen or nitrogen is envisaged and once again both a σ-bond and a weaker π-bond are formed:

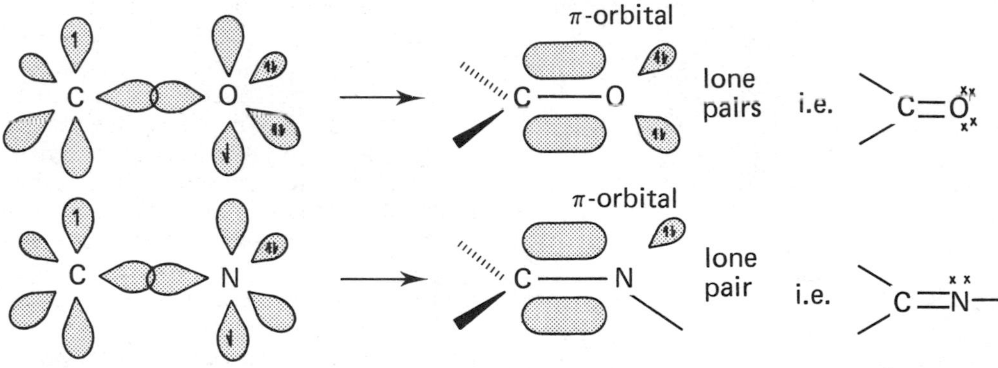

In a molecule containing a triple bond, like ethyne C_2H_2, a third type of hybridization must be invoked. Each carbon forms *two* **sp** hybrid orbitals from the combination of *one* 2s orbital and *one* 2p orbital:

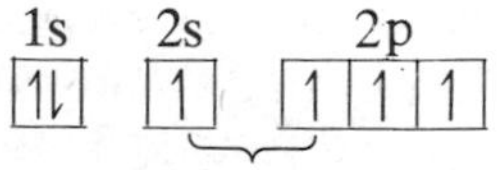

two sp hybrid orbitals

These hybrid orbitals are colinear (i.e. lie in a straight line), and each overlaps with a 1s orbital of hydrogen to form a C—H σ-bond, and with a similar hybrid orbital of the other carbon atom to form a C—C σ-bond.

There remain on each carbon atom two half-filled p orbitals (p_y and p_z) and these overlap, again sideways-on, to form *two* π-bonds:

i.e. $H-C \equiv C-H$

The electron clouds of these π-orbitals effectively envelop the C—C σ-bond in a cylinder of negative charge. The bond energies of the C—C bonds are 346 (σ), 252 (π) and 215 (π) kJ mol^{-1}, and thus the C≡C triple bond is less than three times as strong as a C—C single bond. The distance between the carbon atoms is 0.121 nm.

1.4 Dipoles and intermolecular forces

Elementary bonding theory emphasizes that when two atoms bond together, there normally result particularly stable electronic arrangements, characteristic of the noble-gas elements, namely sets of completely filled energy levels. These are achieved, either by the complete transfer of one or more electrons from one atom to the other, resulting in the formation of **ions** and **ionic** bonding, or else by the sharing of electrons to form electron-pair or **covalent** bonds. Two points need emphasizing, if misunderstandings are not to arise from this oversimplified view. First, although noble-gas electronic arrangements are *frequently* attained on chemical combination of atoms, they are not always present even in stable compounds. For example, neither boron in boron trifluoride BF_3 (covalent), nor copper in copper sulphate $CuSO_4$ (ionic), possesses completely filled energy levels characteristic of a noble-gas element.* Secondly, pure ionic and pure covalent bonds represent two extremes of electronic interaction; there is no clear-cut division between them. In practice, chemical bonds may possess some characteristics of both kinds.

In a covalent bond between two atoms A and B, the pair of bonding electrons will only be shared absolutely equally between them when A and B are identical as, for example, in a hydrogen H—H or chlorine Cl—Cl molecule. In general, one of the atoms will be more electronegative (i.e. will have a greater tendency to acquire one or more electrons) than the other, and will therefore have a greater share of the bonding pair of electrons than the other. This permanent displacement of electrons in a σ-bond is called the **inductive effect**, and it may have

* It is sometimes overlooked that the formation of a cation by the loss of one or more electrons from an atom must always be an endothermic, or energy-absorbing, process and this energy may exceed that released on formation of the anion(s). The formation of noble-gas electron arrangements is not always the driving force behind the reaction; it may be the formation of a stable crystal lattice (cf. lattice energy).

very important consequences. The bond is said to be **polar**. The displacement of the electron-pair is equivalent to the separation of fractional positive and negative charges, analogous to the poles of a magnet, and is described as an electrical **dipole**. Such a dipole exists in a molecule of hydrogen chloride, for example, and is depicted as

$$H \rightarrow\!\!-Cl \quad \text{or} \quad \overset{\delta+}{H}\!\!-\!\!\overset{\delta-}{Cl} \quad \leftrightarrow \text{direction of dipole}$$

The product of the fractional charges on each atom and the distance between them is termed the **dipole moment** and can be measured experimentally. Its magnitude increases with the size of the charges and with the distance between them. While the dipole moment may indicate the polarity of a bond, a molecule may also possess a dipole moment, not on account of any unequal sharing of electron-pairs in covalent bonds, but because of the presence of one or more lone-pairs of electrons. The ammonia molecule, for example, possesses a dipole moment on account of its lone-pair of electrons, rather than on any inductive effect in the N—H bonds. A dipole moment is a vector quantity and therefore depends on molecular geometry. Trichloromethane contains polar bonds and has a net dipole moment (the molecule as a whole can be described as polar); tetrachloromethane also possesses polar bonds but has no dipole moment, as the resultant of the four individual bond moments is zero, and is therefore described as non-polar:

Intermolecular forces are forces which operate between molecules in a liquid or in a molecular crystal, and they influence such physical properties as melting and boiling points, density, solubility and vapour pressure. They result from the presence of permanent or temporary dipoles in the molecules. With polar molecules, the negative end of one dipole may, if correctly oriented, attract the positive end of another and such **dipole–dipole** attractive forces will constitute the main intermolecular forces. Thus, non-polar methane CH_4 boils at $-162°C$, while polar chloromethane CH_3Cl boils at $-24°C$:

Even non-polar molecules must participate in some intermolecular bonding in the liquid or solid state, and this is considered to arise from the existence of **induced dipole** interactions. Slight, periodic displacements of electrons in atoms or molecules can set up temporary dipoles and these can *induce* temporary dipoles in other neighbouring atoms or molecules. Attractive forces can then operate (as with polar molecules) as a result of the **polarization of molecules**. The extent to which this polarization can occur depends very much on molecular size, large molecules being polarized more easily than small ones. Polar molecules, too, of course, are capable of inducing dipoles in neighbouring non-polar molecules. These intermolecular forces, based on dipole interactions, are known collectively as **van der Waals forces** and are very much weaker than covalent bonds, namely $\approx$ 1–2 kJ mol^{-1}.

Another type of bond, the **hydrogen bond**, is intermediate in strength between van der Waals forces and covalent bonds, namely of the order 10–30 kJ mol^{-1}. It arises through the attraction between a fractionally charged hydrogen atom, when it is bonded to a very electronegative element such as fluorine, oxygen or nitrogen in one molecule, and the fractional negative charge which arises in another molecule from the presence of a lone-pair of electrons. It is really a dipole–dipole interaction, made much stronger by the proximity of the very small hydrogen atom, and is *directional*. Thus, compounds which contain hydrogen bonded to such electronegative elements possess stronger intermolecular forces than other compounds of similar size, and hence have higher boiling points.*
For example, the isomers of ethanol and methoxymethane have the same molecular formula C_2H_6O and similarly sized molecules. Ethanol CH_3CH_2OH has a boiling point of 78°C, while methoxymethane CH_3OCH_3 has one of −24°C, a large difference. In methoxymethane, there are no oxygen-attached hydrogen atoms and therefore no hydrogen bonding; only van der Waals forces operate between the molecules:

ethanol   methoxymethane

1.5   Resonance or mesomerism

The permanent displacement of electrons in a σ-bond has been referred to in the previous section and is known as the inductive effect. A similar sort of effect can also operate with π-bonds,

---

* It is, of course, the *intermolecular* forces which have to be overcome in boiling a liquid. *Covalent bonds* would only be broken if the compound decomposed at its boiling point (if the latter is high, this can happen).

In the *carbonyl* group $\diagdown$C=O, which contains both a σ- and a π-bond, it is the π-electrons in particular which are shared unequally between the two atoms and which are drawn towards the more electronegative oxygen atom, because they are more easily polarizable than the σ-electrons. Owing to the limitations of the simple, conventional notation of covalent bonds and the difficulty of depicting electron distributions in structural formulae, the electronic state of the carbonyl group cannot be represented precisely by a single formula. Neither formula (I) below, nor the more extreme representation (II) (which shows the complete transfer of π-electrons from the carbon atom to the oxygen atom), adequately describes the electronic state of the group:

$$\diagdown C = O \qquad\qquad \overset{+}{\diagdown C} - \overset{-}{O}$$

(I)   (II)

The actual electron distribution may be envisaged as lying somewhere in between these two extreme forms. The limiting forms are described as **canonical forms** and the true form is described as a **resonance hybrid**. It should be noted that all the atoms are identically situated in canonical forms—only the electron distribution differs. The phenomenon of **resonance** is also sometimes called **mesomerism** (from the Greek for 'between the parts'). The symbol $\leftrightarrow$ (not to be confused with the equilibrium sign $\rightleftharpoons$) is often used to indicate that a particular structure is represented as a resonance hybrid of the given canonical forms, e.g.

$$\diagdown C = O \longleftrightarrow \overset{+}{\diagdown C} - \overset{-}{O}$$

Other structures commonly depicted as resonance hybrids are the *carboxyl* group —COOH (*see* Section 9.3) and the benzene ring C_6H_6, e.g.

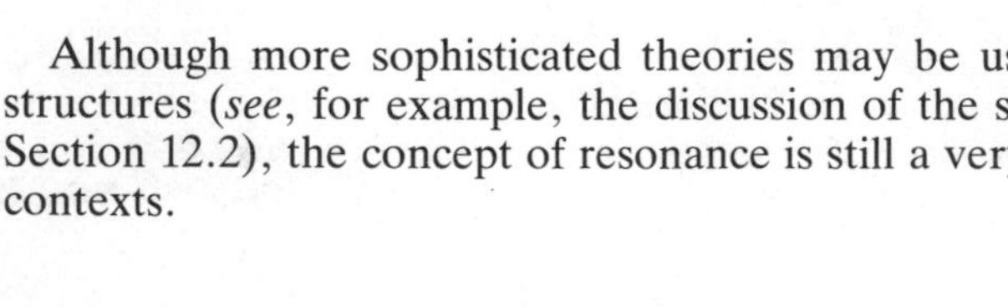

Although more sophisticated theories may be used to describe such structures (*see*, for example, the discussion of the structure of benzene, Section 12.2), the concept of resonance is still a very useful one in many contexts.

1.6   Reactions and mechanisms

Two points concerning organic reactions, as distinct from many inorganic reactions, soon become apparent, once some familiarity with them has been gained. The reactions do not always go to completion, either because they are very slow, or else because a state of dynamic equilibrium is established. Furthermore, many organic compounds are susceptible to attack in more than one way, or by more than one reagent which is present (e.g. the solvent). The actual conditions under which a reaction takes place, such as solvent, temperature, etc., may critically affect the outcome of the reaction in terms of the product(s) formed. For these reasons, organic reactions are frequently accompanied by side-reactions (i.e. those occurring to a lesser degree than one other) and yields of product(s) are commonly much lower than those obtained in inorganic reactions. Therefore, it is not always possible to write a complete stoichiometric equation for an organic reaction; its quantitative significance could be very dubious. Consequently, reactions may sometimes have to be represented as simple transformations without the balancing of atoms on either side of the $\rightarrow$ sign.

As a generalization, *four* major *types* of organic reactions are commonly encountered, namely (i) **substitution**, (ii) **addition**, (iii) **elimination**, and (iv) **rearrangement**. The meanings of these terms will be clarified as they are met with in subsequent chapters.

In principle, when a covalent bond between two atoms or groups A and B breaks, the pair of electrons constituting the bond may be distributed between A and B in any of three ways:

$$
\text{A}^{\times}_{\bullet}\text{B} \quad
\begin{array}{l}
\text{(i)} \\[1em]
\xrightarrow{\text{(ii)}} \\[1em]
\text{(iii)}
\end{array}
\quad
\begin{array}{l}
\text{A}^{\times} + {}_{\bullet}\text{B} \\[1em]
\overset{+}{\text{A}} + {}^{\times}_{\bullet}\bar{\text{B}} \\[1em]
\bar{\text{A}}{}^{\times}_{\bullet} + \overset{+}{\text{B}}
\end{array}
$$

In process (i), each atom or group retains *one* of the electrons from the bond and two neutral atoms or groups are formed (assuming A—B is itself a neutral molecule). This type of fission (breaking) of the bond is known as **homolytic fission** (equal splitting) or **homolysis**. The neutral fragments produced each possess an unpaired electron and are called **free-radicals**.* Generally speaking, a non-polar bond tends to undergo homolytic fission. The process is favoured in the vapour state or in a non-polar solvent and is usually brought about by the action of heat, light or other radicals. The free-radicals so produced are usually highly reactive and rapidly react with other chemical species, or else recombine.

---

* Simple atoms like H or Cl can thus be described as radicals; on the other hand, *stable* molecules like NO_2, although they too contain an unpaired electron, are not usually so described even though they show some properties characteristic of free-radicals (e.g. dimerization $2NO_2 \rightarrow N_2O_4$).

Processes (ii) and (iii) are essentially similar to each other. *Both* the bonding electrons are retained by *one* of the atoms or groups only; which one, A or B, depends on their relative electronegativities. This process is termed **heterolytic fission** (unequal splitting) or **heterolysis**. **Ions** are produced and the process is usually favoured in polar solvents, which can solvate, and hence stabilize, the ions produced. Clearly, polar bonds tend to undergo this type of fission. When a positive charge is located on a carbon atom in a group, the ion is called a **carbocation** (or carbonium ion); when negatively charged, such an ion is called a **carbanion.**

A carbocation would lack two electrons and would thus tend to behave as an **electron-pair acceptor**—such species are referred to by the general term **electrophiles**. A carbanion, on the other hand, would possess an unshared electron-pair and could thus be regarded as an **electron-pair donor**—such species are called **nucleophiles**. Electrophiles in general may possess a formal charge or they may be neutral; e.g. BF_3 or Br^+. This holds, too, for nucleophiles; e.g. NH_3 or I^-.

The precise pathway by which a reaction is thought to take place is referred to as the **mechanism** of the reaction. In theory, a mechanism should provide a complete stereochemical and energetic description of all molecular events occurring during the reaction, but such a hope cannot be realized in practice. Nevertheless, simpler descriptions of the type of species which initiate reaction, the number of elementary steps involved in the total reaction, and any intermediates which may be formed, are commonly referred to as 'mechanisms' and will be discussed under the appropriate headings in subsequent chapters. Evidence for them comes from the application of isotopic, kinetic and stereochemical techniques to a study of the individual reactions. A knowledge of organic reaction mechanisms helps in understanding the distribution of products of a reaction, and the nature of competing side-reactions, and helps, too, in unifying the otherwise very diverse range of known organic reactions.

1.7 Acids and bases

According to the Brønsted–Lowry theory (1923) of acids and bases, an acid may be regarded as a proton donor and a base as a proton acceptor. Hydrogen chloride does not show acidic properties until it is dissolved in a suitable solvent, usually water. Ionization occurs and a proton (namely H^+) is transferred to a water molecule:

$$HCl(g) + H_2O(l) \rightarrow H_3O^+(aq) + Cl^-(aq)$$

In this process, the hydrogen chloride acts as an acid and the water as a base. The resulting solution is acidic since the H_3O^+, or **hydroxonium** ion, is itself a proton donor. Where dissociation into ions in solution is complete, i.e. 100%, the acid is described as a **strong** acid. Organic carboxylic acids are described as **weak** acids as there is only partial dissociation into ions, e.g.

$$CH_3COOH(aq) + H_2O(l) \rightleftharpoons CH_3COO^-(aq) + H_3O^+(aq)$$

The equilibrium constant K_a for this reaction is

$$K_a = \frac{[CH_3COO^-][H_3O^+]}{[CH_3COOH]}$$ (the concentration of water is effectively constant)

where the square brackets signify molar concentrations at equilibrium. The acidity of a weak acid is also often referred to by its pK_a value, where $pK_a = -\log_{10} K_a$. Since the above reaction is reversible, ethanoate ions can behave as a base by accepting a proton from the hydroxonium ion; ethanoate ion is therefore sometimes referred to as the **conjugate base** of the acid, ethanoic (*acetic*) acid.

Organic bases may be treated in a similar fashion. Methylamine CH_3NH_2 dissolves in water to produce an alkaline solution:

$$CH_3NH_2(aq) + H_2O(l) \rightleftharpoons CH_3NH_3^+(aq) + OH^-(aq)$$

This is a **weak** base because there is only partial ionization. The equilibrium constant K_b is defined as

$$K_b = \frac{[CH_3NH_3^+][OH^-]}{[CH_3NH_2]}$$ (the concentration of water is effectively constant)

Since the $CH_3NH_3^+$ ion can donate a proton to OH^- it can be regarded as the **conjugate acid** of methylamine.

The Brønsted–Lowry theory, however, restricts the discussion of acid-base behaviour to that in protic solvents (i.e. where the solvent molecule can donate or accept a proton). The definitions of acids and bases were broadened (1938) by the American chemist G. N. Lewis, who defined an acid as an electron-pair acceptor, and a base as an electron-pair donor. These terms can therefore be applied to reactions in non-protic solvents where no proton transfer is involved, as in the following example:

$$F_3B + :NH_3 \longrightarrow F_3B \leftarrow NH_3$$

The descriptions **Lewis acid** and **Lewis base** are conventionally used in such cases.

By comparison with the previous section, it can be seen that the difference between a Lewis base and a nucleophile is a difference of function and not of electronic structure. A Lewis base donates an electron-pair to hydrogen when reacting with a molecule, the hydrogen being removed as a hydrogen ion, whereas a nucleophile donates an electron-pair to carbon or some element other than hydrogen. An example of a species acting as either, depending on the reaction conditions, is discussed in Section 5.3.3. There is really no distinction to be made between an electrophile and a Lewis acid.

1.8   Functional groups and homologous series

The study of inorganic chemistry is aided by the Periodic Classification of the elements, which helps to summarize and rationalize the trends in the properties of the elements. A study of the immense range of organic compounds is assisted by classifications based on **functional groups** and **homologous series**.

The physical and chemical properties of simple organic compounds usually arise from the presence in the molecules of a recognizable, and characteristic, group or arrangement of atoms or bonds—the **functional group**. As a general rule, the properties of an organic compound may be regarded as the sum of the properties of the individual functional groups within its molecules, although interactions between groups sometimes lead to some modification in properties. Simple organic compounds are therefore classified according to the functional group which they contain.

Compounds which contain the same functional group (and hence have similar chemical properties) can be arranged into families of compounds called **homologous series**, e.g. CH_3OH, CH_3CH_2OH, $CH_3CH_2CH_2OH$, etc. All the members of an individual homologous series can be represented by a general chemical formula and all have closely similar chemical properties. There is, however, a definite gradation in the physical properties of the members with increasing relative molecular mass. As the latter increases, the chemical reactivity tends to decrease somewhat. *Table 1.2* includes some examples of homologous series and the functional groups which they contain. From a knowledge of the reactions of the functional groups and the trends in properties in homologous series, the

Table 1.2

Name of homologous series	Functional group Name	Structure
alkanes	—	—
alkenes		$\diagdown C{=}C \diagup$
alkynes		$-C{\equiv}C-$
halogenoalkanes		$\diagup C{-}X$
	fluoro	X = F
	chloro	X = Cl
	bromo	X = Br
	iodo	X = I
alcohols	hydroxyl	—OH
aldehydes ⎱ ketones ⎰	carbonyl	$\diagdown C{=}O$
carboxylic acids	carboxyl	$-C\diagup^{O}_{\diagdown OH}$
amines	amino	$-NH_2$
	substituted amino	$\left\{ \begin{array}{l} \diagup NH \\ or \\ \diagup N \end{array} \right.$

properties and reactions of familiar compounds can be explained and understood, while those of unfamiliar compounds may be predicted with some confidence. The following chapters set out this knowledge in some detail.

Summary

1   Organic chemistry is the study of the innumerable compounds of carbon.
2   **Isomerism** is the phenomenon of the existence of two or more compounds with the same **molecular formulae** but having one or more different physical or chemical properties. When this arises from differences in structure it is called **structural isomerism**.
3   Bonding and molecular geometry can best be described using the concept of **orbital hybridization**.
4   Many chemical bonds are **polar** in nature and the electron-donating or withdrawing effect of an atom or group in a chemical bond is termed the **inductive effect**.
5   Organic reactions may be considered to be essentially of four types: **substitution, addition, elimination** or **rearrangement**. These reactions may involve **homolytic fission** of bonds when **free-radicals** are initially produced or else **heterolytic fission** when **ions** are initially produced.
6   Chemical species participating in reactions may be described as **electrophiles** if they are electron-pair acceptors or **nucleophiles** if they are electron-pair donors (cf. **Lewis acids** and **Lewis bases**, Section 1.7).
7   Specific arrangements of atoms or bonds which confer characteristic properties on a compound are called **functional groups**. Families of simple compounds containing the same functional group are called **homologous series**.

Questions

1*   Describe *briefly* the hybridization of atomic orbitals with reference to carbon and its compounds.
What is the shape of each of the following species?

C_2H_2 CH_2O $C_6H_5CH:CHCH_3$ CH_3OH $^{\oplus}CH_3$

Comment on the fact that cyclohexene can be prepared and stored over long periods but cyclohexyne is too unstable to be isolated under normal conditions.

Cambridge

2   (a) For each of the following compounds, draw the structures of **three** isomers.
(i) C_5H_{12}; (ii) $C_6H_4Cl_2$; (iii) $C_2H_2Cl_2$
(b) Explain briefly the meaning of the term homologous series.
(c) Give the name and general formula of a homologous series.

JMB

Chapter 2
Alkanes

2.1 Hydrocarbons: introduction

Hydrocarbons—compounds of carbon and hydrogen only—are among the most important organic compounds and occur widely in nature. Some of them are simple and some very complex. The simplest is methane CH_4, which is the predominant constituent of natural gas and is also to be found dissolved in many oil deposits. In the past, it has been called *firedamp* when found in coal mines, in which it constitutes an explosion hazard, and *marsh gas* when it resulted from the bacterial decay of cellulose in vegetable matter under water, as in marshes and swamps. This bacterial decay of cellulose also occurs in the gut of most herbivorous animals and produces methane. Methane is also the major constituent of the atmosphere of the planet Jupiter and may thus have extraterrestrial significance.

Other naturally occurring hydrocarbons can be more complex. The characteristic yellow colour in carrots, butter and egg yolk is due to a hydrocarbon of molecular formula $C_{40}H_{56}$. Many isomers of this compound are known; one of them imparts the red colour to tomatoes, rose-hips and the berries of hawthorn. The smooth, glossy appearance of the leaves of members of the cabbage family is due to the presence of hydrocarbon waxes with formulae such as $C_{29}H_{60}$. Natural rubber is a hydrocarbon with long, chain-like molecules of formula $(C_5H_8)_n$, where n is a whole number of the order of several thousand. Other hydrocarbons, such as limonene $C_{10}H_{16}$ found in lemon and orange oils, are constituents of the pleasant-smelling 'essential oils' found in certain plants.

There are also synthetic or man-made hydrocarbons, the most well known probably being the plastic poly(ethene), known as polythene. Other similar plastics are known by popular names such as polypropylene and polystyrene.

The structures of the three simplest hydrocarbons CH_4, C_2H_4 and C_2H_2 are shown below:

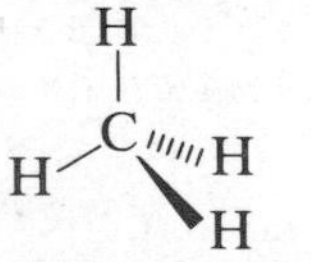

methane CH_4 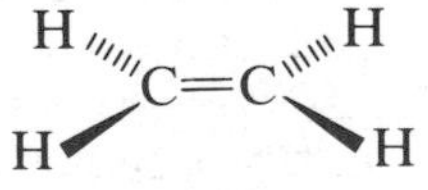   ethene C_2H_4 ethyne C_2H_2

In the molecule of methane, each of the four valency electrons of the
carbon atom is shared with one of a hydrogen atom and thus, in forming
four single bonds, the carbon atom is making use of its maximum combining capacity. Such hydrocarbons are said to be **saturated**. It is clearly
not possible to *add on* any atoms to such a molecule. In the ethene
molecule, a double bond links together the two carbon atoms, while in
the ethyne molecule the corresponding link is a triple bond (*see* Section
1.3). In principle, each of these two molecules would appear to be capable
of adding on more atoms (two in the case of ethene, and four in the case
of ethyne) to give molecules in which carbon once again forms four single
bonds. Such molecules, containing multiple bonds, are said to be
unsaturated.

Methane, ethene and ethyne are the first members of three homologous
series of hydrocarbons called **alkanes**, **alkenes** and **alkynes**, respectively.
This chapter now deals with alkanes.

2.2 Homologous series of alkanes

The alkanes form a homologous series of general formula C_nH_{2n+2}, where
n is a whole number. The first eight straight-chain members are listed in
Table 2.1, together with some physical properties.

The name of each of these alkanes is made up of two parts; one
indicates the number of carbon atoms in the molecule (for historical
reasons, proper systematic naming based on the Greek only starts at C_5
pentane) and the other indicates that it belongs to the homologous series
of alk*anes*. The alkanes listed in *Table 2.1* are all straight-chain compounds
but structural isomerism occurs with all the alkanes above C_3.

There are two possible structures corresponding to a molecular formula
C_4H_{10}:

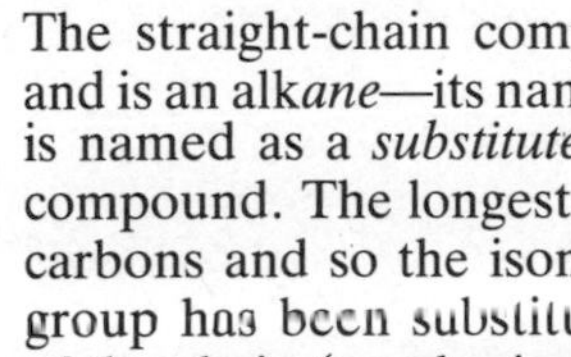

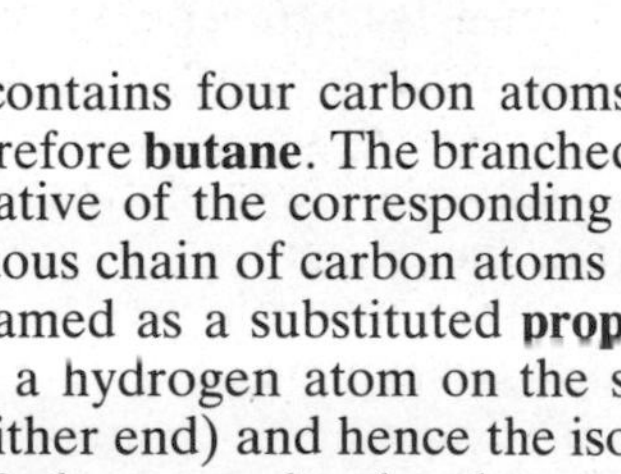

The straight-chain compound contains four carbon atoms (hence *but-*)
and is an alk*ane*—its name is therefore **butane**. The branched-chain isomer
is named as a *substituted* derivative of the corresponding straight-chain
compound. The longest continuous chain of carbon atoms contains three
carbons and so the isomer is named as a substituted **propane**; a methyl
group has been substituted for a hydrogen atom on the second carbon
of the chain (numbering from either end) and hence the isomer's systematic name is **2-methylpropane**. In larger molecules the numbering of the
carbon chain is such that the substituents are attached to the carbon

Table 2.1

Name	Formula	Melting point (°C)	Boiling point (°C)	Density (g cm^{-3})
methane	CH_4	−182	−161	—
ethane	C_2H_6	−172	−89	—
propane	C_3H_8	−188	−42	—
butane	C_4H_{10}	−138	−0.4	—
pentane	C_5H_{12}	−130	36	0.626
hexane	C_6H_{14}	−95	69	0.659
heptane	C_7H_{16}	−91	99	0.684
octane	C_8H_{18}	−57	126	0.703

atoms bearing the lowest numbers. The three isomers of C_5H_{12} are named as follows:

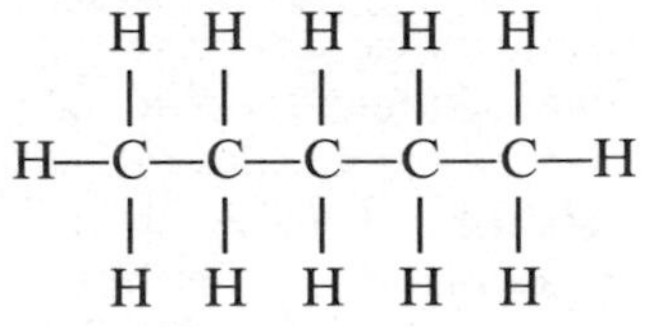

$CH_3CH_2CH_2CH_2CH_3$

pentane
b.p. = 36°C

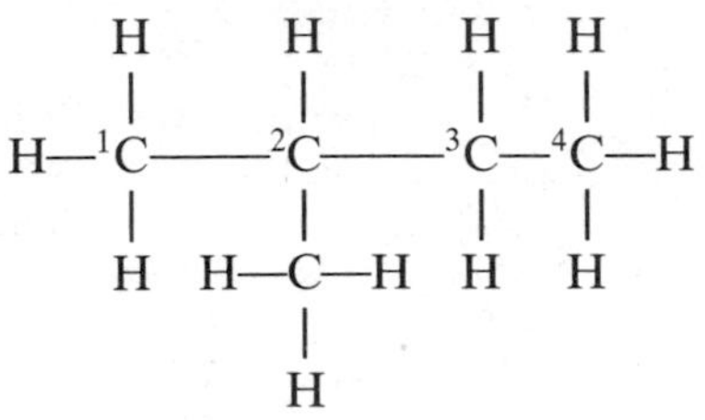

$CH_3CHCH_2CH_3$
|
CH_3

2-methylbutane
(not 3-methylbutane)
b.p. = 28°C

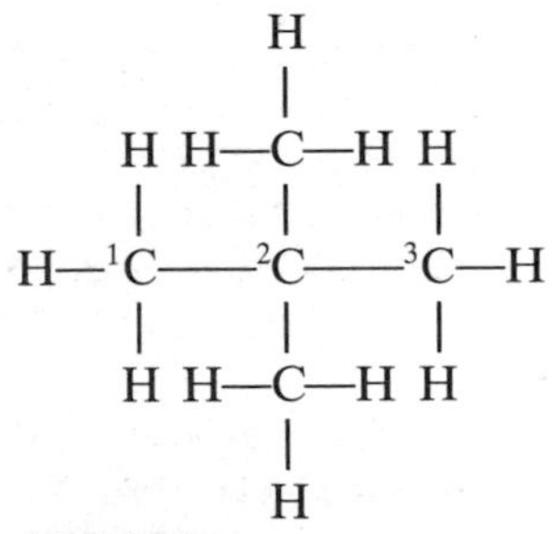

$(CH_3)_4C$

2,2-dimethylpropane
(*di-* indicating two
methyl groups both attached
to carbon atom 2)
b.p. = 9.5°C

The number of isomers increases rapidly with increase in the number of carbon atoms in the molecule. There are five isomers of C_6H_{14}, nine of C_7H_{16} and eighteen of C_8H_{18}.

Cyclic alkanes are also known, but their formulae do not fit the general one for the alkane homologous series. A cyclic alkane must, necessarily, contain two hydrogen atoms fewer than its corresponding straight-chain compound; compare, for example, hexane and *cyclo*hexane (an important hydrocarbon, *see* Section 12.4.1(i)):

$CH_3CH_2CH_2CH_2CH_2CH_3$

hexane C_6H_{14}
b.p. = 69°C

cyclohexane C_6H_{12}
b.p. = 81°C

2.3 Physical properties of alkanes

The gradation in the physical properties of the alkanes is typical of a homologous series. The C_1–C_4 straight-chain compounds are colourless gases at room temperature, the C_5–C_{17} compounds are colourless liquids of increasing viscosity, and the first solid at room temperature appears at C_{18} (m.p. = 28°C). There is an almost linear relationship between the boiling points of the straight-chain alkanes and their relative molecular masses as each CH_2 increment in successive alkanes increases the boiling point by roughly the same amount (particularly true of the higher members of the series)—*see Figure 2.1.*

Alkanes are non-polar and the intermolecular forces are of the weak van der Waals type. As the relative molecular mass increases with increasing number of atoms (and hence electrons), the molecules become more polarizable and the intermolecular forces increase in strength correspondingly (*see* Section 1.4). These forces have to be overcome when the liquid passes into the vapour state, and hence the boiling point increases with increasing relative molecular mass.*

Branched-chain isomers boil at lower temperatures than the corresponding straight-chain ones (*see*, for example, the isomers of pentane C_5H_{12} in Section 2.2), since the intermolecular forces between the branched-chain molecules are slightly weaker. On the other hand, cyclic alkanes tend to boil at slightly higher temperatures than their straight-chain counterparts.

* The effect of increasing relative molecular mass on the boiling point originates mainly from electrostatic forces and *not* gravitational ones (*see* Section 1.4).

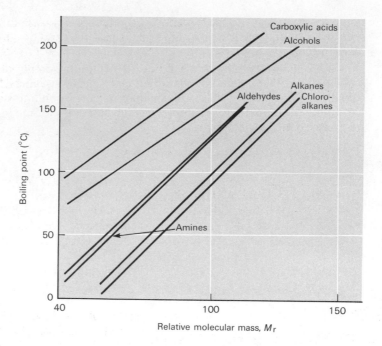

Figure 2.1   A b.p. v. M_r plot for various homologous series

Solid alkanes consist of molecular lattices in which the molecules are held together by van der Waals forces, so the melting points are comparatively low (the melting points of most organic solids are much lower than those of inorganic solids, since the lattices are molecular rather than ionic). The different lattice arrangements preclude any simple relationship between melting point and relative molecular mass.

As the alkanes are non-polar, the liquid hydrocarbons remain totally immiscible with water (forming the top layer in a mixture, since they are less dense than water), which is a highly polar solvent. The hydrocarbons are incapable of hydrogen bonding and hence cannot be accommodated in the extensively hydrogen-bonded water structure. They will, however, dissolve non-polar substances, for which they are reasonably good solvents. As a general rule of thumb, it can be said that like tends to dissolve like.

Straight-chain alkanes can be separated from branched-chain compounds by using crystalline, dehydrated calcium alumino-silicates, called *zeolites*, as **molecular sieves**. Straight-chain isomers can become adsorbed by suitable sieves because they become held in cavities, or pores, in the large cage-like structure, while the branched-chain ones do not. For example, a sieve with a pore size of about 0.5 nm (with 50% of the crystal volume being empty) will adsorb straight-chain alkanes with effective cross-sectional diameters of 0.43 nm, while excluding branched-chain isomers with diameters of 0.5 nm and over. Once the straight-chain isomers have been extracted from a mixture by adsorption by the molecular sieve, the non-adsorbed material is removed and the straight-chain compounds then recovered. The distinction between straight- and branched-chain isomers can be critical in determining certain properties such as the anti-knock ratings of hydrocarbon fuels (*see* Section 2.4.2), the biodegradability of alkylbenzene detergents (*see* Section 19.4) or the bacterial conversion into protein (*see* Section 2.5).

2.4 Chemical properties of alkanes

As they are non-polar, the alkanes are generally inert to ionic reagents such as acids, alkalis and oxidizing agents, under normal conditions. The C—H bond tends to undergo homolytic fission (*see* Section 1.6) and the alkanes therefore undergo free-radical reactions. These usually require some special initiation process, such as the generation of free-radicals through the action of heat or light.

2.4.1 Decomposition

Alkanes decompose when heated strongly, although very high temperatures are needed in the case of the simpler alkanes. Propane, for example, decomposes at 800–850°C to give, in varying amounts, propene, ethene, methane and hydrogen:

$$CH_3CH_2CH_3 \rightarrow CH_3CH{=}CH_2 + H_2$$
$$CH_3CH_2CH_3 \rightarrow CH_2{=}CH_2 + CH_4$$

This type of thermal decomposition is called **pyrolysis** or **thermal cracking** and is of the utmost importance to the petrochemicals industry in providing simple reactive molecules for use as chemical feedstocks (*see* Section 18.3.1). Cracking can also be effected with the aid of catalysts when it is known as **catalytic cracking**; only in thermal cracking are radicals produced.

2.4.2 Oxidation

Although alkanes are resistant to attack by oxidizing agents generally, they will all undergo combustion in air or oxygen, the products of complete oxidation being carbon dioxide and water. For example:

$$CH_4 + 2O_2 \rightarrow CO_2 + 2H_2O \qquad \Delta H^{\ominus} = -890\,kJ$$

In a limited supply of air or oxygen, carbon monoxide or even carbon may be formed. As the number of carbon atoms in a molecule of an alkane increases, the flame with which it burns becomes more luminous. The mechanism of combustion is complex but certainly involves free-radicals.

The gaseous and more volatile alkanes are potentially dangerous as they form explosive mixtures with air or oxygen. Interestingly, the behaviour of straight- and branched-chain alkanes on combustion can critically affect the running of an internal combustion engine. The petrol/air mixture in such an engine is normally ignited at the end of the compression stroke by the passage of a spark from a plug. An explosion occurs and the flame front is steadily propagated through the cylinder during the expansion of the hot gases. Occasionally, the unburned part of the mixture undergoes sudden and violent ignition, before the flame front has reached it, and the result is a violent, metallic rattling or 'knocking' in the cylinder. Under extreme conditions, this can cause cracking of the cylinder head but, in any case, it always results in loss of power and is therefore undesirable. Research has shown that branched-chain alkanes function much more smoothly than straight-chain ones which exacerbate the problem of engine 'knock'.

In 1926 Graham Edgar, an American scientist, introduced the familiar **octane rating** or **number** for petrol. A branched-chain isomer of C_8H_{18} was found to give virtually no knock when used in a single-cylinder test engine, whereas the straight-chain heptane C_7H_{16} caused a considerable amount. These two alkanes were selected to form an arbitrary scale for measuring the anti-knock rating of a petrol:

$$CH_3$$
$$|$$
$$CH_3CCH_2CHCH_3 \qquad\qquad CH_3CH_2CH_2CH_2CH_2CH_2CH_3$$
$$|\quad\ \ \ |$$
$$CH_3\ \ CH_3$$

heptane O.N. = 0

2,2,4-trimethylpentane   O.N. = 100
(formerly known as 'iso-octane')

By making up suitable mixtures of these two alkanes, the anti-knock characteristics of any commercial petrol can be reproduced and the petrol can be assigned an octane number (O.N.), based on the percentage by volume of 'iso-octane' in the mixture. A petrol with an O.N. of 92 (two-star) would have the same anti-knock characteristics as a mixture of 92% 'iso-octane' and 8% heptane. O.N. values can be boosted over the 100 mark by the use of certain additives or by blending (*see* Section 18.2(ii)).

The combustion of gaseous hydrocarbons forms the basis of a method (eudiometry) of determining their molecular formulae. A known volume of hydrocarbon, measured at room temperature and pressure, is burned in an *excess* of oxygen, the reaction being triggered off by the passage of a spark through the gas mixture. When the products are returned to room temperature and pressure the water condenses to a liquid of negligible volume and the residual gas is a mixture of carbon dioxide and unreacted oxygen. If this gas mixture is shaken with a solution of potassium or sodium hydroxide, the resulting contraction in gas volume is a measure of the carbon dioxide which has just been absorbed by the alkali. The remaining gas volume is the volume of unreacted oxygen.

If the formula of the hydrocarbon is represented in general terms as C_xH_y, where x and y are whole numbers (the hydrocarbon may therefore be an alkane, alkene or alkyne), then

$$C_xH_y + \left(x + \frac{y}{4}\right)O_2 \rightarrow xCO_2 + \frac{y}{2}H_2O$$

1 volume $+ \left(x + \dfrac{y}{4}\right)$ volumes $\rightarrow x$ volumes $+ \dfrac{y}{2}$ volumes (as gas)

Volume of CO_2 produced from 1 volume C_xH_y

$= x$ volumes $=$ volume of gas absorbed by alkali

Volume of O_2 used in reacting with 1 volume C_xH_y

$$= \left(x + \frac{y}{4}\right) = \text{initial volume of } O_2 - \text{final volume of gas}$$

Suppose $10\ cm^3$ of hydrocarbon was exploded with $60\ cm^3$ oxygen and that after shaking the gaseous mixture with alkali there was a contraction in volume of $20\ cm^3$, the final volume of gas remaining being $25\ cm^3$ (all volumes measured at room temperature and pressure). Then

$$10\ cm^3\ C_xH_y \rightarrow 20\ cm^3\ CO_2$$

$\therefore$ 1 volume $C_xH_y \rightarrow 2$ volumes CO_2

$\therefore$ $x = 2$

Volume of O_2 used up $= 60\ cm^3 - 25\ cm^3 = 35\ cm^3$

$\therefore$ $10\ cm^3\ C_xH_y$ combined with $35\ cm^3\ O_2$

$\therefore$ 1 volume C_xH_y combined with 3.5 volumes O_2

$\therefore$ $\left(x + \dfrac{y}{4}\right) = 3.5$ and since $x = 2$, y must $= 6$

Hence the molecular formula of the hydrocarbon is C_2H_6

Partial oxidation of alkanes is also used to provide a wide range of simple organic compounds such as ethanoic acid (*acetic acid*):

$$CH_3CH_2CH_2CH_2CH_3 \xrightarrow{O_2} CH_3COOH + HCO_2H + CH_3CH_2CO_2H$$

pentane   ethanoic   methanoic   propanoic
                                     acid   acid   acid

and will be discussed later (*see* Section 9.7).

2.4.3 Halogenation

Alkanes react with the halogens (except iodine) to form **substitution** products. An apparently simple example is the exothermic reaction between methane and chlorine, which is brought about either by heating the mixture to about 400°C, or else by irradiating it with ultraviolet light (the reaction does not take place in the dark). In the reaction, or more correctly series of reactions, hydrogen atoms in the methane molecule are successively replaced by chlorine atoms. The reaction is therefore of the substitution type (*see* Section 1.6):

$$CH_4 \; + \; Cl_2 \; \rightarrow \; CH_3Cl \; + \; HCl$$

$$CH_3Cl \; + \; Cl_2 \; \rightarrow \; CH_2Cl_2 \; + \; HCl$$

$$CH_2Cl_2 \; + \; Cl_2 \; \rightarrow \; CHCl_3 \; + \; HCl$$

$$CHCl_3 \; + \; Cl_2 \; \rightarrow \; CCl_4 \; + \; HCl$$

The organic products are named as substituted products of methane and are chloromethane, dichloromethane, trichloromethane (*chloroform*) and tetrachloromethane (*carbon tetrachloride*), respectively. The reaction is of commercial importance, since these chlorinated hydrocarbons are widely used as solvents (*see also* Section 5.4(i)). All four products are obtained on chlorination but the relative amounts are influenced by the methane:chlorine ratio employed. The first two chloromethanes predominate as products when the ratio is roughly 5:1, but if this is reduced to about 1.7:1 the tri- and tetrachloro products predominate. The mixture of products can be separated by careful fractional distillation, the boiling points of the four products being −24°C, 41°C, 61°C and 79°C, respectively.

  The chlorination of the methane is initiated by the formation of chlorine atoms, or free-radicals; in the presence of ultraviolet light, or at elevated temperatures, these are formed by the dissociation of chlorine molecules (homolytic fission):

$$Cl\!-\!Cl \; \rightarrow \; Cl\cdot \; + \; Cl\cdot \qquad \Delta H^{\ominus} = \; +242 \, kJ$$

These atoms, or free-radicals, each contain an unpaired electron and are highly reactive. As well as recombining to form chlorine molecules they can remove, or abstract, a hydrogen atom which is attached to carbon:

$$CH_4 \; + \; Cl\cdot \; \rightarrow \; CH_3\cdot \; + \; HCl \qquad \Delta H^{\ominus} = -8 \, kJ$$

(The reaction $CH_4 + Cl\cdot \rightarrow CH_3Cl + H\cdot$ is endothermic, $\Delta H^{\ominus} = +91 \, kJ$, and energetically unfavourable.) The highly reactive **methyl radical** so produced is able to attack undissociated chlorine molecules

$$CH_3\cdot \; + \; Cl_2 \; \rightarrow \; CH_3Cl \; + \; Cl\cdot \qquad \Delta H^{\ominus} = -88 \, kJ$$

so generating more chlorine atoms. These may attack molecules of methane or chloromethane and thus another radical species is formed:

$$CH_3Cl + Cl \cdot \;\rightarrow\; \cdot CH_2Cl + HCl$$

This new radical, in turn, is capable of attacking chlorine molecules, so generating more chlorine atoms:

$$\cdot CH_2Cl + Cl_2 \;\rightarrow\; CH_2Cl_2 + Cl \cdot$$

It should now be clear that, once a chlorine atom has been generated, a **chain of reactions** becomes possible leading to all four chloromethanes. For this reason such reactions are called **free-radical chain reactions**. In theory, only one molecule of chlorine needs to dissociate into chlorine atoms, since for every chlorine atom used up in a reaction step another is generated; in practice, however, some recombination of chlorine atoms takes place, and so more than one chlorine atom needs to be generated. In the sequence of reactions, the initial formation of chlorine atoms is called the **initiation step**. The subsequent series of reactions of radicals with molecules to give the products constitute **propagation steps**. Recombination of radicals constitutes **termination steps**, which eventually bring the reaction to a halt, e.g.

$$Cl \cdot + Cl \cdot \;\rightarrow\; Cl_2 \quad \text{or} \quad CH_3 \cdot + Cl \cdot \;\rightarrow\; CH_3Cl$$

An important piece of experimental evidence for the mechanism just outlined is the identification of small amounts of compounds such as ethane C_2H_6 among the products; such a molecule could only have been formed from methane via the recombination of methyl radicals:

$$CH_3 \cdot + CH_3 \cdot \;\rightarrow\; CH_3 - CH_3$$

Essentially similar reactions occur with the other alkanes like ethane, etc., a large number of isomers being formed, even on monochlorination, with the higher members of the homologous series. The C—H bond strengths are not identical in every alkane and it can be seen from the following results of the monochlorination of 2-methylbutane that a hydrogen atom attached to a *tertiary* carbon atom, namely $-\overset{\displaystyle |}{\underset{\displaystyle |}{C}}-H$, is

more easily abstracted than one attached to a *secondary* carbon atom, namely —CH_2—, which, in turn, is more easily abstracted than one attached to a *primary* carbon atom, namely —CH_3. By taking into account the numbers of different types of hydrogen atoms in the molecule, and the ratios of products to be expected on a statistical basis, the figures showed that the relative reactivities of primary, secondary and tertiary hydrogen atoms were 1:2.5:4.

$$CH_3-\underset{\underset{H}{|}}{\overset{\overset{CH_3}{|}}{C}}-CH_2-CH_3 \xrightarrow{300°C} ClCH_2-\underset{\underset{H}{|}}{\overset{\overset{CH_3}{|}}{C}}-CH_2-CH_3 \;+\; CH_3-\underset{\underset{H}{|}}{\overset{\overset{CH_3}{|}}{C}}-CH_2-CH_2Cl \;+$$

$$\underset{34\%}{} \qquad \underset{16\%}{}$$

$$CH_3-\underset{\underset{H}{|}}{\overset{\overset{CH_3}{|}}{C}}-CHCl-CH_3 \;+\; CH_3-\underset{\underset{Cl}{|}}{\overset{\overset{CH_3}{|}}{C}}-CH_2-CH_3$$

$$\underset{28\%}{} \qquad \underset{22\%}{}$$

Bromination takes place less readily and, even then, it is the tertiary hydrogen atoms which tend to be substituted, as the abstraction step

$$-\overset{|}{\underset{|}{C}}-H \;+\; Br\cdot \;\rightarrow\; -\overset{|}{\underset{|}{C}}\cdot \;+\; HBr \qquad \Delta H^{\ominus} \approx +52\,kJ$$

is endothermic, unlike the chlorine case.

Fluorine reacts violently with methane, for example, and with higher alkanes oxidation of C—C bonds tends to take place. Special fluorinating agents are therefore usually used (*see* Section 5.4(v)). Thus, the reactivity of halogens with alkanes is the usual order of chemical reactivity, decreasing from fluorine to iodine (virtually no reaction).

2.4.4 Nitration

Although, under ordinary conditions, concentrated acids have little, if any, effect on alkanes, a substitution reaction does occur between alkanes and nitric acid in the *vapour phase*, at about 400°C and 1 $MN\,m^{-2}$ pressure. Thus, methane yields nitromethane CH_3NO_2:

$$CH_4 + HNO_3 \;\rightarrow\; CH_3NO_2 + H_2O$$

In this reaction, too, free-radicals are involved as is shown by the fact that vapour phase nitration of ethane yields not only nitroethane, but also nitromethane. Nitration of propane yields about 20% nitromethane, 20% nitroethane and 60% nitropropanes. Nitroalkanes find some use as solvents and in the manufacture of amines (*see* Section 11.4).

2.5 Uses of alkanes

The general uses of alkanes may be divided into three categories:

(i) They are used directly as fuels to provide heat and power. In the UK, methane from the extensive natural gas fields under the North Sea is transported by pipeline to the east coast and, after treatment, distributed through the national pipeline system by the British Gas Corporation as a fuel. The natural gas produced from the biggest field, the Leman Field discovered by Shell/Esso in 1966, consists of about 95% methane and 3% ethane, together with small amounts of other substances. Some methane is also imported in liquid form (b.p. = −161°C) from Algeria and is stored in specially insulated tanks, as at Canvey Island for example. The strictest possible precautions have to be taken in storing large amounts of gas because of the fire/explosion hazard.

Mixtures of alkanes such as propane and butane are also liquefied under pressure, when they are known as liquefied petroleum gas (LPG) or 'bottled' gas (for example, 'Calor Gas'), stored in mobile cylinders.

The action of certain anaerobic bacteria on sewage sludge, stored in closed tanks at 35°C, also produces methane, which can be used to provide the power on which to run sewage treatment plants.

The heat, or enthalpy, of combustion of straight-chain alkanes (i.e. the homologous series) *per unit mass* decreases with increasing relative molecular mass because the percentage hydrogen in the molecule is diminishing and the enthalpy of combustion of hydrogen is appreciably greater than that of carbon. However, the density of the alkanes increases with increase in relative molecular mass, and hence liquid hydrocarbons are often a more convenient energy source.

Petrol (gasoline), a complex mixture of hydrocarbons (some of which are C_5–C_9 alkanes), forms an explosive mixture with air or oxygen. In order to achieve maximum power, an internal combustion engine is normally run on an air–fuel mixture which is 10–20% richer in fuel than the stoichiometric mixture (i.e. one in which the reactants are present in their theoretical combining proportions). Consequently, car exhaust emissions contain unburned hydrocarbons in addition to carbon monoxide, carbon dioxide, nitrogen oxides and lead compounds (*see* Section 5.4(ii)). These unburned hydrocarbons can enter into a series of complicated free-radical reactions in the upper atmosphere, under certain climatic conditions, producing extremely unpleasant 'photochemical smogs' which characterize certain cities such as Los Angeles in California. Catalytic devices for car exhausts have been made in order to catalyse the complete combustion of any unburned hydrocarbons in the exhaust, and in the USA all new cars are required by law to be fitted with them in order to satisfy stringent pollution controls.

The combustion of hydrocarbons (gas and oil) and of coal for fuel has led to an ever-increasing carbon dioxide content of the atmosphere (it is about 0.03% by volume). The presence of the carbon dioxide helps to reduce heat loss from the earth's surface (the so-called 'greenhouse effect') and it has been predicted that the ever-increasing carbon dioxide content will lead to a rise in average world temperature. Such a rise would be small and would be unlikely to produce any detrimental effects, however.

(ii) Chemical reactions involving alkanes provide the principal chemical starting materials, or feedstocks, for the petrochemicals manufacturing industry (*see* Section 18.3). In particular, liquid alkanes obtained from

light petroleum distillates are subjected to thermal or catalytic cracking
to provide smaller, more volatile or more reactive compounds for manufacturing routes. The smaller, gaseous alkanes, too, are subjected to such
processes as cracking, oxidation, dehydrogenation and reforming to provide large quantities of a wide range of organic chemicals of industrial
importance. Methane itself is also a major source of hydrogen, which is
produced by the **steam-reforming process**, devised by ICI in the 1950s:

$$CH_4 + H_2O \underset{\text{catalyst}}{\overset{750°C}{\rightleftharpoons}} \underbrace{3H_2 \ + \ CO}_{} \qquad \Delta H^{\ominus} = +205\,kJ$$

synthesis gas

(iii) In 1959, British Petroleum scientists at the Lavéra refinery in the
South of France discovered a way of converting oil into protein, using
microbiological growth. Hydrocarbons provided the carbon source for
the growth of micro-organisms such as bacteria and yeasts, which may
contain 40–70% protein, present mainly as enzymes and in the membranes
of the cells. Oxygen, nitrogen and minerals also have to be present for
cell growth. Although methane itself may be utilized by certain bacteria,
the most readily utilized alkanes seem to be C_{10}–C_{18} compounds from the
gas–oil fraction of oil distillation. After rapid growth of the micro-organisms under fermentation conditions, the cells are recovered and dried and
either used as such, or else extracted for protein (so-called single-cell
protein SCP).

The most commonly used route uses straight-chain alkanes which have
been separated from the original refinery distillate with the aid of molecular sieves (*see* Section 2.3). This feedstock is completely assimilated
by the micro-organisms, so that the product is virtually free from contamination by residual hydrocarbon (a gaseous substrate like methane
would thus appear to be advantageous but there are certain microbiological problems).

In 1971, B.P. commissioned a continuous process plant using straight-
chain alkanes at Grangemouth, Scotland. The plant has a capacity of
4000 tonnes per year of protein, which is designed for use in animal
feedstuffs. Shell operate a process using methane, and ICI's process uses
methanol, made from synthesis gas derived from methane—(*see* paragraph (ii) above).

Thus, although the alkanes are chemically rather unreactive, they play
a crucial role in the chemical industry.

Summary

1   **Hydrocarbons** are compounds containing the elements carbon and hydrogen only.

2   The **alkanes** form a homologous series with a general formula
     C_nH_{2n+2}. They contain no double or triple bonds and are said to be
     saturated.
3   Alkanes are unreactive towards ionic reagents in general, but take part
     in **free-radical reactions** such as **cracking**, **combustion**, **halogenation** and
     nitration (the last two are **substitution** reactions).
4   Alkanes find widespread use as (i) fuels, (ii) a source of petrochemical
     feedstocks, and (iii) a substrate for the microbiological manufacture of
     single-cell protein (SCP).

Questions

1   The following are the boiling points in °C at atmospheric pressure of a
     number of organic compounds:

     butane   −0.5
     pentane   36
     hexane   69
     hexanoic acid   205
     ethyl butanoate   120

     Consider the data carefully and then answer the following questions:
     (a) Why has butane a lower boiling point than pentane?
     (b) Predict the boiling point of heptane.
     (c) Write down the structure of 2,2-dimethylbutane and predict whether
         it would have a higher or lower boiling point than the isomeric
         compound, hexane. Briefly state your reasons.
                                                                          JMB

2   (a) Thermal cracking of propane gives ethene as the main product.
         (i) Describe the type of chemical process involved in cracking.
         (ii) Give the structures of **two** other organic products of this reaction.
     (b) Outline the mechanism of the formation of chloromethane from
         methane and chlorine at 500°C.
     (c) State **two** common features of the reactions in (a) and (b).
                                                                          JMB

Chapter 3
Alkenes

3.1 Homologous series of alkenes

The general formula of this homologous series of **unsaturated** (*see* Section 2.1) hydrocarbons is C_nH_{2n}. An alkene contains a carbon–carbon double bond and this *bond* $\diagdown C{=}C \diagup$ may be regarded as the functional group characteristic of the series. The names of the members of the series correspond to those of the alkane series with the suffix **-ane** replaced by **-ene**, although there can clearly be no C_1 alkene corresponding to methane. The series therefore starts at C_2H_4, ethene, and the first seven alkenes with terminal double bonds (i.e. positioned at the end of a straight chain) are listed in *Table 3.1*, together with some physical properties.

Table 3.1

Name	Formula	Melting point (°C)	Boiling point (°C)	Density (g cm^{-3})
ethene	$CH_2{=}CH_2$	−169	−104	—
propene	$CH_3CH{=}CH_2$	−185	−48	—
but-1-ene	$CH_3CH_2CH{=}CH_2$	−185	−7	—
pent-1-ene	$CH_3(CH_2)_2CH{=}CH_2$	−165	30	0.641
hex-1-ene	$CH_3(CH_2)_3CH{=}CH_2$	−140	64	0.674
hept-1-ene	$CH_3(CH_2)_4CH{=}CH_2$	−119	93	0.698
oct-1-ene	$CH_3(CH_2)_5CH{=}CH_2$	−102	123	0.716

The simple C_1–C_4 alkenes are generally obtained by the cracking of naphtha in the refining of oil (*see* Section 18.3.1), but compounds containing the carbon–carbon double bond are widespread in nature, ranging from cyclic hydrocarbons such as limonene in the 'essential oils' to the unsaturated *fatty acids* obtained from vegetable oils (*see* Section 19.1).

Only one structure is possible for C_2H_4 (ethene) and C_3H_6 (propene), as alkenes (cyclopropane has the same molecular formula as propene, *see* Section 2.2, but is not an alkene), but with C_4H_8 several isomeric alkene structures are possible. The isomerism arises both from branched-chain structures and from alternative positions for the double bond in the chain. For C_4H_8 there are three isomeric alkenes, namely

$$\overset{4}{CH_3}\overset{3}{CH_2}\overset{2}{CH}=\overset{1}{CH_2} \qquad \overset{4}{CH_3}\overset{3}{CH}=\overset{2}{CH}\overset{1}{CH_3} \qquad \overset{3}{CH_3}\overset{\overset{\displaystyle CH_3}{|}}{\underset{2}{C}}=\overset{1}{CH_2}$$

but-1-ene   but-2-ene   2-methylpropene
b.p. = −6°C   b.p. (see below)   b.p. = −7°C

The position of the double bond is indicated by numbering the longest straight chain in the molecule in such a manner that the carbon atom at which the double bond starts has the lowest possible number; thus but-1-ene, rather than but-3-ene. It is unnecessary to indicate the position of the double bond in 2-methylpropene as there is only one position possible.

As well as structural isomerism, **stereoisomerism** is encountered in the homologous series. This is apparent when the geometry of the carbon–carbon double bond is considered. There is no free rotation about this bond (*see* Section 1.3), and therefore the atoms or groups attached to the double bond are fixed in space. In the case of but-2-ene, for example, the two methyl groups attached to the double bond may lie on the same side of the bond or on opposite sides:

$$H_3C_{\ /\!/\!/\!/}\quad\quad_{\!\!\!\!\!\!\backslash\!\backslash\!\backslash}CH_3 \atop \quad\quad C=C \atop H\quad\quad\quad\quad H$$

$$H_3C_{\ /\!/\!/\!/}\quad\quad_{\!\!\!\!\!\!\backslash\!\backslash\!\backslash}H \atop \quad\quad C=C \atop H\quad\quad\quad\quad CH_3$$

cis-but-2-ene   *trans*-but-2-ene
b.p. = 4°C   b.p. = 1°C

The isomers are **geometrical** isomers and are designated *cis*- and *trans*-isomers, respectively. **Geometrical isomerism** (which is one type of stereoisomerism) arises from the different fixed positions which atoms or groups, similarly linked together, may take up in space, and not from the different ways in which they may be linked together, as in structural isomerism. Pairs of geometrical isomers have different physical properties and can differ in some of their chemical properties (*see* Section 16.2).

3.2 Physical properties of alkenes

With two fewer hydrogen atoms in their molecules, the alk-1-enes have very slightly lower boiling points than the corresponding alkanes. Otherwise, the general principles underlying the physical properties of the alkanes (as discussed in Section 2.3) also apply to the homologous series of alkenes. The first three members are colourless gases at room temperature, and the first liquid alkene occurs at C_5, pent-1-ene and its isomers. The first solid member occurs at C_{18} (octadec-1-ene, m.p. = 18°C). The liquids are less dense than water and are immiscible with it.

3.3 Chemical properties of alkenes

The presence of the carbon–carbon double bond confers very considerable chemical activity on the alkenes and consequently they react with a much wider variety of reagents than do the alkanes. The π-bond component of the double bond is weaker than the σ-bond component (see Section 1.3) and the π-electrons are more easily accessible to attacking reagents. Consequently one of the most characteristic features of the alkenes is their participation in **addition** reactions (in contrast to the substitution reactions of alkanes), in which two atoms or groups add on across the double bond to form an addition product:

$$\text{C}=\text{C} \quad + \quad \text{X}-\text{Y} \quad \longrightarrow \quad -\overset{|}{\underset{|}{\text{C}}}-\overset{|}{\underset{|}{\text{C}}}- \\ \qquad\qquad\qquad\qquad\qquad\qquad\quad \text{X} \quad \text{Y}$$

The reactions frequently involve ionic species and are generally heterolytic reactions; the mechanisms involved are discussed later (see Section 3.3.4).

3.3.1 Addition reactions

(i) Reaction with hydrogen
When an alkene is mixed with hydrogen, there is no appreciable reaction, but in the presence of certain noble-metal catalysts such as nickel, platinum or palladium, a fairly rapid reaction occurs leading to the uptake of hydrogen. An addition product is formed in which two hydrogen atoms have been added across the original double bond. The process is referred to as **catalytic hydrogenation** and involves the reduction of the alkene to an alkane:

$$\text{C}=\text{C} \quad + \quad \text{H}_2 \quad \longrightarrow \quad -\overset{|}{\underset{|}{\text{C}}}-\overset{|}{\underset{|}{\text{C}}}- \\ \qquad\qquad\qquad\qquad\qquad\qquad\quad \text{H} \quad \text{H}$$

e.g.

$$CH_2{=}CH_2 + H_2 \quad \rightarrow \quad CH_3CH_3 \qquad \Delta H^{\ominus} = -120 \text{ kJ}$$

The metal is usually in a special, finely divided (i.e. powder-like) form which is highly active (such catalysts can sometimes inflame spontaneously in air, when they are described as being *pyrophoric*). The catalyst rapidly adsorbs molecular hydrogen and hence a large surface to volume ratio, such as exists in a powder, is essential. The catalyst is often supported on an inert material such as charcoal, alumina or silica. Hydrogenation is effected, either by mixing the hydrogen and alkene in the gaseous state, or else by bubbling hydrogen through the liquid alkene (pure, or in solution) in the presence, in either case, of the metal catalyst. In the case of platinum and palladium catalysts, only modest temperatures are required (i.e. about 20–90°C), but slightly higher temperatures are usually necessary with nickel. A measurement of the volume of hydrogen taken up by the alkene allows the number of double bonds present in the molecule to be ascertained since 1 mol of dihydrogen is taken up by 1 mol of double bonds.

Several functional groups (e.g. $-C{\equiv}N$, $\diagdown C{=}O$ and $-NO_2$) can be reduced by catalytic hydrogenation, but alkenes are generally reduced under quite mild conditions. Careful choice and control of such things as temperature, catalyst and solvent can bring about selective hydrogenation. Reduction of an alkene to an alkane is required only infrequently in the laboratory, but catalytic hydrogenation is widely used in the industrial conversion of edible oils and fats into margarine (see Section 19.3) and, to some extent, in their conversion into soap (see Section 19.4).

Dehydrogenation, the removal of hydrogen from an organic compound, can also be carried out using metal catalysts and is sometimes used as an industrial route to alkenes from alkanes, and, more particularly, for making aromatic hydrocarbons (see Section 18.3.2(ii)).

(ii) Reaction with halogens
Chlorine and bromine react readily with alkenes, in the liquid or vapour states, to form dihalogeno addition products:

$$\diagup \!\!\!\! \diagdown C{=}C \diagup \!\!\!\! \diagdown \quad + \quad X-X \quad \rightarrow \quad \begin{array}{c} | \ \ | \\ -C-C- \\ | \ \ | \\ X \ X \end{array}$$

e.g.

$$CH_2{=}CH_2 + Cl_2 \quad \rightarrow \quad ClCH_2CH_2Cl \qquad \Delta H^{\ominus} = -201 \text{ kJ}$$
$$\text{1,2-dichloroethane}$$

1,2-dihalogenoethane is the addition product, with each carbon atom bearing a halogen atom. It is isomeric with the 1,1-dihalogenoethane which can be prepared by other methods. With chlorine, propene gives 1,2-dichloropropane, and a similar reaction occurs with other alkenes. The reaction can be very vigorous, in the case of ethene and chlorine, and has to be moderated by the use of an inert solvent, such as tetrachloromethane or ethanoic acid; otherwise the following oxidation can occur:

$$C_2H_4 + Cl_2 \quad \rightarrow \quad 2C + 4HCl$$

The addition products are colourless and so, if an alkene is shaken with a few drops of a solution of bromine in an inert solvent, the red-brown colour of the bromine disappears. This is a useful, although not on its own conclusive, test for a $\diagup C = C \diagdown$ bond.

The 1,2-dichloro- and dibromoethanes are used as petrol additives to act as lead 'scavengers' (see Section 5.4(ii)), and the dichloro compound is also used to make the extremely important chloroethene $CH_2 = CHCl$ (vinyl chloride), which is the starting material for the manufacture of the plastic PVC (polyvinyl chloride, see Section 20.2.1(iv)). Ethene is allowed to react either with chlorine, or with a mixture of hydrogen chloride and oxygen (oxychlorination), and the addition product subjected to pyrolysis (or cracking):

$$Cl_2 + CH_2 = CH_2$$
$$2HCl + \tfrac{1}{2}O_2 + CH_2 = CH_2$$
$$\left. \right\} \longrightarrow \quad ClCH_2CH_2Cl \quad \xrightarrow[400-500^\circ C]{(-HCl)} \quad CH_2 = CHCl + HCl$$
$$(+ H_2O)$$

Direct reaction of alkenes with fluorine (fluorination) is generally far too vigorous and leads to the oxidation of the double bond. Reaction with iodine scarcely occurs at all, so the order of decreasing reactivity of the halogens with alkenes is fluorine, chlorine, bromine and iodine.

(iii) Reaction with hydrogen halides

Alkenes react, usually on heating, with hydrogen halides (either gaseous or in concentrated solution) to form an addition product:

$$\diagup C = C \diagdown \quad + \quad HX \quad \longrightarrow \quad \begin{matrix} | & | \\ -C - C - \\ | & | \\ H & X \end{matrix}$$

e.g.

$$CH_2 = CH_2 + HCl \quad \rightarrow \quad CH_3CH_2Cl \qquad \Delta H^\ominus = -97\,kJ$$

The product, chloroethane, is used in the manufacture of tetraethyl-lead, TEL (see Section 5.4(ii)), and is made industrially by the hydro-chlorination of ethene at 200°C, in the presence of an aluminium chloride catalyst.

Since the hydrogen halide is an unsymmetrical molecule, there is, in principle, more than one way in which it may add across an unsymmetrical double bond. The reaction with propene illustrates this point:

$$\overset{2}{C}H_3\overset{1}{C}H = CH_2 + HBr \diagup \diagdown \begin{matrix} CH_3CH_2CH_2Br & \text{1-bromopropane} \\ \\ CH_3CHCH_3 & \text{2-bromopropane} \\ | \\ Br \end{matrix}$$

Thus, the hydrogen atom of the HBr molecule may become attached either to carbon atom 1 (and hence the bromine to carbon atom 2, in which case the product is 2-bromopropane), or else to carbon atom 2 (in which case the bromine atom becomes attached to carbon atom 1, and 1-bromopropane is the product). The predominant product is, in fact, 2-bromopropane. A careful survey of the products of such reactions was made by the Russian chemist Vladimir Markovnikov, whose conclusion (1870) can be stated as follows: *when an unsymmetrical molecule adds across an unsymmetrical carbon–carbon double bond, the more electropositive atom or group of the molecule becomes bonded to that carbon atom which is already linked to the greater number of hydrogen atoms, namely the less highly substituted carbon atom of the double bond.* This is known as *Markovnikov's rule*; it was arrived at empirically but it has a sound theoretical basis in modern electronic theory (*see* Section 3.3.4). One product is not necessarily formed exclusively, but the rule enables one to predict with some confidence the likely *predominant* product*.

The order of reactivity of the hydrogen halides with alkenes reflects their relative acid strengths and is (in descending order) hydrogen iodide, hydrogen bromide, hydrogen chloride and hydrogen fluoride (the latter generally reacts only under pressure).

(iv) Reaction with halic(I) acids

The halic(I) acids HOX are weak acids and are formed in an equilibrium reaction when one of the halogens chlorine, bromine or iodine is dissolved in water (fluorine oxidizes water):

$$X_2 + H_2O \rightleftharpoons HOX + HX$$

When X = Cl and Br, the solutions are often referred to as chlorine water and bromine water, respectively. (Very low concentrations of iodic(I) acid are produced because of the low solubility of iodine in water.) When an alkene is shaken with such a solution, mixed products are formed, one of which corresponds to the overall addition of HO and X groups across the double bond:

$$\diagdown C = C \diagup \ + \ HOX \ \longrightarrow \ \underset{\underset{OH}{|}}{-C} \underset{\underset{X}{|}}{-C-}$$

X_2 also adds across the double bond, of course, and gives some of the dihalogeno compound. Thus, for example,

$$CH_3CH{=}CH_2 \ \xrightarrow[\text{HOBr}]{\text{Br}_2} \ \underset{\underset{Br}{|}\ \underset{Br}{|}}{CH_3CH{-}CH_2} \ + \ \underset{\underset{OH}{|}\ \underset{Br}{|}}{CH_3CH{-}CH_2}$$

---

* A reaction can be made to proceed in an anti-Markovnikov fashion by the addition of peroxides to the system; the reaction then proceeds via a free-radical rather than ionic mechanism.

It follows from Markovnikov's rule that, since the bromine atom is less electronegative (or more electropositive) than the OH group, the product with propene is 1-bromopropan-2-ol and *not* 2-bromopropan-1-ol. If an alkene is shaken with a few drops of bromine water, then the red-brown colour disappears immediately, and the reaction can thus be used as a simple, but not conclusive, test for an alkene—*see* (ii) above. If an alkane is shaken with a few drops of bromine water, then the red-brown colour does not disappear, but is extracted into the upper alkane layer (if the alkane is a liquid), as bromine is more soluble in the hydrocarbon than in water.

Industrially, propene is reacted with an aqueous solution of chlorine:

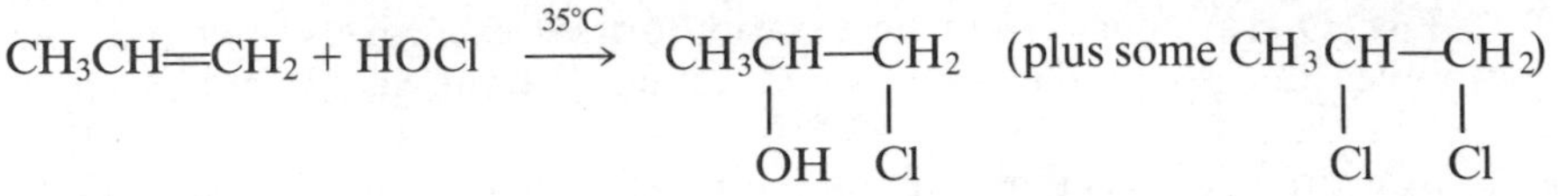

$$CH_3CH{=}CH_2 + HOCl \xrightarrow{35°C} \underset{\underset{OH\quad Cl}{|\quad\;\;|}}{CH_3CH{-}CH_2} \text{ (plus some } \underset{\underset{Cl\quad Cl}{|\quad\;\;|}}{CH_3CH{-}CH_2}\text{)}$$

and the addition product converted by alkali to an **epoxy** compound (*see* Section 20.2.2(ii)):

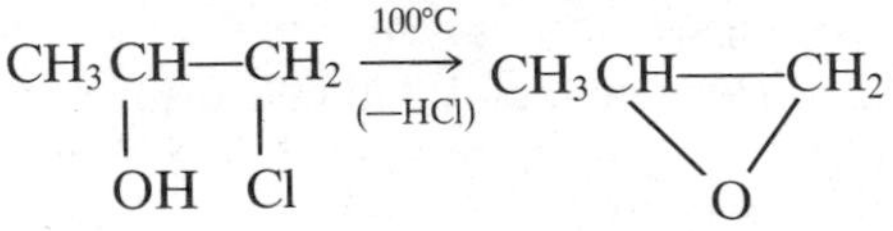

$$\underset{\underset{OH\quad Cl}{|\quad\;\;|}}{CH_3CH{-}CH_2} \xrightarrow[\text{(—HCl)}]{100°C} CH_3CH{-\!\!-\!\!-\!\!-}CH_2$$

epoxypropane

(v) Reaction with concentrated sulphuric acid

Alkenes are slowly absorbed when they are bubbled through, or shaken with, cold concentrated sulphuric acid. The reaction involves the addition of H and HSO_4 groups across the double bond and follows the course predicted by Markovnikov's rule:

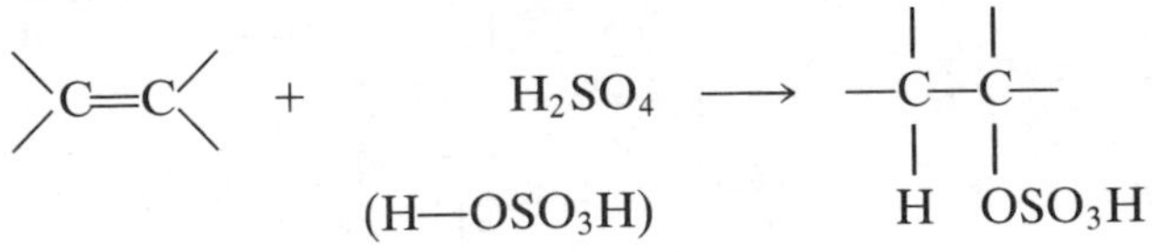

$$\underset{}{\overset{}{>}}C{=}C\overset{}{<} \; + \; \underset{(H{-}OSO_3H)}{H_2SO_4} \longrightarrow \underset{\underset{H\quad OSO_3H}{|\quad\;\;|}}{\overset{\overset{|\quad\;\;|}{}}{-C{-}C{-}}}$$

e.g.

$$CH_3CH{=}CH_2 + H_2SO_4 \rightarrow \underset{\underset{OSO_3H}{|}}{CH_3CHCH_3}$$

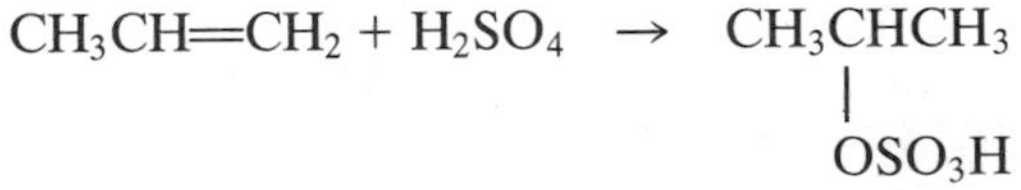

The product is an **ester** (*see* Section 10.1) of sulphuric acid, one of the hydrogen atoms of the acid having been replaced by an alkyl group.

Alkanes do not dissolve in concentrated sulphuric acid, and this reaction is therefore used in the petroleum industry to remove small amounts of alkenes from saturated hydrocarbons. The reaction is also used to manufacture alkyl sulphate detergents (*see* Section 19.4) from C_8 to C_{18} alkenes and sulphuric acid, e.g.

$$CH_3(CH_2)_9 CH{=}CH_2 + H_2SO_4 \ \rightarrow \ CH_3(CH_2)_9 CHCH_3$$
$$|$$
$$OSO_3H$$

Sulphuric acid esters are hydrolysed (i.e. broken down by the action of water) to **alcohols**:

$$\begin{array}{ccc} | \ \ | & & | \ \ | \\ {-}C{-}C{-} & + \ H_2O \ \longrightarrow & {-}C{-}C{-} \ + \ H_2SO_4 \text{ (dilute)} \\ | \ \ | & & | \ \ | \\ H \ \ OSO_3H & & H \ \ OH \end{array}$$

The reaction of an alkene with sulphuric acid, followed by hydrolysis of the ester to an alcohol, is thus equivalent to the overall addition of H and OH groups across the double bond, namely *hydration* of the alkene. Ethanol (*ethyl alcohol*) used to be manufactured from ethene in this way but the route was expensive and inconvenient and a direct, catalytic hydration of ethene is now employed (*see* Section 6.4). However, butan-2-ol is still manufactured from but-2-ene in this way (because direct hydration would lead to some polymerization):

$$CH_3CH{=}CHCH_3 + H_2SO_4 \ \longrightarrow \ CH_3CHCH_2CH_3 \ \xrightarrow{H_2O} \ CH_3CHCH_2CH_3$$
$$\qquad\qquad\qquad\qquad\qquad\quad | \qquad\qquad\qquad\qquad\quad |$$
$$\qquad\qquad\qquad\qquad\qquad OSO_3H \qquad\qquad\qquad\qquad OH$$

(vi) Reaction with metal complexes

This type of reaction was first carried out as long ago as 1830, when the Danish scientist W. C. Zeise passed ethene through a solution of potassium tetrachloroplatinate(II),K_2PtCl_4 and obtained a crystalline salt (Zeise's salt),$K[PtCl_3(C_2H_4)]$. This is now formulated as a π-complex in which the π-electrons of the alkene are donated to the vacant orbitals of the metal but it can be adequately represented thus:

$$\begin{array}{c} CH_2 \\ \| \quad + \ PtCl_4^{2-} \ \longrightarrow \\ CH_2 \end{array} \left[\begin{array}{c} CH_2 \qquad Cl \\ | {-}{-}{-}{\rightarrow} \ Pt \\ CH_2 \qquad\ \ \\ Cl \qquad\quad Cl \end{array} \right]^{-} + \ Cl^{-}$$

This type of complex plays an important role in certain industrial catalytic reactions. For example, when oxygen and ethene are passed into an aqueous solution of palladium(II) chloride and copper(II) chloride, the ethene adds on to the palladium and is then oxidized to an **aldehyde**, ethan*al*. The palladium(II) chloride acts as a catalyst* and the overall reaction is

$$CH_2{=}CH_2 + \tfrac{1}{2}O_2 \xrightarrow[CuCl_2]{PdCl_2} CH_3CHO \qquad \Delta H^{\ominus} = -243\,kJ$$

This route was introduced commercially in 1956 by Wacker-Chemie, and is known as the **Wacker process**. Propene can be converted into the **ketone**, propanone, by the same method:

$$CH_3CH{=}CH_2 + \tfrac{1}{2}O_2 \longrightarrow CH_3CCH_3$$
$$\qquad\qquad\qquad\qquad\qquad \overset{\|}{\underset{O}{}}$$

In the presence of ethanoic acid, ethene forms ethenyl ethanoate (*see also* Section 20.2.1(iv)):

$$CH_2{=}CH_2 + \tfrac{1}{2}O_2 + CH_3CO_2H \rightarrow CH_2{=}CHOCOCH_3 + H_2O$$

In the **OXO reaction**, an alkene is reacted with carbon monoxide and hydrogen at elevated temperatures and pressures, in the presence of a cobalt catalyst such as $Co_2(CO)_8$, and the net reaction is the addition of H and CHO groups across the double bond (a cobalt–alkene complex is formed as an intermediate):

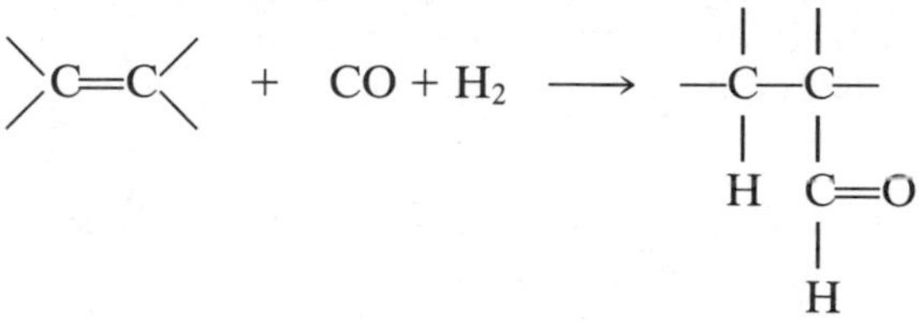

$$\diagup\!\!\!\!C{=}C\diagdown + \quad CO + H_2 \longrightarrow \underset{\substack{H \quad C=O \\ | \\ H}}{-C-C-}$$

e.g.

$$CH_3CH{=}CH_2 + CO + H_2 \xrightarrow[20\text{-}25\ MN\ m^{-2}]{150\text{-}200°C} CH_3CH_2CH_2CHO \quad \text{and} \quad CH_3\underset{\underset{CH_3}{|}}{CHCHO}$$

3.3.2 Oxidation reactions

(i) Combustion

The gaseous and volatile alkenes form explosive mixtures with air or oxygen and, like other hydrocarbons, will burn in excess air or oxygen to form carbon dioxide and water. For example,

$$C_2H_4 + 3O_2 \rightarrow 2CO_2 + 2H_2O \qquad \Delta H^{\ominus} = -1410\,kJ$$

* $PdCl_4^{2-} + C_2H_4 \rightarrow [PdCl_3(C_2H_4)]^- + Cl^-$
$[PdCl_3(C_2H_4)]^- + H_2O \rightarrow CH_3CHO + Pd + 3Cl^- + 2H^+$
$Pd + CuCl_2 \rightarrow PdCl_2 + 2CuCl$
$2CuCl + 2HCl + \tfrac{1}{2}O_2 \rightarrow 2CuCl_2 + H_2O$

Alkenes contain a higher proportion of carbon than do the corresponding alkanes, and tend to burn in air with a luminous, sometimes smoky, flame, owing to their incomplete combustion to carbon.

(ii) Reaction with potassium manganate(VII)

Alkanes are highly resistant to chemical oxidizing agents under normal conditions. In contrast, *alkenes* are quite sensitive to oxidation by several oxidizing agents. If an alkene is shaken with a few drops of a dilute, acidified solution of potassium manganate(VII) the purple solution is rapidly decolorized. The alkene is *oxidized* to an addition product, in which two hydroxyl groups have been added across the carbon–carbon double bond; the product is a *di*-alcohol or **diol**:

$$>C=C< \longrightarrow \underset{\underset{OH}{|}}{-C-}\underset{\underset{OH}{|}}{C-}$$

e.g.

$$CH_2{=}CH_2 \rightarrow HOCH_2CH_2OH$$
$$\text{ethane-1,2-diol}$$

The purple manganate(VII) ion is *reduced* to the virtually colourless Mn^{2+} ion. The actual mechanism of this reaction is complex, but the reaction may be summarized by the equation

$$5 >C=C< + 6H^+ + 2MnO_4^- + 2H_2O \longrightarrow 5 \underset{\underset{OH}{|}\;\underset{OH}{|}}{>C{-}C<} + 2Mn^{2+}$$

If the reaction is carried out under alkaline conditions, then the purple colour first changes to the green colour of manganate(VI) ions, and this is followed by further reduction to a brown precipitate of manganese(IV) oxide. The alkene is oxidized to the diol again. The colour changes characteristic of this oxidation, under acid or alkaline conditions, distinguish alkenes from alkanes, but the reaction is not a conclusive test for alkenes, as alkynes and other compounds which contain easily oxidizable functional groups also react.

Under more vigorous conditions, manganate(VII) causes fission of the carbon–carbon double bond and a variety of products such as aldehydes, ketones and carboxylic acids are formed, e.g.

$$\underset{\underset{CH_3C=CH_2}{}}{\overset{CH_3}{|}} \longrightarrow \underset{\underset{OH\;\;OH}{|\;\;\;|}}{\overset{CH_3}{\overset{|}{CH_3C{-}CH_2}}} \longrightarrow \underset{O}{\overset{\|}{CH_3CCH_3}} + \underset{O}{\overset{\|}{HCH}}$$
$$\downarrow$$
$$HCO_2H$$

(iii) Reaction with trioxygen (ozonolysis)

The oxygen allotrope trioxygen O_3 (*ozone*) is produced by the passage of an electric discharge through oxygen O_2. If the mixture of gases is bubbled through a solution of an alkene at a low temperature, an unstable product known as an **ozonide** is formed. Ozonides are unstable and are readily hydrolysed to form aldehydes or ketones:

$$\begin{array}{ccc} \ce{>C{=}C<} + O_3 & \longrightarrow & \text{ozonide} & \xrightarrow{H_2O} & \ce{>C=O} + \ce{O=C<} + H_2O_2 \\ \text{fission} & & & & \end{array}$$

Since hydrogen peroxide is also formed in the hydrolysis, and since aldehydes may be oxidized by it to **carboxylic acids**, the hydrolysis is usually performed in the presence of zinc dust which destroys any peroxide formed. Alternatively, the ozonide may be decomposed by catalytic hydrogenation. (Note that, although trioxygen must initially add across the double bond of the alkene, *complete* fission of the double bond occurs.)

This reaction with trioxygen is of particular importance because, by examining the hydrolysis products of the ozonide (they may be separated by fractional distillation or by chromatography), the precise position of the double bond in the original alkene can be located. Such a procedure would, for example, distinguish but-1-ene from but-2-ene:

$$CH_3CH_2CH{=}CH_2 + O_3 \rightarrow CH_3CH_2CH \underset{O{-}O}{\overset{O}{\diamond}} CH_2 \xrightarrow[Zn]{H_2O}$$

$$\underset{H}{\overset{CH_3CH_2}{>}}C{=}O + O{=}C\underset{H}{\overset{H}{<}}$$

propanal   methanal

$$CH_3CH{=}CHCH_3 + O_3 \rightarrow CH_3CH \underset{O{-}O}{\overset{O}{\diamond}} CHCH_3 \xrightarrow[Zn]{H_2O}$$

$$\underset{H}{\overset{CH_3}{>}}C{=}O + O{=}C\underset{H}{\overset{CH_3}{<}}$$

2 mol ethanal

Natural rubber contains carbon–carbon double bonds and hence suffers from the serious drawback of perishing when in contact with trioxygen.

(iv) Reaction with peroxoacids

The reaction between a carboxylic acid and a concentrated solution of hydrogen peroxide H_2O_2 produces a **peroxoacid***; for example, ethanoic

---

* A peroxoacid contains the peroxo linkage —O—O—, thus:

$$R{-}C\underset{O{-}O{-}H}{\overset{O}{<}}$$

acid CH_3CO_2H gives peroxoethanoic acid $CH_3C\overset{\displaystyle O}{\underset{\displaystyle O-OH}{}}$. If an alkene is reacted with the resulting solution an **epoxy** compound is produced, an oxygen atom having added on across the double bond:

$$\text{>C=C<} + RCO_3H \longrightarrow \overset{\displaystyle O}{\text{>C---C<}} + RCO_2H$$

(R = an alkyl group, H or C_6H_5)

e.g.

$$CH_2{=}CH_2 + CH_3CO_3H \longrightarrow \overset{\displaystyle O}{CH_2{-}CH_2} + CH_3CO_2H$$

Epoxy compounds are hydrolysed by water, on heating, to diols.

(v) Catalytic oxidation

Alkenes undergo a variety of oxidations in the presence of suitable catalysts. These reactions often form the basis of important manufacturing routes in large-scale organic chemicals production.

For example, ethene is converted into epoxyethane (*see* Section 7.3(iv)), by reaction with air or oxygen at 250°C and 1–3 MN m^{-2} pressure, in the presence of a *silver catalyst*:

$$CH_2{=}CH_2 + \tfrac{1}{2}O_2 \longrightarrow \overset{\displaystyle O}{CH_2{-}CH_2} \qquad \Delta H^{\ominus} = -146 \text{ kJ}$$

Epoxyethane is an important commercial product, as it is converted by hydrolysis into ethane-1,2-diol, which is the major component of anti-freeze fluids and which is also used in the making of detergents and the synthetic fibre Terylene (*see* Section 20.3(ii)).

The catalytic oxidation of alkenes to aldehydes and ketones by the Wacker process has already been mentioned (*see* Section 3.3.1(vi)).

As well as oxidation of the double bond, oxidation of the carbon atom adjacent to the double bond can occur and this is illustrated by the reaction used in the manufacture of propenoic acid from propene (N.B. epoxypropane is *not* formed, *cf.* Section 3.3.1(iv)):

$$CH_2{=}CHCH_3 \xrightarrow[\substack{400-500°C}]{\substack{O_2, \text{ catalyst}}} CH_2{=}CHCO_2H$$

The acid (formerly known as *acrylic acid*) is an important intermediate in the manufacture of paints, plasticizers and adhesives.

3.3.3 Polymerization reactions

In the presence of certain catalysts alkene molecules can add on *to each other* in a *'head-to-tail'* fashion to form long-chain molecules of very high relative molecular mass, i.e. **macromolecules**. These macromolecules are built up by innumerable repetitions of a small molecular unit, the **monomer** (in this case the alkene molecule), and are called **polymers** (from the Greek *polus* meaning *many*, and *meros* meaning *parts*). Since the units are linked together, in this case, by an addition reaction, such polymers are termed **addition polymers**.

The substances derived from the polymerization of alkenes all have the property of 'mouldability' and are examples of **plastics** (*see* Section 20.2). The most well known and extensively manufactured example is poly(ethene), or polythene, made by polymerizing ethene:

$$n\,(CH_2{=}CH_2) \;\rightarrow\; {+}CH_2{-}CH_2{+}_{\overline{n}} \qquad \text{where } n \text{ is of the}$$
monomer   polymer   order of many thousands

If ethene is polymerized in the presence of a trace of oxygen, as initiator, at a pressure of about 200 MN m^{-2} and a temperature of 100–300°C, then a low-density (LD) form of the polymer is produced. The polymerization under these conditions involves a free-radical chain reaction (*see* Section 20.2.1(i)). The alternative, high-density (HD) form is prepared either by the Phillips process, which uses a chromium(III) oxide catalyst at 7 MN m^{-2} pressure and 125–175°C, or else by the Ziegler method which employs organometallic catalysts ('Ziegler catalysts') at 0.7 MN m^{-2} pressure and 100–170°C.

Propene, too, can be polymerized using 'Ziegler catalysts':

$$\phantom{n\,(CH_3CH{=}CH_2) \longrightarrow {+}} \overset{\displaystyle CH_3}{\underset{\displaystyle |}{}}$$
$$n\,(CH_3CH{=}CH_2) \longrightarrow {+}CH{-}CH_2{+}_{\overline{n}}$$

(There are three different kinds of poly(propene), *see* Section 20.2.1(ii).) Other important polymerizations are shown below:

$$n(CH_2{=}CHCl) \longrightarrow {+}CH_2{-}CH{+}_{\overline{n}}$$
$$\phantom{n(CH_2{=}CHCl) \longrightarrow {+}CH_2{-}} |$$
$$\phantom{n(CH_2{=}CHCl) \longrightarrow {+}CH_2{-}} Cl$$
chloroethene   poly(chloroethene) (also known as PVC)

$$n(CH_2{=}CHCN) \longrightarrow {+}CH_2{-}CH{+}_{\overline{n}}$$
$$\phantom{n(CH_2{=}CHCN) \longrightarrow {+}CH_2{-}} |$$
$$\phantom{n(CH_2{=}CHCN) \longrightarrow {+}CH_2{-}} CN$$
propenonitrile   poly(propenonitrile)

$$n(C_6H_5CH{=}CH_2) \longrightarrow {+}CH{-}CH_2{+}_{\overline{n}}$$
$$\phantom{n(C_6H_5CH{=}CH_2) \longrightarrow {+}} |$$
$$\phantom{n(C_6H_5CH{=}CH_2) \longrightarrow {+}} C_6H_5$$
phenylethene   poly(phenylethene) (also known as *polystyrene*)

$$n(CF_2{=}CF_2) \longrightarrow {+}CF_2{-}CF_2{+}_{\overline{n}}$$
tetrafluoroethene   poly(tetrafluoroethene) (also known as PTFE)

These important plastics are discussed in more detail in Chapter 20.

3.3.4   Mechanism of alkene addition reactions

The addition reactions of alkenes with halogens, hydrogen halides, halic(I) acids and sulphuric acid all follow essentially the same type of mechanism, involving ions.

In a carbon–carbon double bond the π-bond is weaker than the σ-bond (*see* Section 1.3), and the π-electrons are generally accessible to other reagents and available for sharing with other atoms, i.e. for electron-pair, or covalent, bond formation. The type of reagent which tends to attack the double bond of alkenes is therefore commonly an **electrophile**, i.e. an electron-seeking species. The mechanism involves shifts of electron-pairs (denoted by an arrow ⌢) and is ionic.

Taking the addition of bromine to an alkene as an example, stereo-chemical and other evidence suggests that the bromine molecule does *not* add on as a whole, in a single step, but that the reaction occurs in stages. The close approach of the bromine molecule to the π-cloud of the alkene induces polarization of the bromine–bromine bond, followed by heter-olysis of the bond. In the process, a positively charged organic ion and a bromide ion are formed.

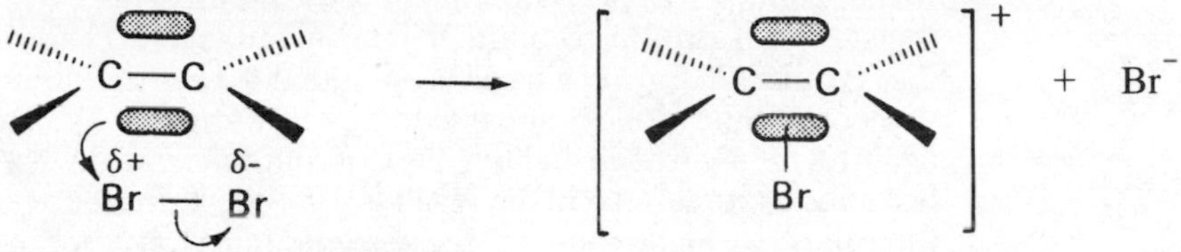

The positively charged ion may be formulated in several different ways, e.g.

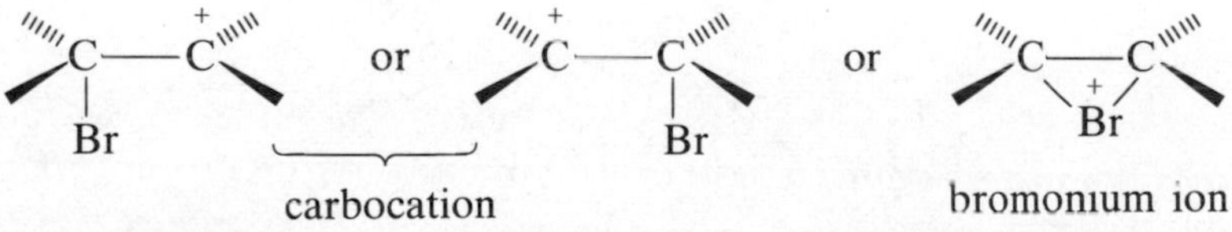

The bromonium ion representation is perhaps the most satisfactory. Reaction is completed by the combination of a bromide ion (which provides *two* electrons to form a covalent bond and acts as a nucleophile) with the bromonium ion:

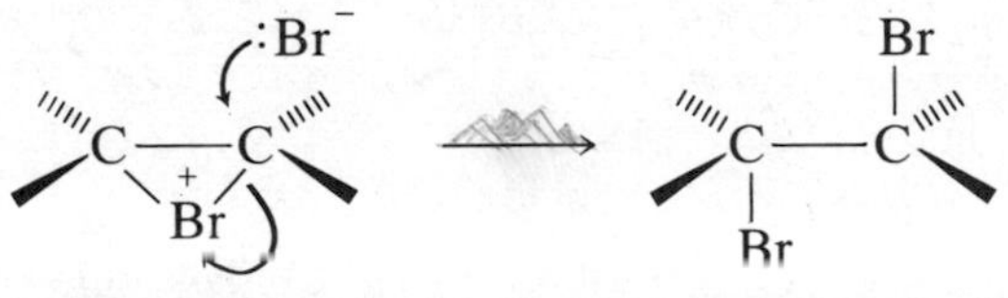

There is ample evidence from experiments (*see,* for example, Section 16.4.2) that the two bromine atoms add on across the double bond from *opposite sides* of the alkene, and this observation is most conveniently

explained in terms of the bromonium ion intermediate, rather than a
carbocation one, since the bulky bromine atom would be expected to
shield one side of the ion from attack by bromide ion.

When an alkene reacts with a solution of a halic(I) acid, it is usually
treated with a solution of the halogen in water, e.g. bromine water. In
this case, a bromonium ion is once again thought to be formed but it will
be susceptible to attack by either bromide ion, or by water molecules,
and hence gives mixed products (*see* Section 3.3.1(iv)):

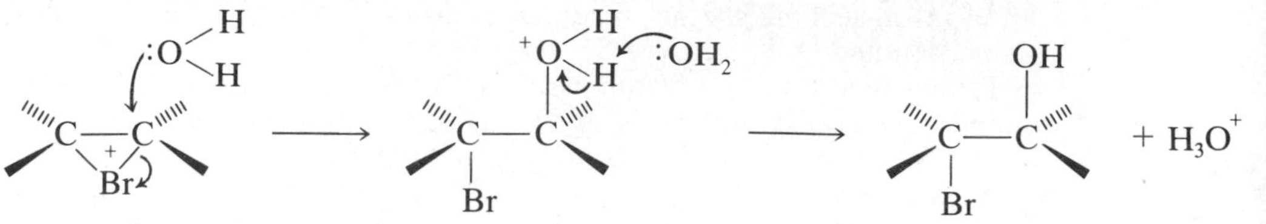

The oxygen atom of a solvent water molecule donates a pair of electrons
to form a covalent bond with the carbon atom, and the positive charge
is therefore transferred to the oxygen atom. This is followed by the
transfer of a proton to another solvent molecule. The overall effect is
that Br and OH groups have been added across the double bond.

Electrophilic attack on alkenes by protons is also involved in the
addition of hydrogen halides and of sulphuric acid. The double bond
becomes protonated and the result is a carbocation; there is no alternative
formulation, analogous to the bromonium ion, because the proton
(hydrogen ion H^+) has no lone-pair of electrons associated with it. The
reaction is completed by the attack of a halide ion:

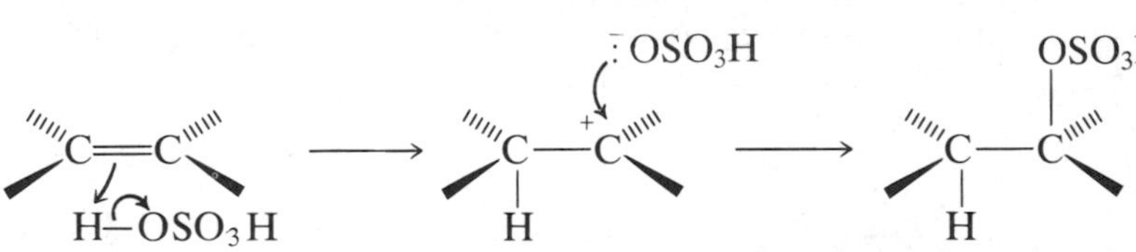

In the case of sulphuric acid the mechanism may be formulated as

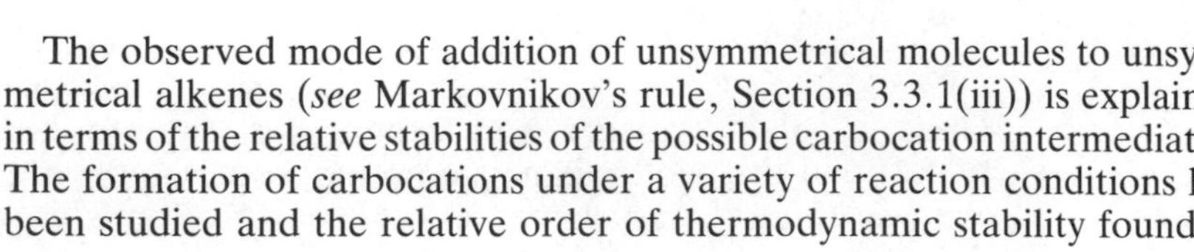

The observed mode of addition of unsymmetrical molecules to unsymmetrical alkenes (*see* Markovnikov's rule, Section 3.3.1(iii)) is explained
in terms of the relative stabilities of the possible carbocation intermediates.
The formation of carbocations under a variety of reaction conditions has
been studied and the relative order of thermodynamic stability found to

be **tertiary** more stable than **secondary**, which is more stable than a **primary** one (*see also* Section 1.2 for meaning of tertiary, etc.), e.g.

$$
\begin{array}{ccc}
\overset{\displaystyle CH_3}{\underset{\displaystyle CH_3}{\overset{|}{\underset{|}{CH_3-C^+}}}} & > & \overset{\displaystyle CH_3}{\underset{\displaystyle H}{\overset{|}{\underset{|}{CH_3-C^+}}}} & > & \overset{\displaystyle H}{\underset{\displaystyle H}{\overset{|}{\underset{|}{CH_3-C^+}}}} \\
\text{tertiary} & & \text{secondary} & & \text{primary}
\end{array}
$$

Thus, when hydrogen bromide adds to propene, a secondary carbocation intermediate is preferred to the primary one, and the product is 2-bromopropane and not the 1-isomer:

$$
CH_3CH=CH_2 \quad \underset{\underset{H-Br}{\overset{\delta+}{}\overset{\delta-}{}}}{\nearrow\searrow}
\begin{array}{l}
CH_3\overset{+}{C}HCH_3 \xrightarrow{\;Br^-\;} CH_3CHCH_3 \\
| \\
Br \\
\\
CH_3CH_2CH_2^+ \xrightarrow{\;^-Br^-\;} CH_3CH_2CH_2Br
\end{array}
$$

The electron-donating tendency (inductive effect) (*see* Section 1.4) of alkyl groups, e.g. $CH_3{\rightarrow}C{<}$, albeit rather weak, helps to stabilize the positive charge on a carbon atom; hence the observed order of stabilities of carbocations. It should be emphasized that these are *relative* stabilities; even a tertiary carbocation is highly reactive towards nucleophilic attack and normally has little more than a transient existence during the course of the reaction.

In the cationic polymerization of alkenes, as in that of 2-methylpropene with an aluminium chloride/water catalyst, a similar electrophilic mechanism is thought to be involved:

$$
AlCl_3 + H_2O \rightarrow [AlCl_3(OH)]^- + H^+
$$

$$
\underset{CH_3}{\overset{CH_3}{{>}}}C=CH_2 \xrightarrow{} CH_3-\underset{\underset{CH_3}{|}}{\overset{\overset{CH_3}{|}}{C^+}} + CH_2=C\underset{CH_3}{\overset{CH_3}{{<}}} \xrightarrow{} CH_3-\underset{\underset{CH_3}{|}}{\overset{\overset{CH_3}{|}}{C}}-CH_2-\underset{\underset{CH_3}{|}}{\overset{\overset{CH_3}{|}}{C^+}} \text{ etc.}
$$

Thus, most of the addition reactions of alkenes involve an initial **electrophilic** attack on the alkene, reaction being completed by subsequent nucleophilic attack. The reactions are therefore referred to as **electrophilic addition** reactions.

However, it is worth noting that the addition of hydrogen across a double bond does *not* involve an ionic mechanism. Both *atoms* of hydrogen and alkene molecules are adsorbed on the metal catalyst surface and the hydrogen atoms add across the double bond from the *same side,* i.e. the addition is *cis-*.

3.4 Dienes

Unsaturated hydrocarbons containing two double bonds per molecule are
known as **dienes**. They generally display properties typical of alkenes,
but when the double bonds are **conjugated**, i.e. separated by only one
carbon–carbon single bond, there are some modifications.

Buta-1,3-diene is a typical conjugated diene. Although it undergoes
addition reactions with hydrogen and halogen very readily, it is possible
to stop the reaction after the uptake of 1 mol of reagent, when it is found
that 1,4-addition has taken place with the rearrangement of a double
bond. Thus, buta-1,3-diene reacts with a halogen or hydrogen halide:

$$CH_2{=}CH{-}CH{=}CH_2 + X_2 \quad \rightarrow \quad X{-}CH_2CH{=}CHCH_2{-}X$$

$$\text{(plus a little } CH_2{=}CH{-}CH{-}CH_2X)$$
$$|$$
$$X$$

$$CH_2{=}CH{-}CH{=}CH_2 + HX \quad \rightarrow \quad CH_3\,CH{=}CHCH_2\,X$$

$$\text{(plus some } CH_2{=}CH{-}CHCH_3)$$
$$|$$
$$X$$

Polymerization is also an important feature of diene chemistry. Natural
rubber is, in fact, a polymer of methylbuta-1,3-diene (*isoprene*), the latter
being the product of the distillation of natural rubber:

$$\left(\begin{array}{c} CH_2 \\ \diagdown \\ \quad C{=}CH \quad CH_2 \\ \diagup \\ CH_3 \end{array}\right)_n \xrightarrow{\;heat\;}$$

$$CH_2{=}C{-}CH{=}CH_2$$
$$|$$
$$CH_3$$

methylbuta-1,3-diene

b.p. = 34°C

Buta-1,3-diene is obtained from the C_4 products of the cracking of
naphtha (*see* Section 18.3.1) and is an important starting material for the
synthetic rubber industry (*see* Section 20.4).

3.5 Manufacture and uses of alkenes

The simpler alkenes are obtained industrially from the cracking of the
naphtha fraction from oil distillation, and are some of the most versatile
and widely used chemical feedstocks in the petrochemicals industry. A
large amount of ethene is polymerized to poly(ethene) and it also finds
extensive use in the manufacture of such important chemicals as PVC,
ethanol, epoxyethane (for anti-freeze, detergents and synthetic fibres),
poly(phenylethene) and ethanoic acid. Some of the reactions are summarized in *Figure 3.1*.

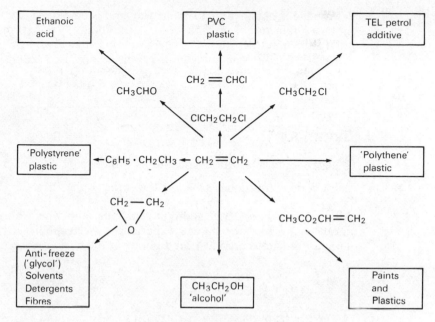

Figure 3.1   Some synthetic routes from ethene

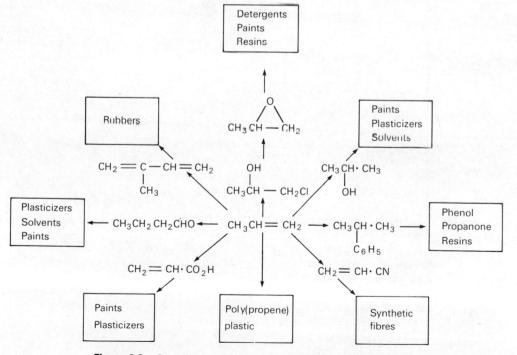

Figure 3.2   Some synthetic routes from propene

Propene is obtained in a similar way and is employed, for example, in the manufacture of phenol, propanone, poly(propene) and poly(propenonitrile). Some of the reactions are shown in *Figure 3.2*.

The various butenes and buta-1,3-diene are used extensively in the manufacture of synthetic rubbers (*see* Section 20.4).

Summary

1   The **alkenes** form a homologous series with a general formula C_nH_{2n}. The characteristic functional group of the series is the $\diagup\!\!\!C\!=\!C\diagdown$ *bond*.

2   Alkenes readily undergo **addition** reactions, and where an ionic mechanism is involved the alkene experiences **electrophilic** attack. Typical addition reactions are illustrated below in the case of propene:

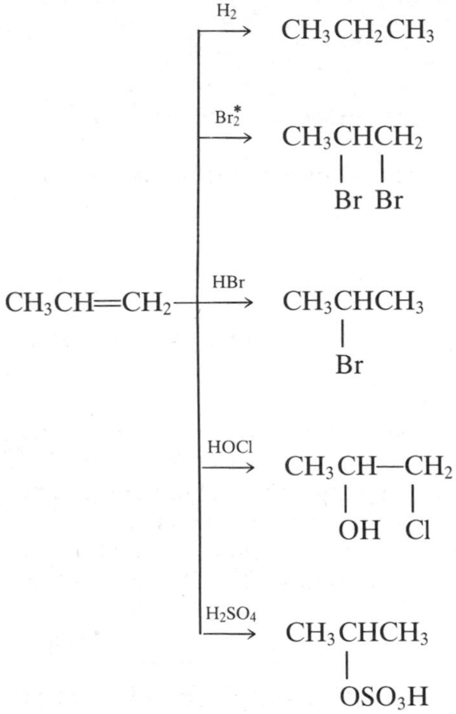

$$CH_3CH{=}CH_2$$

$\xrightarrow{\;H_2\;}$ $CH_3CH_2CH_3$

$\xrightarrow{\;Br_2^{*}\;}$ CH_3CHCH_2
$\qquad\qquad\;\;$ | |
$\qquad\qquad\;\;$ Br Br

$\xrightarrow{\;HBr\;}$ CH_3CHCH_3
$\qquad\qquad\;\;$ |
$\qquad\qquad\;\;$ Br

$\xrightarrow{\;HOCl\;}$ $CH_3CH{-}CH_2$
$\qquad\qquad\;\;$ | |
$\qquad\qquad\;\;$ OH Cl

$\xrightarrow{\;H_2SO_4\;}$ CH_3CHCH_3
$\qquad\qquad\;\;$ |
$\qquad\qquad\;\;$ OSO_3H

* These reactions form the basis of useful methods for distinguishing between alkenes and alkanes.

3   Alkenes are also sensitive to **oxidation** by a variety of reagents. Examples
    are shown for propene:

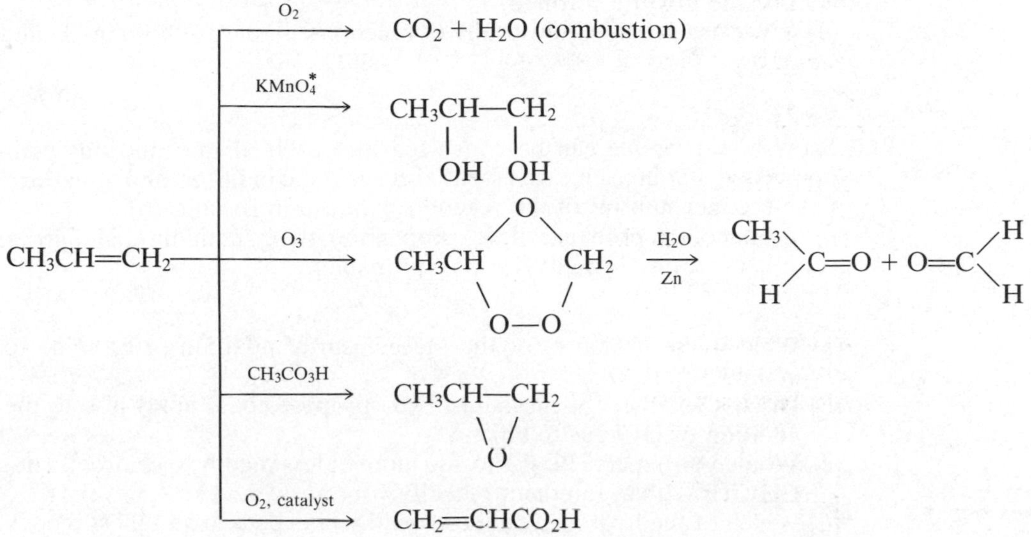

$$CH_3CH{=}CH_2$$

$\xrightarrow{\ O_2\ }$ $CO_2 + H_2O$ (combustion)

$\xrightarrow{\ KMnO_4^*\ }$

$$CH_3CH{-}CH_2$$
$$\quad | \quad\ \ |$$
$$\quad OH\ \ OH$$

$\xrightarrow{\ O_3\ }$

$$CH_3CH \overset{O}{\underset{O-O}{\diagup\diagdown}} CH_2$$

$\xrightarrow[Zn]{\ H_2O\ }$

$$\underset{H}{\overset{CH_3}{>}}C{=}O + O{=}C\underset{H}{\overset{H}{<}}$$

$\xrightarrow{\ CH_3CO_3H\ }$

$$CH_3CH{-}CH_2$$
$$\quad \diagdown\,O\,\diagup$$

$\xrightarrow{\ O_2,\ catalyst\ }$

$$CH_2{=}CHCO_2H$$

4   Alkenes can add on to each other to form **addition polymers** which are
    important plastics.
5   Alkenes find their major use as reactive chemical feedstocks for the
    petrochemicals industry.

Questions

1   (a) Write an equation for the electrophilic reaction of hydrogen bromide
        with propene.
    (b) Give the mechanism for the reaction.
    (c) Give the structural formula for an isomer of the product of this
        reaction.
    (d) Explain briefly why this isomer is not a major product in the reaction
        of hydrogen bromide with propene under these conditions.
    (e) Write a structure for the product that you would predict to be formed
        in the reaction of propene with nitrosyl chloride, NOCl (which reacts
        as NO^+Cl^-).

                                                                          JMB

2   (a) Write an equation for the principal reaction which takes place when
        propene is bubbled into 80% sulphuric acid at room temperature and
        give the structural formula of the organic product.
    (b) Outline the mechanism of the reaction leading to the formation of
        this product.
    (c) How does this mechanism explain the formation of one product rather
        than another when sulphuric acid reacts with propene?

(d) Write an equation for the reaction which occurs when the product of the reaction between sulphuric acid and propene is added to water and the mixture warmed.

(e) What use is made industrially of reactions similar to those in (a) and (d)?

JMB

3* Show how propene can be converted into each of the following compounds, giving in each case essential reagents, conditions and equations. Give the mechanisms of the reactions you use in (b) and (d).
(a) propane; (b) propan-2-ol; (c) propanone; (d) 2-methylpropanonitrile; (e) ethanoic (*acetic*) acid; (f) epoxypropane.

JMB

4* (a) What is the evidence for the mechanism of addition of bromine to ethene?

(b) Discuss whether the mechanism you propose could apply also to the addition of HCl gas to ethene.

(c) Would you expect HCl gas to add more or less readily to chloroethene, CH_2CHCl, than to ethene? Justify your answer.

(d) Which of the hydrogen halides would you expect to add most readily to ethene? Justify your answer.

(e) Suggest a reason why the addition of HCl to propene normally produces more 2-chloropropane than 1-chloropropane.

Nuffield

5* State the Markovnikov rule for addition reactions of HBr to alkenes, and give a mechanistic explanation of the rule.
Bromine water was added to an alkene A, C_4H_8, to give a compound **B**. Acid-catalysed dehydration of **B** resulted in the formation of two isomeric bromoalkenes **C** and D, C_4H_7Br. When **C** was cleaved with trioxygen (*ozone*), methanal was identified as one of the products. **D** gave no methanal with trioxygen, but propanone was formed in this reaction. Identify compounds **A,B,C** and **D**, and explain the observations described above.

JMB

6** Compound M is a gaseous hydrocarbon and on combustion yields carbon dioxide and water in the molar ratio 4:3. It reacts readily with chlorine, forming compound N of empirical formula C_2H_3Cl. Compound N, on ozonolysis, gives a single compound P which may be further oxidized to chloroethanoic acid.
Identify compounds M, N and P and explain the reactions leading from M to the final product. In particular, discuss the chlorination reaction in which compound N is formed.

London

7* A compound **A**, C_6H_{10}, takes up only **one** mole of hydrogen on hydrogenation over a nickel catalyst at room temperature. **A** reacts with bromine in tetrachloromethane and with dilute neutral potassium permanganate solution. Ozonolysis of **A** (reaction with ozone) gives, after treatment with zinc dust and water, methanal (*formaldehyde* CH_2O) and a second carbonyl compound of formula C_5H_8O. Deduce at least one possible structure for **A**, and write structures for the products derived from **A** in the above-mentioned reactions.

Oxford

8   Discuss the detailed mechanism of the reaction of hydrogen bromide with (i) ethene and (ii) propene.
Show how the use of addition reactions involving ethene leads to the industrial production of each of the following compounds:

$$\underset{\Large CH_2\text{——}CH_2}{\overset{\Large O}{\diagup \diagdown}}$$

ethanol, epoxyethane $\overset{O}{CH_2\text{——}CH_2}$, ethane-1,2-diol $HOCH_2CH_2OH$, and chloroethene $CH_2{=}CHCl$.
Describe briefly the importance of these compounds.

JMB

Chapter 4
Alkynes

4.1 Homologous series of alkynes

The alkynes constitute a homologous series of highly unsaturated hydro-carbons of general formula C_nH_{2n-2}. The functional group of the series may be regarded as the carbon–carbon *triple bond* —C≡C—. The alkynes are named by retaining the prefix indicating the number of carbon atoms in the molecule and employing the suffix *-yne*. Thus, the simplest member of the series C_2H_2 is known as ethyne (formerly called *acetylene*). Some physical properties of the first seven members of the series of 1-alkynes (i.e. those in which the triple bond is to be found at the end of the chain, or in the *terminal* position as it is sometimes referred to) are summarized in *Table 4.1*.

Table 4.1

Name	Formula	Melting point (°C)	Boiling point (°C)	Density (g cm^{-3})
ethyne	HC≡CH	−81*	−84	—
propyne	$CH_3C≡CH$	−103	−23	—
but-1-yne	$CH_3CH_2C≡CH$	−126	8	—
pent-1-yne	$CH_3(CH_2)_2C≡CH$	−106	40	0.690
hex-1-yne	$CH_3(CH_2)_3C≡CH$	−132	72	0.716
hept-1-yne	$CH_3(CH_2)_4C≡CH$	−81	100	0.733
oct-1-yne	$CH_3(CH_2)_5C≡CH$	−79	126	0.746

Ethyne is obtained from the cracking of methane or naphtha and is the only alkyne which is widely used. A range of compounds containing the carbon–carbon triple bond is to be found in natural plant sources.

4.2 Physical properties of alkynes

The series of 1-alkynes shows the usual gradation of physical properties, as exemplified by melting and boiling points and densities. On the whole,

* On cooling in liquid air, ethyne solidifies to a wax-like solid which melts at a temperature just above ethyne's boiling point at normal pressure.

the boiling points of the alkynes tend to be 10–20°C higher than those of the corresponding alkanes and alkenes. Ethyne itself is a colourless gas which is slightly soluble in water. But-2-yne $CH_3C{\equiv}CCH_3$ (b.p. = 27°C) and the C_5 alkynes are the first of the series to be liquids at room temperature.

4.3 Chemical properties of alkynes

The carbon–carbon triple bond —C≡C— consists of one σ-bond and two weaker π-bonds, so it is not surprising that **addition** reactions are a characteristic feature of this functional group. Despite its high formal unsaturation, the triple bond does not usually react so vigorously as a carbon–carbon double bond, as in alkenes, and it is therefore sometimes possible to stop the addition reaction after the uptake of 1 mol of reactant, rather than 2 mol which would be possible in principle, e.g.

$$-C{\equiv}C- \ + \ X-Y \ \longrightarrow \ \underset{\underset{X \ \ Y}{| \ \ |}}{-C{=}C-} \ \xrightarrow{X-Y} \ -CX_2-CY_2-$$

The reactions frequently involve ionic species and are generally heterolytic ones, but the mechanism may involve initial attack by electrophiles (like alkenes, *see* Section 4.4) or by nucleophiles (unlike alkenes).

Since the alkynes are hydrocarbons, they would be expected to undergo **combustion** and **oxidation**; since they are highly unsaturated they would be expected to undergo **addition** and maybe **polymerization reactions**.

4.3.1 Addition reactions

(i) Reaction with hydrogen
As would be expected, alkynes can be hydrogenated catalytically, taking up 2 mol of hydrogen per mol of alkyne to form the corresponding saturated molecule, an alkane:

$$-C{\equiv}C- \ + \ 2H_2 \ \rightarrow \ -CH_2-CH_2-$$

e.g.

$$HC{\equiv}CH \ + \ 2H_2 \ \rightarrow \ CH_3-CH_3$$

Careful control of the hydrogenation can stop the reaction at the alkene stage, since the subsequent addition is a little slower, but use of a **Lindlar catalyst** (palladium treated with certain additives, and introduced by H. Lindlar in 1952) is more effective since this catalyses only the first step. Experimental evidence shows once again that the hydrogen atoms are added on from the same side of the triple bond, so that disubstituted alkynes yield the *cis* geometrical isomer of the alkene when reduced in this way, e.g.

$$CH_3-C\equiv C-CH_3 + H_2 \longrightarrow$$

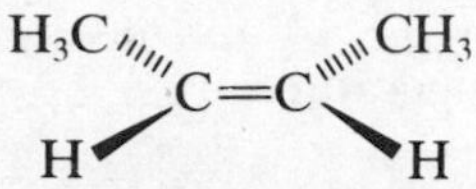

b.p. = 27°C   b.p. = 4°C
but-2-yne   *cis*-but-2-ene

(ii) Reaction with halogens
Ethyne itself is oxidized by chlorine; on mixing of the two gases, there is an explosive reaction liberating hydrogen chloride and substantial amounts of soot (carbon):

$$C_2H_2 + Cl_2 \rightarrow 2HCl + 2C$$

The reaction can be modified, however, so that an addition reaction takes place, and this may proceed with the uptake of 2 mol of chlorine per mol of alkyne, e.g.

$$HC\equiv CH + Cl_2 \xrightarrow[\text{FeCl}_3]{80°C} CHCl=CHCl \xrightarrow{Cl_2} Cl_2CH-CHCl_2$$

The ultimate product is a useful solvent for oils and resins but is also used to make another very widely used solvent, trichloroethene:

$$Cl_2CH-CHCl_2 \xrightarrow[(-HCl)]{220-320°C} Cl_2C=CHCl$$

1,1,2,2-tetrachloroethane   trichloroethene
b.p. = 147°C   b.p. = 87°C

Bromine, too, reacts in a similar manner on addition, but there is no reaction with iodine. Thus alkynes, as well as alkenes, etc., decolorize an aqueous solution of bromine (*see* Section 3.3.1(ii)).

(iii) Reaction with hydrogen halides
The addition of hydrogen chloride to ethyne is catalysed by a mixture of mercury(II) chloride and activated charcoal:

$$HC\equiv CH + HCl \xrightarrow[\text{HgCl}_2]{200°C} CH_2=CHCl \qquad \Delta H^\ominus = -184 \text{ kJ}$$

ethyne   chloroethene
                                                  b.p. = 57°C

The product is very important for the manufacture of PVC plastic but is now more usually manufactured from ethene (*see* Section 3.3.1(ii)). Another molecule of hydrogen chloride may be taken up, when the unsymmetrical 1,1-dichloroethane is produced. This addition follows Markovnikov's rule:

$$CH_2{=}CHCl + HCl \rightarrow CH_3CHCl_2$$

The reaction of alkynes with hydrogen halides generally proceeds more slowly than with alkenes.

(iv) Reaction with hydrogen cyanide
This is not a general reaction, but hydrogen cyanide HCN (usually manufactured from methane and ammonia) adds on to ethyne, in the presence of a copper(I) chloride catalyst, to give propenonitrile. This substance is polymerized to form an important plastic (*see* Section 3.3.3, and Section 20.3(iii)), but is made at the present time almost entirely from propene (but *see* Section 4.5):

$$HC{\equiv}CH + HCN \xrightarrow[\text{CuCl/HCl}]{80-90^\circ C} CH_2{=}CHCN$$

ethyne   propenonitrile
                                                 b.p. = 78°C

(v) Reaction with water
In the presence of dilute sulphuric acid and a mercury(II) sulphate catalyst at about 60°C, water adds across a triple bond and the alkyne is said to be *hydrated*. The initial addition product, however, immediately isomerizes to an aldehyde or ketone (*see also* Section 8.3.6):

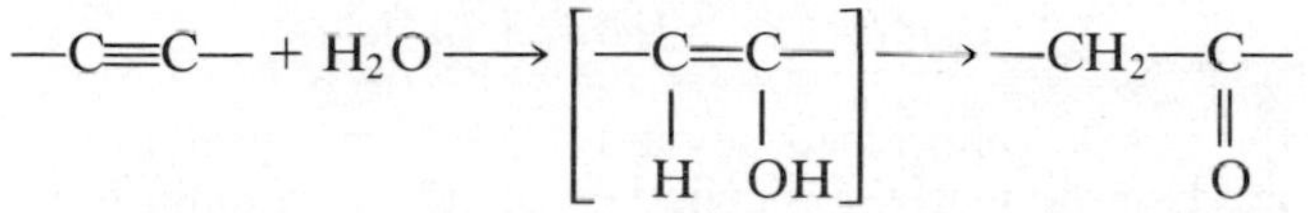

$$-C{\equiv}C- + H_2O \longrightarrow \left[\begin{array}{c} -C{=}C- \\ | \quad | \\ H \quad OH \end{array} \right] \longrightarrow -CH_2{-}\underset{O}{\overset{\parallel}{C}}{-}$$

Thus ethyne gives an aldehyde, ethanal, as the product:

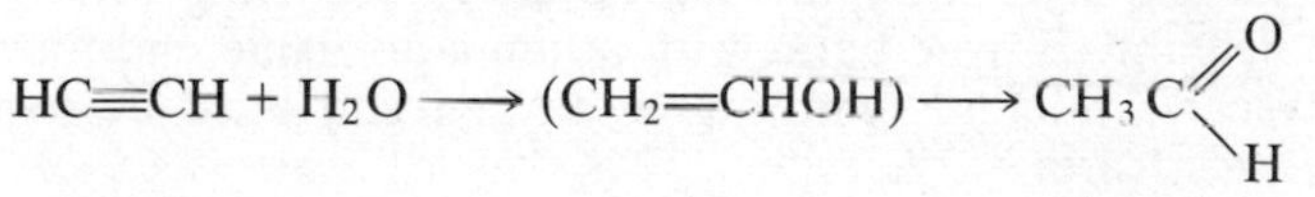

$$HC{\equiv}CH + H_2O \longrightarrow (CH_2{=}CHOH) \longrightarrow CH_3C{\overset{\displaystyle O}{\underset{\displaystyle H}{<}}}$$

ethanal
b.p. = 21°C

This reaction has been of some commercial importance for the manufacture of ethanal, but has now been superseded by other routes starting from ethene or ethanol.

Propyne gives a ketone, propanone, on hydration so the addition clearly follows Markovnikov's rule:

$$CH_3C{\equiv}CH + H_2O \longrightarrow \left(\begin{array}{c} CH_3C{-}CH_2 \\ | \\ OH \end{array} \right) \longrightarrow CH_3{-}\underset{O}{\overset{\parallel}{C}}{-}CH_3$$

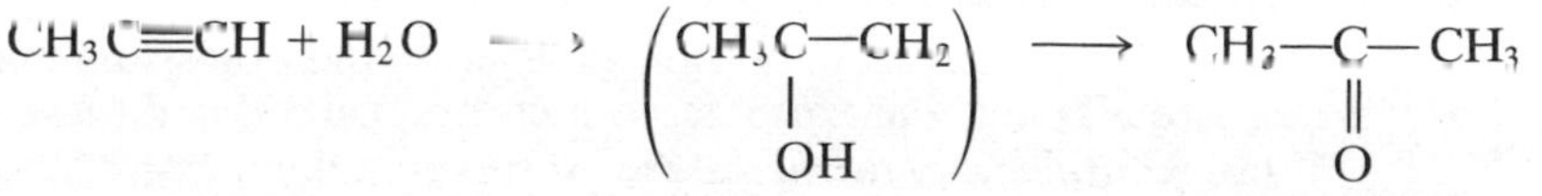

propanone
b.p. = 56°C

(vi) Reaction with ethanoic acid

Here again, this is not a general reaction but is the basis of an important commercial reaction with ethyne. In the presence of a zinc or cadmium ethanoate catalyst, supported on charcoal, ethyne and ethanoic acid react in the vapour state, at a temperature of about 200°C, to produce the ester, ethenyl ethanoate (formerly known as *vinyl acetate*). This product finds important uses, in polymerized form, in adhesives and emulsion paints (*see* Section 20.2.1(iv)):

$$HC\equiv CH + CH_3C\overset{O}{\underset{OH}{\diagdown}} \longrightarrow CH_3C\overset{O}{\underset{OCH=CH_2}{\diagdown}}$$

ethenyl ethanoate
b.p. = 73°C

4.3.2   Oxidation reactions

(i) Combustion

The alkynes, as hydrocarbons, will undergo combustion to carbon dioxide and water in excess air or oxygen, but in a limited supply they undergo incomplete combustion to carbon, and hence burn with a sooty flame (an indication of their high degree of unsaturation, or high carbon content).

Ethyne has a positive standard enthalpy of formation (i.e. it is an *endothermic* compound), and is thermodynamically unstable with respect to its constituent elements:

$$2C + H_2 \quad \rightarrow \quad C_2H_2 \qquad \Delta H^\ominus = +227 \text{ kJ}$$

If the gas is compressed beyond 0.2 MN m^{-2} pressure, it explodes violently. Consequently, gas cylinders of ethyne contain solutions of the gas, under pressure of about 1.2 MN m^{-2}, in propanone, the latter being absorbed in a finely divided naturally occurring form of silica called kieselguhr. Ethyne burns with a luminous flame and forms dangerously explosive mixtures with air or oxygen and must be handled with the greatest care:

$$2C_2H_2 + 5O_2 \quad \rightarrow \quad 4CO_2 + 2H_2O \qquad \Delta H^\ominus = -1300 \text{ kJ}$$

Partly owing to the endothermic nature of ethyne, its enthalpy of combustion is high and is responsible for the high flame temperatures reached (up to 3000°C) when ethyne/oxygen mixtures are burned (as in 'oxyacetylene' equipment used in welding).

(ii) Other oxidations

Alkynes are oxidized by trioxygen/water and potassium manganate(VII), as are alkenes, but various products are formed and these reactions have no general preparative value. However, ethyne for example, will decolorize acidified manganate(VII)—*see* Section 3.3.2(ii). Alkynes are much less sensitive to oxidation by peroxoacids than alkenes.

4.3.3   Polymerization reactions

Polymerization, strictly speaking, implies the linking together of a very large number of molecules to produce a macromolecule. When only a few molecules are linked together, the process should be referred to as **oligomerization** and the products described as **oligomers**. In the specific case of *two* molecules linking together, the process is described as **dimerization**.

In the presence of acid/copper(I) chloride as catalyst, ethyne dimerizes to form but-1-ene-3-yne:

$$2HC\equiv CH \quad \rightarrow \quad CH_2\!\!=\!\!CH\!\!-\!\!C\equiv CH$$
$$\text{but-1-ene-3-yne}$$
$$\text{b.p.} = 5°C$$

Addition of hydrogen chloride to this compound, in the presence of the same catalyst at 30–60°C, yields 2-chloro-1,3-butadiene (*chloroprene*):

$$CH_2\!\!=\!\!CHC\equiv CH + HCl \quad \longrightarrow \quad CH_2\!\!=\!\!CHC\!\!=\!\!CH_2$$
$$\qquad\qquad\qquad\qquad\qquad\qquad\qquad\qquad | $$
$$\qquad\qquad\qquad\qquad\qquad\qquad\qquad\qquad Cl$$

which when polymerized yields the important synthetic *neoprene* rubbers (*see* Section 20.4).

Extensive researches carried out by the German chemist Walter Reppe have shown that **trimerization** of ethyne to benzene (for discussion of structure, *see* Section 12.2) can be achieved using Ziegler–Natta organometallic catalysts (*see* Section 20.2.1(i)):

$$3HC\equiv CH \quad \rightarrow \quad C_6H_6$$

Using substituted alkynes, substituted benzenes may be prepared.

Higher oligomers may also be prepared, often using nickel(II) complexes as catalysts:

$$4HC\equiv CH \quad \longrightarrow$$

cyclo-octatetraene

Oligomerizations of alkynes, alkenes and dienes, using catalysts of this sort, have been extensively investigated in recent years and are beginning to play a more important role in industrial manufacturing routes.

4.3.4   Metal derivatives

The hydrogen atom attached to the triple bond in 1-alkynes can be replaced by metals under a variety of conditions; this C—H bond is evidently weaker than a normal C—H bond in an alkane, or in an alkene. With a very strong base, such as sodium amide (*sodamide*) $NaNH_2$ in liquid ammonia, a proton is removed and the alkyne can then be considered to behave as a very weak acid. Simple sodium derivatives of alkynes can be made by passing the alkyne through a solution of sodium, or better $NaNH_2$, in liquid ammonia (b.p. $= -34°C$), e.g.

$$HC\equiv CH + 2Na \xrightarrow{NH_3} 2HC\equiv C^- Na^+ + H_2$$

Sodium amide acts simply as a strong base, which permits the transfer of a proton from the alkyne to the base:

$$HC\equiv C\overset{\frown}{-}H + :NH_2^- \xrightarrow{NH_3} HC\equiv C^- + NH_3$$

Copper(I) and silver(I) salts of 1-alkynes may be prepared by passing the alkyne through an ammoniacal solution of either copper(I) chloride, or of silver nitrate. In the case of ethyne, the copper(I) salt is brick-red in colour and the silver salt is white; both are precipitated from the ammonia-cal solution:

$$HC\equiv CH + 2Cu(NH_3)_2^+ \rightarrow CuC\equiv CCu + 2NH_4^+ + 2NH_3$$
$$HC\equiv CH + 2Ag(NH_3)_2^+ \rightarrow AgC\equiv CAg + 2NH_4^+ + 2NH_3$$

The salts are called copper(I) dicarbide and silver dicarbide, and neither should be handled when dry as they can then be dangerously explosive. The formation of these salts is the basis of a simple laboratory test for 1-alkynes (non-terminal alkynes cannot, of course, form such salts). Ethene and ethyne would be readily distinguished by such a test.

Ionic dicarbides are not explosive but are easily decomposed by water. Calcium dicarbide can be prepared by heating coke and calcium oxide to a temperature of about 2000°C in an electric furnace:

$$CaO + 3C \rightarrow CaC_2 + CO$$

Water decomposes the dicarbide immediately into ethyne and this reaction can be used to generate small amounts of the gas in the laboratory, although the ethyne produced is not pure:

$$Ca^{2+} \; ^-C\equiv C^- + 2H_2O \rightarrow HC\equiv CH + Ca(OH)_2$$

Metal dicarbides are considered by some chemists to have been a possible source of the hydrocarbons found in natural oil and gas deposits, although most of these are undoubtedly of biological origin.

4.4   Mechanism of alkyne addition reactions

The carbon–carbon triple bond consists of one σ-bond and two π-bonds, and is therefore effectively sheathed in a cylindrical distribution of electron charge (*see* Section 1.3); this can act as a source of electron-pairs which are used to form new bonds in ionic reactions. Undoubtedly, some reactions involve **electrophilic** addition. The addition reaction between alkyne and halogen, or hydrogen halide, for example, is thought to proceed in an analogous fashion to that for alkenes:

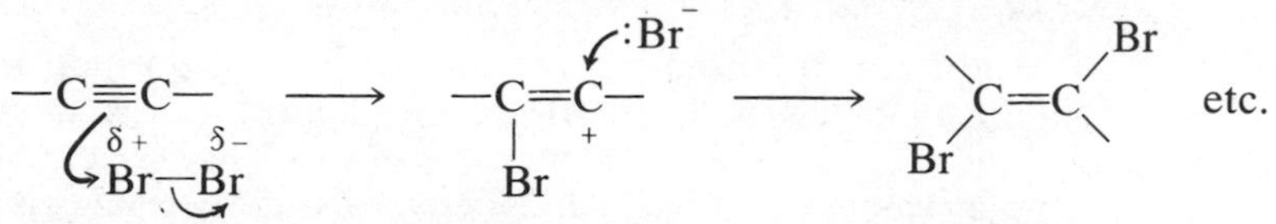

$$-C\equiv C- \longrightarrow -C=C- \longrightarrow \;\;C=C\;\; \text{etc.}$$

The product of the addition of 1 mol of halogen is predominantly the *trans* geometrical isomer. Similarly, with hydrogen halides:

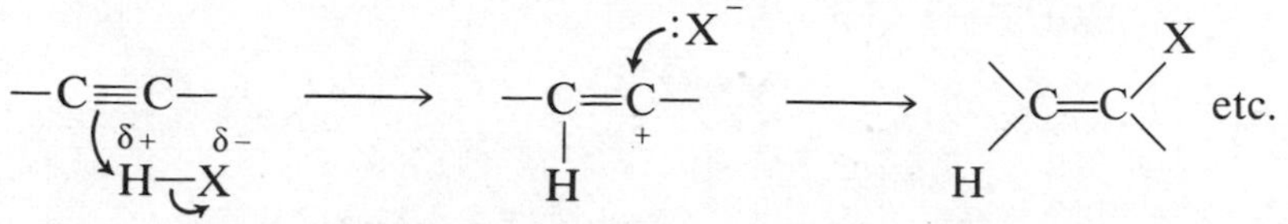

$$-C\equiv C- \longrightarrow -C=C- \longrightarrow \;\;C=C\;\; \text{etc.}$$

Some of the reactions of alkynes are not so easy to rationalize or explain in mechanistic terms. For example, despite the electron distribution of the triple bond, alkynes tend to react *more slowly* than alkenes. Some reactions, too, are thought to proceed via **nucleophilic** addition. Metal ions, such as those of copper(I) and mercury(II), can form complexes with the triple bond and catalyse nucleophilic additions. Therefore, the hydration of alkynes is thought to proceed thus:

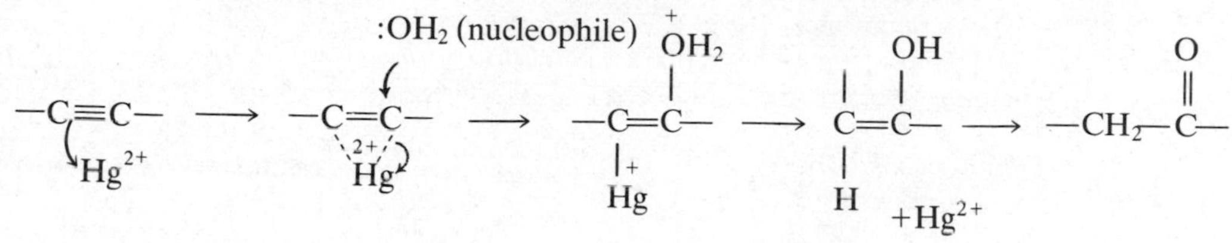

$$-C\equiv C- \longrightarrow -C=C- \longrightarrow -C=C- \longrightarrow C=C- \longrightarrow -CH_2-C-$$

The copper(I) catalysed addition of hydrogen cyanide to ethyne is also thought to involve nucleophilic addition.

Catalytic hydrogenation, on the other hand, undoubtedly proceeds via the addition of *hydrogen atoms* to the triple bond. When molecular hydrogen is adsorbed on the catalyst surface, the covalent bond is broken and the hydrogen atoms form chemical bonds with metal atoms. When alkyne molecules approach the catalyst surface, these metal–hydrogen bonds break and new carbon–hydrogen bonds are formed. The hydrogen atoms therefore add on to the alkyne molecule from the *same side* and the *cis* geometrical isomer is thus the predominant product in certain cases (namely disubstituted alkynes $RC\equiv CR$, where R = an alkyl group):

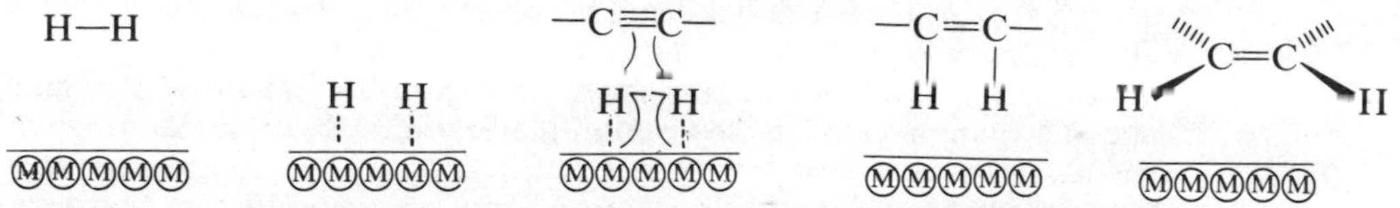

The same mechanism applies to the catalytic hydrogenation of alkenes.

4.5   Manufacture and uses of alkynes

Ethyne is an important alkyne and is used on an industrial scale. It is obtained industrially by the cracking of hydrocarbons, in particular naphtha and methane. The temperature of cracking is momentarily allowed to reach values of 1200–1500°C (being an endothermic compound, ethyne is less unstable at high temperatures) before the gas mixture is rapidly cooled to prevent any further decomposition, e.g.

$$2CH_4 \ \rightarrow \ HC\equiv CH + 3H_2$$

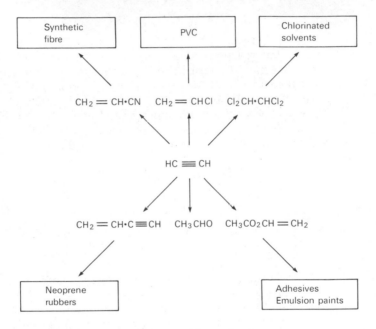

Figure 4.1   Some synthetic routes from ethyne

Many of the manufacturing routes which once used ethyne (e.g. the route to PVC plastic) have now been superseded by ones utilizing ethene, since the latter is readily and cheaply available from the cracking of naphtha (*see* Section 18.3.1). Also, many of the addition reactions of ethyne are exothermic and difficult to control. Nevertheless, some routes based on ethyne are still operated by some companies and some possible transformations of ethyne are shown in *Figure 4.1*. These routes may become economically viable again in the future, if there is a shortage of oil and hence of ethene feedstock. Ethyne may be considered as a coal-based chemical, as well as an oil-based one, since it can be manufactured from calcium carbide, itself derived from coke (*see* Section 4.3.4).

Summary

1 The **alkynes** form a homologous series with a general formula C_nH_{2n-2}.
2 The characteristic functional group of the series is the *triple bond* $-C{\equiv}C-$.
3 Alkynes are unsaturated hydrocarbons and take part in **addition** reactions, although in general they react more slowly than alkenes. Some typical reactions are summarized below:

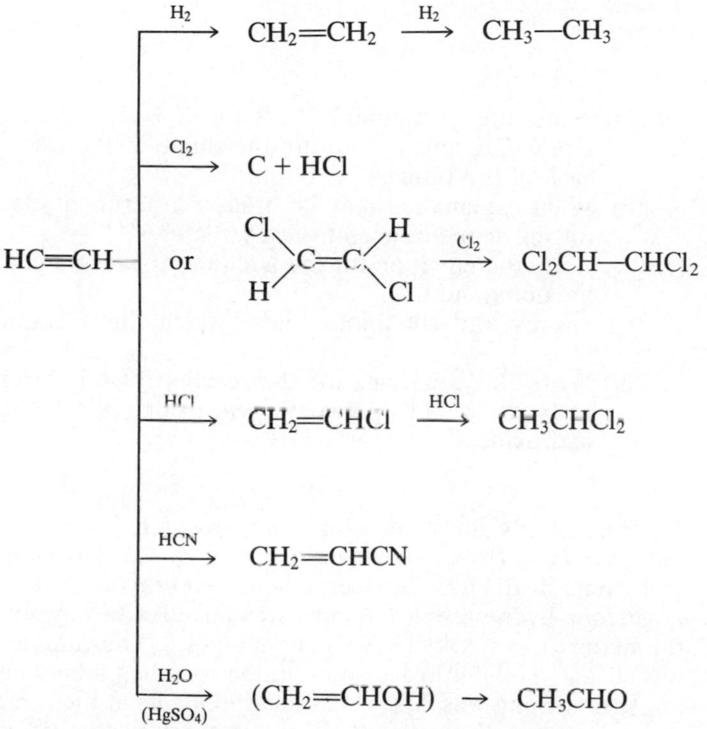

$$HC{\equiv}CH$$

$$\xrightarrow{H_2} CH_2{=}CH_2 \xrightarrow{H_2} CH_3{-}CH_3$$

$$\xrightarrow{Cl_2} C + HCl$$

$$\begin{array}{c} Cl \\ \diagdown \\ H \diagup \end{array} C{=}C \begin{array}{c} H \\ \diagup \\ \diagdown Cl \end{array} \xrightarrow{Cl_2} Cl_2CH{-}CHCl_2$$

or

$$\xrightarrow{HCl} CH_2{=}CHCl \xrightarrow{HCl} CH_3CHCl_2$$

$$\xrightarrow{HCN} CH_2{=}CHCN$$

$$\xrightarrow[(HgSO_4)]{H_2O} (CH_2{=}CHOH) \rightarrow CH_3CHO$$

4 Alkynes are **oxidized** by reagents such as ozone and potassium manganate(VII), but various products are formed and these reactions have little general preparative value. Alkynes burn in excess air or oxygen to carbon dioxide and water.
5 Alkynes can be **oligomerized** in the presence of suitable metal catalysts.
6 1-Alkynes form ionic metal derivatives when treated with a very strong base in a non-aqueous solvent. The formation of the insoluble brick-red copper(I) dicarbide and white silver dicarbide, when ethyne is bubbled through ammoniacal solutions of copper(I) chloride and silver nitrate, respectively, is a useful test for distinguishing between ethyne and ethene.
7 Ethyne is an alkyne of considerable industrial importance and is used as a chemical feedstock.

Questions

1   The structural formulae of three isomers having the same molecular formula C_4H_6 are shown below:

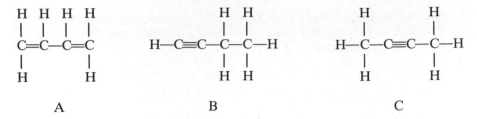

   A   B   C

(a) Name the three compounds A, B and C.
(b) (i) Draw diagrams to indicate the shape of the carbon skeleton in each of the isomers A, B and C.
 (ii) What explanation may be offered in terms of electronic theory for the shape of **one** of these structures?
(c) (i) Write the equation for the addition of excess hydrogen bromide to compound C.
 (ii) Suggest the conditions under which this reaction would take place.
 (iii) Write the equations for the reaction of this bromo compound with (a) aqueous potassium hydroxide, (b) alcoholic potassium hydroxide.

London

2   Describe briefly how you would prepare **both** (a) a sample of ethyne (*acetylene*) **and** (b) an alkene of your own choice. Give **one** chemical test that would distinguish between ethene and ethyne.
A gaseous hydrocarbon (10 cm³) was mixed with oxygen (80 cm³) and the mixture was sparked at room temperature. The volume after sparking was 60 cm³. Sodium hydroxide solution was then added and the volume of gas remaining was 20 cm³ (all measurements at the same temperature and pressure). Write the structures of **three** hydrocarbons that fit this information.

Oxford

3   (a) Outline two industrial manufacturing routes to ethyne indicating starting materials and reaction conditions.
(b) What is the significance of the extremely high temperature used in the modern manufacturing route to ethyne?
(c) Name two factors which contribute to the high cost of ethyne as compared with that of ethene.
(d) Under what circumstances might the older route become viable again?
(e) Why is ethyne more difficult to handle and store than is ethene?
(f) Compare the manufacturing routes to propenonitrile starting from (i) ethyne, (ii) propene.

Chapter 5
Halogenoalkanes

5.1 Homologous series of halogenoalkanes

Halogen-substituted derivatives of alkanes are very commonly encountered in the laboratory, particularly in synthetic work where they can be converted into a variety of classes of compounds. Chlorinated hydrocarbons have become important as industrial solvents, dry-cleaning liquids and pesticides, and in the past thirty years or so there has been a tremendous growth in the production of fluorinated compounds. There are few naturally occurring halogenoalkanes, although substantial amounts of bromomethane CH_3Br are produced by a certain type of seaweed.

Simple monohalogenated alkanes have the general formula $C_nH_{2n+1}X$, where X is a halogen atom, and form homologous series. Some physical properties for the 1-chloroalkane series are listed in *Table 5.1*.

From chloropropane onwards, in the series given in *Table 5.1*, structural isomers exist. There are, for example, four isomers of C_4H_9Cl. They are named as chlorine-substituted alkanes (based on the longest straight carbon chain), the positions of substituents being indicated by appropriate numbering of the chain. Thus:

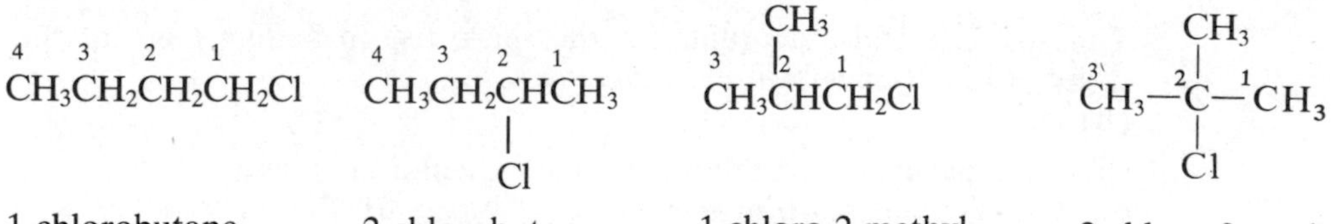

$$\overset{4}{C}H_3\overset{3}{C}H_2\overset{2}{C}H_2\overset{1}{C}H_2Cl \qquad \overset{4}{C}H_3\overset{3}{C}H_2\overset{2}{C}H\overset{1}{C}H_3$$

$$\underset{\displaystyle Cl}{|}$$

1-chlorobutane
b.p. = 77°C

2-chlorobutane
b.p. = 68°C

$$\begin{array}{c} CH_3 \\ | \\ \overset{3}{C}H_3\overset{2}{C}H\overset{1}{C}H_2Cl \end{array} \qquad \begin{array}{c} CH_3 \\ | \\ \overset{3'}{C}H_3-\overset{2}{C}-\overset{1}{C}H_3 \\ | \\ Cl \end{array}$$

1-chloro-2-methyl
propane
b.p. = 69°C

2-chloro-2-methyl
propane
b.p. = 51°C

Compounds in which there is only *one* alkyl group attached to the carbon atom bearing the halogen atom (as in the first and third isomers above) are known as **primary** halogenoalkanes; if *two* or *three* alkyl groups are attached (as in the second and fourth isomers) the compounds are referred to as **secondary** and **tertiary** halogenoalkanes, respectively.

Table 5.1

Name	Formula	Melting point (°C)	Boiling point (°C)	Density (g cm^{-3})
chloromethane	CH_3Cl	−97	−24	—
chloroethane	CH_3CH_2Cl	−139	12	—
1-chloropropane	$CH_3CH_2CH_2Cl$	−123	46	0.889
1-chlorobutane	$CH_3CH_2CH_2CH_2Cl$	−123	77	0.886
1-chloropentane	$CH_3(CH_2)_3CH_2Cl$	−99	105	0.882
1-chlorohexane	$CH_3(CH_2)_4CH_2Cl$	−83	134	0.878
1-chloroheptane	$CH_3(CH_2)_5CH_2Cl$	−70	159	0.877
1-chlorooctane	$CH_3(CH_2)_6CH_2Cl$	−58	182	0.875

5.2 Physical properties of halogenoalkanes

The halogenoalkanes possess polar C—X bonds which can confer a dipole moment on the molecule. The intermolecular forces between molecules consist of *permanent dipole–permanent dipole* interactions, as well as the universal *induced dipole* interactions. However, these forces do not lead to any very appreciable decrease in volatility in comparison with alkanes of corresponding relative molecular mass (there are some exceptions), a fact illustrated in *Figure 2.1*.

As the chain length of 1-chloroalkanes, for example, increases, the halogen's influence on the intermolecular forces diminishes and the boiling points become virtually identical to those of alkanes. With increase in relative molecular mass, the boiling points of halogenoalkanes increase, as shown in *Figure 2.1*. With increasing size of halogen atom, the molecular size increases, with an accompanying increase in mutual polarization forces; hence the boiling points increase:

CH_3CH_2Cl	CH_3CH_2Br	CH_3CH_2I
chloroethane	bromoethane	iodoethane
b.p. = 12°C	b.p. = 38°C	b.p. = 73°C

This size effect also accounts for the sharp rise in boiling point as the degree of halogen substitution increases, e.g.

CH_3Cl	CH_2Cl_2	$CHCl_3$
chloromethane	dichloromethane	trichloromethane
b.p. = −24°C	b.p. = 40°C	b.p. = 61°C

CCl_4	CH_3CH_2Br	$BrCH_2CH_2Br$
tetrachloromethane	bromoethane	1,2-dibromoethane
b.p. = 77°C	b.p. = 38°C	b.p. = 131°C.

The boiling points of isomeric halogenoalkanes also vary in the order primary > secondary > tertiary isomer, as illustrated by the isomers of chlorobutane previously discussed.

The majority of halogenoalkanes are covalent liquids, which are virtually unattacked by cold water and are immiscible with it. The lower chloro compounds are less dense than water, but the bromo and iodo compounds are generally denser and form the bottom layer when mixed with water.

Halogenoalkanes have low flammability, in general, but their careless use can be hazardous. The vapour of the chloro and iodo compounds, in particular, can be toxic and the compounds are often absorbed through the skin. Inhalation of the vapour of the very familiar trichloromethane and tetrachloromethane (formerly known as *chloroform* and *carbon tetrachloride*, respectively) must always be avoided.

5.3 Chemical properties of halogenoalkanes

The relative bond strengths (C—F 485, C—Cl 339, C—Br 284 and C—I 213 kJ mol^{-1}) indicate that the iodo compounds are likely to be the most reactive, and these are frequently used in synthetic work in the laboratory. The C—F bond is certainly the most inert and so the fluoro compounds find special use where lack of reactivity is an important factor (*see* Section 5.4(v)).

In the main, being saturated compounds, the halogenoalkanes undergo two characteristic types of reaction, namely **substitution** and **elimination**. They also form important compounds by reactions with metals.

5.3.1 Substitution reactions

(i) Reaction with water or alkali
The carbon–halogen bond is polarized, the bonding electron-pair being shared unequally between the two atoms as shown below:

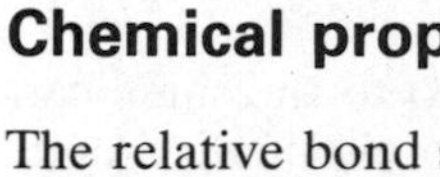

Consequently, the carbon is electropositive and is an *electrophilic* centre, so that it should be susceptible to attack by a *nucleophilic* species (either a negative ion or neutral molecule, possessing one or more lone-pairs electrons). Water is such a reagent. Halogenoalkanes are only slowly attacked, or hydrolysed, by water but reaction takes place readily when the halogenoalkane is warmed with alkali. Usually, aqueous ethanol ('alcohol') is employed as a solvent, in order to dissolve both the covalent halogenoalkane and the ionic alkali. The product of alkaline hydrolysis is an **alcohol**, formed by the **substitution** of an —OH group for an X atom:

R—X + OH$^-$ → R—OH + X$^-$ (where R represents an alkyl group)

e.g.

CH$_3$CH$_2$I + OH$^-$ → CH$_3$CH$_2$OH + I$^-$
iodoethane   ethanol
b.p. = 73°C   b.p. = 78°C

As reaction proceeds, the halogen atom is displaced as a halide ion. If the solution is then acidified with dilute nitric acid and silver nitrate solution added, a precipitate of the silver halide is produced. Unlike an *ionic* halide, the halogenoalkane itself will give no precipitate, since the halogen atom is *covalently* bonded to carbon. This procedure can therefore be used as a test for simple halogenoalkanes.

A disubstituted alkane follows a similar course of reaction, e.g.

$$BrCH_2CH_2Br + 2OH^- \rightarrow HOCH_2CH_2OH + 2Br^-$$
1,2-dibromoethane   ethane-1,2-diol
b.p. = 131°C   b.p. = 197°C

but if the two halogen atoms are linked to the same carbon atom, then substitution is immediately followed by elimination and an aldehyde or ketone is produced, e.g.

$$CH_3CHCl_2 + 2OH^- \longrightarrow CH_3-C\underset{\displaystyle H}{\overset{\displaystyle O}{\diagup}} + 2Cl^- + H_2O$$

1,1-dichloroethane   ethanal
b.p. = 57°C   b.p. = 21°C

Trichloromethane (*chloroform*) is hydrolysed by boiling alkali to the salt of a carboxylic acid:

$$CHCl_3 + 4OH^- \longrightarrow HC\underset{\displaystyle O^-}{\overset{\displaystyle O}{\diagup}} + 3Cl^- + 2H_2O$$

(ii) Reaction with alkoxides
Alcohols react with metallic sodium to form ionic salts called **alkoxides**, e.g.

$$2ROH + 2Na \rightarrow 2RO^-Na^+ + H_2$$

When a halogenoalkane is heated with a solution of the alkoxide in its corresponding alcohol, substitution takes place and an **ether** is formed, the halogen atom having been substituted by an **alkoxy** group OR:

$$R-X + {}^-OR \rightarrow R-O-R + X^-$$

e.g.

$$CH_3CH_2I + {}^-OCH_3 \rightarrow CH_3CH_2OCH_3 + I^-$$
methoxyethane
b.p. = 11°C

This reaction is the basis of the **Williamson ether synthesis** (*see* Section 15.6.1) but, under certain conditions, products of an elimination reaction, as well as a substitution reaction, are obtained (*see* Section 5.3.3).

(iii) Reaction with cyanide ion, CN⁻

A halogen atom can be substituted by a *cyano* group —CN by heating under reflux the halogenoalkane with potassium cyanide in aqueous, alcoholic solution. The greatest possible care must be taken when handling potassium cyanide since both the salt and hydrogen cyanide, which may be formed by salt hydrolysis, are highly toxic:

$$R—X \; + \; CN⁻ \; \rightarrow \; R—CN + X⁻$$

e.g.

$$CH_3CH_2I + CN⁻ \; \rightarrow \; CH_3CH_2CN + I⁻$$

$$\text{propanonitrile}$$
$$\text{b.p.} = 97°C$$

The product is an organic cyanide which is usually called a **nitrile**. Since an additional carbon atom has been introduced into the molecule, the name of the nitrile is based on the alkane which contains the same number of carbon atoms, with *nitrile* as a suffix. This reaction is a very useful one in synthetic work because nitriles can be reduced to amines or hydrolysed to carboxylic acids (*see* Section 10.5). The reaction also provides a useful way of introducing another carbon atom into an organic molecule.

(iv) Reaction with ammonia

When a halogenoalkane is heated with an alcoholic solution of ammonia under pressure, the immediate product is an alkyl-substituted ammonium salt:

$$R—X + :NH_3 \; \rightarrow \; R—NH_3^+ + X⁻$$

This can exchange a proton with unreacted ammonia, acting as a base, to form an **amine** in which the *amino* group —NH₂ has been substituted for the halogen atom overall:

$$R—NH_3^+ + :NH_3 \; \rightleftharpoons \; R—NH_2 + NH_4^+$$

In any case, however, treatment of the substituted ammonium salt with alkali will liberate the free amine:

$$R—NH_3^+ + OH⁻ \; \rightarrow \; R—NH_2 + H_2O$$

Unfortunately, a series of further substitutions take place and these severely limit the synthetic value of such a reaction. Competing with the ammonia as nucleophile are the amine products themselves, so that a series of further substitutions takes place leading to a mixture of products:

$$R—X + H_2N—R \; \rightarrow \; R_2NH_2^+ + X⁻ \; \overset{NH_3}{\rightleftharpoons} \; R_2NH + NH_4^+ + X⁻$$

$$R—X + HNR_2 \; \rightarrow \; R_3NH^+ + X⁻ \; \overset{NH_3}{\rightleftharpoons} \; R_3N \; + NH_4^+ + X⁻$$

Reaction ceases at the formation of a *quaternary* ammonium salt (i.e.

one in which all four hydrogen atoms of the ammonium ion have been substituted by alkyl groups):

$$R—X + NR_3 \;\rightarrow\; R_4N^+ + X^-$$

Some of these competing reactions may be minimized by employing a large excess of ammonia in the reaction but, except in a few cases, the method is not a very satisfactory one for making amines.

(v) Reaction with silver salts

If a halogenoalkane is heated with an alcoholic solution of the *silver* salt of a carboxylic acid, then the carboxylate anion substitutes for the halogen atom since it can behave as a nucleophile:

$$R—X + RCO_2^- Ag^+ \longrightarrow RC{\overset{\displaystyle O}{\underset{\displaystyle OR}{\diagup}}} + AgX$$

An **ester** is formed and the silver halide which is the other product precipitates out of the solution. The reason for using a *silver* salt is given later (*see* Section 5.3.2). The following example makes it clear that, in the above equation, R represents any alkyl group and these are not necessarily identical in the two reactants, e.g.

$$CH_3CH_2I + CH_3CO_2^- Ag^+ \;\rightarrow\; CH_3CO_2CH_2CH_3 + AgI$$
$$\text{ethyl ethanoate}$$
$$\text{b.p.} = 77°C$$

There are other more widely used methods for making esters but this one finds an application when mild reaction conditions are necessary.

(vi) Reaction with metal derivatives of 1-alkynes

The metal derivatives are obtained by reacting the 1-alkyne with sodium amide (*sodamide*) in liquid ammonia (*see* Section 4.3.4). The anion of the salt is a nucleophile and substitutes for a halogen atom when a halogenoalkane is added to the solution:

$$R—X + Na^+ {}^- C{\equiv}CR \;\rightarrow\; R—C{\equiv}C—R + NaX$$

e.g.

$$CH_3CH_2I + Na^+ {}^- C{\equiv}CCH_3 \;\rightarrow\; CH_3CH_2—C{\equiv}C—CH_3 + NaI$$
$$\text{pent-2-yne}$$
$$\text{b.p.} = 56°C$$

This reaction is very important for making disubstituted alkynes from 1-alkynes.

5.3.2 Mechanism of substitution reactions

The carbon atom forming the polar carbon–halogen bond in a halogenoalkane is electropositive and is prone to nucleophilic attack by an electron-pair donor. This results in the displacement of the halogen atom as a halide ion and its substitution by a new group. These reactions are termed **nucleophilic substitution** reactions and involve ionic mechanisms. The factors influencing these reactions are complex, but it should be noted that the chemical nature of both the attacking nucleophile and the group being displaced (called the leaving group) plays an important role; the nucleophilic power of the attacking reagent and the ability of the leaving group to form a stable anion or molecule must be considered. On the basis of carbon–halogen bond strengths, the iodoalkanes should be the most reactive of the halogeno compounds towards nucleophilic substitution* and in most circumstances they are.

A wide range of experimental evidence, much of it derived from a study of the rates of nucleophilic substitution reactions, shows that there are two main pathways, or mechanisms, by which such reactions take place.

The simplest mechanism involves a single step. The nucleophile attacks the halogenoalkane from behind the electrophilic carbon atom, bearing the halogen atom, and as the new bond begins to form, the existing carbon–halogen bond stretches and weakens until it finally breaks and halide ion is displaced, e.g.

$$HO: + CH_3\overset{\delta+}{-}CH_2\overset{\delta-}{-}I \;\rightarrow\; CH_3-CH_2-OH + I^-$$

In what is called the **transition state**, a state corresponding to an energy maximum in the process of the reaction, both nucleophile and halogen are partially bonded to the carbon atom; the remaining atoms or groups temporarily adopt an approximately planar configuration about the carbon atom. As the substitution is completed, the original stereochemical arrangement of groups about the carbon atom becomes **inverted** (*see* Section 16.4.1). The whole process, as depicted below, is a synchronous one:

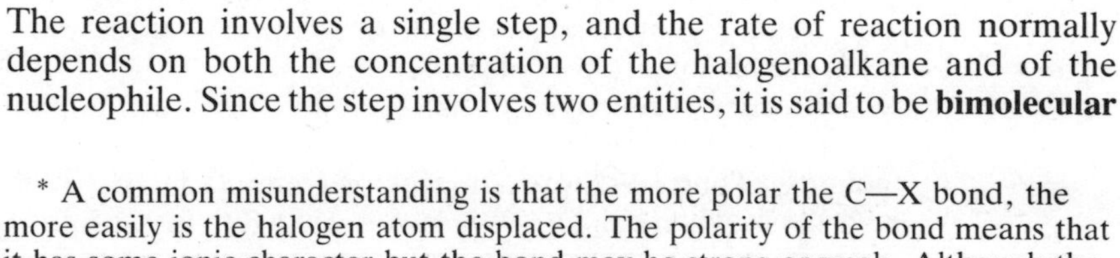

$$HO^- \cdots C - X \;\rightarrow\; \overset{\delta-}{HO}\cdots \overset{\delta+}{C}\cdots \overset{\delta-}{X} \;\rightarrow\; HO - C \cdots H + X^-$$

transition state

The reaction involves a single step, and the rate of reaction normally depends on both the concentration of the halogenoalkane and of the nucleophile. Since the step involves two entities, it is said to be **bimolecular**

---

* A common misunderstanding is that the more polar the C—X bond, the more easily is the halogen atom displaced. The polarity of the bond means that it has some ionic character but the bond may be strong or weak. Although the C—F bond is very polar, it is also very strong and not easily broken.

and so this substitution mechanism, which is very common, is referred to as the S_{N2} mechanism—Substitution, Nucleophilic and bimolecular (2).

With some halogenoalkanes, however, kinetic studies show that the nucleophilic substitution reactions must occur via a different mechanism, or pathway, because the rate of the reaction is found to depend only on the concentration of the halogenoalkane and to be quite independent of the concentration of the nucleophile. Therefore, the reaction pathway must involve more than one step and the slowest, or **rate-determining**, step must involve only the halogenoalkane and not the nucleophile. This step is thought to be the slow heterolysis (ionic fission) of the carbon–halogen bond to produce the halide ion and a highly reactive carbocation:

$$
\begin{array}{ccc}
& CH_3 & & CH_3 \\
& | & & | \\
CH_3{-}C{-}X & \xrightarrow{\text{slow}} & CH_3{-}C^+ & +\ X^- \\
& | & & | \\
& CH_3 & & CH_3
\end{array}
$$

The carbocation produced is a highly reactive electrophile and undergoes very fast attack by the nucleophile to complete the reaction:

$$
\begin{array}{ccc}
CH_3 & & CH_3 \\
| & & | \\
CH_3{-}C^+ +\ :OH^- & \xrightarrow{\text{fast}} & CH_3{-}C{-}OH \\
| & & | \\
CH_3 & & CH_3
\end{array}
$$

The carbocation is planar and hence, theoretically, nucleophilic attack is possible from either side of the ion.

The rate-determining step in this reaction involves *one* entity only, the halogenoalkane molecule, and is therefore a **unimolecular** step. This type of mechanism is referred to as the S_{N1} mechanism—Substitution, Nucleophilic, unimolecular (1).

The relative stabilities of carbocations have already been referred to (*see* Section 3.3.4). The least stable are the primary ones, so that they are not favoured as intermediates in an S_{N1} mechanism. Consequently, primary halogenoalkanes usually react via an S_{N2} mechanism. Tertiary halogenoalkanes can yield tertiary carbocations and hence their nucleophilic substitution reactions usually follow the S_{N1} mechanism. Secondary halogenoalkanes may follow either, or both, mechanisms during reaction. The solvent also affects the mechanism; ions are stabilized by solvation and hence S_{N1} mechanisms will only operate in certain polar solvents where such solvation is possible.

The order of reactivity of halogenoalkanes in nucleophilic substitutions is tertiary > secondary > primary, and iodo > bromo > chloro compounds. Thus, 2-chloro-2-methylpropane $(CH_3)_3CCl$ undergoes rapid hydrolysis when shaken with water, giving an instant precipitate of silver chloride when shaken with an ethanolic solution of silver nitrate in the cold. Secondary and primary halides require warming. Iodo compounds precipitate silver iodide under similar conditions, whereas the chlorides and bromides generally require warming.

These mechanisms are general ones for nucleophilic substitution reactions and apply to the cases discussed in Section 5.3.1. When the nucleophile is not a strong one, the ease of displacement of halide ion may be aided sometimes by employing a specific reagent. Thus, the use of the *silver* salt of a carboxylic acid, in forming an ester from the halogenoalkane, is dictated by the need to increase the rate of reaction by complexing the silver ion Ag^+ with a lone-pair of electrons on the halogen atom, making it a better leaving group.

5.3.3   Elimination reactions

The actual reaction conditions under which reagents are allowed to participate in organic reactions can critically affect the course of the reaction and thereby the products formed. The reaction between a halogenoalkane and a base affords a good example. When the reaction is carried out at room temperature, or on warming, in a dilute aqueous alcohol solution, nucleophilic substitution takes place; the hydroxide ion acts as a nucleophile. For example,

$$HO:^-$$

$$CH_3-CH-CH_3 \longrightarrow CH_3-CH-CH_3 + I^-$$
$$\qquad\quad | \qquad\qquad\qquad\qquad\quad |$$
$$\qquad\quad I \qquad\qquad\qquad\qquad\quad OH$$

2-iodopropane   propan-2-ol
b.p. = 89°C   b.p. = 82°C

However, if a *hot, concentrated alcoholic* solution of base and halogenoalkane is employed, then an **elimination**, rather than substitution, reaction occurs and an alkene is formed, e.g.

$$CH_3-CH-CH_3 + OH^- \longrightarrow CH_3CH{=}CH_2 + H_2O + I^-$$
$$\qquad\qquad |\qquad\qquad\qquad\qquad\qquad propene$$
$$\qquad\qquad I$$

Such an elimination reaction is the converse of an addition reaction, the elements of a hydrogen halide having been removed or eliminated from the molecule to give an unsaturated product. In such a reaction, the hydroxide ion behaves, not as a nucleophile, but as a *base* (*see* Section 1.7) by removing a proton from the molecule thus:

$$H :OH^-$$

$$CH_3-CH-CH_2 \longrightarrow CH_3-CH{=}CH_2 + H_2O + I^-$$
$$\qquad\qquad |$$
$$\qquad\qquad I$$

Depending on the structure of the halogenoalkane, elimination may often compete with substitution if the nucleophile can act as a base, and so the reaction yield may be reduced. Elimination is favoured at higher temperatures and with tertiary halogenoalkanes. Thus, even at room temperature some elimination occurs with the following tertiary compound:

$$
\underset{\begin{array}{c}\mid\\CH_3\end{array}}{\overset{\begin{array}{c}CH_3\\\mid\end{array}}{CH_3-C-Cl}} \longrightarrow \quad \underset{CH_3}{\overset{CH_3}{>}}C=CH_2 \quad + \quad \underset{\begin{array}{c}\mid\\CH_3\end{array}}{\overset{\begin{array}{c}CH_3\\\mid\end{array}}{CH_3-C-OH}}
$$

$$
17\% \qquad\qquad\qquad 83\%
$$

When the corresponding bromo compound is warmed with alcoholic hydroxide, almost 100% elimination takes place.

When an unsymmetrical halogenoalkane undergoes elimination, the more highly substituted alkene is the predominant product, e.g.

$$
\underset{\begin{array}{c}\mid\\I\end{array}}{\overset{\begin{array}{c}CH_3\\\mid\end{array}}{CH_3-C-CH_2-CH_3}} \xrightarrow{OH^-} \underset{}{\overset{\begin{array}{c}CH_3\\\mid\end{array}}{CH_2=C-CH_2CH_3}} + \underset{}{\overset{\begin{array}{c}CH_3\\\mid\end{array}}{CH_3-C=CHCH_3}}
$$

2-iodo-2-methylbutane   20%   80%
b.p. = 125°C   2-methylbut-1-ene   2-methylbut-2-ene
                       b.p. = 31°C   b.p. = 39°C

In the Williamson ether synthesis (*see* Section 5.3.1(ii)) the alkoxide ion, too, can act as a base as well as a nucleophile, and hence some alkene will be produced as an elimination product.

An elimination reaction is used to prepare various chlorinated solvents industrially, including the important degreasing solvent trichloroethene:

$$
CHCl_2-CHCl_2 \xrightarrow{(-HCl)} CHCl=CCl_2
$$

1,1,2,2-tetrachloroethane   trichloroethene
b.p. = 147°C   b.p. = 87°C

Disubstituted halogeno compounds may also eliminate 2 mol of hydrogen halide per mol of halogenoalkane. Thus:

$$
BrCH_2CH_2Br \xrightarrow[OH^-]{(-2HBr)} HC\equiv CH
$$

1,2-dibromoethane   ethyne
b.p. = 131°C

If a 1,2-dibromoalkane is heated with zinc powder in an alcohol solvent then 1 mol of bromine is eliminated and an alkene is produced. Thus:

$$
BrCH_2CH_2Br \xrightarrow[heat]{Zn} CH_2=CH_2
$$

                 ethene

5.3.4   Reaction with metals

Under the appropriate conditions, halogenoalkanes will react with a variety of metals such as zinc, magnesium and lithium, to form organometallic compounds, in which the metal atom becomes bonded to a carbon atom. The products are usually very sensitive to air oxidation, and to hydrolysis, but their reactivity can be put to good use in synthetic reactions.

One of the most important classes of organometallic compounds obtained from halogenoalkanes is the organomagnesium compounds called **Grignard reagents**. These are named after the Nobel Prize-winning French chemist Victor Grignard, who introduced them in 1900. Grignard reagents are prepared by gradually adding a totally dry halogenoalkane to clean, dry magnesium turnings, suspended in a dry ether solvent. The reaction often has to be started by adding a crystal of iodine (which reacts with the surface of magnesium and exposes a pure metal surface free of any oxide) or by heating; once the reaction has commenced, it usually continues without further heating, since the reaction is exothermic (it can become very vigorous). The Grignard reagent is stabilized by the ether solvent and cannot be isolated; it is used *in situ* (i.e. in the medium in which it has been prepared):

$$R—X + Mg \rightarrow R—Mg—X \qquad e.g. \ CH_3I + Mg \rightarrow CH_3MgI$$

There has been much controversy over the actual structures of the Grignard reagents, which are likely to be complex, but their behaviour can be satisfactorily explained by representing the structure thus, $\overset{\delta-}{R}—\overset{\delta+}{Mg}—\overset{\delta-}{X}$; the magnesium may be considered to be bonded to carbon and halogen by two polar bonds. These Grignard reagents are thus a source of nucleophilic carbon ($\delta-$) and take part in important nucleophilic additions with aldehydes and ketones (*see* Section 8.3.1(v)). Their reaction with water provides a method of reducing a halogenoalkane to an alkane. Here, effectively, the $R^{\delta-}$ residue acts as a base and removes a proton from the water molecule.

$$R—Mg—X + H_2O \rightarrow RH + Mg(OH) X$$

Carbon dioxide, either as the gas or in the solid 'dry ice' form, reacts readily with Grignard reagents and this reaction provides a useful synthetic route to carboxylic acids:

$$R—Mg—X + CO_2 \rightarrow RCO_2MgX \xrightarrow{H_3O^+} RCO_2H$$

Grignard reagents will also react with metal chlorides to form organometallic compounds, e.g.

$$4CH_3MgBr + 2PbBr_2 \rightarrow (CH_3)_4Pb + Pb + 4MgBr_2$$
$$CH_3MgBr + SiCl_4 \rightarrow CH_3SiCl_3 + MgBrCl$$

Similar reactions to those involving Grignard reagents may be performed with the more reactive organolithium compounds, prepared by the reaction of lithium with a halogenoalkane in a dry ether solution and an inert nitrogen atmosphere, e.g.

$$CH_3CH_2CH_2CH_2Cl + 2Li \rightarrow CH_3CH_2CH_2CH_2Li + LiCl$$

5.4 Manufacture and uses of halogenoalkanes

Halogenoalkanes have found extensive use in a wide range of products and industries.

(i) Solvents

Chlorinated hydrocarbons are widely used as solvents; some of the common ones are chloroalkanes and others chloroalkenes. They combine the desirable properties of good solvent power, high vapour densities (low evaporation) and reasonably low boiling points with very low flammability. Even monochlorinated hydrocarbons are much less flammable than the corresponding alkanes since, on strong heating, they lose chlorine atoms which react with radicals which would otherwise propagate chain reactions on combustion. Tetrachloromethane (*carbon tetrachloride*) has been used in fire-extinguishers* but is really to be avoided for this purpose, since it can become oxidized to the poisonous gas carbonyl chloride (*phosgene*) $COCl_2$, as can the familiar trichloromethane (*chloroform*). Both tetrachloromethane and tetrachloroethene are widely used drycleaning solvents and trichloroethene is a very effective degreasing solvent. Some of the manufacturing routes are shown below, using chlorination (substitution and addition), hydrochlorination (addition) and dehydrochlorination (elimination):

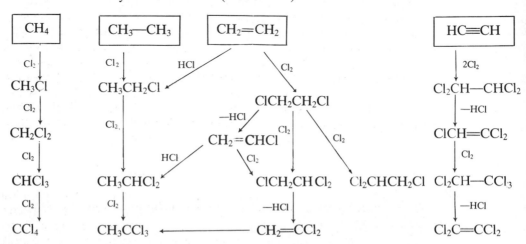

* $CBrClF_2$ (BCF) is the most widely used vaporizing liquid in fire-extinguishers.

Chlorination by substitution also produces hydrogen chloride, which is wasteful of chlorine, so that frequently oxychlorination is used, i.e. reaction with a mixture of hydrogen chloride and oxygen, using copper catalysts.

(ii) Petrol additives

Chloroethane is used in the manufacture of tetra-ethyl lead, which is added to petrol to improve its octane rating:

$$4CH_3CH_2Cl + 4Na/Pb \rightarrow (CH_3CH_2)_4Pb + 4NaCl + 3Pb$$
$$\text{(alloy)}$$

Such organolead compounds are banned as petrol additives in some countries, and carefully controlled in others (including the UK), owing to the dangers of atmospheric pollution by toxic lead compounds emanating from car exhaust emissions. 1,2-dichloro- and dibromo-ethanes are also used in petrol as 'lead scavengers'. During combustion, the lead forms volatile halides and oxyhalides which are vented out of car exhausts, so preventing the accumulation of lead in the engine cylinders.

(iii) Pesticides

Until recently, chlorinated hydrocarbons have been extensively used in agriculture as pesticides (see, for example, DDT, Section 13.2.2) but their persistence in the environment has led to restricted use. Most of them have not been halogenoalkanes, but bromomethane is still used as an effective all-round pesticide and is one of the most effective nematocides known (i.e. is very active against nematodes or roundworms). Both 1,2-dibromoethane and 1,2-dibromo-3-chloropropane are widely used in soil fumigation.

(iv) Anaesthetics

Both chloromethane and chloroethane have been used as local anaesthetics, being sprayed on the skin, but the most well-known example is trichloromethane (chloroform). This is now rarely used as it can prove toxic to the liver and has other unpleasant side-effects. One of the most widely used anaesthetics in present use is halothane (ICI trade-name Fluothane) which was introduced by C. W. Suckling and J. Raventós, working for ICI, in 1956. Its formula is $CF_3CHBrCl$. It possesses outstanding properties of high volatility, non-flammability, chemical stability (important in closed-circuit anaesthesia) and high potency.

(v) Fluorocarbons

The development of fluorine chemistry and technology in the separation of uranium isotopes during the Second World War has led to an enormous growth in organofluorine chemistry in the post-war period. Direct fluorination of alkanes is not possible, since the reactions are violently exothermic. Instead, chloroalkanes, particularly tetrachloromethane, are treated with anhydrous hydrogen fluoride and an antimony trifluoride/pentachloride catalyst, either in the liquid phase or gas phase, e.g.

$$CHCl_3 + 2HF \rightarrow CHClF_2 + 2HCl$$
$$CCl_4 + HF \rightarrow CCl_3F + HCl$$
$$CCl_4 + 2HF \rightarrow CCl_2F_2 + 2HCl$$

The non-toxic, inert, non-flammable, odourless and volatile nature of a fluorine compound like CF_2Cl_2 was first exploited by Thomas Midgley Jnr. (who also had the vital discovery of tetraethyl-lead as an anti-knock agent to his credit) and A. L. Henne in 1930, when it was introduced as a replacement for ammonia as a refrigerant gas. Such compounds are now widely used for this purpose, being known in the USA as Freons (DuPont) and in the UK as Arctons (ICI). The thermal decomposition of chlorodifluoromethane is used to manufacture tetrafluoroethene, the starting material for PTFE plastic (*see* Section 20.2.1(vi)).

$$2CHClF_2 \xrightarrow{800°C} CF_2{=}CF_2 + 2HCl$$
(Arcton 22)

A further development in the 1950s was the use of such fluorinated compounds as propellant gases in aerosol dispensers, the most commonly used being CCl_3F (b.p. = 25°C) and CCl_2F_2 (b.p. = -30°C). In the 1970s there were fears that their widespread use might lead to the depletion of the essential 'ozone layer' which filters out the more harmful ultraviolet components of sunlight (chlorine atoms from these compounds can react with 'ozone'), and they are likely to be replaced gradually in the future. The USA banned 'non-essential uses' of such compounds from 1979.

Summary

1   Mono**halogenoalkanes** form homologous series of general formula $C_nH_{2n+1}X$, where X = halogen. The characteristic functional group is the carbon–halogen bond $-\overset{\diagdown}{\underset{\diagup}{C}}-X$.

2   The halogenoalkanes are saturated compounds but, possessing a polar carbon–halogen bond, they undergo **nucleophilic substitution** reactions which are of great use in synthetic work:

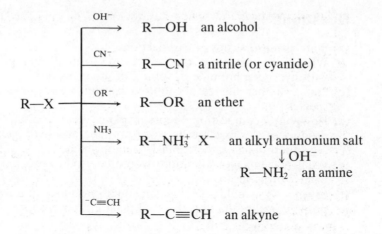

3   Under different reaction conditions halogenoalkanes can also undergo
     elimination reactions to form *unsaturated* compounds, e.g.

$$CH_3—CH—CH_3 \xrightarrow{OH^-} CH_3—CH=CH_2$$
$$\quad\quad\quad |$$
$$\quad\quad\quad X$$

$$CH_3—CH—CH_2 \xrightarrow{OH^-} CH_3—C\equiv CH$$
$$\quad\quad\quad |\quad\; |$$
$$\quad\quad\quad X\quad X \searrow_{Zn}$$
$$\quad\quad\quad\quad\quad\quad\quad CH_3—CH=CH_2$$

4   The halogenoalkanes are used to form various important classes of organo-
     metallic compounds, including alkyllithium reagents and Grignard
     reagents:

$$R—X + 2Li \;\rightarrow\; RLi + LiX$$
$$R—X + Mg \;\rightarrow\; R\;Mg—X$$

     Grignard reagents react with water and carbon dioxide and may be used
     to convert halogenoalkanes to alkanes and carboxylic acids, respectively.
5   Halogenoalkanes find important uses as general synthetic reagents in the
     laboratory. Industrially, they find uses as solvents, anaesthetics,
     refrigerant gases, aerosol propellants and, to a lesser degree, pesticides
     and petrol additives.

Questions

1 (a) State **two** properties of carbocations.
   (b) What is the formula of the carbocation formed when propene reacts
       with hydrogen bromide to form 2-bromopropane?
   (c) State whether this carbocation is a primary, secondary or tertiary
       species.
   (d) How does carbocation theory explain the fact that the principal
       product of the reaction between propene and hydrogen bromide is
       2-bromopropane rather than 1-bromopropane?
   (e) What reagents and conditions would be used to prepare propan-2-ol
       from 2-bromopropane?
   (f) In what way would you alter the conditions in (e) in order to prepare
       propene rather than propan-2-ol from 2-bromopropane?

                                                                         JMB

2 (a) The characteristic reactions of halogenoalkanes may be described as
       nucleophilic substitutions. Explain briefly the meaning of this term.
   (b) What is the significant feature of halogenoalkanes which makes them
       susceptible to reactions of this kind?
   (c) Name **two** classes of organic compounds, other than halogenoalkanes,
       which react readily with nucleophilic reagents.
   (d) State what would be observed if bromoethane were warmed for a
       few minutes with an ethanolic solution of silver nitrate, and explain
       why your observation(s) would be different if bromobenzene were
       used instead of bromoethane.

                                                                         JMB

3 When 1-bromobutane is treated with alkali under different conditions,
   the product may be one or more of the following: (i) an alcohol; (ii) an
   alkene; (iii) an ether.
   Explain this behaviour in terms of the *reaction mechanism* and give
   structural formulae for possible products. Under what conditions might
   each of the products be favoured?

                                                                St. Andrews

4* State the reagents and give the equations summarizing the preparation
   of each of the following from bromoethane:
   (a) ethanol; (b) ethyl ethanoate (*acetate*); (c) ethoxypropane; (d) 1-
   aminopropane (two steps are needed).

                                                                         JMB

5* Bromobutanes react with hot, ethanolic potassium hydroxide to produce
   gaseous butenes, by elimination of hydrogen bromide. The rates of
   reaction can be studied by timing the collection of the gaseous butenes
   in a gas syringe. 0.004 mol of a liquid bromobutane is injected into the
   flask through a rubber cap using a hypodermic syringe. The hypodermic
   syringe is weighed before and after injection. The volume of gas produced
   is recorded at suitable time intervals and the final volume noted.

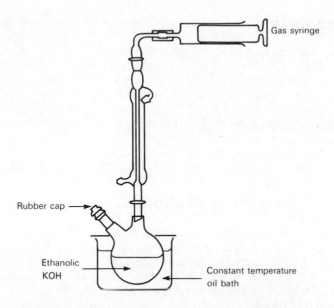

From 1-bromobutane only 8 cm^3 of gas are obtained, but from 2-bromo-
2-methylpropane 80 cm^3 of gas are obtained at room temperature.
Molar volume at room temperature = 24 dm^3.
(a) What is the purpose of the reflux condenser?
(b) Give *one* **chemical test** which would show that the gas collected was
     an alkene.
(c) Draw the structural formulae of 1-bromobutane, 2-bromo-2-methyl-
     propane and the corresponding alkenes which would be obtained
     from them.
(d) Draw the structural formulae of *two* additional isomers of 1-bromo-
     butane which might similarly be investigated.
(e) (i)   Calculate the number of moles of butene obtained in the 1-
          bromobutane experiment and hence the percentage of 1
          bromobutane which participated in the elimination reaction.
    (ii) What reaction might be occurring as well as the elimination
          reaction?
    (iii) Draw the structural formula of the product you predict for the
          alternative reaction in (e) (ii).
(f) Kinetic studies show that the reaction producing an alkene from 2-
     bromo-2-methylpropane is second order overall. What form would
     you predict for the rate expression of the reaction?

                                                                    Nuffield

6   Write an essay on the monohalogenoalkanes.
    Your answer should include reference to methods of synthesis, physical
    and chemical properties and the commercial importance of halogeno-
    alkanes in the manufacture of organic chemicals.

                                                                    JMB

Chapter 6
Alcohols

6.1 Homologous series of alcohols

Alcohols are compounds containing the *hydroxyl group* —OH bonded to an alkyl group. The hydroxyl group is the functional group of the series. In some respects, alcohols resemble water, in which the hydroxyl group is bonded to a hydrogen atom. By far the best-known alcohol is ethanol, otherwise known as *ethyl alcohol*, or simply 'alcohol'. Its formula is CH_3CH_2OH and it is a product of many natural fermentations. Another commonly encountered alcohol is methanol CH_3OH, sometimes known as *methyl alcohol*, or 'wood alcohol', which is the simplest alcohol of all.

The simple alcohols are named by adding the suffix *-ol* to the root of the name of the corresponding alkane. They form a homologous series of general formula $C_nH_{2n+1}OH$. The first eight members of a series of alcohols, in which the hydroxyl group is attached to the end of the chain, are listed in *Table 6.1*, together with some physical data.

Isomerism occurs and there are two structures for C_3H_7OH, the position of the hydroxyl group being indicated by the usual numbering scheme:

$$CH_3CH_2CH_2OH \qquad CH_3{-}CH{-}CH_3$$
$$| $$
$$OH$$

propan-1-ol   propan-2-ol
b.p. = 97°C   b.p. = 83°C

Table 6.1

Name	Formula	Melting point (°C)	Boiling point (°C)	Density (g cm^{-3})
methanol	CH_3OH	−98	65	0.793
ethanol	CH_3CH_2OH	−117	78	0.789
propanol	$CH_3CH_2CH_2OH$	−126	97	0.804
butanol	$CH_3(CH_2)_2CH_2OH$	−89	118	0.810
pentanol	$CH_3(CH_2)_3CH_2OH$	−78	138	0.815
hexanol	$CH_3(CH_2)_4CH_2OH$	−44	157	0.820
heptanol	$CH_3(CH_2)_5CH_2OH$	−34	177	0.822
octanol	$CH_3(CH_2)_6CH_2OH$	−16	195	0.826

There are four isomers of C_4H_9OH, arising from variations in the position of the hydroxyl group and the branching of the carbon chain:

$$CH_3CH_2CH_2CH_2OH \qquad \underset{\underset{OH}{|}}{CH_3CH_2CHCH_3} \qquad \underset{\underset{CH_3}{|}}{CH_3CHCH_2OH} \qquad \underset{\underset{OH}{|}}{\overset{\overset{CH_3}{|}}{CH_3-C-CH_3}}$$

butan-1-ol   butan-2-ol   2-methylpropan-1-ol   2-methylpropan-2-ol
b.p. = 118°C   b.p. = 100°C   b.p. = 108°C   b.p. = 83°C
                                                                         (m.p. = 26°C)

In the first and third isomers, the hydroxyl group is attached to a carbon atom which is linked to two hydrogen atoms and *one* alkyl group. Such an alcohol contains the —CH_2OH group and is said to be a **primary** alcohol (cf. halogenoalkanes, Section 5.1). In the second isomer, the hydroxyl group is attached to a carbon atom which is linked to one hydrogen atom and *two* alkyl groups $\diagdown$CHOH and the alcohol is described as a **secondary** alcohol. In the fourth isomer, the hydroxyl group is attached to a carbon atom linked to *three* alkyl groups $\overset{\diagdown}{\underset{\diagup}{-}}$C—OH and the alcohol is then described as a **tertiary** one.

Alcohols which contain one hydroxyl group only in the molecule are called **monohydric** alcohols. Alcohols such as ethane-1,2-diol (*glycol*), which contain two hydroxyl groups per molecule, are called **dihydric** alcohols, while those like propane-1,2,3-triol (*glycerol*, or *glycerine*), which contain three such groups, are called **trihydric** alcohols:

$$HOCH_2CH_2OH \qquad\qquad \underset{\underset{OH}{|}}{HOCH_2CHCH_2OH}$$

ethane-1,2-diol   propane-1,2,3-triol
b.p. = 197°C   b.p. = 290°C

In the triol, two of the hydroxyl groups are primary ones and one is a secondary one. Alcohols containing more than three hydroxyl groups per molecule are called **polyhydric** alcohols. Generally speaking, compounds in which two hydroxyl groups are attached to the same carbon atom cannot be isolated; their formation would be followed by the spontaneous elimination of the elements of water to form a *carbonyl* group (but *see* Section 8.3.1(iii), and also —$CHCl_2$ → —CHO, Section 5.3.1(i)):

$$\underset{\diagup}{\overset{\diagdown}{}}C\underset{OH}{\overset{OH}{\diagup}} \quad \xrightarrow{-H_2O} \quad \underset{\diagup}{\overset{\diagdown}{}}C{=}O$$

6.2   Physical properties of alcohols

The alcohols display the usual trends in physical properties, characteristic of a homologous series. Methanol and ethanol are colourless, volatile liquids; higher members are viscous liquids and the first solid alcohol occurs at C_{11}, although 2-methylpropan-2-ol has a melting point (26°C) very close to room temperature. The boiling points of the alcohols increase by a nearly regular amount as the relative molecular mass increases, but the actual values are substantially higher than those of alkanes, or halogenoalkanes, of corresponding relative molecular mass (*see Figure 2.1*). The extensive hydrogen bonding (*see* Section 1.4) which is present in alcohols accounts for this:

The hydrogen bonds are stronger than the universal van der Waals forces, and more energy is required to overcome these intermolecular forces. Alcohols therefore have higher boiling points and enthalpies of vaporization than, for instance, halogenoalkanes of corresponding relative molecular mass. As the number of hydroxyl groups in an alcohol molecule increases, so does the degree of hydrogen bonding, and this is reflected in higher boiling points and viscosities, as illustrated by a comparison of these properties for ethanol, ethane-1,2-diol and propane-1,2,3-triol.

The presence of the hydroxyl group also confers water solubility on the simple alcohols, as they can readily fit into the loosely hydrogen-bonded water structure. Thus, methanol and ethanol mix freely with water; heat is evolved on mixing and there is a contraction in total volume. These are indications of hydrogen bond formation between molecules. Propanols, too, mix with water but as the chain length of the alcohol increases, so the effect of the hydroxyl group on the physical properties of the molecule diminishes, and this is reflected in the diminishing solubility of the higher alcohols in water (as well as in the boiling points, which begin to approach those of alkanes of corresponding relative molecular mass). Butan-1-ol, for example, is only partially soluble in water.

Ethanol forms a constant-boiling mixture with water, making the separation of the anhydrous alcohol from aqueous solution by fractional distillation impossible. The constant-boiling mixture has a boiling point of 78.2°C at 0.1 MN m^{-2} pressure and a composition of 95.6% ethanol, 4.4% water by mass. In this form, it is sometimes known as *rectified spirit*. Pure ethanol may be obtained from it by removing the water with calcium oxide, or by azeotropic distillation of the alcohol with benzene (*see* Section 21.2.2(ii)). It is then known as *absolute alcohol* but may still contain up to about 0.5% water. Completely anhydrous ethanol, which is very hygroscopic indeed, may be obtained from absolute alcohol by reaction with magnesium (*see* Section 6.3.4).

The so-called *methylated spirit* normally consists of about 90% rectified spirit, 9% methanol, a little pyridine and a violet dye. Methanol is a very dangerous poison* which can cause blindness when a small amount is ingested and is added, together with the poisonous pyridine and the dye, to deter people from drinking 'meths'. It is illegal to attempt to obtain pure ethanol from mixtures without possession of a Customs and Excise permit. *Surgical spirit* contains about 95% rectified spirit and 5% methanol. A little Oil of Wintergreen is sometimes added to it to give it a characteristic smell.

The solubility of ethane-1,2-diol (formerly known as *ethylene glycol* or just *glycol*) in water is exploited in its use as a component of 'anti-freeze' fluids. A 60% solution in water lowers the freezing point by about 40°C. Certain types of insect contain in their bodies propane-1,2,3-triol which, by depressing the freezing point of body water, prevents tissue damage resulting from ice formation in body cells when the insects are exposed to low temperatures. This triol has also been used for protecting animal sperm against damage during thawing, after it has been preserved at low temperatures for use in artificial insemination in animal breeding.

6.3 Chemical properties of alcohols

Alcohols are saturated compounds containing the functional group —OH. They can undergo **nucleophilic substitution** reactions, in which the hydroxyl group is displaced by other groups, or they can undergo **elimination** reactions, involving the loss of the elements of water when treated with a dehydrating agent. Primary and secondary alcohols also undergo **oxidation** (equivalent to dehydrogenation of the alcohol grouping) and the hydroxyl group shows acidic properties when treated with reactive metals (cf. water).

6.3.1 Substitution reactions

(i) Alcohols undergo few substitution reactions because the hydroxyl group is not easily displaced. Under a variety of conditions, and using a variety of reagents, alcohols may be converted into halogenoalkanes, however. Phosphorus pentachloride, for example, will react with an alcohol, forming a chloroalkane. The evolution of hydrogen chloride fumes which occurs during the reaction indicates the presence of the hydroxyl group. This is present as a simple group in an alcohol and as part of the more complicated carboxyl group (*see* Section 9.1) in a carboxylic acid:

$$R—OH + PCl_5 \rightarrow R—Cl + HCl + POCl_3$$

---

* In the body, methanol is oxidized, via methanal, to methanoic acid and this causes intense acidosis (cf. ethanol in Section 6.3.3).

The organic product is not always easily separated from the unpleasantly hygroscopic phosphorus trichloride oxide (b.p. = 105°C), and a much 'cleaner' and more efficient reagent is sulphur dichloride oxide (*thionyl chloride*). This reacts smoothly with an alcohol, in an anhydrous solvent, to form the chloroalkane which is already substantially free of the other (gaseous) reaction products. Unreacted sulphur dichloride oxide (b.p. = 77°C) may be distilled off:

$$R{-}OH + SOCl_2 \ \rightarrow \ R{-}Cl + SO_2 + HCl$$

e.g.

$$CH_3CH_2CH_2OH + SOCl_2 \ \rightarrow \ CH_3CH_2CH_2Cl + SO_2 + HCl$$
propan-1-ol   1-chloropropane
b.p. = 97°C   b.p. = 46°C

Bromo compounds can be made from alcohols by the action of phosphorus tribromide, which is conveniently prepared *in situ* by the action of bromine on red phosphorus:

$$3CH_3CH_2OH + PBr_3 \ \rightarrow \ 3CH_3CH_2Br + H_3PO_3$$
ethanol   bromoethane
                                               b.p. = 38°C

An analogous reaction can be used to prepare iodo compounds.

(ii) Alternatively, halogenoalkanes can be prepared from alcohols using hydrogen halides:

$$R{-}OH + HX \ \rightarrow \ R{-}X + H_2O$$

The ease of reaction depends both on the nature of the hydrogen halide (the order of reactivity is HI > HBr > HCl), and on the structure of the alcohol. The bromo compounds are most conveniently prepared by heating the alcohol with a mixture of sodium bromide and concentrated sulphuric acid (which generates hydrogen bromide and helps absorb the water formed), or else with a constant-boiling solution of hydrobromic acid. However, yields may be poor with some secondary alcohols, and certainly with tertiary alcohols, since some dehydration (*see* Section 6.3.2) will also occur. The action of a constant-boiling solution of hydriodic acid will bring about the corresponding formation of iodo compounds. Concentrated hydrochloric acid readily reacts with tertiary alcohols at room temperature to form the chloro compound, but the conversion of primary alcohols requires heating with the concentrated acid in the presence of anhydrous zinc chloride. Secondary chloro compounds are probably best made by alternative methods, although the $HCl/ZnCl_2$ method may be used:

$$CH_3CH_2CH_2CH_2OH + HCl \xrightarrow{\ ZnCl_2\ } CH_3CH_2CH_2CH_2Cl + H_2O$$
butan-1-ol   1-chlorobutane
b.p. = 118°C   b.p. = 77°C

$(CH_3)_3COH + HCl \rightarrow (CH_3)_3CCl + H_2O$
2-methylpropan-2-ol   2-chloro-2-methylpropane
b.p. = 83°C   b.p. = 50°C

$$CH_3CH_2CH_2CH_2OH + HBr \xrightarrow{H_2SO_4} CH_3CH_2CH_2CH_2Br + H_2O$$
1-bromobutane
b.p. = 101°C

$$CH_3CHCH_3 + HI \longrightarrow CH_3CHCH_3 + H_2O$$
| |
OH   I
propan-2-ol   2-iodopropane
b.p. = 83°C   b.p. = 89°C

The mechanism of reaction between alcohols and hydrogen halides is an ionic one, and involves nucleophilic attack by halide ion at the carbon atom bearing the hydroxyl group. The hydroxyl group is not easily displaced as hydroxide ion but, when it is protonated, the carbon–oxygen bond is weakened, e.g.

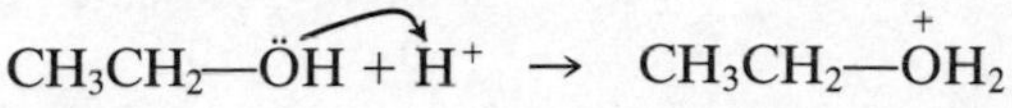

$$CH_3CH_2-\overset{..}{O}H + H^+ \rightarrow CH_3CH_2-\overset{+}{O}H_2$$

In the case of primary and usually secondary alcohols, the halide anion then displaces a neutral water molecule from the protonated alcohol, so substituting for the hydroxyl group:

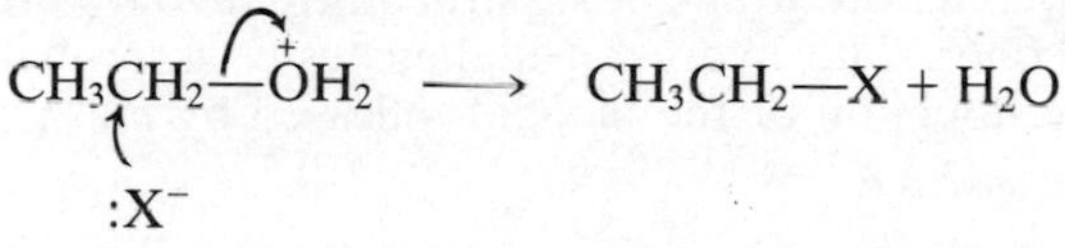

$$CH_3CH_2-\overset{+}{O}H_2 \longrightarrow CH_3CH_2-X + H_2O$$
:X⁻

In the case of a protonated tertiary alcohol, the carbon–oxygen bond undergoes spontaneous heterolytic fission to form a carbocation, which immediately combines with a halide ion to form the halogenoalkane:

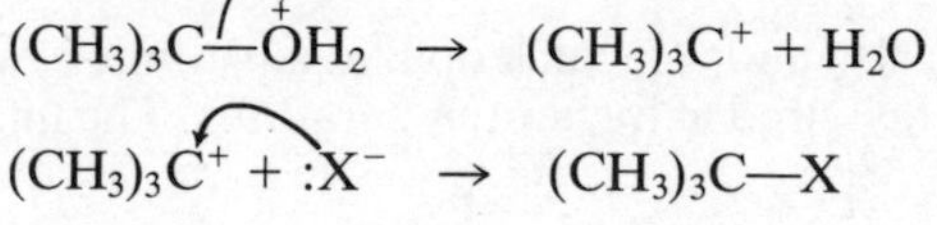

$$(CH_3)_3C-\overset{+}{O}H_2 \rightarrow (CH_3)_3C^+ + H_2O$$

$$(CH_3)_3\overset{+}{C} + :X^- \rightarrow (CH_3)_3C-X$$

It is easy to see how the reaction with a tertiary alcohol may give rise to some alkene formation; the intermediate carbocation is capable of losing a proton to form an alkene as well as undergoing nucleophilic attack, e.g.

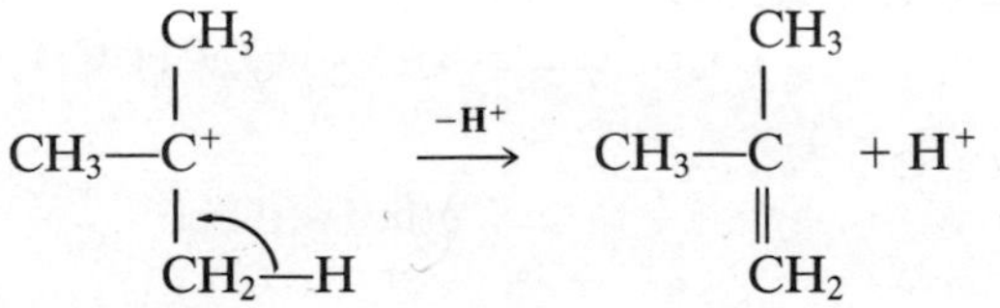

$$
\begin{array}{ccc}
CH_3 & & CH_3 \\
| & & | \\
CH_3-C^+ & \xrightarrow{-H^+} & CH_3-C \quad + H^+ \\
| & & \| \\
CH_2-H & & CH_2 \\
\end{array}
$$

Since the order of stabilities of carbocations is tertiary > secondary > primary, it is usually tertiary alcohols only which react by this mechanism, although secondary alcohols may do so on occasion.

(iii) The products of the reaction between alcohols and concentrated sulphuric acid depend very much on the reaction conditions employed (*see also* Section 6.3.2), but they may result from a substitution reaction. Ethanol, for example, reacts when mixed with concentrated sulphuric acid at room or slightly elevated temperatures, to form an *ester* of sulphuric acid, ethyl hydrogensulphate:

$$CH_3CH_2OH + H_2SO_4 \rightleftharpoons CH_3CH_2OSO_3H + H_2O$$

The sulphuric acid absorbs the water formed and the position of equilibrium therefore favours the product. If the mixture of acid and alcohol is heated under reduced pressure, and the product removed by distillation, then diethyl sulphate is formed:

$$CH_3CH_2OSO_3H + CH_3CH_2OH \xrightarrow{100°C} CH_3CH_2OSO_2OCH_2CH_3 + H_2O$$
diethyl sulphate
b.p. = 210°C

Dimethyl and diethyl sulphates should be handled with the greatest care, since they are very poisonous and are harmful by skin absorption.

At more elevated temperatures, i.e. about 140°C, another displacement or substitution can occur and it is particularly favoured by the presence of excess alcohol. The hydrogensulphate ester is probably formed, as before, by protonation of the alcohol followed by nucleophilic substitution, e.g.

$$CH_3CH_2{-}OH + H_2SO_4 \rightleftharpoons CH_3CH_2{-}\overset{+}{O}H_2 + \overset{-}{O}SO_3H$$

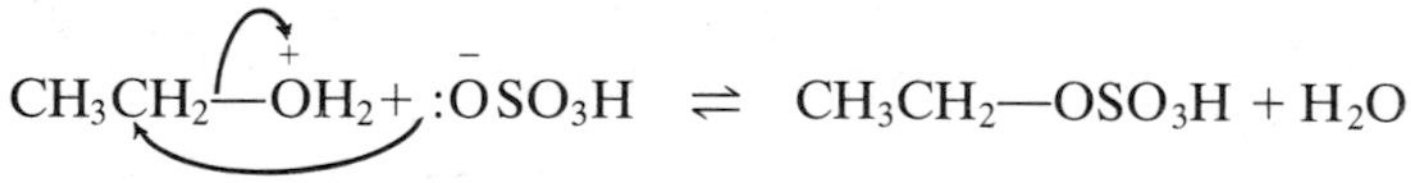

$$CH_3CH_2{-}\overset{+}{O}H_2 + :\overset{-}{O}SO_3H \rightleftharpoons CH_3CH_2{-}OSO_3H + H_2O$$

The ester itself can also undergo nucleophilic substitution by unprotonated alcohol and this is favoured at higher temperatures. The final step involves proton transfer:

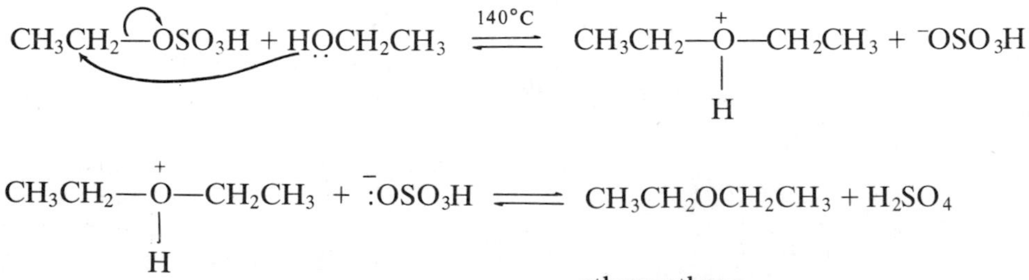

$$CH_3CH_2{-}OSO_3H + HOCH_2CH_3 \xrightarrow{140°C} CH_3CH_2{-}\overset{+}{O}{-}CH_2CH_3 + {}^-OSO_3H$$
$$|$$
$$H$$

$$CH_3CH_2{-}\overset{+}{O}{-}CH_2CH_3 + :^-OSO_3H \rightleftharpoons CH_3CH_2OCH_2CH_3 + H_2SO_4$$
$$|$$
$$H$$

ethoxyethane
b.p. = 37°C

The product is called an **ether** (*see also* Section 5.3.1(ii) and Chapter 7).

In these substitution reactions, the protonated alcohols, and their esters of inorganic acids, have behaved in a similar manner to halogenoalkanes.

When propane-1,2,3-triol is treated with a mixture of concentrated sulphuric and nitric acids, the nitric acid ester, propane-1,2,3-triyl trinitrate (formerly known as *nitroglycerine*), is formed:

$$
\begin{array}{ll}
CH_2OH & CH_2ONO_2 \\
| & | \\
CHOH + 3HONO_2 \longrightarrow & CHONO_2 + 3H_2O \\
| & | \\
CH_2OH & CH_2ONO_2
\end{array}
$$

The first patent for its use as a commercial explosive was filed by the Swedish scientist Alfred Nobel (originator of the Nobel Prizes) in 1863. When the nitrate ester is absorbed in 'kieselguhr', a natural finely divided form of silicon(IV) oxide, dynamite is produced which is much safer to handle. A modified form is known as 'gelignite'. On detonation of the nitrate ester, by ignition or shock, a rapidly expanding volume of gas is produced:

$$4C_3H_5N_3O_9 \rightarrow 12CO_2 + 10H_2O + 6N_2 + O_2 \qquad \Delta H^{\ominus} = -7230\,kJ$$

Alcohols also form esters with organic carboxylic acids and these reactions are discussed in Section 9.3.2.

6.3.2   Elimination reactions

As alcohols contain the hydroxyl group —OH, they may undergo the elimination of the elements of water, i.e. dehydration, when treated with a suitable reagent under appropriate conditions. Concentrated sulphuric acid has a very strong affinity for water and behaves as a dehydrating agent towards alcohols. When heated with sulphuric acid, alcohols can be dehydrated to alkenes:

$$
-\overset{|}{\underset{|}{C}}-\overset{|}{\underset{|}{C}}-OH \longrightarrow ^{>}C{=}C^{<} + H_2O
$$

The ease of dehydration of an alcohol depends markedly on its structure; tertiary alcohols readily dehydrate under mildly acid conditions at moderate temperatures, secondary alcohols are intermediate in reactivity, while primary alcohols generally require concentrated acid and relatively high temperatures. The ease of dehydration is thus tertiary > secondary > primary, as illustrated:

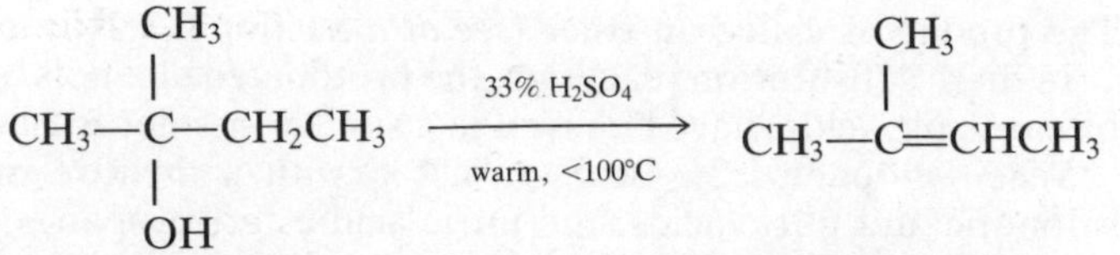

$$\underset{\underset{OH}{|}}{\overset{\overset{CH_3}{|}}{CH_3-C-CH_2CH_3}} \xrightarrow[\text{warm, <100°C}]{\text{33\% H}_2\text{SO}_4} \underset{}{\overset{\overset{CH_3}{|}}{CH_3-C=CHCH_3}}$$

2-methylbutan-2-ol   2-methylbut-2-ene
b.p. = 102°C   b.p. = 39°C

$$\underset{\underset{OH}{|}}{CH_3-CH-CH_2CH_2CH_3} \xrightarrow[\text{warm, <100°C}]{\text{50\% H}_2\text{SO}_4} CH_3-CH=CH-CH_2CH_3$$

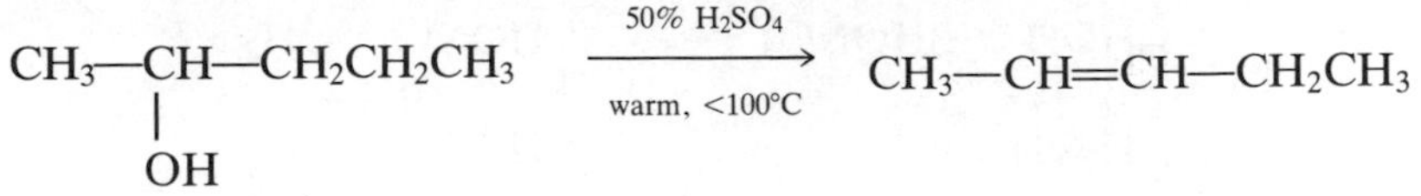

pentan-2-ol   pent-2-ene
b.p. = 120°C   b.p. = 37°C

$$CH_3CH_2OH \xrightarrow[\text{170° C}]{\text{98\% H}_2\text{SO}_4} CH_2=CH_2$$

ethene

The reaction with primary alcohols may involve the intermediate formation of a hydrogensulphate ester, which at higher temperatures loses the elements of sulphuric acid to form the alkene (*see* Section 6.3.1(iii)).

$$CH_3CH_2OH + H_2SO_4 \rightleftharpoons CH_3CH_2OSO_3H + H_2O \rightleftharpoons CH_2=CH_2 + H_2SO_4 + H_2O$$

An alternative mechanism involves the action of the hydrogensulphate ion as a *base* on the protonated alcohol:

$$\underset{\underset{-:\bar{O}SO_3H}{\overset{|}{H}}}{CH_2-CH_2-\overset{+}{O}H_2} \rightleftharpoons CH_2=CH_2 + H_2SO_4 + H_2O$$

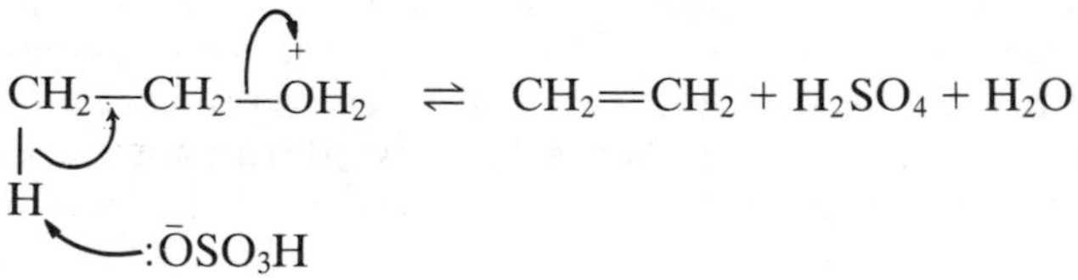

Overall, of course, the reaction is the converse of the sulphuric acid-catalysed hydration of alkenes (*see* Section 3.3.1(v)).

The mechanism of dehydration of tertiary alcohols undoubtedly involves heterolysis of the protonated alcohol to form a relatively stable carbocation, which then loses a proton to form an alkene (*see* Section 6.3.1(ii)). It should be clear from the previous examples that, where elimination can take place in more than one way, the predominant alkene product is the most highly substituted one (cf. Section 5.3.3).

Concentrated sulphuric acid is an oxidizing agent and therefore dehydration of alcohols with this acid sometimes leads to some oxidation of organic material to carbon ('charring'), and the production of sulphur dioxide as a reduction product of the acid. Such problems may often be overcome by the use of 85% phosphoric(V) acid, which is non-oxidizing although it does not absorb the water formed as sulphuric acid does. Cyclohexanol can be conveniently dehydrated to the liquid cyclohexene by this method:

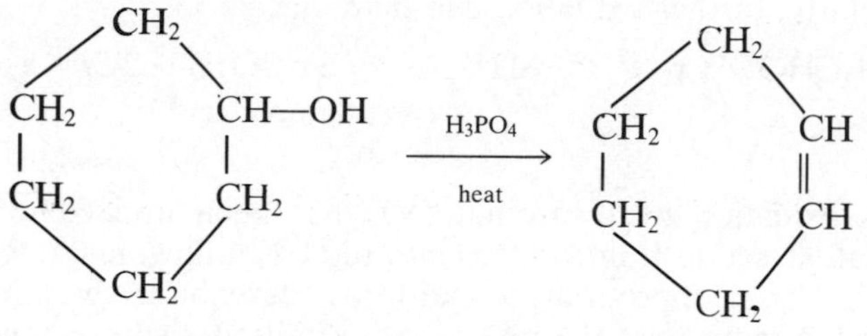

cyclohexanol
b.p. = 161°C

cyclohexene
b.p. = 83°C

more conveniently written as

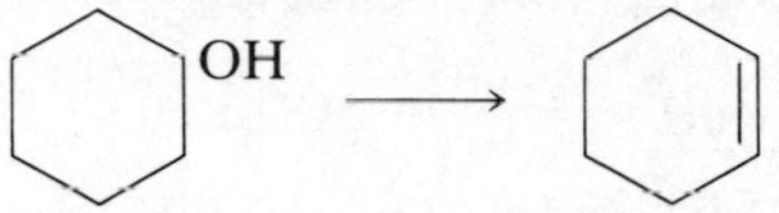

Alcohols can also be dehydrated by passing the vapour over strongly heated aluminium oxide or silicon(IV) oxide, e.g.

$$CH_3CH_2OH \xrightarrow[375°C]{Al_2O_3} CH_2{=}CH_2$$

6.3.3   Oxidation reactions

Alcohols react with strong oxidizing agents, such as acidified solutions of sodium or potassium dichromate(VI), to give products which depend on the structure of the alcohol.

A primary alcohol gives first an **aldehyde**; this is usually isolated by distilling it from the hot reaction mixture as it is formed. If the aldehyde is left in contact with the oxidizing agent, further oxidation occurs to give a **carboxylic acid**:

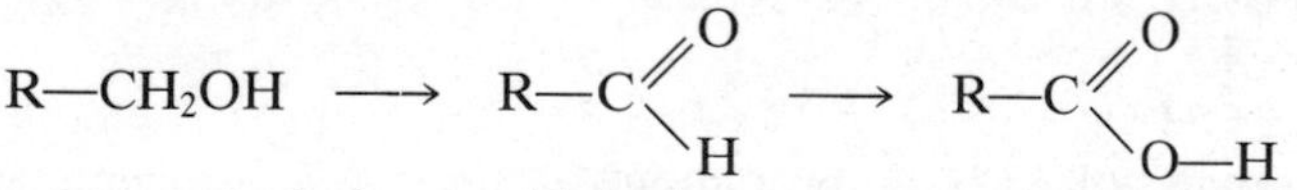

primary alcohol   aldehyde   carboxylic acid

During the oxidation the orange solution containing dichromate(VI) ions changes its colour to green as reduction to chromium(III) ions takes place, e.g.

$$3CH_3CH_2OH - 6e^- \rightarrow 3CH_3CHO + 6H^+ \quad \text{oxidation}$$
$$Cr_2O_7^{2-} + 14H^+ + 6e^- \rightarrow 2Cr^{3+} + 7H_2O \quad \text{reduction}$$

i.e. overall

$$3CH_3CH_2OH + Cr_2O_7^{2-} + 8H^+ \xrightarrow{warm} 3CH_3CHO + 2Cr^{3+} + 7H_2O$$

ethanol   ethanal
b.p. = 78°C   b.p. = 21°C

Similarly, further oxidation can take place:

$$3CH_3CHO + Cr_2O_7^{2-} + 8H^+ \rightarrow 3CH_3CO_2H + 2Cr^{3+} + 4H_2O$$

ethanoic acid
b.p. = 118°C

This oxidation by dichromate(VI) has been utilized in the screening 'breathalyser' test, introduced into the UK following the Road Safety Act 1967. Under appropriate conditions, the exhaled breath of a motorist who has more than the maximum permitted 80 mg ethanol per 100 cm³ of blood leads to this colour change, and the test may be regarded as positive.

If acidified potassium manganate(VII) is used as oxidizing agent then, in the presence of a limited amount of it, the colour change accompanying oxidation is purple to colourless (Mn^{2+} ions), e.g.

$$5CH_3CH_2OH + 2MnO_4^- + 6H^+ \xrightarrow{warm} 5CH_3CHO + 2Mn^{2+} + 8H_2O$$

In the body, ethanol is oxidized in the liver, first to ethanal, then to ethanoate ions, and finally to carbon dioxide and water. In the aversion therapy treatment of alcoholics, a chemical may be given which prevents the oxidation of the ethanal. The accumulation of this in the blood causes the patient to feel very ill.

Secondary alcohols are oxidized to **ketones** by either dichromate(VI) or manganate(VII), and these products are relatively stable towards further oxidation. The colour changes accompanying oxidation are as just described.

secondary alcohol   ketone

e.g.

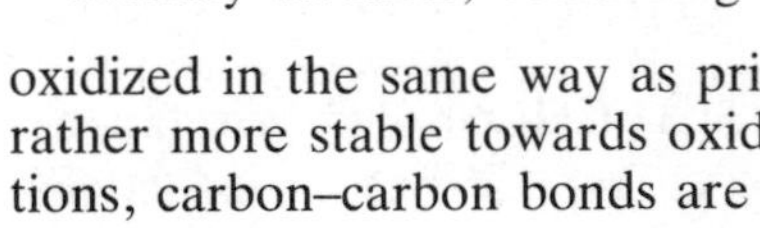

$$CH_3CHCH_3 \longrightarrow CH_3CCH_3$$

$$| \qquad\qquad\qquad ||$$

OH   O

propan-2-ol   propanone
b.p. = 83°C   b.p. = 56°C

Tertiary alcohols, containing the $\diagdown\!C\!\!-\!OH$ group, clearly cannot be oxidized in the same way as primary and secondary alcohols. They are rather more stable towards oxidizing agents but, under vigorous conditions, carbon–carbon bonds are broken and the alcohol is broken down into smaller fragments which can include ketones, carboxylic acids and even carbon dioxide.

Since there are simple tests to distinguish aldehydes and ketones (*see* Section 8.3.7), a study of the behaviour of an alcohol on oxidation, under the conditions described, is a useful way of distinguishing the three types of alcohol. The colour change accompanying oxidation readily distinguishes the primary and secondary alcohols from tertiary ones.

Diols and triols, too, are oxidized but fission of the carbon–carbon bond also takes place if vigorous conditions are used, e.g.

$$HOCH_2{-}CH_2OH \rightarrow HO_2C{-}CO_2H \rightarrow 2CO_2 + H_2O \quad \text{(via various intermediate oxidation products)}$$

ethane-1,2-diol   ethanedioic acid
                        (*oxalic acid*)

Dehydrogenation, brought about by passing the alcohol vapour over heated metal catalysts, such as copper, or metal compounds, is used industrially to convert alcohols into aldehydes or ketones, e.g.

$$CH_3{-}\underset{\underset{\displaystyle OH}{|}}{CH}{-}CH_3 \xrightarrow[\text{400–500°C}]{\text{ZnO or Cu}} CH_3{-}\underset{\underset{\displaystyle O}{\|}}{C}{-}CH_3 + H_2$$

propan-2-ol   propanone
b.p. = 83°C   b.p. = 56°C

A little dehydration may also accompany this reaction and, under these conditions, tertiary alcohols are dehydrated, not dehydrogenated.

More generally, oxidative dehydrogenation is employed, as in the manufacture of methanal (*formaldehyde*) from methanol in the vapour phase:

$$CH_3OH + \tfrac{1}{2}O_2 \xrightarrow[\text{500–650°C}]{\text{Ag}} HCHO + H_2O$$

6.3.4   Reaction with metals

Because alcohols contain hydroxyl groups, they show some resemblance to water. For example, methanol and ethanol form solvated salts with calcium chloride, namely $CaCl_2.4CH_3OH$ and $CaCl_2.3CH_3CH_2OH$, respectively (cf. $CaCl_2.6H_2O$), so that anhydrous calcium chloride cannot be used to dry either of these alcohols.

More important, however, the alcohols react with metals like sodium and potassium (although less violently than water does) to form ionic compounds called **alkoxides**, e.g.

$$2ROH + 2Na \rightarrow 2RO^-Na^+ + H_2$$

For example, if sodium is added to ethanol there is a brisk evolution of hydrogen as the metal dissolves. If the remaining ethanol is evaporated, a white crystalline solid, sodium ethoxide, is obtained:

$$2CH_3CH_2OH + 2Na \rightarrow 2CH_3CH_2O^-Na^+ + H_2$$
$$\text{sodium ethoxide}$$

Alkoxides can be used as organic bases and will also act as nucleophiles (cf. their reaction with halogenoalkanes, Section 5.3.1(ii)).

Under suitable conditions, alcohols will also react with less reactive metals such as aluminium and magnesium. Magnesium finds a particular use in the preparation of very pure 'super-dry' ethanol from the commercial *absolute alcohol*. Sodium cannot be used to remove all the water from the ethanol, owing to the establishment of the following equilibrium:

$$CH_3CH_2O^-Na^+ + H_2O \rightleftharpoons CH_3CH_2OH + NaOH$$

The complete removal of up to 1% water can be accomplished by distilling the ethanol from magnesium (plus a trace of iodine to activate the metal). Magnesium ethoxide is first formed and this then reacts with the water to form the almost totally insoluble magnesium hydroxide:

$$2CH_3CH_2OH + Mg \rightarrow (CH_3CH_2O^-)_2Mg^{2+} + H_2$$
$$(CH_3CH_2O^-)_2Mg^{2+} + 2H_2O \rightarrow 2CH_3CH_2OH + Mg(OH)_2$$

6.4 Manufacture and uses of alcohols

The simpler alcohols are manufactured in large amounts, since they are required for a variety of purposes.

Methanol is manufactured by the reduction of carbon monoxide, from synthesis gas (*see* Section 2.5(ii)). Synthesis gas is obtained either by partial oxidation of methane, or by the 'steam-reforming' process

$$CH_4 + O_2 \rightarrow CO + 2H_2 \qquad \Delta H^{\ominus} = -36\,kJ$$

or

$$CH_4 + H_2O \rightleftharpoons CO + 3H_2 \qquad \Delta H^{\ominus} = +205\,kJ$$

The conversion to methanol involves a zinc oxide/chromium oxide catalyst,* at temperatures between 350°C and 450°C and a pressure of about 30 MN m^{-2}:

$$CO + 2H_2 \rightarrow CH_3OH \qquad \Delta H^{\ominus} = -109\,kJ$$

Less than 20% of the gases are converted to methanol, which is removed by condensation, allowing the gas stream to be recycled.

* If a metal which forms a carbonyl, e.g. nickel, is used as a catalyst, then methane and water are formed instead of methanol.

Large amounts of methanol are used to make methanal, by oxidation (*see* Section 6.3.3 and Section 8.4). Increasing amounts of methanol are being employed, by ICI for example, in the production of single-cell protein (SCP, *see also* Section 2.5(iii)), as it is a relatively cheap and readily available feedstock which is cheaper and easier to ferment than hydrocarbons. Methanol has also been studied as a potential fuel for use in fuel cells.

Ethanol can be made by fermentation of carbohydrates or from ethene. Some is still manufactured from the latter by reaction with sulphuric acid followed by hydrolysis (*see* Section 3.3.1(v)), but in the main direct hydration is now employed. This involves passing ethene over a phosphoric(V) acid catalyst at about 300°C and 7 MN m^{-2} pressure, together with steam. The reaction is an equilibrium one. In principle, formation of ethanol is favoured by low temperatures and high pressures:

$$CH_2{=}CH_2 + H_2O \; \rightleftharpoons \; CH_3CH_2OH \qquad \Delta H^{\ominus} = -46\,kJ$$

In practice, a temperature of about 300°C is necessary to ensure a reasonably fast reaction. Consequently, the yield of ethanol is low (about 5% per pass) but, after the product has been condensed as a 10–20% aqueous solution, the gas stream is purified and recycled (excess ethene is employed), so the process is a continuous one.

The major use of ethanol today is as a solvent employed in a very wide variety of formulations. Its importance as a chemical feedstock has declined somewhat as direct routes from other materials such as ethene have been developed.

Higher alkenes can be converted into higher alcohols via the OXO reaction (*see* Section 3.3.1(vi)), the aldehyde products being then hydrogenated to alcohols.

Ethane-1,2-diol, which is widely used as a component of 'anti-freeze' mixtures and in the manufacture of Terylene (*see* Section 20.3(ii)), is prepared by hydrolysis of epoxyethane, derived from the oxidation of ethene (*see* Section 3.3.2(v)):

$$CH_2{=}CH_2 + \tfrac{1}{2}O_2 \xrightarrow[\substack{250°C \\ 1-3\ MN\ m^{-2}}]{Ag} CH_2{-}CH_2 \xrightarrow{H_2O} HOCH_2{-}CH_2OH$$
$$\diagdown O \diagup$$

Propane-1,2,3-triol (*glycerol*) used to be obtained mainly as one of the hydrolysis products of esters in the vegetable oils and animal fats (*see* Section 19.2.1) employed in soap making, but much of it is now manufactured from propene. One route involves the following:

$$CH_3CH{=}CH_2 \xrightarrow[\substack{350°C \\ 0.2\ MN\ m^{-2}}]{CuO/O_2} O{=}CHCH{=}CH_2 \xrightarrow[\substack{ZnO/MgO \\ 400°C}]{(CH_3)_2CHOH}$$

$$HOCH_2CH{=}CH_2 \xrightarrow[\text{catalyst}]{aq.H_2O_2} HOCH_2\underset{\underset{OH}{|}}{C}HCH_2OH$$

Propane-1,2,3-triol is used as a component of 'anti-freeze' fluids, as a moistening agent in food and cosmetics, and in the manufacture of dynamite.

Summary

1   An **alcohol** is a compound containing one or more **hydroxyl** groups in a molecule. Alcohols can be classified according to the number of hydroxyl groups per molecule (e.g. **monohydric, dihydric,** etc.) and the number of alkyl groups bonded to the carbon atom bearing the hydroxyl group (primary —CH_2OH; secondary $\diagdown CHOH$; tertiary —$\overset{\diagdown}{\underset{\diagup}{C}}$—OH). Monohydric alcohols form a homologous series of general formula $C_nH_{2n+1}OH$.

2   The characteristic functional group of alcohols is the **hydroxyl** group —OH. Alcohols will normally undergo **nucleophilic substitution** reactions, **elimination** reactions and **oxidations**. Although extremely weak, the acidic nature of the hydroxyl group also means that metal derivatives can be formed.

(i) Nucleophilic substitution

R—OH

$\xrightarrow{\text{PCl}_5}$ R—Cl

$\xrightarrow{\text{SOCl}_2}$ R—Cl

$\xrightarrow{\text{PBr}_3}$ R—Br

$\xrightarrow{\text{HBr}}$ R—Br

$\xrightarrow{\text{HI}}$ R—I

$\xrightarrow[\text{low temp.}]{\text{H}_2\text{SO}_4}$ R—OSO_3H

$\xrightarrow[\text{higher temp.}]{\text{H}_2\text{SO}_4}$ R—O—R

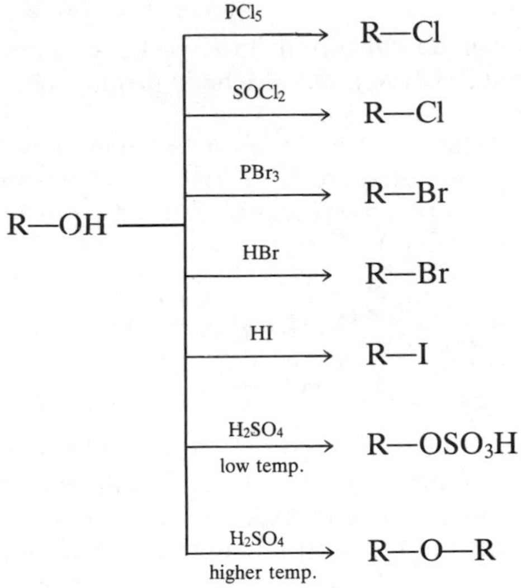

(ii) Elimination

$$-\overset{|}{\underset{|}{C}}-\overset{|}{\underset{|}{C}}- \quad \xrightarrow[\text{H}_3\text{PO}_4]{\overset{\text{H}_2\text{SO}_4}{\text{or}}} \quad \overset{\diagup}{\diagdown}C{=}C\overset{\diagdown}{\diagup} + H_2O$$
$$\;\;\;\;\text{H}\;\;\;\text{OH}$$

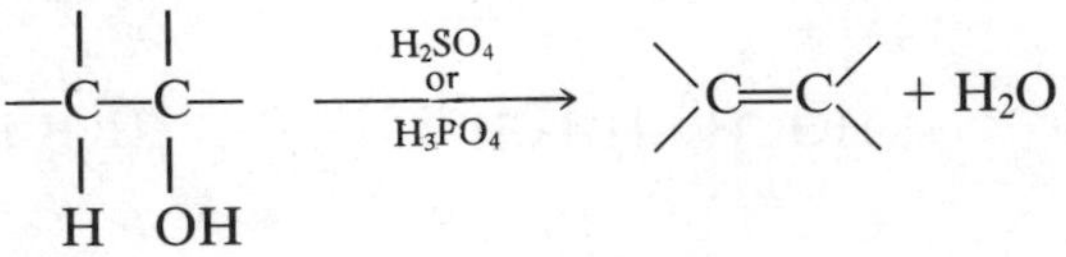

(iii) Oxidation

On reaction with oxidizing agents such as acidified solutions of potassium dichromate(VI) or manganate(VII):

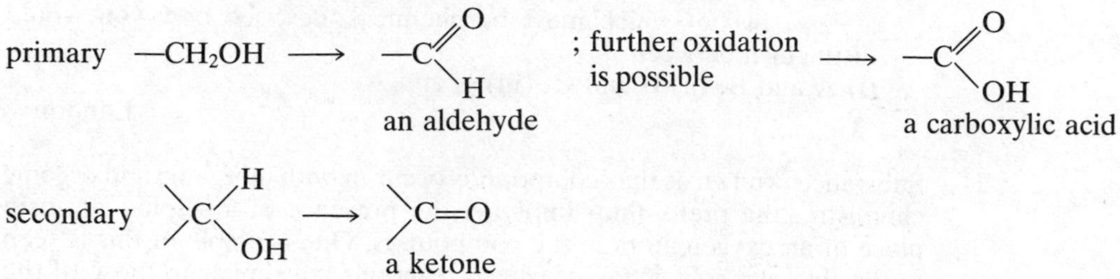

primary —CH$_2$OH $\longrightarrow$ —C$\overset{\displaystyle O}{\underset{\displaystyle H}{\big\langle}}$; further oxidation is possible $\longrightarrow$ —C$\overset{\displaystyle O}{\underset{\displaystyle OH}{\big\langle}}$

an aldehyde   a carboxylic acid

secondary $\overset{\displaystyle H}{\underset{\displaystyle OH}{>\!C\!<}}$ $\longrightarrow$ $>\!C\!=\!O$

a ketone

tertiary $>\!C\!-\!OH$ not affected under mild conditions. Vigorous oxidation breaks down the molecule, giving products containing fewer carbon atoms per molecule.

Alternatively, **dehydrogenation**, oxidative or otherwise, can be used. Primary alcohols give aldehydes; secondary alcohols give ketones.

(iv) Reaction with metals

The hydrogen atom of the hydroxyl group is very weakly acidic (*see also* Section 9.3.1), for example pK_a of ethanol is ≈ 18, and can be replaced by reaction of the alcohol with a reactive metal such as potassium or sodium. Alkoxides are formed:

2R—OH + 2Na $\rightarrow$ 2R—O$^-$Na$^+$ + H$_2$

3   The alcohols generally are widely used as solvents. Methanol, ethanol, ethane-1,2-diol and propane-1,2,3-triol are some of the most important to be manufactured industrially, starting (with the exception of methanol) from the corresponding alkene.

Questions

1* The structural formulae of three alcohols are shown below:

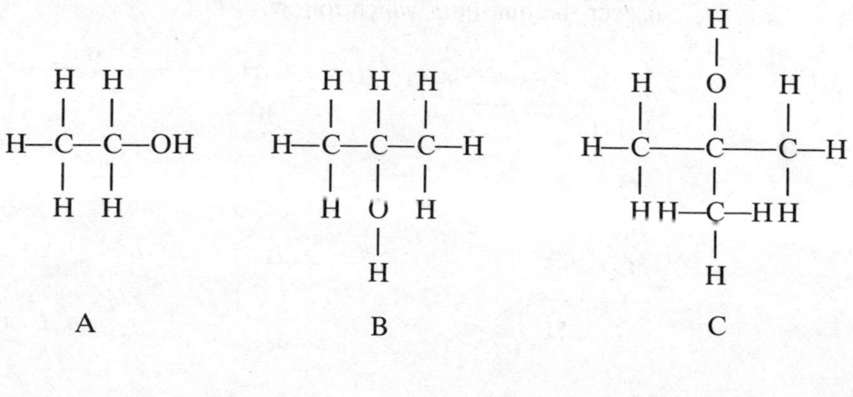

A   B   C

(a) Give the systematic names of B and C.
(b) Give a reaction scheme, i.e. reagents, conditions of reaction and structures of intermediates, for the formation of A from iodomethane.
(c) Using *three* different methods (one for each of (i), (ii) and (iii) below), at least *two* of which must be chemical, describe how you would distinguish between
(i) A and B, (ii) B and C, (iii) C and A.

London

2*   Substances known as thio-compounds occur in both inorganic and organic chemistry, the prefix thio- implying the presence of a sulphur atom in place of an oxygen atom in the compounds. One example of this is seen in the thio-alcohols, many of whose reactions are similar to those of the corresponding alcohols.
The questions which follow are about thioethanol (C_2H_5SH). Assume that its reactions are analogous to those of ethanol.
(a) (i) Write a balanced equation for the reaction of thioethanol with sodium.
(ii) Give the formula of the anion formed when thioethanol reacts with sodium.
(b) Name the products from the complete combustion of thioethanol.
(c) How would you expect the boiling point of thioethanol to compare with that of ethanol? Give a reason.
(d) How would you expect the solubility in water of thioethanol to compare with that of ethanol? Give a reason.
(e) (i) When an alcohol containing a large proportion of ^{18}O atoms is reacted with a carboxylic acid, the ^{18}O atoms are found in the organic product and not in the water produced during the reaction. State briefly what can be deduced about the mechanism of the reaction.
(ii) Draw a diagram of the structural formula of the probable product of the reaction between thioethanol and ethanoic acid (*acetic acid*).
(iii) What would be a suitable catalyst for this reaction?

Nuffield

3*   Compound **A** is an alcohol. Examine the reaction scheme below and answer the questions which follow.

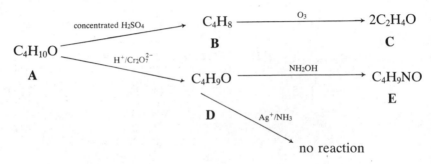

(a) Give the structural formula of **each** of the compounds **A** to **E**.
(b) Give the mechanism for the conversion of **A** into **B**, showing clearly how the formation of an isomeric product, C_4H_8, arises.

JMB

4*   Deduce the structures of **A,B,C** and **D** and explain the significance of the italicized phrases in the following paragraph.
The compound **A**, C_3H_8O, *acts as a Lewis base* in its reaction with concentrated hydrobromic acid to give **B**, C_3H_7Br. On treatment with concentrated sulphuric acid, **A** *undergoes an elimination reaction* to give the hydrocarbon **C**, C_3H_6. When **C** is reacted with concentrated hydrobromic acid, the resultant product **D**, C_3H_7Br, is formed in accordance with the *Markovnikov rule* and is isomeric with **B**.

JMB

5   Show, by means of equations, how (a) ethanol, and (b) ethane-1,2-diol (*ethylene glycol*), are obtained on an industrial scale. How would you obtain anhydrous ethanol from a 50% aqueous solution?
Give **two** reactions which demonstrate the presence of a hydroxyl group in ethanol.
How would you demonstrate that phenol is more acidic than cyclohexanol $(C_6H_{11}OH)$?

Oxford & Cambridge

6*   Ethanol for industrial use was formerly made by fermentation, but now it is made almost entirely by the reaction:

$$C_2H_4\,(g) + H_2O\,(g) \;\rightleftharpoons\; C_2H_5OH\,(g) \qquad \Delta H^\circ = -50\,kJ\,mol^{-1}$$

In a typical industrial plant, a mixture of ethene (*ethylene*) and steam passes over a catalyst consisting of phosphoric acid on a porous, inert support at a temperature of about 570 K. About 5% of the ethene is converted into ethanol and the reaction is, in fact, well short of the position of equilibrium but the ethanol is separated and the unreacted ethene is recycled.
(a) Why do you think that ethene has become the chief industrial source of ethanol?
(b) Write an expression for the equilibrium constant K_p of the reaction.
(c) Why do you think that the conversion of ethene into ethanol in the typical industrial plant stops well short of the position of equilibrium?
(d) Suggest a method by which the ethanol could be separated in a nearly pure state from unchanged reactants.
(e) Why is the phosphoric acid on a 'porous support'?
(f) Suppose the operating temperature were reduced to 540 K, all other factors remaining the same. What effect might this have on the percentage conversion of ethene to ethanol? Justify your answer.
(g) Apart from altering the temperature, propose *two* different ways in which the process might be modified in order to increase the percentage conversion of ethene to ethanol.

Nuffield

7*   (a) Write down structural formulae for the isomers of the alcohol,
         $C_5H_{11}OH$.
         An alcohol of formula $C_5H_{11}OH$ gives a ketone $C_5H_{10}O$ on oxidation
         with chromic acid. When the alcohol is dehydrated by passing over
         heated aluminium oxide and the product is treated with ozonized
         oxygen and subsequently with water, a mixture of a ketone and an
         acid results.
         What is the structure of this alcohol?
         Give balanced equations for the dehydration and the two oxidations.
     (b) What is the reaction of aqueous sodium hydroxide with
         (i) propanal CH_3CH_2CHO,
         (ii) trimethylethanal $(CH_3)_3CCHO$?

                                                                                                   JMB

Chapter 7
Ethers

7.1 Homologous series of ethers

The characteristic functional group of this class of compounds is the $>C-O-C<$ structure, the oxygen atom linking two alkyl groups. By far the best-known, and most commonly encountered, member of the series is $CH_3CH_2OCH_2CH_3$ formerly known as *diethyl ether*, or just *ether*. Its anaesthetic properties were first used in surgery in 1842. The ethers are named as alkoxy-substituted alkanes, so that *ether* becomes ethoxyethane. The simplest ether is methoxymethane, CH_3OCH_3.

As well as the symmetrical ethers named above there are, of course, unsymmetrical ones and cyclic ones, as illustrated below:

$CH_3CH_2OCH_2CH_3$ $CH_3OCH_2CH_2CH_3$

ethoxyethane
b.p. = 35°C

methoxypropane
(or propoxymethane)
b.p. = 39°C

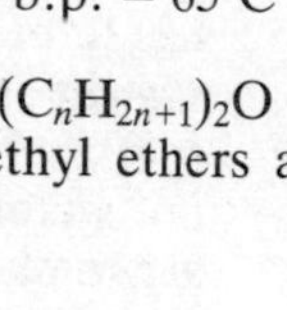

$$\begin{array}{c} CH_2-CH_2 \\ | \qquad | \\ CH_2 \quad CH_2 \\ \diagdown \diagup \\ O \end{array}$$

b.p. = 65°C

The simple ethers have the general formula $(C_nH_{2n+1})_2O$ and form a homologous series. Some physical data for methyl ethers are given in *Table 7.1*.

Table 7.1

Name	Formula	Melting point (°C)	Boiling point (°C)	Density (g cm^{-3})
methoxymethane	CH_3OCH_3	−140	−25	0.661
methoxyethane	$CH_3OCH_2CH_3$	−139	8	0.697
methoxypropane	$CH_3OCH_2CH_2CH_3$	—	39	0.738
methoxybutane	$CH_3O(CH_2)_3CH_3$	−116	70	0.744
methoxypentane	$CH_3O(CH_2)_4CH_3$	—	99	0.761
methoxyhexane	$CH_3O(CH_2)_5CH_3$	—	126	0.772

The simple ethers are isomeric with corresponding monohydric alcohols; isomerism also arises as a result of the position of the oxygen atom in the chain and of the branching of chains. The simplest ethers are gases at room temperature but the commonly encountered ones like ethoxyethane are volatile, colourless liquids with a characteristic 'ether' smell.

7.2 Physical properties of ethers

The volatility of ethers, in comparison with isomeric alcohols, is the most striking physical property. Whereas the hydrogen atom of the hydroxyl group of an alcohol is bonded to the electronegative oxygen atom, so permitting hydrogen bonding (see Section 6.2), the hydrogen atoms in ethers are bonded to carbon as part of alkyl groups. Consequently, the lack of intermolecular hydrogen bonding means that the boiling points of ethers are very similar to those of alkanes of corresponding relative molecular mass and very much lower than those of isomeric alcohols (see Section 5.2).

The lack of hydrogen bonding also means that the volatile, liquid ethers are generally immiscible with water, although with ethoxyethane and water, for example, each is very slightly soluble in the other. Absolutely dry ethoxyethane can be obtained by adding a very small amount of sodium to it.

The ethers are chemically unreactive, possess excellent solvent properties for a wide range of organic compounds, are virtually immiscible with water and are volatile and thus easily evaporated. This combination of properties has led to the widespread use of ethoxyethane as an extracting solvent for the isolation of organic compounds from aqueous solutions (see also Section 21.2.3). Its major drawback is its quite remarkable flammability; its flash-point of $-45°C$ (the minimum temperature at which an air/ethoxyethane mixture can be ignited by spark) and ignition temperature of $180°C$ (the minimum temperature to which such a mixture has to be heated to spontaneously inflame) means that under no circumstances should the ether be handled in the presence of a naked flame or sparks. Ether distillations must always be carried out with the aid of a hot water bath or electric mantle, and precautions taken to prevent the accumulation of any vapour in the air.

A particularly useful class of ethers are the mono-alkyl ethers of diols, particularly the methyl, ethyl and butyl ethers of ethane-1,2-diol:

$CH_3OCH_2CH_2OH$ $CH_3CH_2OCH_2CH_2OH$ $CH_3(CH_2)_2CH_2OCH_2CH_2OH$
b.p. = 124°C   b.p. = 135°C   b.p. = 171°C

These are obtained commercially by treating epoxyethane (see Section 7.3(iv)) with the appropriate alcohol and are marketed as Cellosolve solvents. Because they contain both an ether linkage and a hydroxyl group, these compounds possess excellent solvent properties and are miscible with water. Another widely used solvent, with a much lower volatility, is sold as Diglyme and is also an ether:

$CH_3OCH_2CH_2OCH_2CH_2OCH_3$
Diglyme
b.p. = 161°C

7.3 Chemical properties of ethers

The $\overset{/}{\underset{}{>}}C-O-C\overset{}{\underset{}{<}}$ linkage is particularly strong so that the ëthers are chemically unreactive. This feature makes them suitable solvents for general use. Examples of three types of chemical behaviour are shown below.

(i) Oxidation
The ease with which the volatile ethers undergo combustion poses a considerable hazard when air/ether mixtures are handled, and suitable precautions must be taken (*see* Section 7.2). The final products of combustion are, of course, carbon dioxide and water, e.g.

$CH_3CH_2OCH_2CH_3 + 6O_2 \rightarrow 4CO_2 + 5H_2O$

(ii) Nucleophilic properties
Ethers are generally too weakly nucleophilic to participate in typical nucleophilic substitution reactions (although they can behave as Lewis bases, see (iii) below), but such character is displayed in certain cases. Thus, ethoxyethane forms a stable complex (an *etherate*) with boron trifluoride, the oxygen atom of the ether donating a lone-pair of electrons to the electron-deficient boron atom:

$$\begin{matrix} CH_3CH_2 \\ CH_3CH_2 \end{matrix}\!\!\!\overset{\frown}{\ddot{O}} + BF_3 \longrightarrow \begin{matrix} CH_3CH_2 \\ CH_3CH_2 \end{matrix}\!\!\!\ddot{O}{\rightarrow}BF_3$$

In analogous fashion, it is thought that the use of ether solvents for the preparation of the organomagnesium compounds called Grignard reagents (*see* Section 5.3.4) allows the highly reactive reagents to be stabilized by solvation, the oxygen atom of the ether donating an electron-pair to the magnesium, e.g.

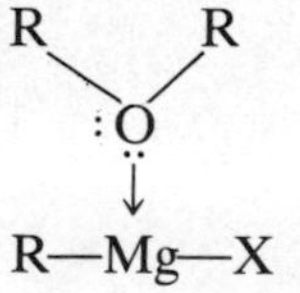

$$\begin{matrix} R \qquad\quad R \\ \diagdown \qquad \diagup \\ :\!O \\ \downarrow \\ R-Mg-X \end{matrix}$$

(iii) Cleavage of the ether link
Ethers do not react with metallic sodium, nor, in the cold, with phosphorus

pentachloride; this indicates that there is no hydroxyl group in the molecule (cf. alcohols, Section 6.3.1(i), and Section 6.3.4). When an ether is heated with the pentachloride, however, fission of the ether linkage occurs:

$$R—O—R + PCl_5 \rightarrow 2R—Cl + POCl_3$$

With an unsymmetrical ether, two chloroalkanes will be formed.

A more widely used method of cleaving ethers, however, involves reacting them with hot, concentrated hydriodic acid:

$$R—O—R + 2HI \rightarrow 2R—I + H_2O$$

Ethers can behave as weak Lewis bases; for example, ethoxyethane dissolves appreciably in concentrated hydrochloric acid to form a soluble **oxonium** salt:

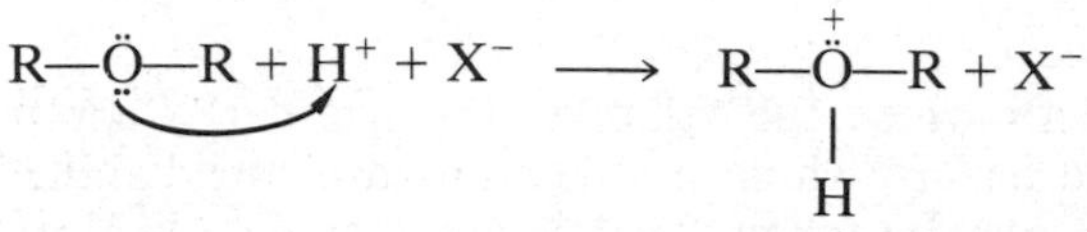

$$R—\overset{..}{\underset{..}{O}}—R + H^+ + X^- \longrightarrow R—\overset{+}{\overset{..}{O}}—R + X^-$$
$$|$$
$$H$$

oxonium ion

In the case of hydriodic acid, the iodide ion is sufficiently nucleophilic to attack the oxonium ion in a nucleophilic substitution reaction (cf. Section 6.3.1(ii)) and the alcohol so produced reacts with hydriodic acid to form another molecule of iodoalkane:

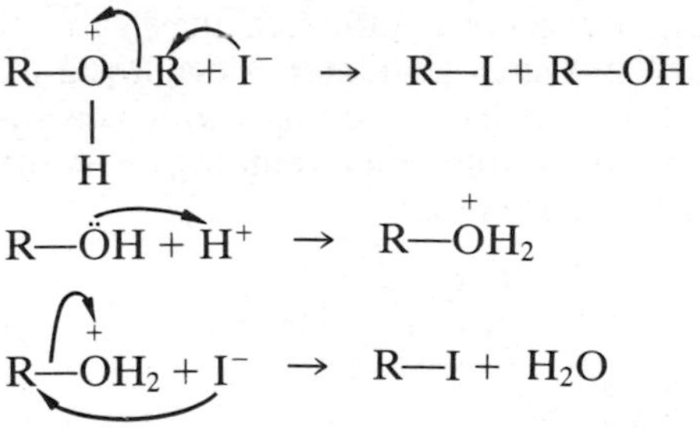

$$R—\overset{+}{O}—R + I^- \rightarrow R—I + R—OH$$
$$|$$
$$H$$

$$R—\overset{..}{O}H + H^+ \rightarrow R—\overset{+}{O}H_2$$

$$R—\overset{+}{O}H_2 + I^- \rightarrow R—I + H_2O$$

(iv) Cyclic ethers

The simple cyclic ether epoxyethane is an exception to the general rule that ethers are chemically unreactive. In this case the ether linkage is readily broken, owing to the steric strain in the three-membered ring, for example when the ether is treated with water or an alcohol:

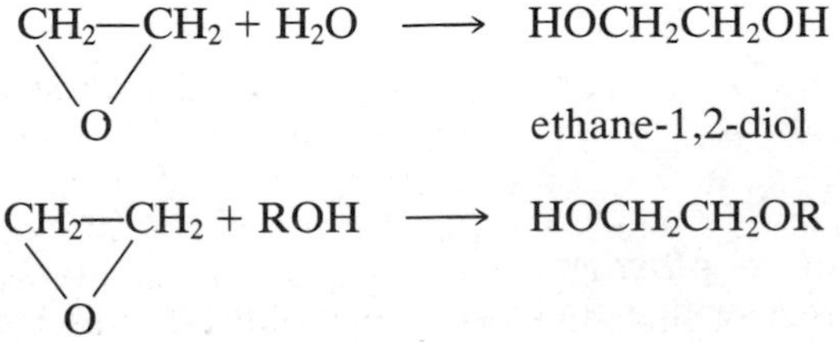

$$CH_2—CH_2 + H_2O \longrightarrow HOCH_2CH_2OH$$
$$\diagdown \diagup$$
$$O$$

ethane-1,2-diol

$$CH_2—CH_2 + ROH \longrightarrow HOCH_2CH_2OR$$
$$\diagdown \diagup$$
$$O$$

The breaking of a carbon–oxygen bond is assisted by employing an acid which protonates the oxygen atom and the reaction then involves a nucleophilic substitution:

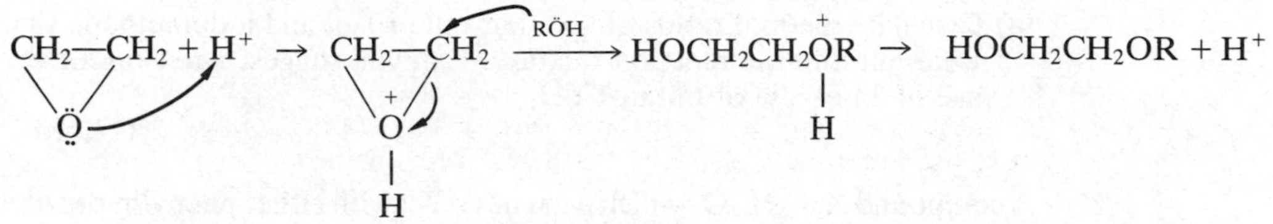

$$CH_2\!-\!CH_2 + H^+ \rightarrow CH_2\!-\!CH_2 \xrightarrow{\text{R\"OH}} HOCH_2CH_2\overset{+}{O}R \rightarrow HOCH_2CH_2OR + H^+$$

7.4 Manufacture and uses of ethers

By far the most commonly used ether is ethoxyethane. This ether is an excellent solvent for a wide variety of organic compounds, including fats and resins. Ethoxyethane is the product of a side-reaction in the commercial manufacture of ethanol by the direct hydration route (*see* Section 6.4):

$$2CH_3CH_2OH \rightleftharpoons CH_3CH_2OCH_2CH_3 + H_2O$$

Epoxyethane is manufactured from ethene (*see* Section 3.3.2(v)), and is an important starting material in the manufacture of 'anti-freeze' (*see* Section 6.4), Terylene (*see* Section 20.3(ii)), non-ionic detergents (*see* Section 19.4) and ethers like Diglyme.

Summary

1 **Ethers** contain the characteristic linkage C—O—C. The simple ethers form a homologous series with general formula $(C_nH_{2n+1})_2O$.
2 The ethers are in general highly flammable, volatile liquids and burn to carbon dioxide and water. They are chemically unreactive but the ether linkage can be cleaved by the action of hot, concentrated hydriodic acid:

$$R\!-\!O\!-\!R + 2HI \rightarrow 2RI + H_2O$$

Questions

1* 1 Methoxybutane $(CH_3(CH_2)_3OCH_3)$ can be prepared in the following way. Dissolve sodium (1 g) in warm butan-1-ol (20 g). Add iodomethane (6.15 g) and gently reflux the solution for 2 h. Separate the ether (b.p. 70–71°C) from the excess of butan-1-ol (b.p. 118°C) by distillation.
  (a) Suggest the identity of the precipitate that separates during the 2 h heating period and state how you could test your suggestion.

(b) Write a balanced equation for the main reaction.
(c) Suggest a likely mechanism for the main reaction.
(d) The mass of ether obtained was 2.7 g. What is its percentage yield based on iodomethane?
(e) Calculate the mol ratios of sodium, butan-1-ol and iodomethane that were put into the reaction mixture. Can you suggest a reason for the use of an excess of butan-1-ol?

Oxford

2*   A compound X, $C_8H_{14}O$, which does not react with either phenylhydrazine or ethanoic (*acetic*) anhydride, on catalytic hydrogenation gives rise to Y, $C_8H_{18}O$. Oxidation of X gives 2 mol of ethanoic (*acetic*) acid and 1 mol of an acid Z (Found: C, 35.8; H, 4.5%, relative molecular mass 134). Suggest formulae for X, Y, Z. How may Y be synthesized from butan-1-ol? (H = 1, C = 12, O = 16.)

Cambridge

Chapter 8
Aldehydes and ketones

8.1 Homologous series of aldehydes and ketones

The functional group discussed in this chapter is the **carbonyl** group, $\diagdown C{=}O$, present in both aldehydes and ketones. If the carbonyl group in a molecule is situated between two carbon atoms, then the compound is a **ketone**. If a hydrogen atom is directly attached to the carbonyl group, then the compound is an **aldehyde**. The simplest aldehyde has two hydrogen atoms attached to the carbonyl group. Aldehydes are named by adding the suffix -*al* to the root of the corresponding alkane, ketones by adding the suffix -*one*. The simplest aldehyde is methanal (*formaldehyde*) and the simplest ketone is propanone (*acetone*):

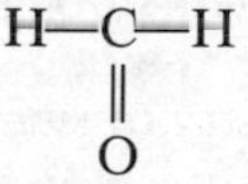

$$
\begin{array}{cc}
\text{H--C--H} & \text{CH}_3\text{--C--CH}_3 \\
\parallel & \parallel \\
\text{O} & \text{O}
\end{array}
$$

methanal   propanone
b.p. = $-21°C$ b.p. = $56°C$

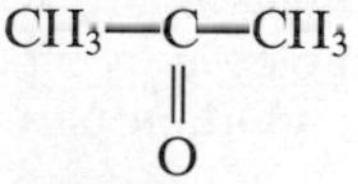

Some physical data are provided in *Table 8.1*.

Table 8.1

Name	Formula	Melting point (°C)	Boiling point (°C)	Density (g cm^{-3})
Aldehydes:				
methanal	HCHO	-92	-21	0.815
ethanal	CH_3CHO	-121	21	0.795
propanal	CH_3CH_2CHO	-81	49	0.797
butanal	$CH_3(CH_2)_2CHO$	-99	76	0.817
pentanal	$CH_3(CH_2)_3CHO$	-92	104	0.819
hexanal	$CH_3(CH_2)_4CHO$	-56	129	0.834
Ketones:				
propanone	CH_3COCH_3	-95	56	0.792
butanone	$CH_3COCH_2CH_3$	-86	80	0.805
pentan-2-one	$CH_3CO(CH_2)_2CH_3$	-78	102	0.812
hexan-2-one	$CH_3CO(CH_2)_3CH_3$	-57	127	0.830

Isomerism occurs within the series of aldehydes as a result of chain-branching, and within the series of ketones as a result of variations in the positions of the carbonyl group in the chain as well, e.g.

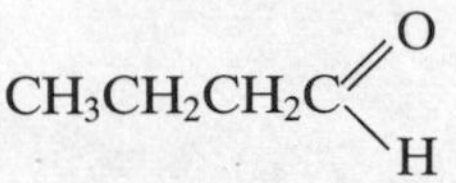

$$CH_3CH_2CH_2C\diagdown\begin{matrix}O\\H\end{matrix}$$

$$CH_3{-}\underset{\underset{CH_3}{|}}{CH}{-}C\diagdown\begin{matrix}O\\H\end{matrix}$$

butanal
b.p. = 76°C

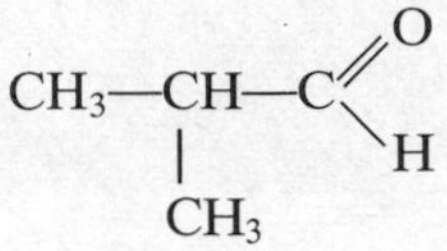

2-methylpropanal
b.p. = 61°C

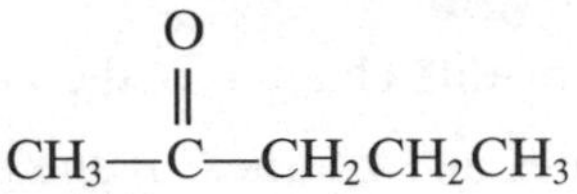

$$CH_3{-}\overset{\overset{O}{\|}}{C}{-}CH_2CH_2CH_3$$

$$CH_3CH_2{-}\overset{\overset{O}{\|}}{C}{-}CH_2CH_3$$

pentan-2-one
b.p. = 102°C

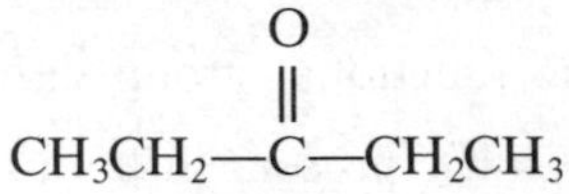

pentan-3-one
b.p. = 102°C

Aldehydes and ketones can also be isomeric, of course; propanal and propanone are examples.

Aldehydes and ketones are widely distributed throughout the plant kingdom, for example, as components of the fragrant 'essential oils'. They also play an important part in the chemistry of sugars (*see* Section 17.2).

8.2 Physical properties of aldehydes and ketones

The carbonyl group is polarized (*see* Section 8.3), but there is no hydrogen attached to the electronegative oxygen and hence no hydrogen bonding in either aldehydes or ketones. Nevertheless, the boiling points are higher (typically, about 50°C higher, *see Figure 2.1*) than those of alkanes of corresponding relative molecular mass and must be a reflection of the dipole–dipole intermolecular forces; they are lower than those of corresponding alcohols in which hydrogen bonding is prevalent. The simplest aldehyde, methanal, is a gas at room temperature and is usually used as a 40% aqueous solution (*formalin*), particularly for preserving biological specimens. Otherwise, the most commonly encountered aldehyde is ethanal, a colourless, volatile liquid with a very pungent smell. The first few members of the series are miscible with water, but after C₅ they become sparingly soluble or insoluble.

The simplest ketone is propanone, a colourless, volatile liquid with a characteristic smell. It, too, like ethanal, mixes readily with water, ethanol and ethoxyethane; it is frequently used as a polar solvent.

8.3   Chemical properties of aldehydes and ketones

In a carbon–*carbon* double bond $\diagdown$C$=$C$\diagup$, both the σ- and π-electron clouds are symmetrically distributed between the two carbon atoms. In the carbon–*oxygen* double bond (i.e. the carbonyl group) $\diagdown$C$=$O, both the σ- and π-bonds are polarized, the electrons being distributed unequally between the two atoms. Owing to the greater electronegativity of the oxygen atom, this atom has the greater share of the bonding electrons and the distortion of the π-electron density can be represented in several ways (*see also* Section 1.5):

 or $\overset{\delta+}{\diagup}C=\overset{\delta-}{O}$ which is midway between the extreme forms: $\diagup$C$=$O $\leftrightarrow$ $\overset{+}{\diagup}$C$-\overset{-}{O}$

As a result of the *permanent* polarization of the bond, the carbon atom is electrophilic in character and is therefore susceptible to **nucleophilic** attack, i.e. attack by reagents which are able to donate an electron-pair to the carbon atom and form a new bond. Conversely, the oxygen atom is susceptible to **electrophilic** attack, i.e. attack by an electron-pair acceptor, but this usually involves the comparatively simple step of protonation (i.e. reaction with H^+).

This is all in marked contrast to the situation with $\diagdown$C$-$C$\diagup$, where the π-electron pair of the double bond is shared with an electrophile in a process of electrophilic attack on the double bond. While the $\diagdown$C$=$C$\diagup$ is weaker than two σ-bonds (*see* Section 1.3), the $\diagdown$C$=$O bond is stronger than two C$-$O σ-bonds (about 752 kJ mol^{-1} as compared with 2×357 kJ mol^{-1}), but despite this it is reactive, owing to the polarization. Thus, whereas alkenes undergo *electrophilic* addition reactions, aldehydes and ketones undergo **nucleophilic addition** reactions, e.g.

$$\diagup C=O \quad + \quad \overset{\delta-}{X}-\overset{\delta+}{Y} \quad \rightarrow \quad \diagup \underset{X \quad Y}{C-O}$$

In these reactions, aldehydes are generally more reactive than ketones because the electrophilic character of the carbon atom of the carbonyl group is reduced a little in ketones, owing to the weak inductive effect of *two* alkyl groups:

$$CH_3 \rightarrow \overset{\delta+}{\underset{\underset{O^{\delta-}}{\|}}{C}} \leftarrow CH_3 \qquad cf. \qquad CH_3 \rightarrow \overset{\delta+}{\underset{\underset{O^{\delta-}}{\|}}{C}}-H$$

Many of the reactions of the carbonyl group are shared by both alde-hydes and ketones—particularly nucleophilic additions. Sometimes the initial addition products spontaneously **eliminate** the elements of water and the combined addition–elimination sequence is referred to as a **condensation** reaction. Other characteristic reactions depend on the fact that hydrogen attached to carbon adjacent to the carbonyl group is very weakly acidic. Finally, the C—H bond of aldehydes is susceptible to oxidation and therefore aldehydes, but not ketones, show **reducing properties**.

8.3.1 Nucleophilic addition reactions

(i) Reaction with hydrogen cyanide

The addition reaction with hydrogen cyanide is base-catalysed and involves nucleophilic attack by cyanide ion:

$$HCN + OH^- \rightleftharpoons CN^- + H_2O$$

$$\begin{array}{c}\\ \end{array}C{=}O + CN^- \rightleftharpoons \begin{array}{c} O^- \\ C \\ CN \end{array} \xrightarrow{H_2O} \begin{array}{c} OH \\ C \\ CN \end{array} + OH^-$$

i.e. overall

$$\begin{array}{c}\\ \end{array}C{=}O + HCN \rightleftharpoons \begin{array}{c}\\ \end{array}C(OH)CN$$

e.g.

$$CH_3COCH_3 + HCN \rightleftharpoons CH_3C(OH)CNCH_3$$

The optimum pH for the reaction is about 9. At low pH values, the dissociation of the weak hydrocyanic acid is so suppressed that the con-centration of free cyanide ion is very small; at high values, protonation of the initial addition product is inhibited and a product cannot be isolated. The reaction is usually performed by adding mineral acid to a mixture of the aldehyde or ketone and potassium cyanide (*care!*) so that the solution is buffered at the correct pH. The product of this reaction is sometimes known as a cyanohydrin and is of particular synthetic import-ance because it can be readily hydrolysed by mineral acid to a hydroxy carboxylic acid, e.g.

$$\begin{array}{c} CH_3 \quad OH \\ C \\ H \quad CN \end{array} \xrightarrow{H_3O^+} \begin{array}{c} CH_3 \quad OH \\ C \\ H \quad CO_2H \end{array}$$

2-hydroxypropanonitrile   2-hydroxypropanoic acid

and can also be dehydrated (*see* Section 8.4), e.g.

$$CH_3CH(OH)CN \xrightarrow[600-700°C]{H_3PO_4} CH_2{=}CHCN$$

✓*(ii) Reaction with sodium hydrogensulphite*

When an aldehyde or ketone is mixed with a fresh, aqueous solution of sodium hydrogensulphite (conveniently prepared by passing sulphur dioxide through a cold saturated solution of sodium carbonate), a white crystalline product usually separates out after shaking of the solution:

$$\text{\Large$>$}C{=}O + NaHSO_3 \quad \rightleftharpoons \quad \text{\Large$>$}C\big\langle {}^{O^-Na^+}_{SO_3H} \quad \rightleftharpoons \quad \text{\Large$>$}C\big\langle {}^{OH}_{SO_3^-Na^+}$$

$$(Na^+, {}^-SO_3H)$$

(N.B. It is the S which bonds to the carbon)

Since the aldehyde or ketone can easily be regenerated by heating this addition product with dilute sulphuric acid, the reaction is sometimes used as a means of separating a carbonyl compound from other compounds. However, as a rule, ketones other than those containing the CH_3CO- group fail to give this reaction.

Not done in class. ✗ *(iii) Reaction with alcohols and with water*

In the presence of an anhydrous acid such as dry hydrogen chloride* an alcohol will react with an aldehyde to produce first, a **hemi-acetal**:

$$\text{\Large$>$}C{=}O + ROH \quad \rightleftharpoons \quad \text{\Large$>$}C\big\langle {}^{OH}_{OR}$$

e.g.

$$CH_3CHO + CH_3OH \quad \rightleftharpoons \quad CH_3CH(OH)OCH_3$$

ethanal   1-methoxyethanol

In the presence of excess alcohol a more stable **acetal** is produced:

$$\text{\Large$>$}C\big\langle {}^{OH}_{OR} + ROH \quad \rightleftharpoons \quad \text{\Large$>$}C\big\langle {}^{OR}_{OR} + H_2O$$

e.g. overall

$$CH_3CHO + 2CH_3OH \quad \rightleftharpoons \quad CH_3CH(OCH_3)_2$$
$$1,1\text{-dimethoxyethane}$$
$$\text{b.p.} = 64°C$$

The corresponding derivatives of ketones are called hemi-ketals and ketals but are generally not easily formed. Acetals and ketals contain ether linkages but their formation from carbonyl compounds is reversible and they are easily hydrolysed by aqueous acid; they are, however, stable to alkali and they are therefore used to 'protect' the carbonyl group from

* Carbonyl compounds undergo various self-condensation reactions with acids.

reaction in alkaline solution. They play an important part in the chemistry of sugars (*see* Section 17.2).

The hydration of carbonyl compounds involves a similar sort of reaction. Methanal forms a relatively stable hydrate when dissolved in water:

$$
\begin{array}{c}
H \\
\diagdown \\
\diagup \\
H
\end{array}
C{=}O + H_2O \;\rightleftharpoons\;
\begin{array}{c}
H \quad OH \\
\diagdown\; \diagup \\
C \\
\diagup\; \diagdown \\
H \quad OH
\end{array}
$$

With aldehydes in general, the hydrate is present in the equilibrium mixture in very small proportions but, exceptionally, hydrates can sometimes be isolated. One such example is the hydrate of trichloroethanal (formerly known as *chloral hydrate* and used as a soporific) $Cl_3CCH(OH)_2$, a white crystalline solid with m.p. = 52°C.

(iv) Reaction with complex metal hydrides

Aldehydes and ketones are very rapidly reduced by complex metal hydrides, such as the highly reactive $LiAlH_4$, in anhydrous solvents. The product is obtained after acidification of the reaction mixture:

$$
\begin{array}{c}
\diagdown \\
\diagup
\end{array}
C{=}O + H^- \;\longrightarrow\;
\begin{array}{c}
\diagdown \quad O^- \\
C \\
\diagup \quad H
\end{array}
\;\xrightarrow{H_3O^+}\;
\begin{array}{c}
\diagdown \quad OH \\
C \\
\diagup \quad H
\end{array}
$$

e.g.

$$
CH_3COCH_3 + H^- \;\longrightarrow\;
\begin{array}{c}
CH_3CHCH_3 \\
| \\
O^-
\end{array}
\;\xrightarrow{H_3O^+}\;
\begin{array}{c}
CH_3CHCH_3 \\
| \\
OH
\end{array}
$$

propanone   propan-2-ol
b.p. = 56°C   b.p. = 83°C

Aldehydes are reduced to primary alcohols and ketones are reduced to secondary alcohols. There is a variety of reducing agents available for these reactions; they are discussed later (Section 8.3.5), since they do not all involve nucleophilic addition.

(v) Reaction with Grignard reagents

Grignard reagents are a source of nucleophilic carbon (*see* Section 5.3.4) and they react rapidly with aldehydes and ketones in dry ether solvents to form organomagnesium complexes which undergo hydrolysis, on the addition of water or acid, to yield alcohols:

$$
\begin{array}{c}
\diagdown \\
\diagup
\end{array}
C{=}O + RMgX \;\longrightarrow\;
\begin{array}{c}
\diagdown \quad OMgX \\
C \\
\diagup \quad R
\end{array}
\;\xrightarrow{H_2O}\;
\begin{array}{c}
\diagdown \quad OH \\
C \\
\diagup \quad R
\end{array}
\; + \; Mg(OH)X
$$

e.g.

$$
CH_3CHO + CH_3CH_2MgI \rightarrow CH_3CH(OH)CH_2CH_3 + Mg(OH)I
$$

ethanal   ethylmagnesium   butan-2-ol
b.p. = 21°C   iodide   b.p. = 100°C

With methanal, primary alcohols are formed:

$$HCHO + RMgX \;\rightarrow\; RCH_2OMgX \xrightarrow{H_2O} RCH_2OH + Mg(OH)X$$

These reactions are of great synthetic importance.

8.3.2   Mechanism of nucleophilic addition reactions

In each of the previous examples, the addition reaction is thought to involve nucleophilic attack on the carbon atom of the carbonyl group. The oxygen atom is subsequently protonated to give a stable product. Some of the reactions involve an equilibrium but this generally favours the product. Thus:

(i)

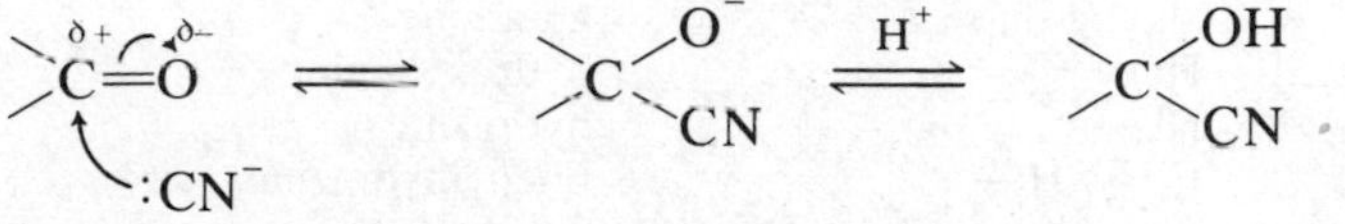

(ii)

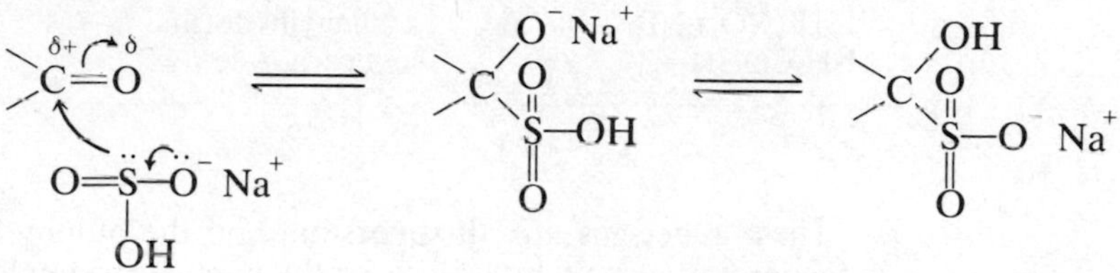

(iii)

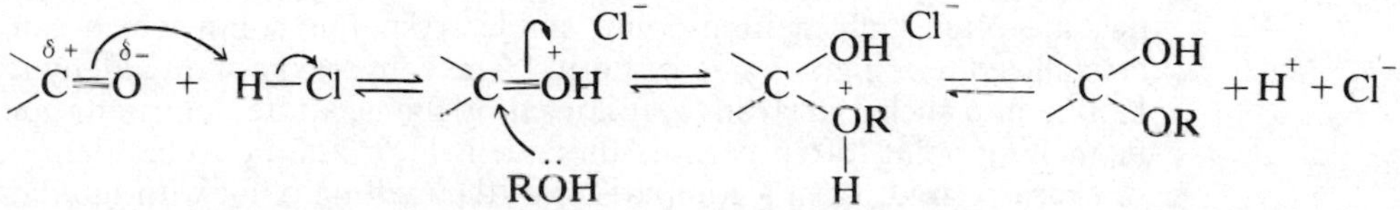

Here, protonation of the oxygen first renders the carbon more susceptible to attack. Then

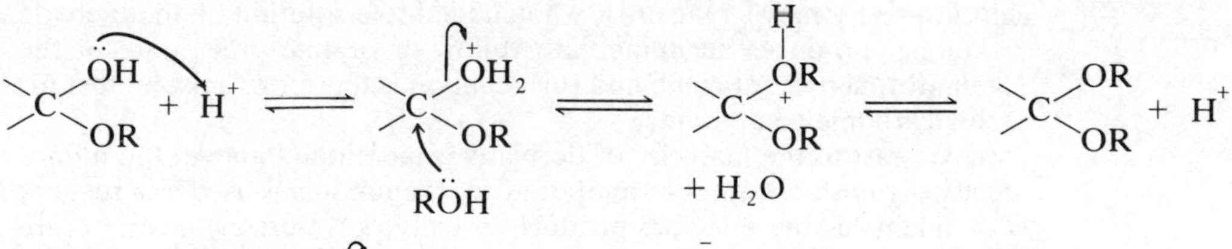

(iv)

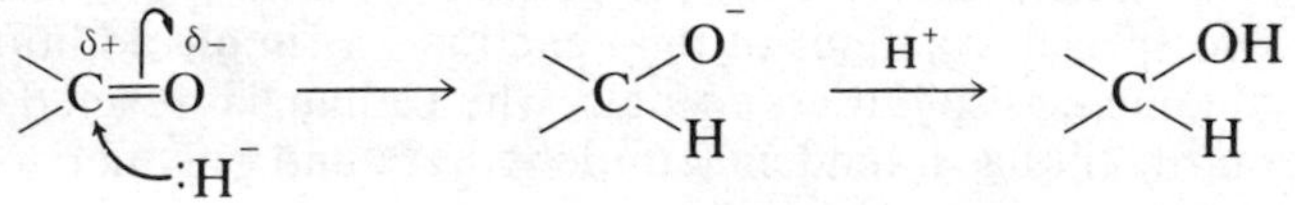

(v)

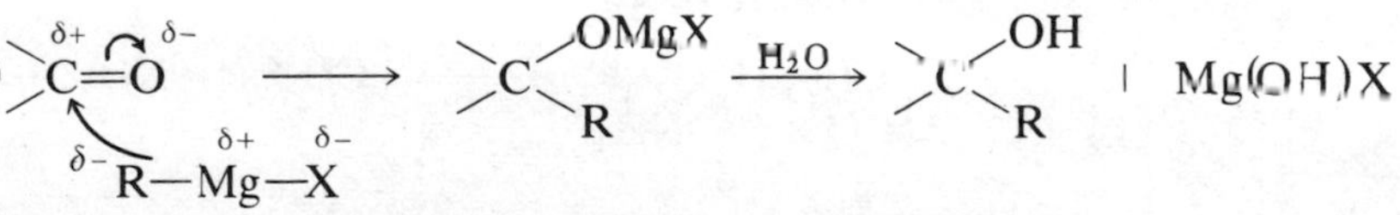

(For a more precise mechanism for (iv) *see* Section 8.3.5(iii)).

8.3.3 Addition–elimination reactions

With certain reagents, nucleophilic *addition* to the carbonyl group of aldehydes and ketones is followed by the spontaneous *elimination* of the elements of water and the overall reaction is therefore sometimes referred to as a **condensation** reaction (*Table 8.2*):

$$\text{>C=O} + \text{H}_2\ddot{\text{N}}\text{—R} \rightleftharpoons \text{>C=N—R} + \text{H}_2\text{O}$$

Table 8.2

R =	Name of reagent	Type of product
alkyl group	an amine	an imine
HO—	hydroxylamine	an oxime
NH₂—	hydrazine	a hydrazone
C₆H₅NH—	phenylhydrazine	a phenylhydrazone
	2,4-dinitro-	a 2,4-dinitro-
C₆H₃(NO₂)₂NH—	phenylhydrazine	phenylhydrazone
NH₂CONH—	semicarbazide	a semicarbazone

These reactions are all reversible and the aldehyde or ketone can be regenerated by acid hydrolysis of the condensation products. These products are often high melting-point, stable crystalline solids which can be crystallized to a high degree of purity. The conversion of an aldehyde or ketone into such a **derivative**, followed by the accurate determination of its melting point, often permits the aldehyde or ketone to be identified, or **characterized,** from a comparison of the melting point with tabulated, authentic values for known compounds. A mixed melting point may also be a confirmatory test (*see* Section 21.1.1).

A solution of the orange 2,4-dinitrophenylhydrazine in dilute hydrochloric acid ('Brady's reagent'), when added to a solution of an aldehyde or ketone, produces an immediate yellow or orange precipitate of the 2,4-dinitrophenylhydrazone and this reaction is therefore a useful test for such a carbonyl compound.

In contrast to the majority of derivatives mentioned above, the imines are usually unstable at room temperature. If ammonia is used as a reagent with aldehydes the unstable products usually polymerize, but there are two special examples of this reaction. A simple addition product can be obtained, as colourless crystals, when ethanal is reacted with dry ammonia but it still has a tendency to dehydrate and polymerize:

$$\text{CH}_3\text{CHO} + \text{NH}_3 \longrightarrow \text{CH}_3\text{CH}\begin{array}{c}\text{OH}\\\text{NH}_2\end{array} \xrightarrow{-\text{H}_2\text{O}} \text{CH}_3\text{CH=NH} \longrightarrow \text{polymer}$$

m.p. = 97°C

Methanal condenses with ammonia to produce a fairly complex cyclic structure; the reaction can be brought about by warming a solution containing methanal and ammonia:

$$6HCHO + 4NH_3 \rightarrow (CH_2)_6N_4 + 6H_2O$$
$$\text{m.p.} > 230°C$$

This product was once used in medicine but is now of interest because when heated with concentrated nitric acid, it is converted into a product known by the name of Cyclonite, or RDX, and widely used as a military explosive:

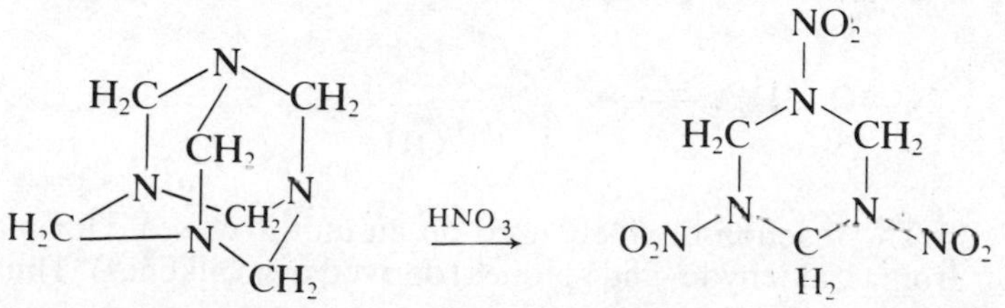

$$\text{m.p.} = 203°C$$

8.3.4   Mechanism of addition–elimination reactions

The first step in these reversible reactions involves nucleophilic addition, but the product then undergoes an *acid-catalysed* elimination reaction. Too low a pH value for the reaction medium will inhibit nucleophilic attack by R—NH$_2$, because this is basic and will become protonated in acid solution to R—NH$_3^+$, which cannot act as a nucleophile. If the pH value is too high, then the acid-catalysed elimination step will not occur at any appreciable rate. An optimum pH value of about 5 is therefore maintained and the buffering of the solution is usually attained by adding a calculated quantity of R—NH$_2$ in the form of its hydrochloride salt R—NH$_3^+$Cl$^-$:

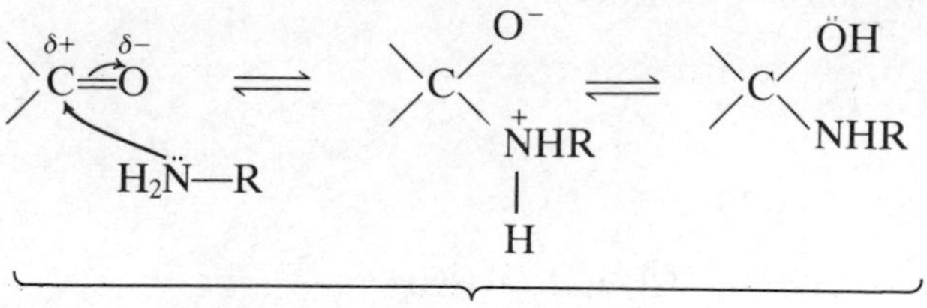

nucleophilic addition

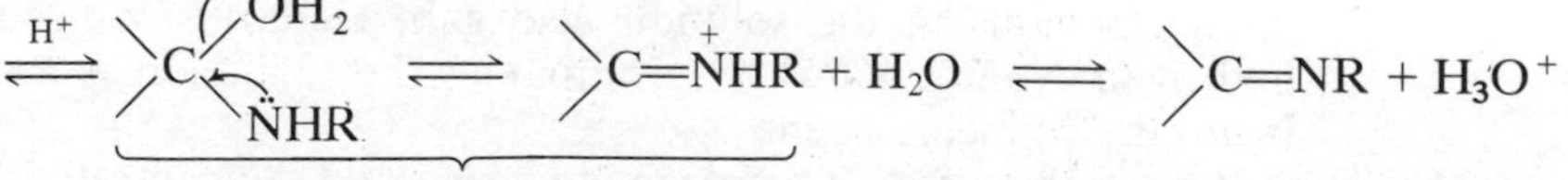

acid-catalysed elimination

8.3.5 Reduction of aldehydes and ketones

Aldehydes and ketones can be reduced under a variety of reaction conditions and by a variety of reducing agents.

(i) Hydrogenation
Like alkenes, they can be reduced by catalytic hydrogenation (the catalyst usually being Ni, Pt or Pd). The reduction product is an alcohol but reduction of other functional groups can also be effected by the same reagent, so it is rather unselective:

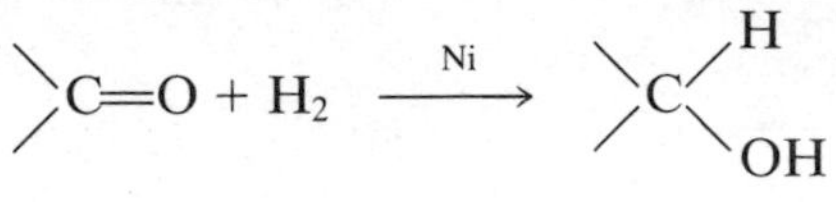

$$\text{\Large$\diagdown$C}{=}\text{O} + \text{H}_2 \xrightarrow{\text{Ni}} \text{\Large$\diagdown$C}\diagup^{\text{H}}_{\text{OH}}$$

The reaction is widely used on an industrial scale for making butanols from C_4 aldehydes and ketones (derived from alkenes). Thus, for example:

$$CH_3CH{=}CH_2 + CO + H_2 \rightarrow CH_3CH_2CH_2CHO + (CH_3)_2CHCHO$$

$$\downarrow H_2 \qquad\qquad\qquad\qquad \downarrow H_2$$

$$CH_3CH_2CH_2CH_2OH \quad (CH_3)_2CHCH_2OH$$

(ii) 'Dissolving metal' reductions
These generally involve a combination of a metal and a protic solvent, e.g. sodium/ethanol or zinc/ethanoic acid. The usual reduction product is an alcohol. The real reducing agent is the electron. As the metal dissolves, it gives up electrons and these are usually transferred to the carbonyl group to give an anion radical, which subsequently undergoes protonation to the alcohol:

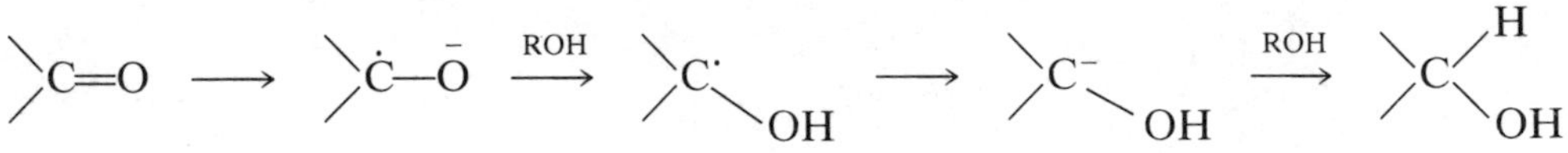

$$\text{\Large$\diagdown$C}{=}\text{O} \longrightarrow \text{\Large$\diagdown$\dot{C}}{-}\bar{\text{O}} \xrightarrow{\text{ROH}} \text{\Large$\diagdown$C}^{\cdot}{\diagdown_{\text{OH}}} \longrightarrow \text{\Large$\diagdown$C}^{-}{\diagdown_{\text{OH}}} \xrightarrow{\text{ROH}} \text{\Large$\diagdown$C}\diagup^{\text{H}}_{\text{OH}}$$

e.g.

$$CH_3COCH_3 \rightarrow CH_3CH(OH)CH_3$$

or, overall,

$$CH_3COCH_3 + 2H^+ + 2e^- \rightarrow CH_3CH(OH)CH_3$$

Hydrogen ions in the solution also gain electrons, of course, so that hydrogen evolution always accompanies the reduction, but the hydrogen is *not* the reducing agent.

If a combination of amalgamated zinc and concentrated hydrochloric acid is employed, in a method introduced by Clemmensen in 1913, then the carbonyl group, particularly in ketones, is reduced to —CH₂—:

$$>C=O \xrightarrow[\text{HCl}]{\text{Zn/Hg}} >CH_2$$

e.g.

$$CH_3CO(CH_2)_5CH_3 \rightarrow CH_3CH_2(CH_2)_5CH_3$$

octan-2-one   octane
b.p. = 173°C   b.p. = 125°C

(iii) Complex metal hydride reductions
By far the most widely used laboratory reagents for the reduction of aldehydes and ketones have been the complex metal hydrides (a simple metal hydride like sodium hydride acts as a very strong base, rather than as a reducing agent). The most powerfully reducing complex hydride is lithium tetrahydridoaluminate, prepared by reduction of anhydrous aluminium chloride with lithium hydride and readily available commercially. Reduction is usually performed in a *dry* ethoxyethane solution, in which the hydride is quite soluble; all reagents must be absolutely dry, since the hydride is violently decomposed by water with the evolution of hydrogen:

$$LiAlH_4 + 4H_2O \rightarrow LiOH + Al(OH)_3 + 4H_2$$

In reacting with an aldehyde or ketone, the complex acts as a hydride ion donor (nucleophile) and hence, in theory, 1 mol of reducing agent will reduce 4 mol of aldehyde or ketone. In practice, excess hydride is employed. The initial product of reduction is an organoaluminium complex and this, together with any excess hydride, is very cautiously decomposed by the addition of water. Subsequent acidification of the mixture dissolves the metal hydroxides and the alcohol product is extracted from the mixture with ethoxyethane:

$$4 >C=O + Li^+AlH_4^- \rightarrow \left(\begin{array}{c} | \\ -C-O \\ | \\ H \end{array} \right)_4 Al^-Li^+ \xrightarrow{4H_2O} 4 >C \begin{array}{c} OH \\ H \end{array}$$

$$+ Al(OH)_3 + LiOH$$

e.g.

$$4CH_3CHO + LiAlH_4 \rightarrow (CH_3CH_2O)_4AlLi \xrightarrow{4H_2O} 4CH_3CH_2OH + Al(OH)_3 + LiOH$$

This hydride is a very versatile reducing agent and will reduce esters, carboxylic acids, aldehydes and ketones to alcohols. It does not normally attack carbon–carbon double bonds.

An alternative reagent is sodium tetrahydridoborate. This is not quite so reactive a reagent and has the advantage that it can be used in aqueous, or aqueous alcohol, solutions. Once again, a complex is formed and this is decomposed by the addition of acid:

$$4 \;\diagdown\!\!\!\!\diagup C{=}O + Na^+BH_4^- \;\rightarrow\; \left(-\overset{|}{\underset{\underset{H}{|}}{C}}{-}O\right)_4 B^-Na^+ \;\xrightarrow{4H_2O}\; 4 \;\diagup\!\!\!\!\diagdown C \diagdown\!\!\!\diagdown \begin{matrix} OH \\ H \end{matrix} + NaH_2BO_3 + H_2O$$

This reagent does not reduce carboxylic acids or their esters.

The mechanism of these reductions is thought to involve transfer of a hydride ion from the complex to the carbonyl group in a nucleophilic attack, which is aided by co-ordination of the oxygen with the aluminium or boron atom:

$$\overset{\delta+}{\underset{\underset{H}{|}}{C}}{=}O^{\delta-}\;{}^-\!AlH_3 \;\longrightarrow\; -\overset{|}{\underset{\underset{H}{|}}{C}}{-}O{-}\overset{-}{A}lH_3 \;\xrightarrow{etc.}$$

$$\left(-\overset{|}{\underset{\underset{H}{|}}{C}}{-}O\right)_4 Al^-Li^+ \;\xrightarrow{4H_2O}\; 4 \;\diagup\!\!\!\!\diagdown C \diagdown\!\!\!\diagdown \begin{matrix} OH \\ H \end{matrix} + Al(OH)_3 + LiOH$$

8.3.6 Reactions not involving the carbonyl group

Although most of the characteristic reactions of aldehydes and ketones involve the carbonyl group, the carbonyl group itself also affects the strength of bonds between hydrogen and the carbon atoms immediately adjacent to the carbonyl group. In a solution of an aldehyde or ketone, there is an equilibrium between two distinct species, although there is usually only a small percentage of the second species present:

$$\underset{\textbf{keto}\text{ form}}{-CH_2-\underset{\underset{O}{\|}}{C}-} \;\rightleftharpoons\; \underset{\textbf{enol}\text{ form}}{-CH{=}\underset{\underset{OH}{|}}{C}-}$$

(The enol form takes its name from the combination of an alk*ene*

$$\diagdown\!\!\!\!\diagup C{=}C\diagdown\!\!\!\!\diagup \text{ and an alcoh}ol \;\;-OH \text{ group.)}$$

These forms arise from a redistribution of electrons within the molecule, and the transfer of a very small atom (in this case a hydrogen atom) from one site to another; the phenomenon is termed **tautomerism**. As a result of this, the hydrogen attached to carbon becomes very weakly acidic because, when a proton is removed, the remaining negative charge can be spread out, or **delocalized**, over three atoms, so that the anion is stabilized. The energy change for carbanion formation is therefore favourable:

$$-\underset{\underset{\displaystyle H}{|}}{C}H-\underset{\underset{\displaystyle O}{\|}}{C}- \quad \rightleftharpoons \quad -\overset{-}{C}H-\underset{\underset{\displaystyle O}{\|}}{C}- \quad \leftrightarrow \quad -CH=\underset{\underset{\displaystyle O^-}{|}}{C}- \quad \text{i.e.} \quad -\underset{}{C}H-\underset{\underset{\displaystyle O}{\|}}{C}-$$

$$:OH^- \qquad\qquad + H_2O$$

If an aldehyde is treated with mild alkali, such as a very dilute solution of sodium hydroxide or a solution of sodium carbonate, the resulting carbanion can attack another molecule of aldehyde in a nucleophilic manner; the overall reaction has been called an **aldol** (the product is both an *ald*ehyde and an alcoh*ol*) **condensation**, e.g.

$$CH_3CHO + OH^- \quad \rightleftharpoons \quad {}^-CH_2CHO + H_2O$$

$$CH_3-\overset{\displaystyle O}{\underset{\displaystyle H}{C}} \qquad\qquad \rightleftharpoons \quad CH_3\underset{\underset{\displaystyle O^-}{|}}{C}HCH_2CHO$$

$$:CH_2CHO$$

$$CH_3\underset{\underset{\displaystyle O^-}{|}}{C}HCH_2CHO + H_2O \quad \rightleftharpoons \quad CH_3\underset{\underset{\displaystyle OH}{|}}{C}HCH_2CHO + OH^-$$

i.e. overall

$$2CH_3CHO \quad \rightleftharpoons \quad CH_3\underset{\underset{\displaystyle OH}{|}}{C}HCH_2CHO$$

ethanal   3-hydroxybutanal
                           b.p. $= 80°C$ (at 1.6 kN m^{-2})

If the aldol product is warmed, or treated with mineral acid, then it eliminates the elements of water very easily:

$$CH_3\underset{\underset{\displaystyle OH}{|}}{C}HCH_2CHO \quad \longrightarrow \quad CH_3CH{=}CHCHO + H_2O$$

                           but-2-enal
                           b.p. $= 104°C$

Ketones do not undergo this reaction very easily, but mixed aldol condensations between an aldehyde and a ketone, or between two different aldehydes, can be carried out.

When aldehydes are treated with more concentrated alkali, then repeated condensations lead to the gradual precipitation of a white resinous polymer.* Ketones do not generally react with alkali.

---

* Under acid conditions, ethanal forms both a trimer (b.p. $= 124°C$) and a tetramer (at low temperature). Methanal forms a polymer, poly(methanal), and a trimer (m.p. $= 62°C$).

If an aldehyde lacks any hydrogen atoms bonded to the carbon atom adjacent to the carbonyl group, then it clearly cannot undergo an aldol condensation; instead it undergoes the **Cannizzaro reaction**, discovered by the Italian chemist Stanislao Cannizzaro in 1853. This reaction involves disproportionation of the aldehyde, i.e. its simultaneous oxidation and reduction. Thus, methanal, for example, when treated with alkali, forms an alcohol (the reduction product) and (after acidification) a carboxylic acid (the oxidation product):

$$2HCHO + NaOH \ \rightarrow \ CH_3OH + HCOONa \ (\xrightarrow{H+} \ HCOOH)$$

$$\begin{array}{ll} \text{methanol} & \text{methanoic acid} \\ \text{b.p.} = 65°C & \text{b.p.} = 101°C \end{array}$$

Aldehydes and ketones can be halogenated quite readily at the carbon atoms adjacent to the carbonyl group (where the hydrogen–carbon bonds are weaker), the overall reaction being one of **substitution**:

$$-CH_2-\underset{\underset{O}{\|}}{C}- \ + \ X_2 \longrightarrow \ -CHX-\underset{\underset{O}{\|}}{C}- \ + \ HX \xrightarrow{X_2} \ -CX_2-\underset{\underset{O}{\|}}{C}- \ + \ HX$$

If chlorine is bubbled through ethanal, for example, in the presence of calcium carbonate (to remove the acid), then a colourless, oily liquid, trichloroethanal, is formed (this forms an unusual hydrate, *see* Section 8.3.1(iii)):

$$CH_3CHO + 3Cl_2 \ \rightarrow \ Cl_3CCHO + 3HCl$$

$$\begin{array}{l} \text{trichloroethanal} \\ \text{b.p.} = 98°C \end{array}$$

Bromination and iodination of carbonyl compounds can be brought about by shaking an aqueous solution of the halogen with the carbonyl compound. The halogenation is both acid- and base-catalysed. In the acid-catalysed reaction, monohalogenation may be performed quite easily, as subsequent halogenation takes place much more slowly. Thus:

$$CH_3COCH_3 + Br_2 \ \rightarrow \ BrCH_2COCH_3 + HBr$$

This is not the case in the base-catalysed reaction, and hence complete substitution can occur if there is an excess of halogen. Note that substitution occurs only at the carbon atoms adjacent to the carbonyl group, e.g.

$$CH_3COCH_2CH_3 + 5Br_2 \ \rightarrow \ CBr_3COCBr_2CH_3 + 5HBr$$

Under alkaline conditions, however, the halogenated product can undergo hydrolysis. If, for example, propanone is shaken with an aqueous solution of iodine, and sodium hydroxide solution then added until the iodine colour disappears, a yellow precipitate of triiodomethane (*iodoform*) gradually forms.

$$CH_3COCl_3 + H_2O \xrightarrow{OH^-} CH_3COOH + CHI_3$$

(as its salt)   triiodomethane
                     m.p. = 119°C

The yellow crystalline solid is easily recognized by its characteristic, rather 'antiseptic'-like smell and by its melting point. The appearance of the solid under these conditions constitutes a useful test (the *iodoform* test) for carbonyl compounds containing the group CH_3CO—, namely methyl ketones and the aldehyde ethanal. In addition, compounds containing the group $CH_3CH(OH)$— also give a positive result because they are oxidized, under the reaction conditions, to methyl ketones. Thus, ethanol and propan-2-ol will give a positive test because they are oxidized to ethanal and propanone, respectively, whereas methanol and propan-1-ol will not.

The halogenation of a carbonyl compound is an interesting reaction to study in the laboratory, because the rate of halogenation is found to be independent of the concentration of halogen. The rate *is*, however, dependent on the concentration of carbonyl compound and on the concentration of acid or base, used as catalyst. This is interpreted as meaning that the slowest, or rate-determining, step of the mechanism is the acid- or base-catalysed formation of the enol, which subsequently reacts rapidly with halogen:

acid-catalysed case:

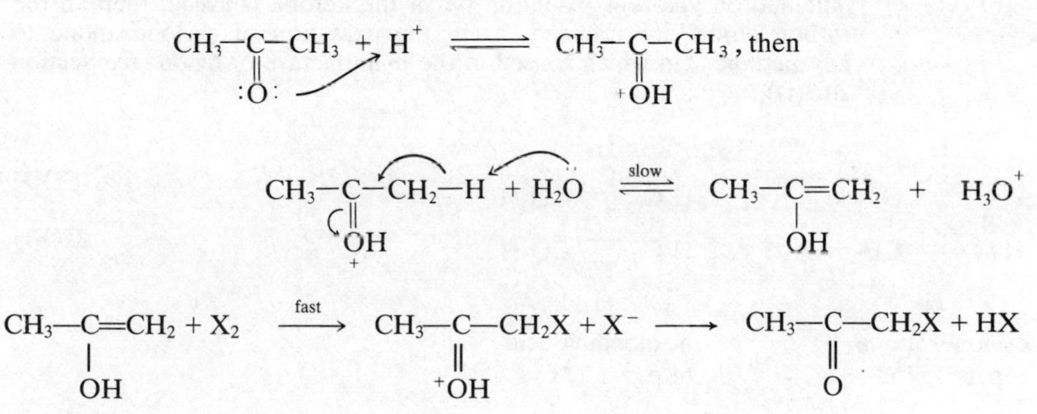

$$CH_3-\overset{\|}{\underset{:O:}{C}}-CH_3 + H^+ \rightleftharpoons CH_3-\overset{\|}{\underset{+OH}{C}}-CH_3 \text{, then}$$

$$CH_3-\overset{\|}{\underset{+OH}{C}}-CH_2-H + H_2O \underset{slow}{\rightleftharpoons} CH_3-\underset{OH}{C}=CH_2 + H_3O^+$$

$$CH_3-\underset{OH}{C}=CH_2 + X_2 \xrightarrow{fast} CH_3-\overset{\|}{\underset{+OH}{C}}-CH_2X + X^- \longrightarrow CH_3-\overset{\|}{\underset{O}{C}}-CH_2X + HX$$

base-catalysed case:

$$CH_3-\overset{\|}{\underset{O}{C}}-CH_2-H + :OH^- \underset{slow}{\rightleftharpoons} CH_3-\overset{\|}{\underset{O}{C}}-CH_2^- \longleftrightarrow CH_3-\underset{O^-}{C}=CH_2$$

$$CH_3-\underset{O^-}{C}=CH_2 + X_2 \xrightarrow{fast} CH_3-\overset{\|}{\underset{O}{C}}-CH_2X + X^-$$

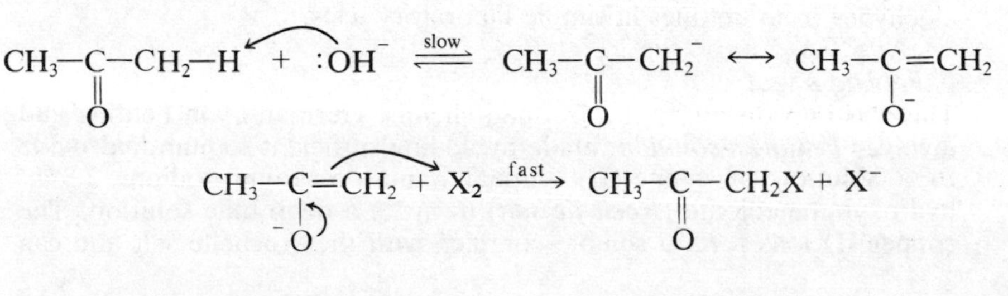

In this second case, the trihalogenated ketone undergoes alkaline hydrolysis as in the *iodoform* reaction:

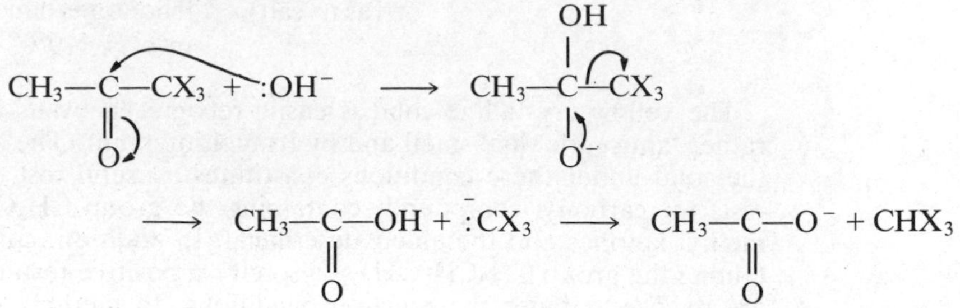

$$CH_3-C-CX_3 + :OH^- \longrightarrow CH_3-\underset{O^-}{\overset{OH}{C}}-CX_3$$

$$\longrightarrow CH_3-\underset{O}{C}-OH + :CX_3 \longrightarrow CH_3-\underset{O}{C}-O^- + CHX_3$$

8.3.7 Chemical differences between aldehydes and ketones

A major difference is that the aldehyde group —CHO is sensitive to oxidation, being oxidized to a carboxylic acid group, and thus aldehydes display **reducing properties**; ketones do not. Under strongly oxidizing conditions, ketones are oxidized but cleavage of the bonds adjacent to the carbonyl group takes place to give smaller molecular fragments as products. Thus:

$$CH_3COCH_3 \rightarrow CH_3CO_2H \text{ and } CO_2$$

A carboxylic acid containing fewer carbon atoms than the ketone is obtained on vigorous oxidation but if the ketone is cyclic, then all the carbon atoms are retained, as in the oxidation of cyclohexanone to hexanedioic acid which is used in the manufacture of nylon (*see* Section 20.3(i)):

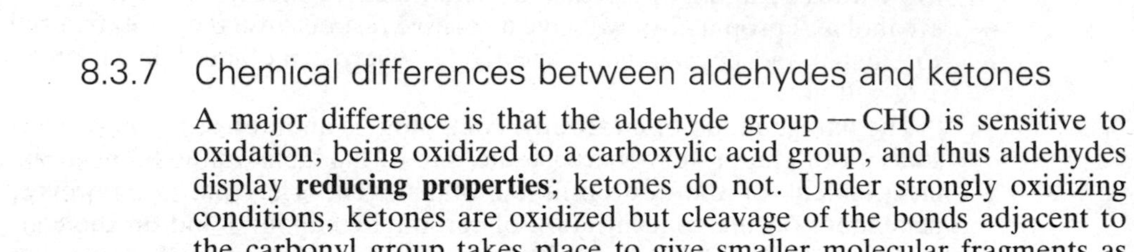

cyclohexanone
b.p. = 157°C

hexanedioic acid
m.p. = 152°C

There are three simple reactions which readily serve to distinguish aldehydes from ketones in simple laboratory tests.

(i) Fehling's test
This was introduced by the German chemist Hermann von Fehling and involves *Fehling's solution*, made by adding sufficient sodium hydroxide to a solution of copper(II) sulphate and potassium sodium 2,3-dihydroxybutanedioate (*Rochelle salt*) to make a deep blue solution. The copper(II) ions form a soluble complex with the Rochelle salt and can

then be reduced under alkaline conditions, without precipitation of any copper(II) hydroxide. On adding an aldehyde and warming the solution, an orange-red precipitate of hydrated copper(I) oxide is formed. The aldehyde is oxidized to a carboxylic acid:

$$R\!-\!CHO + 2Cu^{2+} + 5OH^- \;\rightarrow\; R\!-\!COO^- + Cu_2O + 3H_2O$$

This test is also widely used for reducing-sugars (*see* Section 17.2.2(iii)) and, since one or two other compounds (such as trichloromethane) also give a positive result, the results of such a test are not, by themselves, conclusive evidence of an aldehyde. Methanal is the most powerfully reducing aldehyde and gives rise to a precipitate of copper(I) oxide without warming; with warming and excess of methanal, metallic copper is precipitated, sometimes as a 'mirror'.

(ii) Tollens's test
This test was introduced by the German chemist Bernhard Tollens and involves warming the aldehyde with an ammoniacal solution of silver nitrate (made by adding ammonia solution to silver nitrate solution until the silver oxide precipitate just dissolves). The silver(I) ions are reduced to metallic silver and, if the reaction is performed in a scrupulously clean test-tube, the metal is deposited as a shining silver 'mirror' on the walls of the tube:

$$R\!-\!CHO + 2Ag(NH_3)_2^+ + OH^- \;\rightarrow\; R\!-\!COO^- + 2Ag + 2NH_4^+ + 2NH_3$$

The aldehyde is again oxidized to a carboxylic acid, but in neither this test nor the Fehling's test is it practicable to try to isolate the organic product. This test is not always reliable (ammoniacal silver nitrate occasionally yields a silver 'mirror' on its own) and a control should be used during the test.

(iii) Oxidation
Aldehydes are readily oxidized by conventional oxidizing agents such as warm, acidified solutions of potassium dichromate(VI) (colour change, orange to green) or potassium manganate(VII) (colour change, purple to colourless); the oxidation product in each case is a carboxylic acid, e.g.

$$3CH_3CHO + Cr_2O_7^{2-} + 8H^+ \;\rightarrow\; 3CH_3CO_2H + 2Cr^{3+} + 4H_2O$$
$$\text{ethanoic acid}$$
$$\text{b.p.} = 118°C$$

$$5CH_3CHO + 2MnO_4^- + 6H^+ \;\rightarrow\; 5CH_3CO_2H + 2Mn^{2+} + 3H_2O$$

The colour changes accompanying these reactions, in the presence of excess aldehyde, readily distinguish an aldehyde from a ketone (which does not react), but once again they are not exclusive to aldehyde oxidations. Alcohols, if primary or secondary, give the same overall result (*see* Section 6.3.3).

8.4 Manufacture and uses of aldehydes and ketones

The most widely used aldehydes and ketones are methanal, ethanal and propanone.

Methanal is manufactured from methanol by air oxidation in the presence of a silver catalyst:

$$CH_3OH + \tfrac{1}{2}O_2 \xrightarrow[500-650°C]{Ag} HCHO + H_2O \qquad \Delta H^\ominus = -158 \text{ kJ}$$

Its main use is in the manufacture of thermosetting plastics obtained by condensing methanal with either phenol (which gives Bakelite) or carbamide (*see* Section 20.2.2(i)).

Ethanal is made principally by the Wacker process, in which ethene is oxidized in the presence of a palladium(II) catalyst (*see* Section 3.3.1(vi)):

$$CH_2{=}CH_2 + \tfrac{1}{2}O_2 \rightarrow CH_3CHO \qquad \Delta H^\ominus = -243 \text{ kJ}$$

It can also be obtained by the hydration of ethyne (*see* Section 4.3.1(v)). Most of the ethanal produced goes to make organic compounds like ethanoic acid, ethanoic anhydride and butan-1-ol.

Propanone used to be made by the dehydrogenation of propan-2-ol (derived from propene), but most is manufactured commercially as a joint product, with phenol, from (1-methylethyl)benzene (*cumene, see* Section 13.3.2). It is widely used as a solvent. Propanone is also converted into its cyanohydrin (*see* Section 8.3.1(i)) which is hydrolysed and then esterified with methanol to give methyl 2-methylpropenoate. This is polymerized into the important plastic Perspex (*see* Section 20.2.1(v)):

$$CH_3COCH_3 + HCN \rightarrow CH_3C(OH)CNCH_3$$

$$CH_3C(OH)CNCH_3 + H_2SO_4 \longrightarrow \underset{\underset{CH_3}{|}}{CH_2{=}C{-}CONH_3^+} \ HSO_4^-$$

$$\underset{\underset{CH_3}{|}}{CH_2{=}C{-}CONH_3^+} \ HSO_4^- + CH_3OH \longrightarrow \underset{\underset{CH_3}{|}}{CH_2{=}CCOOCH_3} + NH_4^+ + HSO_4^-$$

Other aldehydes are made by the OXO process (*see* Section 3.3.1(vi)) and other ketones by dehydrogenation of alcohols.

Summary

1 **Aldehydes** contain the functional group $-C{\Large\diagdown}^{\textstyle O}_{\textstyle H}$ and **ketones** the group $\diagup^{\diagdown}C{=}O$. The simple members of each class form a homologous series.

2   Since both aldehydes and ketones possess the **carbonyl** group —C—

they have many chemical properties in common.   ‖
                                                                          O

(i) Nucleophilic addition reactions

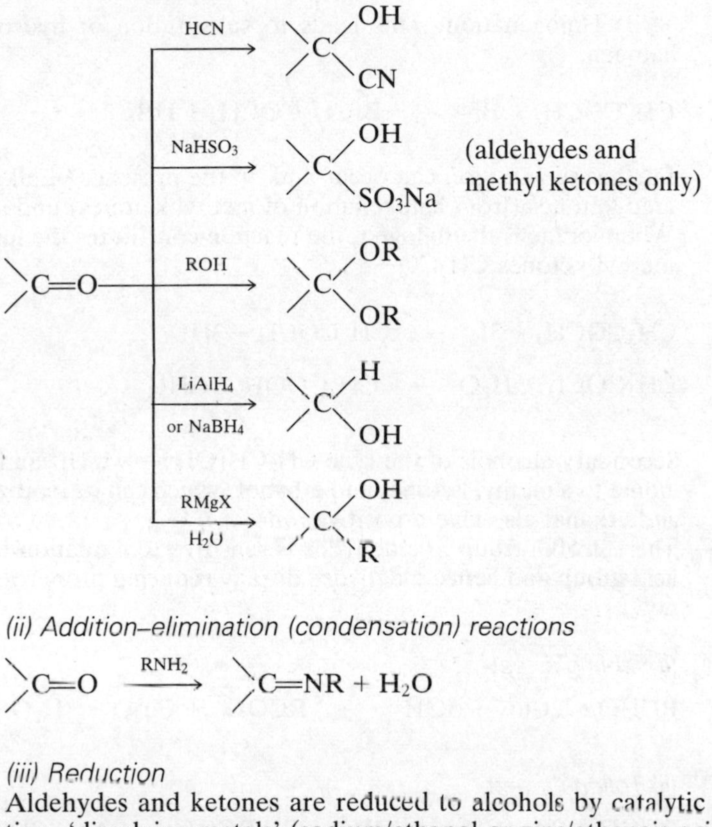

$\ce{>C=O}$

HCN → $\ce{>C<^{OH}_{CN}}$

NaHSO₃ → $\ce{>C<^{OH}_{SO_3Na}}$ (aldehydes and methyl ketones only)

ROH → $\ce{>C<^{OR}_{OR}}$

LiAlH₄ or NaBH₄ → $\ce{>C<^{H}_{OH}}$

RMgX H₂O → $\ce{>C<^{OH}_{R}}$

(ii) Addition–elimination (condensation) reactions

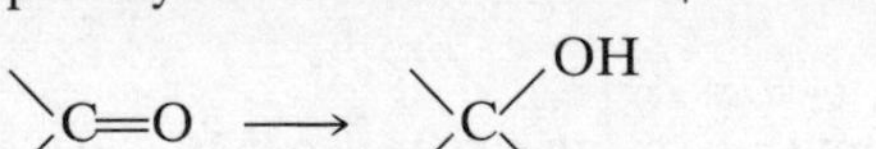

$$\ce{>C=O} \quad \xrightarrow{\text{RNH}_2} \quad \ce{>C=NR + H_2O}$$

(iii) Reduction
Aldehydes and ketones are reduced to alcohols by catalytic hydrogenation, 'dissolving metals' (sodium/ethanol or zinc/ethanoic acid), or complex hydrides such as LiAlH₄ or NaBH₄:

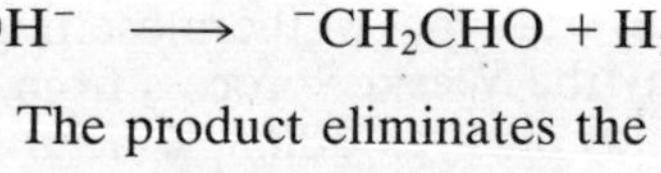

$$\ce{>C=O} \quad \longrightarrow \quad \ce{>C<^{OH}_{H}}$$

(iv) Reactions at carbon adjacent to the carbonyl group
    (a) Aldol condensation, e.g.

$$\ce{CH_3CHO + OH^- \longrightarrow {}^-CH_2CHO + H_2O} \xrightarrow{\text{CH}_3\text{CHO}} \ce{CH_3CH(OH)CH_2CHO}$$

The product eliminates the elements of water easily:

$$\ce{CH_3CH(OH)CH_2CHO} \xrightarrow{-\text{H}_2\text{O}} \ce{CH_3CH=CHCHO}$$

With more concentrated alkali, polymerization to a resin occurs. Methanal cannot undergo aldol condensation but undergoes disproportionation in the Cannizzaro reaction instead:

$$2HCHO + NaOH \rightarrow CH_3OH + HCOONa$$

(b) Halogenation. This leads to substitution of hydrogen atoms by halogen, e.g.

$$CH_3COCH_3 + Br_2 \rightarrow BrCH_2COCH_3 + HBr$$

Further substitution can occur and, in the presence of alkali, trihalogenated ketones (from halogenation of methyl ketones) undergo hydrolysis. When iodine is the halogen, the reaction constitutes the **iodoform test** for methyl ketones CH_3CO—:

$$CH_3COCH_3 + 3I_2 \rightarrow CH_3COCI_3 + 3HI$$

$$CH_3COCI_3 + H_2O \rightarrow CH_3COOH + CHI_3$$

Secondary alcohols of the type $CH_3CH(OH)$— which can be oxidized by iodine to a methyl ketone, and ethanol (which can be oxidized to ethanal) and ethanal also give a positive *iodoform* test.

3 The —CHO group in aldehydes is sensitive to oxidation to a carboxylic acid group and hence aldehydes display **reducing** properties. Ketones do not.

(i) Fehling's test

$$RCHO + 2Cu^{2+} + 5OH^- \rightarrow RCOO^- + Cu_2O + 3H_2O$$

(ii) Tollens's test

$$RCHO + 2Ag(NH_3)_2^+ + OH^- \rightarrow RCOO^- + 2Ag + 2NH_4^+ + 2NH_3$$

(iii) Oxidation

$$3RCHO + Cr_2O_7^{2-} + 8H^+ \rightarrow 3RCO_2H + 2Cr^{3+} + 4H_2O$$

$$5RCHO + 2MnO_4^- + 6H^+ \rightarrow 5RCO_2H + 2Mn^{2+} + 3H_2O$$

4 The most widely used carbonyl compounds are methanal, ethanal and propanone. Methanal is manufactured by the oxidation of methanol and is used in the manufacture of several types of thermosetting plastics such as Bakelite. Ethanal is made by the Wacker process from ethene and is converted into other compounds such as ethanoic acid. Propanone is made as a co-product, with phenol, of the oxidation of (1-methylethyl)benzene (*cumene*) and is used as a solvent.

Questions

1   The relative molecular mass of hex-1-ene, $CH_3CH_2CH_2CH_2CH{=}CH_2$, is 84 and that of pentan-2-one, $CH_3CH_2CH_2COCH_3$, is 86.
    (a) State, giving a reason, which of these two compounds you would expect to have the higher boiling point.
    (b) (i) State one reagent which will form an addition product with hex-1-ene but not with pentan-2-one.
        (ii) Write an equation for the reaction which takes place (mechanism not required here).
    (c) (i) State one reagent which will form an addition product with pentan-2-one but not with hex-1-ene.
        (ii) Write an equation for the reaction which takes place (mechanism not required here).
    (d) Indicate the mechanism of the reaction in **either** (b)(ii) **or** (c)(ii).
    (e) Explain briefly why reagents which form addition products with carbonyl compounds will not, in general, form addition products with alkenes.

                                                                            JMB

2   An organic substance **X** of molecular formula C_5H_8O is found to react readily with the following: bromine, hydroxylamine (NH_2OH), and sodium tetrahydridoborate ($NaBH_4$).
    (a) For each of the reagents listed name **one** functional group within **X** which would give a positive reaction with it.
    (b) Bearing in mind the above reactions explain which functional group must be present in **X** if it gives a carboxylic acid when treated with chromic(VI) acid.
    (c) What can you deduce from the observation that propanone is formed when **X** is suitably treated with trioxygen (*ozone*)?
    (d) Suggest a structure for **X**.

                                                                            JMB

3   The reaction of propanone (*acetone*) with hydrogen cyanide is imperceptibly slow in the absence of a suitable catalyst. What substance would you use to catalyse this reaction? How does the catalyst function?
    What is the reaction of propanone with
    (a) 2,4-dinitrophenylhydrazine, (b) lithium tetrahydridoaluminate ($LiAlH_4$), (c) aqueous ammoniacal silver nitrate?
    What mass of tri-iodomethane (*iodoform*) is obtained by treating 1.00 g of propanone with an excess of iodine and alkali?

                                                            Oxford & Cambridge

4* With ethanol as the *sole* carbon source, design a synthesis for each of the following:

    (a) CH_3CH_2Br, (b) CH_3CHO, (c) CHI_3, (d) $CH_3CH(OCH_2CH_3)_2$,
    (e) $CH_3CH{=}CHCHO$, (f) $CH_3\underset{\substack{\|\\O}}{C}CH_2CH_3$.

Give the inorganic reagents required and the necessary conditions but not experimental procedures.

Imperial College

5   When pentan-2-ol, $CH_3CH_2CH_2CH(OH)CH_3$, is heated with an acidified aqueous solution of potassium dichromate(VI), the colour of the solution changes from orange to green and the alcohol is oxidized to the corresponding ketone.

(a) What is the oxidation state of the chromium species responsible for the green colour?

(b) Name one other reagent which could have been used instead of potassium dichromate(VI) to oxidize the alcohol to the ketone.

(c) What is the structural formula of the ketone formed by oxidizing pentan-2-ol?

(d) Show, by means of an equation in each case, how this ketone reacts with
   (i) 2,4-dinitrophenylhydrazine, (ii) hydroxylamine.

(e) How might you use the 2,4-dinitrophenylhydrazine derivatives of two isomeric ketones in order to distinguish between them?

JMB

6* The following is a list of carbonyl compounds, each of which is indicated by a capital letter:

A CH_3CHO;   B CH_3COCH_3;   C $C_2H_5COC_2H_5$;   D C_6H_5CHO;
E $C_6H_5COCH_3$

(a) Give the names and structural formulae of *two* ketones which are isomeric with C.

(b) Give the names and structural formulae of *two* aldehydes which are isomeric with E.

(c) Give *one* chemical test in each case to distinguish between
   (i) A and B, (ii) B and C.

(d) (i)   Write an equation for the reaction of D with sodium hydroxide.
   (ii) How would you isolate benzenecarboxylic acid (*benzoic acid*) from the reaction mixture?
   (iii) This reaction may be described as a *disproportionation*. Explain the meaning of this term as applied to this reaction.

London

Chapter 9
Carboxylic acids

9.1 Homologous series of carboxylic acids

Carboxylic acids are weak organic acids which contain the **carboxyl** group,

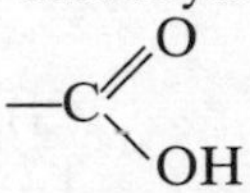

$$-C\underset{\displaystyle OH}{\overset{\displaystyle O}{}}$$

as functional group. The acids are named by adding the suffix
-*oic* to the roots of the names of the alkanes which possess the same
number of carbon atoms. Thus, the most widely encountered carboxylic
acid (*acetic acid*), present in vinegar, is called ethanoic acid, CH_3CO_2H.
Table 9.1 lists some physical data for the first eight straight-chain members
of the homologous series.

Table 9.1

Name	Formula	Melting point (°C)	Boiling point (°C)	Density (g cm^{-3})
methanoic acid	HCO_2H	8	101	1.220
ethanoic acid	CH_3CO_2H	17	118	1.049
propanoic acid	$CH_3CH_2CO_2H$	−22	141	0.992
butanoic acid	$CH_3(CH_2)_2CO_2H$	−5	163	0.959
pentanoic acid	$CH_3(CH_2)_3CO_2H$	−35	186	0.939
hexanoic acid	$CH_3(CH_2)_4CO_2H$	−2	205	0.929
heptanoic acid	$CH_3(CH_2)_5CO_2H$	−11	224	0.922
octanoic acid	$CH_3(CH_2)_6CO_2H$	16	237	0.910

Some of the higher members of the series (systematic name, alkanoic
acids) are referred to as *fatty acids* because they are constituents, in the
form of their esters, of naturally-occurring vegetable oils and animal fats,
although only those acids containing an even number of carbon atoms,
e.g. octadecanoic acid $C_{17}H_{35}CO_2H$ (*stearic acid*), are normally to be
found in this source (*see* Section 19.1).

Some of the middle members of the homologous series have very
disagreeable odours; butanoic acid, for example, is a minor constituent
of perspiration and the unpleasant smell of rancid butter and strong
cheese is due to the same acid, set free by bacterial action.

Isomerism occurs owing to variations in the alkyl group attached to the carboxyl group, e.g.

$$CH_3CH_2CH_2CO_2H$$

$$\overset{3}{C}H_3\!-\!\overset{2}{C}H\!-\!\overset{1}{C}O_2H$$
$$|$$
$$CH_3$$

butanoic acid
b.p. = 163°C

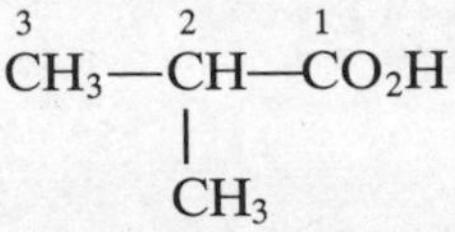

2-methylpropanoic acid
b.p. = 154°C

Acids which contain two carboxyl groups per molecule are **dibasic** acids; examples are ethanedioic acid (*oxalic acid*) which is present as its mono potassium salt in the leaves and roots of rhubarb and other plants and is very poisonous, and 2,3-dihydroxybutanedioic acid (*tartaric acid*), occurring in many berries and fruits and probably the most widely distributed acid in the vegetable kingdom:

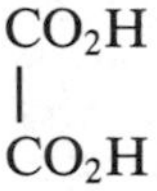

$$CO_2H$$
$$|$$
$$CO_2H$$

$$HO\!-\!CH\!-\!CO_2H$$
$$|$$
$$HO\!-\!CH\!-\!CO_2H$$

$$CH_2\!-\!CO_2H$$
$$|$$
$$HO\!-\!C\!-\!CO_2H$$
$$|$$
$$CH_2\!-\!CO_2H$$

ethanedioic acid
m.p. = 187°C

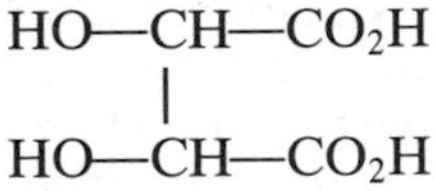

2,3-dihydroxybutanedioic acid
m.p. = 206°C

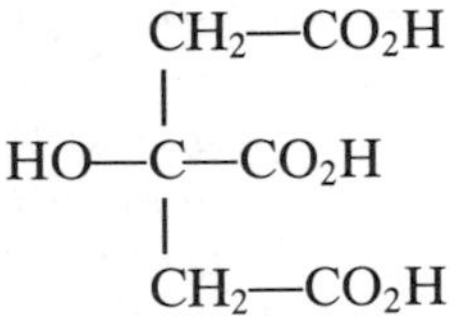

2-hydroxypropane-1,2,3-
tricarboxylic acid
m.p. = 153°C

Acids which contain three carboxyl groups per molecule are more uncommon; an example is 2-hydroxypropane-1,2,3-tricarboxylic acid (*citric acid*), familiar to biochemists from the *Krebs* or *citric acid cycle*, and found in citrus fruits such as lemons.

9.2 Physical properties of carboxylic acids

The lowest members of the homologous series of *monocarboxylic* acids are colourless liquids with pungent smells, and are freely soluble in water. In the liquid state, there is extensive hydrogen bonding between carboxyl groups of different molecules, and the boiling points are substantially higher than those of alkanes of corresponding relative molecular mass (*see Figure 2.1*). The molecules are said to be extensively **associated** by hydrogen bonding. In non-polar solvents such as hydrocarbons the simple acids exist as associated pairs, or **dimers**, as revealed by relative molecular mass determinations, e.g.

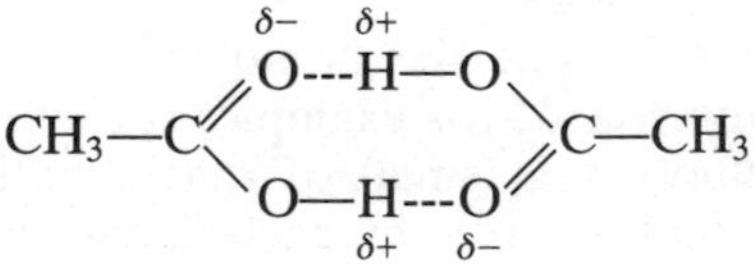

apparent $M_r = 120$
M_r of acid = 60

As the carbon chain length of an acid increases, so its solubility in water decreases, the effect of the carboxyl group on the physical properties of the molecule becoming less significant. Those acids containing more than nine carbon atoms per molecule are oily or waxy solids; they are insoluble in water and, because of their low volatility, possess practically no smell.

Methanoic acid (*formic acid*) is a colourless, corrosive liquid with a pungent smell. Ethanoic acid was one of the earliest known acids, being formed by the oxidation of ethanol in the presence of various *acetobacter* micro-organisms, when beer and light wines turn sour. When pure, it is a colourless liquid with a strong pungent smell. Since it freezes at 17°C, it often solidifies in cold weather to an ice-like solid. For this reason, the pure anhydrous acid (which is quite hygroscopic, i.e. it absorbs water from the atmosphere) is sometimes called *glacial* ethanoic acid.

9.3 Chemical properties of carboxylic acids

Although the carboxyl group apparently contains a carbonyl group linked to a hydroxyl group, carboxylic acids rarely display properties of either ketones or alcohols. The chemical properties of carboxylic acids suggest that this formulation is an oversimplification. Two structures (canonical forms, *see* Section 1.5) can be written in conventional formula notation, but the real structure is thought to lie somewhere between these two extreme representations, i.e. the distribution of electrons in the carboxyl group is not precisely described by either formula:

$$R-C\begin{array}{c}{}^{\displaystyle O}\\[-2pt]\diagdown\\[-4pt]\ddot{O}-H\end{array} \longleftrightarrow R-C\begin{array}{c}{}^{\displaystyle O^{-}}\\[-2pt]\diagdown\\[-4pt]\overset{+}{O}-H\end{array}$$

Consequently, the carbon atom of the carboxyl group is not really electrophilic and not very sensitive to nucleophilic attack (but *see* Section 9.3.2). The carboxyl group displays acidic properties in water. It also influences the strength of C—H bonds adjacent to the group.

9.3.1 Acidic properties

Although they are weak acids by comparison with mineral acids, the soluble carboxylic acids form distinctly acidic solutions in water, unlike alcohols:

$$R-C\begin{array}{c}{}^{\displaystyle O}\\[-2pt]\diagdown\\[-4pt]OH\end{array} + H_2O \rightleftharpoons R-C\begin{array}{c}{}^{\displaystyle O}\\[-2pt]\diagdown\\[-4pt]O^{-}\end{array} + H_3O^+$$

The fact that dissociation occurs is attributed to the stabilization of the negative charge on the anion. This is achieved by the spreading out, or

delocalization, of the electronic charge over the carbon atom and both oxygen atoms:

$$R-C\overset{O}{\underset{O^-}{\Bigl\langle}} \longleftrightarrow R-C\overset{O^-}{\underset{O}{\Bigl\langle}} \qquad \text{i.e.} \quad R-C\overset{O}{\underset{O}{\Bigl\langle}}$$

In contrast, alcohols only display acidic properties on reaction with very strong bases, because the formation of the corresponding anion, the alkoxide ion, is thermodynamically unfavourable (no delocalization of negative charge on the alkoxide ion is possible). In fact, alkoxides immediately react with water to form the alcohol and hydroxide ions (*see* Section 6.3.4).

The strength of a weak acid can be expressed by a pK_a value (*see* Section 1.7); the lower the value, the stronger the acid (cf. pH values). The presence of an electron-withdrawing atom or group adjacent to the carboxyl group in an acid leads to the weakening of the O—H bond (the *inductive effect, see* Section 1.4) and enhanced acidity, e.g.

$$CH_3-C\overset{O}{\underset{OH}{\Bigl\langle}} \qquad\qquad Cl\!\leftarrow\!CH_2\!\leftarrow\!C\overset{O}{\underset{OH}{\Bigl\langle}} \qquad\qquad Cl\!\leftarrow\!\overset{\displaystyle Cl}{\underset{\displaystyle Cl}{C}}\!\leftarrow\!C\overset{O}{\underset{OH}{\Bigl\langle}}$$

$pK_a = 4.8$ $pK_a = 2.9$ $pK_a = 0.7$

The inductive effect can be transmitted along a chain of atoms but it falls off rapidly with increasing distance, e.g.

$$Cl\!\leftarrow\!CH_2\!\leftarrow\!CH_2\!\leftarrow\!C\overset{O}{\underset{OH}{\Bigl\langle}}$$

$pK_a = 4.1$

Conversely, the presence of an electron-donating atom or group in the molecule strengthens the O—H bond and the acid strength diminishes, e.g.

$$H-C\overset{O}{\underset{OH}{\Bigl\langle}} \qquad\qquad CH_3\!\rightarrow\!C\overset{O}{\underset{OH}{\Bigl\langle}}$$

$pK_a = 3.8$ $pK_a = 4.8$

Thus, ethanoic acid is a weaker acid than methanoic acid. In the pure anhydrous state, of course, ethanoic acid does not display typical acidic properties, because the degree of ionization is so very small.

(i) Reaction with alkali
Carboxylic acids react with alkali to form salts and water. However,

solutions of salts of carboxylic acids in water are not neutral but alkaline owing to salt hydrolysis:

$$\left.\begin{array}{ll} RCO_2H + OH^- & \rightarrow \quad RCO_2^- + H_2O \\ RCO_2^- + H_2O & \rightarrow \quad RCO_2H + OH^- \end{array}\right\} \quad \text{i.e.} \quad RCO_2H + OH^- \rightleftharpoons RCO_2^- + H_2O$$

Thus, if a solution of a carboxylic acid is to be titrated against alkali (in order to determine its relative molecular mass, for example), an indicator which changes colour in the alkaline region of pH, such as *phenolphthalein*, must be used to indicate the end-point.

Even when carboxylic acids are insoluble in water, their salts are usually soluble, so that acids can conveniently be extracted from organic mixtures with alkali or sodium carbonate solution—*see* (iii) below.

(ii) Reaction with metals
Fairly reactive metals dissolve in aqueous solutions of carboxylic acids, displacing hydrogen, e.g.

$$2RCO_2H + Mg \rightarrow (RCO_2)_2Mg + H_2$$

(iii) Reaction with carbonates and hydrogencarbonates
The reaction between a solution of a carboxylic acid and a carbonate or hydrogencarbonate is accompanied by the vigorous evolution of carbon dioxide:

$$2RCO_2H + CO_3^{2-} \rightarrow 2RCO_2^- + CO_2 + H_2O$$

This reaction conveniently distinguishes carboxylic acids from neutral compounds, such as alcohols, but it cannot be the basis of a specific test, since other organic acids such as sulphonic acids (*see* Section 12.4.2(ii)) react similarly.

9.3.2   Esterification

Carboxylic acids react with alcohols to form neutral compounds called **esters** (*see* Section 6.3.1(iii) and Section 10.1). The reaction is usually very slow and does not go to completion, a state of equilibrium being eventually reached:

$$\underset{\displaystyle RC-OH + R'OH}{\overset{\displaystyle O}{\overset{\displaystyle \|}{}}} \rightleftharpoons \underset{\displaystyle RC-OR' + H_2O}{\overset{\displaystyle O}{\overset{\displaystyle \|}{}}}$$

The water produced as a result of the reaction between the acid and alcohol contains oxygen, which has come from the carboxyl group; the OR' of the ester comes from the alcohol. (This has been shown by 'labelling' the oxygen of the alcohol with the heavy isotope ^{18}O; all of the ^{18}O appears in the ester, after reaction, and none in the water produced.) The reaction is termed **esterification**. Esters are named from the acid from which they are derived, with a prefix indicating the alkyl group supplied by the alcohol. Thus:

$$CH_3CO_2H + CH_3CH_2OH \rightleftharpoons CH_3\overset{\overset{\displaystyle O}{\|}}{C}-OCH_2CH_3 + H_2O$$

ethanoic acid   ethanol   ethyl ethanoate
b.p. = 118°C   b.p. = 78°C   b.p. = 77°C

$$CH_3CH_2CO_2H + CH_3OH \rightleftharpoons CH_3CH_2\overset{\overset{\displaystyle O}{\|}}{C}-OCH_3 + H_2O$$

propanoic acid   methanol   methyl propanoate
b.p. = 141°C   b.p. = 65°C   b.p. = 80°C

Note that, in the two examples given above, the esters are isomeric and are derived from different acids by esterification with different alcohols. The process of esterification is catalysed by mineral acid and one method of preparing esters involves heating the acid and alcohol with a small amount of concentrated sulphuric acid. The sulphuric acid not only catalyses the reaction, but increases the equilibrium yield of ester by absorbing the water formed, so displacing the position of equilibrium in favour of the product. The equilibrium constant for the ethyl ethanoate reaction is about 3.8 at 25°C, corresponding to about 66% conversion to ester, using an equimolar mixture of acid and alcohol. The yield can be improved by distilling the ester out of the reaction mixture as the ester is formed, and by employing an excess of one of the reagents (usually the cheaper one). This method of esterification is not always the best one, since the concentrated sulphuric acid can lead to dehydration of the alcohol, particularly if it is a tertiary one.

The alternative method of esterification, using an acid and alcohol, is called the **Fischer–Speier** method, in which dry hydrogen chloride is passed through the reaction mixture (which usually contains excess of one of the reagents) to act as acid catalyst.

The mechanism of acid-catalysed esterification is thought to involve protonation of the carboxyl group of the acid to produce a more electrophilic carbon atom (*see* introduction to Section 9.3), able to undergo nucleophilic attack by the alcohol:

$$R-C\overset{\ddot{O}:}{\underset{OH}{\diagdown}} + H^+ \rightleftharpoons R-C\overset{+OH}{\underset{OH}{\diagup}} \xrightarrow{R'OH} R-\overset{\overset{\displaystyle OH}{|}}{\underset{\underset{\displaystyle OH}{|}}{C}}-\overset{+}{O}\overset{R'}{\underset{H}{\diagdown}}$$

The reaction is completed by the transfer of a proton, followed by the elimination of water and regeneration of the acid catalyst:

$$R-\underset{\underset{\displaystyle OH}{|}}{\overset{\overset{\displaystyle OH}{|}}{C}}-\overset{R'}{\underset{H}{O}} \rightleftharpoons R-\underset{\underset{\displaystyle :OH}{|}}{\overset{\overset{\displaystyle +OH_2}{|}}{C}}-OR' \rightleftharpoons R-\underset{\displaystyle +OH}{\overset{\displaystyle ||}{C}}-OR' + H_2O \rightleftharpoons R-\underset{\displaystyle O}{\overset{\displaystyle ||}{C}}-OR' + H_2O + H^+$$

9.3.3 Substitution

(i) Occasionally the —OH component of the carboxyl group behaves in a similar fashion to the same group in an alcohol (*see* introduction to Section 9.3), and a substitution reaction takes place. Treatment of a carboxylic acid, usually dissolved in a dry solvent to moderate the reaction, with either phosphorus pentachloride or (better) sulphur dichloride oxide (*thionyl chloride*), yields an **acid chloride** (*see* Section 10.2):

$$RCOOH + PCl_5 \quad \rightarrow \quad RCOCl + POCl_3 + HCl$$
$$RCOOH + SOCl_2 \quad \rightarrow \quad RCOCl + SO_2 \quad + HCl$$

e.g.

$$CH_3COOH + SOCl_2 \quad \rightarrow \quad CH_3COCl + SO_2 + HCl$$
ethanoic acid   ethanoyl chloride
b.p. = 118°C   b.p. = 52°C

(ii) Hydrogen atoms bonded to the carbon atom adjacent to a carboxyl group can be substituted by chlorine by passing chlorine into the boiling acid in the presence of ultraviolet light or a *halogen carrier*, such as iodine or red phosphorus.* Successive substitutions can take place but they can be controlled by regulating the uptake of chlorine, until the calculated increase in mass has been achieved, e.g.

$$CH_3CO_2H + Cl_2 \quad \rightarrow \quad ClCH_2CO_2H + HCl$$

ethanoic acid   chloroethanoic acid
b.p. = 118°C   m.p. = 61°C

$$ClCH_2CO_2H + Cl_2 \quad \rightarrow \quad Cl_2CHCO_2H + HCl$$

                              dichloroethanoic acid
                              b.p. = 194°C

$$Cl_2CHCO_2H + Cl_2 \quad \rightarrow \quad Cl_3CCO_2H + HCl$$

                              trichloroethanoic acid
                              m.p. = 58°C

* ICl_3 may well be the effective chlorinating agent when iodine is used; when phosphorus is used, halogenation proceeds via attack on the corresponding acid halide RCOX. This is formed by reaction with PX_3 (from $P + X_2$) and,

owing to the presence of the $-C\overset{\displaystyle O}{\underset{\displaystyle X}{\diagdown}}$ group, contains an *activated* —CH$_2$—
group.

Substitution mainly takes place at the carbon atom immediately adjacent to the carboxyl group but can occur at other carbon atoms in the molecule. Bromination can also be carried out in a similar fashion, in the presence of a small quantity of red phosphorus, but bromination takes place exclusively at the carbon atom adjacent to the carboxyl group.

9.3.4 Reduction

Carboxylic acids are readily reduced by lithium tetrahydridoaluminate $LiAlH_4$ to primary alcohols:

$$RCO_2H \quad \rightarrow \quad RCH_2OH$$

So, too, are esters, acid chlorides and acid anhydrides (*see* Chapter 10). In contrast, sodium tetrahydridoborate $NaBH_4$ will *not* effect reduction.

9.4 Unusual properties of methanoic acid

Methanoic acid, the simplest member of the homologous series, displays some properties not shown by other members of the series.

(i) The arrangement of atoms in a molecule of the acid are characteristic both of a carboxylic acid, and an aldehyde, and therefore methanoic acid shows reducing properties:

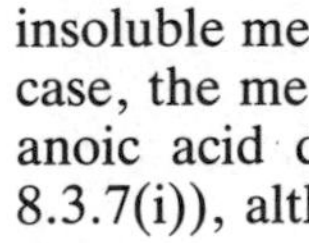

$$\begin{array}{c} H-C-OH \\ \parallel \\ O \end{array}$$

The acid and its salts reduce ammoniacal silver nitrate solution (Tollens's test, *see* Section 8.3.7(ii)), manganate(VII) and dichromate(VI) solutions and, like methanal, reduce mercury(II) chloride solution, first to white insoluble mercury(I) chloride and then to grey metallic mercury. In each case, the methanoic acid is oxidized to carbon dioxide and water. Methanoic acid does *not* reduce Fehling's solution, however (*see* Section 8.3.7(i)), although its methyl and ethyl esters do to some slight extent.

(ii) When treated with concentrated sulphuric acid, methanoic acid and its salts evolve carbon monoxide, and this reaction is a useful method for preparing the gas:

$$HCO_2H \xrightarrow{H_2SO_4} H_2O + CO$$

(iii) Methanoic acid does *not* form an acid chloride, but on treatment with phosphorus pentachloride yields carbon monoxide and hydrogen chloride:

$$HCO_2H + PCl_5 \quad \rightarrow \quad POCl_3 + CO + 2HCl$$

(iv) When an alkali metal salt of methanoic acid is strongly heated, hydrogen is evolved and a salt of ethanedioic acid is formed:

$$2HCO_2^- \rightarrow C_2O_4^{2-} + H_2$$

9.5 Salts of carboxylic acids

The metallic salts of carboxylic acids are ionic compounds which are generally soluble in water (even the lead(II) salt of ethanoic acid is soluble and is used in the normal laboratory test for hydrogen sulphide). In *neutral* aqueous solution, the salts usually give a deep-red coloration with *neutralized* iron(III) chloride solution (cf. phenols, Section 13.3.1(iii)).

When the salts are treated with mineral acid, the carboxylic acid is liberated and can be isolated by distillation, or by extraction with suitable solvents.

On strong heating with soda-lime, the acids or their salts are **decarboxylated**, i.e. they lose the carboxyl group as carbon dioxide:

$$RCO_2H \rightarrow RH + CO_2$$

Electrolysis of an aqueous, methanolic solution of a salt of a carboxylic acid produces a hydrocarbon, a reaction discovered by the German chemist Hermann Kolbe in 1849 and sometimes called the **Kolbe synthesis**. The carboxylate anion undergoes oxidation at the platinum anode:

$$2RCO_2^- - 2e^- \rightarrow R\!-\!R + 2CO_2$$

This is a useful method for preparing higher symmetrical alkanes from carboxylic acids.

When an *ammonium* salt of a simple carboxylic acid is heated strongly, the corresponding **acid amide** is produced (*see* Section 10.4):

$$R\!-\!COO^-NH_4^+ \rightarrow RCONH_2 + H_2O$$

9.6 Dibasic acids

Acids which contain two carboxyl groups per molecule are **dibasic** acids, e.g.

$HO_2C\!-\!CO_2H$

ethanedioic acid
m.p. = 189°C

$HO_2C\!-\!CH_2\!-\!CO_2H$

propanedioic acid
m.p. = 136°C (dec)

$HO_2C\!-\!CH_2CH_2\!-\!CO_2H$

butanedioic acid
m.p. = 185°C

In most reactions, the dicarboxylic acids behave in a similar fashion to those already discussed. However, the acids named above show some individual properties.

(i) Decarboxylation

$$HO_2C—CO_2H \xrightarrow{150°C} HCO_2H + CO_2$$

$$HO_2C—CH_2—CO_2H \xrightarrow{150°C} CH_3CO_2H + CO_2$$

(ii) Dehydration
This reaction is particularly characteristic of dibasic acids in which both carboxylic acid groups are attached to the same carbon atom, e.g.

$$HO_2C—CO_2H \xrightarrow[\text{conc. } H_2SO_4]{\text{heat}} CO + CO_2 + H_2O$$

$$HO_2C—CH_2—CO_2H \xrightarrow[\text{conc. } H_2SO_4]{\text{heat}} O{=}C{=}C{=}C{=}O + 2H_2O$$

$$\text{tricarbon dioxide}$$

(iii) Anhydride formation
When strongly heated in ethanoic anhydride, butanedioic acid forms a cyclic anhydride (for anhydrides *see* Section 10.3):

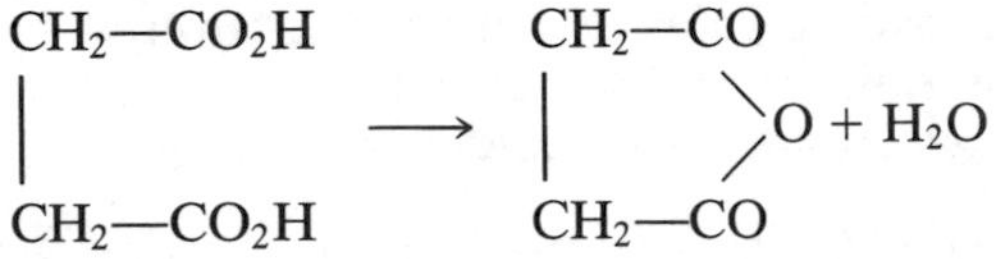

$$\begin{array}{c} CH_2—CO_2H \\ | \\ CH_2—CO_2H \end{array} \longrightarrow \begin{array}{c} CH_2—CO \\ | \quad\quad \rangle O + H_2O \\ CH_2—CO \end{array}$$

butanedioic anhydride
m.p. = 120°C

(iv) Oxidation
Ethanedioic acid (*oxalic acid*) and its salts are oxidized to carbon dioxide by warm, acidified potassium manganate(VII) solution:

$$5C_2O_4^{2-} + 2MnO_4^- + 16H^+ \rightarrow 10CO_2 + 2Mn^{2+} + 8H_2O$$

9.7 Manufacture and uses of carboxylic acids

The major acid manufactured commercially is ethanoic acid. Some is produced by the oxidation of ethanal, derived from ethene by the Wacker process (*see* Section 3.3.1(vi)), in the liquid phase with a manganese(II) ethanoate catalyst:

$$CH_3CHO + \tfrac{1}{2}O_2 \xrightarrow{50-70°C} CH_3CO_2H$$

Some ethanal is also oxidized directly to ethanoic anhydride (*see* Section 10.3):

$$2CH_3CHO + O_2 \rightarrow (CH_3CO)_2O + H_2O$$

Increasingly, ethanoic acid is being manufactured by a process developed by BP Chemicals. It involves the continuous liquid-phase air oxidation of naphtha, a light hydrocarbon fraction (containing mainly C_5–C_7 alkanes) which is obtained from petroleum (*see* Section 18.2). The oxidation is a free-radical reaction and provides methanoic, ethanoic and propanoic acids in the approximate ratio, by mass, of 15%, 77% and 8%, as the major products:

$$\text{naphtha + air} \xrightarrow[\text{3–9 MN m}^{-2}]{\text{150–200°C}} HCO_2H + CH_3CO_2H + CH_3CH_2CO_2H$$

Ethanoic acid is used primarily for the manufacture of cellulose ethanoate (*see* Section 17.6), ethenyl ethanoate (*see* Section 4.3.1(vi)) for paints, and for making ester solvents. Various applications have been developed for the co-products, and both methanoic and propanoic acids are now being used as fungicides for the protection of animal fodder (silage) and the latter acid as a preservative for bread.

An alternative route to ethanoic acid has been developed by the German company BASF and by Monsanto. This utilizes a continuous liquid-phase reaction between methanol and carbon monoxide in the presence of a catalyst:

$$CH_3OH + CO \rightarrow CH_3CO_2H$$

This route has the advantage of deriving its starting materials from coal via synthesis gas (*see* Section 18.6.2(iii)) and might therefore become more widely adopted in the future, in the event of major oil or naphtha shortages.

Summary

1 **Carboxylic acids** are weak acids containing the **carboxyl** group $-C\underset{\displaystyle \text{OH}}{\overset{\displaystyle O}{\diagup\kern-0.6em\diagdown}}$ as functional group. Acids containing two groups per molecule are called **dibasic** acids.

2 (i) Carboxylic acids display typical **acid properties**, e.g.

$$RCO_2H + OH^- \rightarrow RCO_2^- + H_2O$$

$$2RCO_2H + Mg \rightarrow (RCO_2)_2Mg + H_2$$

$$2RCO_2H + CO_3^{2-} \rightarrow 2RCO_2^- + CO_2 + H_2O$$

(ii) Carboxylic acids react with alcohols to give *esters* in an equilibrium reaction called **esterification**:

$$RCO_2H + R'OH \rightleftharpoons RCOOR' + H_2O$$

(iii) Carboxylic acids also undergo **substitution** reactions at the carboxyl group, e.g.

$RCOOH + SOCl_2 \rightarrow RCOCl + SO_2 + HCl$

and at the alkyl group, e.g.

$CH_3CO_2H + Cl_2 \rightarrow ClCH_2CO_2H + HCl$
                          and eventually Cl_3CCO_2H

(iv) Carboxylic acids are reduced by lithium tetrahydridoaluminate to primary alcohols:

$RCO_2H \rightarrow RCH_2OH$

3   Methanoic acid is unusual in that it shows **reducing** properties, being oxidized to carbon dioxide and water by MnO_4^-, $Cr_2O_7^{2-}$ and $Ag(NH_3)_2^+$:

$HCO_2H \xrightarrow{\text{oxidation}} CO_2 \text{ and } H_2O$

It also takes part in the following reactions:

$HCO_2H \xrightarrow{\text{conc. } H_2SO_4} H_2O + CO$

$2HCO_2^- \xrightarrow{\text{heat}} C_2O_4^{2-} + H_2$

The acid does *not* form an acid chloride but reacts with PCl_5 thus:

$HCO_2H + PCl_5 \rightarrow POCl_3 + CO + 2HCl$

4   Salts of simple carboxylic acids (and the acids themselves) are **decarboxylated**, when heated strongly with soda-lime:

$RCO_2H \rightarrow RH + CO_2$

The electrolysis of these salts in aqueous methanolic solution yields hydrocarbons at a platinum anode; the reaction is known as the **Kolbe synthesis**:

$2RCO_2^- - 2e^- \rightarrow R—R + CO_2$

When ammonium salts are heated the corresponding *acid amide* is produced:

$RCO_2^-NH_4^+ \rightarrow RCONH_2 + H_2O$

5   Ethanoic acid is the most important carboxylic acid commercially and much of it is now made by the oxidation of naphtha, the reaction also producing methanoic and propanoic acids as important co-products. Ethanoic acid is used for making cellulose ethanoate (*see* Section 17.6), paints and solvents.

Questions

1* The table below lists the dissociation constants K_a of a number of carboxylic acids:

Acid	Formula	K_a (mol l^{-1})
methanoic	HCO_2H	1.8×10^{-4}
ethanoic	CH_3CO_2H	1.8×10^{-5}
propanoic	$CH_3CH_2CO_2H$	1.4×10^{-5}
butanoic	$CH_3CH_2CH_2CO_2H$	1.5×10^{-5}
chloroethanoic	CH_2ClCO_2H	1.6×10^{-3}
bromoethanoic	CH_2BrCO_2H	1.4×10^{-3}
iodoethanoic	CH_2ICO_2H	7.5×10^{-4}
2-chlorobutanoic	$CH_3CH_2CHClCO_2H$	1.4×10^{-3}
3-chlorobutanoic	$CH_3CHClCH_2CO_2H$	8.9×10^{-5}
4-chlorobutanoic	$CH_2ClCH_2CH_2CO_2H$	3.0×10^{-5}

Use the data to deduce patterns in which the dissociation constants of the acids change in relation to the length of their carbon chains and to the type and position of substituents. Suggest explanations for these patterns and point out any anomalies which, you feel, are present.

London

2* (a) Two compounds, A and B, have the same molecular formula, $C_4H_4O_4$, and show the following reactions:
    (i)   In aqueous solution both react with magnesium, liberating 1 mol of hydrogen gas for every mol of A or of B.
    (ii) Both compounds react with bromine dissolved in tetrachloromethane.
    (iii) On complete reduction, both compounds yield butane.
    Use the above information to deduce the possible structures of compounds A and B.
    (b) The following table lists some of the properties of A and B:

Property	Compound A	Compound B
melting point (°C)	130	287
solubility in water	fairly soluble	slightly soluble
temperature at which dehydration occurs (°C)	about 160	230

From this additional information, deduce the identity of A and of B. Fully explain your reasoning.

London

3*   Chloroethanoic acid may be prepared by mixing glacial ethanoic acid with
a small quantity of red phosphorus and then passing chlorine gas into this
mixture, in the presence of sunlight. After completion of the reaction,
chloroethanoic acid may be separated from the reaction mixture by
fractional distillation at about 180°C.

The stoichiometric equation is

$$CH_3CO_2H + Cl_2 \rightarrow CH_2ClCO_2H + HCl$$

(a) What substances other than chloroethanoic acid and hydrogen chloride would you expect to be formed in the above preparation? Give
formulae and names.
(b) What might be the role of the small quantity of red phosphorus added
to the reaction mixture?
(c) In the absence of sunlight, the formation of chloroethanoic acid
proceeds extremely slowly. How do you account for this?
(d) The progress of the chlorination reaction may be followed by period-
ically weighing the reaction mixture to determine the increase in mass
resulting from the addition of chlorine to the acid.
In an actual experiment, 25 g of glacial ethanoic acid were used. After
chlorination for 3 h, an increase in mass of 5.0 g was determined.
What yield of chloroethanoic acid had been obtained at that stage?

London

4   Ethanoic acid used to be manufactured by the air oxidation of ethanal,
derived from ethene either by direct oxidation or via ethanol with subsequent catalytic dehydrogenation.
This method is now being replaced by the continuous liquid-phase air
oxidation of alkanes (C_5-C_7). The condensate from the reactor is separated
by fractional distillation into ethanoic acid and the by-products methanoic
and propanoic acids.
(a) Write equations for the reactions involved in the ethanal route.
(b) What is the source of both ethene and the C_5-C_7 alkanes and how are
they obtained?
(c) What advantages does the new process have over the older route?
(d) What mechanism is involved in the new route?
(e) What is a major industrial use for ethanoic acid and to what use are
the by-products of the new route put?

Chapter 10
Derivatives of carboxylic acids

The classes of compounds discussed in this chapter are all derived from carboxylic acids. Their names and functional groups are as follows:

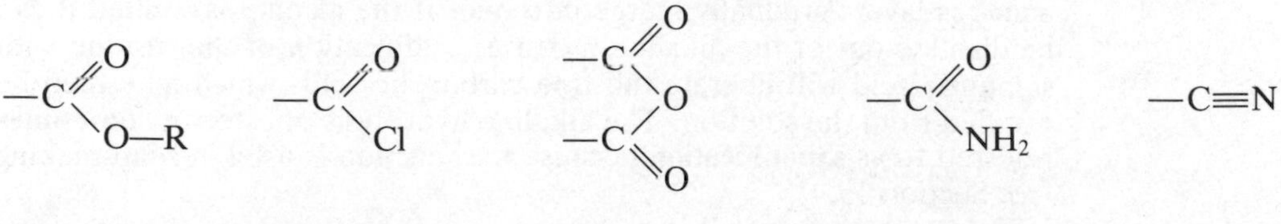

ester	acid chloride	acid anhydride	acid amide	nitrile

10.1 Chemistry of esters

Esters are neutral compounds derived from reaction between a carboxylic acid and an alcohol (*see* Section 9.3.2). The esters of any individual carboxylic acid form a homologous series whose lower members are volatile liquids with pleasant, often fruit-like odours. As the relative molecular mass increases, volatility decreases and the esters gradually become viscous, oily liquids or waxy solids. They are immiscible with water and ethyl ethanoate is widely used as an organic solvent. They are polar compounds with boiling points well below those of carboxylic acids of corresponding relative molecular mass, since there is no intermolecular hydrogen bonding:

$CH_3CO_2CH_2CH_3$

ethyl ethanoate ($M_r = 88$)
b.p. = 77°C

$CH_3CH_2CO_2CH_3$

methyl propanoate ($M_r = 88$)
b.p. = 80°C

$CH_3CH_2CH_2CO_2H$

butanoic acid ($M_r = 88$)
b.p. = 163°C

Esters occur naturally. Those derived from certain long-chain carboxylic acids (*fatty acids*) and the alcohol propane-1,2,3-triol (*glycerol*) are important constituents of vegetable oils and animal fats (*see* Section 19.1). Ester polymers, called **polyesters**, are very important synthetic materials; an example is the synthetic polyester fibre Terylene (*see* Section 20.3(ii)).

The following reactions are typical of esters.

10.1.1   Hydrolysis

The hydrolysis of an ester to carboxylic acid and alcohol is the reverse of the esterification process and involves an equilibrium:

$$RCOOR' + H_2O \rightleftharpoons RCOOH + R'OH$$

The composition of the equilibrium mixture is the same, no matter whether one has started from an equimolar mixture of acid and alcohol, or ester and water. Like esterification, the rate of hydrolysis is very slow but hydrolysis is catalysed by acid or alkali. *Alkaline* hydrolysis is usually the more effective method because the salt of the carboxylic acid, rather than the free acid, is obtained, so that the equilibrium position is pushed so much towards the products that reaction is virtually complete. The hydrolysis is normally effected by heating the ester with sodium hydroxide solution under reflux. As hydrolysis proceeds, the separate ester and aqueous layers gradually merge into one. If the alcohol is wanted it can be distilled out of the alkaline mixture; acidification of the residue with sulphuric acid will liberate the free carboxylic acid, which may then be distilled from the solution. The alkaline hydrolysis of esters is sometimes referred to as **saponification** because the reaction is used in soap-making (*see* Section 19.4).

The mechanism of alkaline hydrolysis involves nucleophilic attack on the ester; the final step is irreversible and hence the reaction goes to completion:

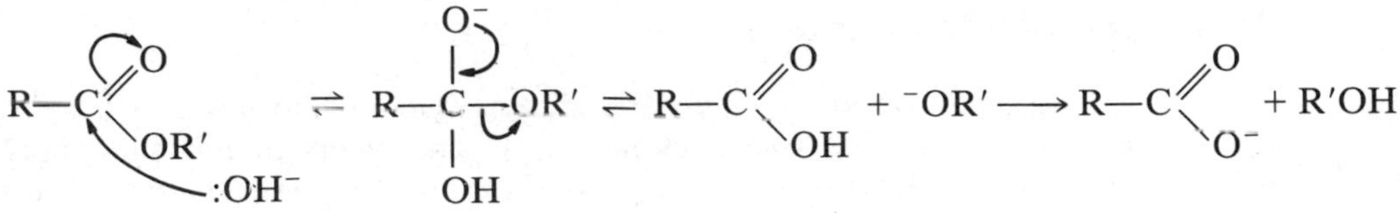

10.1.2   Reduction

Esters are reduced by lithium tetrahydridoaluminate to primary alcohols corresponding to the acid from which the ester is derived. The other organic product is the alcohol corresponding to the alcohol component of the ester:

$$RCOOR' \rightarrow RCH_2OH + R'OH$$

e.g.

$$CH_3COOCH_3 \rightarrow CH_3CH_2OH + CH_3OH$$

methyl ethanoate   ethanol   methanol

$$CH_3COOCH_2CH_3 \rightarrow 2CH_3CH_2OH$$

ethyl ethanoate   ethanol

10.1.3   Uses of esters

Esters are widely used as solvents and as plasticizers (*see* Section 20.2.1(iv)). Several are important as flavouring essences. For example:

3-methylbutyl ethanoate	$CH_3CO_2CH_2CH_2CHCH_3$ $\quad\quad\quad\quad\quad\quad\quad\mid$ $\quad\quad\quad\quad\quad\quad\quad CH_3$	pear-drop
3-methylbutyl propanoate	$CH_3CH_2CO_2CH_2CH_2CHCH_3$ $\quad\quad\quad\quad\quad\quad\quad\quad\quad\mid$ $\quad\quad\quad\quad\quad\quad\quad\quad\quad CH_3$	apricot/ plum
methyl butanoate	$CH_3CH_2CH_2CO_2CH_3$	apple
ethyl butanoate	$CH_3CH_2CH_2CO_2CH_2CH_3$	pineapple
3-methylbutyl butanoate	$CH_3CH_2CH_2CO_2CH_2CH_2CHCH_3$ $\quad\quad\quad\quad\quad\quad\quad\quad\quad\quad\quad\mid$ $\quad\quad\quad\quad\quad\quad\quad\quad\quad\quad\quad CH_3$	plum

Dimethyl benzene-1,4-dicarboxylate is an important ester used in the production of the synthetic polyester fibre Terylene (*see* Section 20.3(ii)).

10.2 Chemistry of acid chlorides

Acid chlorides are very reactive compounds and hence are not found in nature. They are conveniently prepared by the action of sulphur dichloride oxide (*thionyl chloride*) on a carboxylic acid (*see* Section 9.3.3(i)); since the other products of the reaction are gases, the acid chloride is not always isolated but is sometimes used *in situ*. Acid chlorides are generally volatile liquids and find applications in many small-scale laboratory syntheses. The following reactions, in which the chlorine is replaced by other groups, are typical of acid chlorides.

10.2.1 Hydrolysis

Acid chlorides are vigorously hydrolysed on contact with water, heat and fumes of hydrogen chloride being evolved. The acid chlorides even fume on contact with moist air, and hence must always be protected from moisture during storage. For the same reason, reactions involving acid chlorides must be carried out with water-free reagents, and if a solvent is used, it must also be dry. Care must always be exercised when handling them.

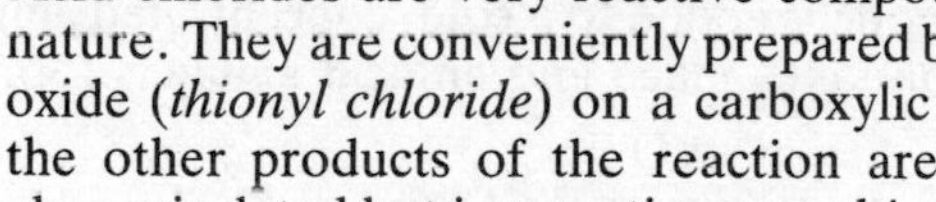

$$RC \overset{\displaystyle O}{\underset{\displaystyle Cl}{\big\backslash}} + H_2O \longrightarrow RC \overset{\displaystyle O}{\underset{\displaystyle OH}{\big\backslash}} + HCl$$

e.g.

$$CH_3COCl + H_2O \longrightarrow CH_3COOH + HCl$$

ethanoyl chloride   ethanoic acid
b.p. = 55°C   b.p. = 118°C

The acid chlorides are named according to the corresponding carboxylic acid, the suffix *-oic* being replaced by *-oyl*; thus ethan*oic* acid forms the acid chloride, ethan*oyl* chloride.

10.2.2 Reaction with alcohols

Acid chlorides react with alcohols, although slightly less vigorously than with water, to form esters:

$$RCOCl + R'OH \rightarrow RCOOR' + HCl$$

e.g.

$$CH_3COCl + CH_3CH_2OH \rightarrow CH_3COOCH_2CH_3 + HCl$$

ethyl ethanoate
b.p. = 77°C

This is a particularly useful method of making esters because the reaction goes to completion (cf. Section 9.3.2), and hence high yields may be expected.

10.2.3 Reaction with ammonia and amines

When an acid chloride is added to a cold, concentrated ammonia solution, a vigorous reaction takes place and an **acid amide** is formed:

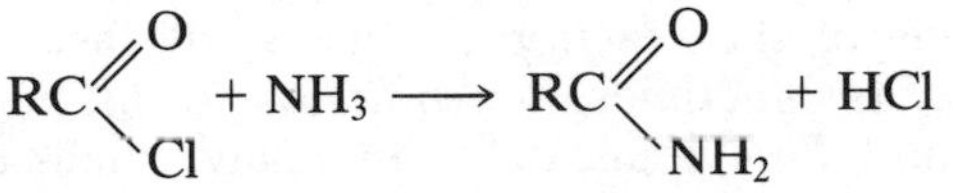

$$RC\underset{Cl}{\overset{O}{<}} + NH_3 \longrightarrow RC\underset{NH_2}{\overset{O}{<}} + HCl$$

e.g.

$$CH_3COCl + NH_3 \rightarrow CH_3CONH_2 + HCl$$

ethanoyl chloride   ethanamide
b.p. = 55°C   m.p. = 82°C

A similar reaction occurs with an amine and the reaction is often performed in a dry ether solvent to moderate its vigour. The product is an *N-substituted amide*, i.e. an amide in which hydrogen attached to the nitrogen atom has been replaced by an alkyl group:

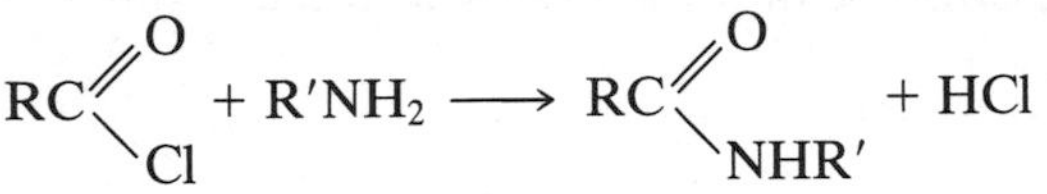

$$RC\underset{Cl}{\overset{O}{<}} + R'NH_2 \longrightarrow RC\underset{NHR'}{\overset{O}{<}} + HCl$$

e.g.

$$CH_3COCl + CH_3CH_2NH_2 \rightarrow CH_3CONHCH_2CH_3 + HCl$$

ethylamine   N-ethylethanamide
b.p. = 19°C   b.p. = 205°C

A secondary amine will also react in a similar fashion but a tertiary amine will not (*see* Section 11.3.2(iii)).

The reaction of an acid chloride with an amine is used commercially in the manufacture of the very important range of semi-synthetic penicillins, first produced by The Beecham Group in 1959. By varying the acid chloride, various groups can be introduced into the essential penicillin structure to produce new penicillin antibiotics with important pharmacological properties not possessed by natural penicillin.

10.2.4 Reaction with carboxylic acid salts

The addition of an acid chloride to the powdered, anhydrous sodium salt of a carboxylic acid produces an exothermic reaction in which an **acid anhydride** is formed (*see* Section 10.3):

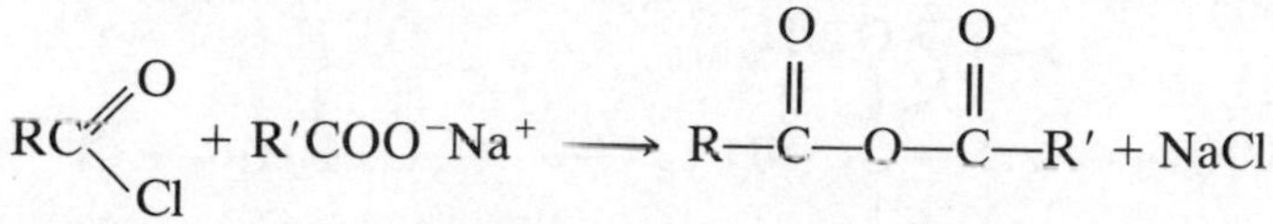

$$RC{\overset{O}{\underset{Cl}{\diagdown}}} + R'COO^-Na^+ \longrightarrow R-\overset{\overset{\displaystyle O}{\|}}{C}-O-\overset{\overset{\displaystyle O}{\|}}{C}-R' + NaCl$$

e.g.

$$CH_3COCl + CH_3COONa \rightarrow CH_3COOCOCH_3 + NaCl$$

ethanoic anhydride
b.p. = 137°C

Symmetrical and unsymmetrical anhydrides can obviously be made by this method.

10.2.5 Reduction

Acid chlorides can be reduced by catalytic hydrogenation or, more conveniently in the laboratory, by lithium tetrahydridoaluminate. The reduction product is a primary alcohol (*see also* reduction of carboxylic acids and esters, Section 9.3.4):

$$RCOCl \rightarrow RCH_2OH$$

10.2.6 Friedel–Crafts reaction

Acid chlorides enter into a substitution reaction with *aromatic* compounds to form ketones. This reaction is known as the Friedel–Crafts reaction and is discussed in Section 12.4.2(iv)).

10.2.7 Mechanism of reactions involving acid chlorides

In acid chlorides, the presence of the electronegative (electron-withdrawing) chlorine atom bonded to the carbon atom of the carbonyl group increases the electrophilic character of that carbon atom, and therefore makes it prone to nucleophilic attack. Following this, the chlorine atom is usually displaced as the chloride ion, with the result that, overall, the chlorine has been replaced by the nucleophile. Experimental evidence suggests that the reaction involves essentially two steps, i.e. addition followed by elimination, and not the direct, concerted displacement of chlorine by the nucleophile:

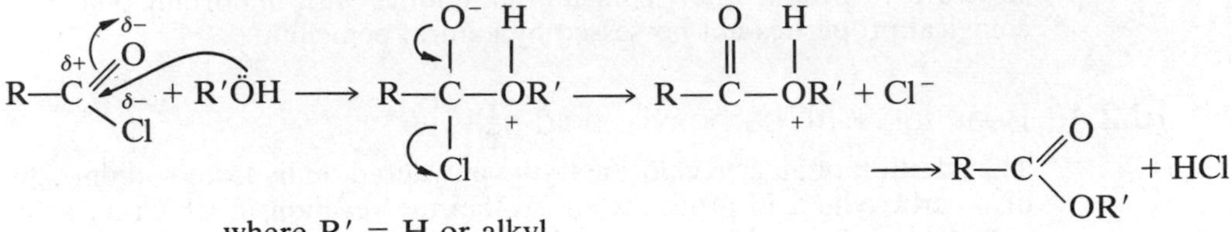

$$R-C\overset{\delta-}{\underset{Cl}{\overset{O}{\diagdown}}} + R'\ddot{O}H \longrightarrow R-\underset{\underset{Cl}{|}}{\overset{\overset{O^-}{|}}{C}}-\underset{+}{\overset{H}{\underset{|}{O}R'}} \longrightarrow R-\underset{+}{\overset{\overset{O}{\|}}{C}}-\overset{H}{\underset{|}{O}}R' + Cl^-$$

$$\longrightarrow R-C\overset{O}{\underset{OR'}{\diagup}} + HCl$$

where R' = H or alkyl

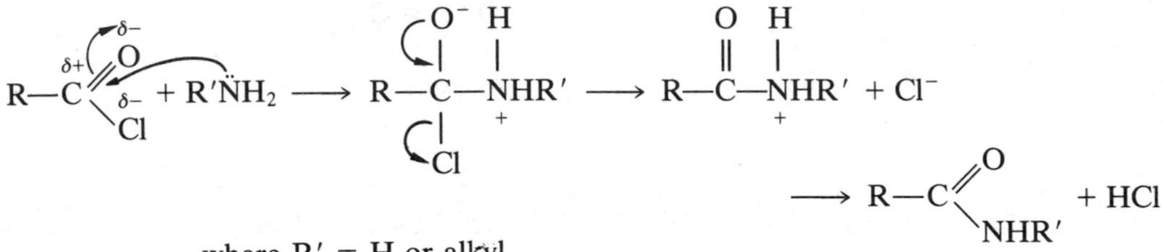

$$R-C\overset{\delta-}{\underset{Cl}{\overset{O}{\diagdown}}} + R'\ddot{N}H_2 \longrightarrow R-\underset{\underset{Cl}{|}}{\overset{\overset{O^-}{|}}{C}}-\underset{+}{\overset{H}{\underset{|}{N}HR'}} \longrightarrow R-\underset{+}{\overset{\overset{O}{\|}}{C}}-\overset{H}{\underset{|}{N}}HR' + Cl^-$$

$$\longrightarrow R-C\overset{O}{\underset{NHR'}{\diagup}} + HCl$$

where R' = H or alkyl

$$R-C\overset{\delta+}{\underset{Cl}{\overset{O}{\diagdown}}} + R'COO^- \longrightarrow R-\underset{\underset{Cl}{|}}{\overset{\overset{O^-}{|}}{C}}-OCOR' \longrightarrow R-C\overset{O}{\underset{OCOR'}{\diagup}} + Cl^-$$

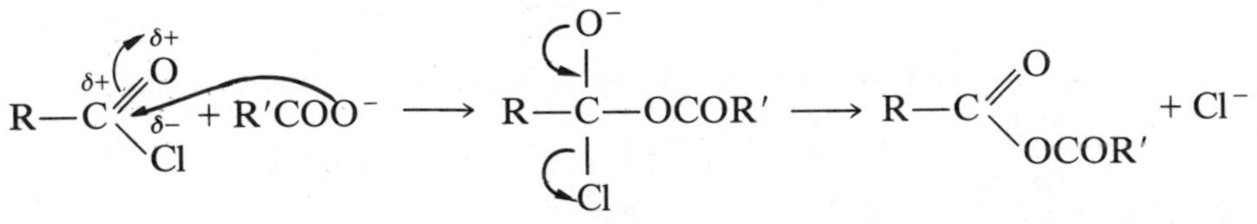

10.3 Chemistry of acid anhydrides

Acid anhydrides are derived from the elimination of water from 2 molecules of carboxylic acids. They are usually liquids with boiling points higher than those of other derivatives, such as esters, of comparable relative molecular mass. Ethanoic anhydride (anhydrides are simply named from the acids from which they are derived) is a colourless liquid with a boiling point of 137°C. The anhydrides resemble the acid chlorides in chemical behaviour, but they are not so reactive and are therefore often more convenient to use in the laboratory.

10.3.1 Hydrolysis

Anhydrides are slowly hydrolysed by water and dissolve rapidly in warm water and in alkaline solutions:

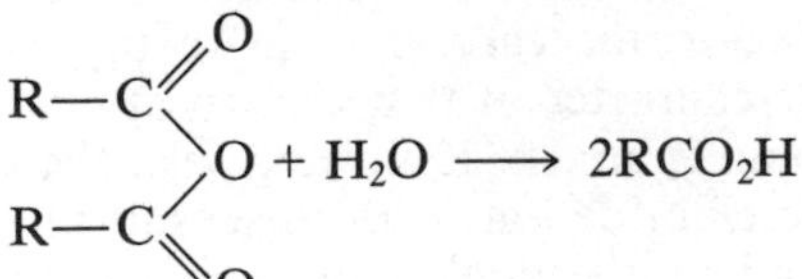

$$\begin{matrix} R-C\overset{O}{\diagdown} \\ \qquad\;\; O \\ R-C\diagdown \\ \qquad\;\; O \end{matrix} + H_2O \longrightarrow 2RCO_2H$$

The addition of ethanoic anhydride to ethanoic acid is a convenient way of removing any traces of water from the acid; the product of the reaction is, of course, ethanoic acid.

10.3.2 Reaction with alcohols

Anhydrides react with alcohols to form an ester and the free carboxylic acid:

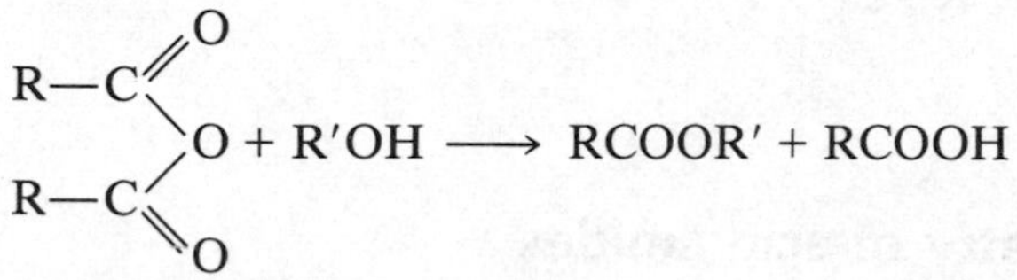

$$
\begin{array}{c}
R-C{\overset{\displaystyle O}{\big\backslash}} \\
\quad\quad\;\; O + R'OH \longrightarrow RCOOR' + RCOOH \\
R-C{\underset{\displaystyle O}{\big\diagup}}
\end{array}
$$

10.3.3 Reaction with ammonia and amines

Amides are formed on reaction of anhydrides with ammonia or amines:

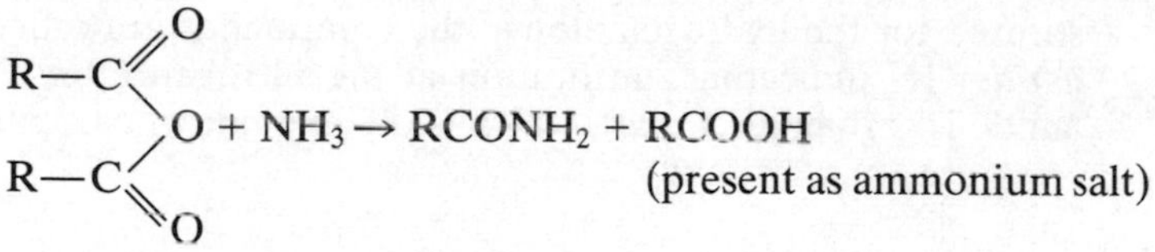

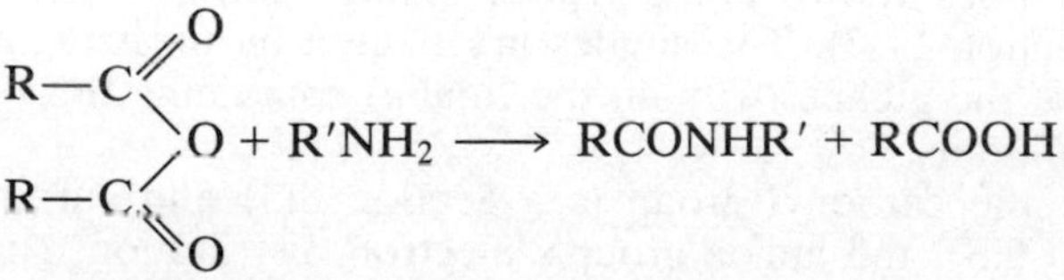

$$
\begin{array}{c}
R-C{\overset{\displaystyle O}{\big\backslash}} \\
\quad\quad\;\; O + NH_3 \rightarrow RCONH_2 + RCOOH \\
R-C{\underset{\displaystyle O}{\big\diagup}}
\end{array}
$$
(present as ammonium salt)

$$
\begin{array}{c}
R-C{\overset{\displaystyle O}{\big\backslash}} \\
\quad\quad\;\; O + R'NH_2 \longrightarrow RCONHR' + RCOOH \\
R-C{\underset{\displaystyle O}{\big\diagup}}
\end{array}
$$

Both ethanoyl chloride and ethanoic anhydride introduce the **ethanoyl** group CH_3CO- into the compounds with which they react, and can be described as **ethanoylating agents** (they are still commonly referred to as *acetylating* agents, the older name for the CH_3CO- group being *acetyl*). The reaction with amines can be particularly important in synthetic work, when amines can be *protected* by an ethanoyl group (hence being temporarily converted into a N-substituted amide) during a particular reaction (*see* Section 14.2.3). The protecting ethanoyl group is subsequently removed by hydrolysis (*see* Section 10.4.1).

Ethanoic anhydride is manufactured commercially, mainly by the oxidation of ethanal using a copper/cobalt ethanoates catalyst:

$$2CH_3CHO + O_2 \rightarrow (CH_3CO)_2O + H_2O$$

10.3.4   Mechanism of reactions involving acid anhydrides

The reactions are mechanistically very similar to those of the acid chlorides, a carboxylate anion being displaced rather than a chloride ion, e.g.

$$R—C\overset{O}{\underset{O}{\diagdown}} \quad R—C\overset{O}{\diagdown} \quad + H_2\ddot{O} \longrightarrow R—\overset{O^-}{\underset{\underset{R—C\diagdown}{O}}{\overset{|}{C}}}—\overset{+}{O}H_2 \longrightarrow R—C\overset{O}{\underset{\overset{+}{O}H_2}{\diagdown}} + \quad + RCOO^- \longrightarrow 2RCOOH$$

10.4   Chemistry of acid amides

With the exception of methanamide $HC\overset{O}{\underset{NH_2}{\diagup}}$ (b.p. $= 193°C$) and its simple N-alkyl derivatives, most simple amides are white, crystalline solids. They form a homologous series containing the **amide** group —$CONH_2$ as functional group. When one or two alkyl groups are substituted for the hydrogen atoms, the compounds are called *N-substituted amides* (N indicating substitution at the nitrogen atom). An important amide is carbamide (*urea*), which is the *di*amide of carbonic acid, namely

$$H_2N—\overset{\overset{\textstyle O}{\|}}{C}—NH_2$$ (the *mono*amide is not known). The linkage —CONH— is sometimes known as the **peptide** linkage and is present in all proteins (*see* Section 17.7). The amides are named by prefixing *amide* with the name of the alkane (without the final *e*) containing the same number of carbon atoms.

Like the carbonyl group (*see* Section 8.3) and carboxyl group (*see* Section 9.3), the amide group's electron distribution is difficult to represent accurately by a single structure, and is best regarded as lying somewhere between the following two extreme forms (**canonical forms**), i.e. as a **resonance hybrid**:

$$—\overset{\overset{\textstyle O}{\|}}{C}—NH_2 \longleftrightarrow —\overset{\overset{\textstyle O^-}{|}}{C}=\overset{+}{N}H_2 \quad \text{namely} \quad —\overset{\overset{\textstyle O}{|}}{C}—NH_2$$

The lone-pair of electrons on the nitrogen atom is **delocalized** over three atoms, and hence is not readily available for sharing with other atoms. Consequently, most amides are extremely weak bases, unlike amines R—NH_2 which are decidedly basic (*see* Section 11.3.1). It also follows that amides display little carbonyl group reactivity (cf. carboxylic acids, Section 9.3). Carbamide is exceptional in some of its basic properties, as, for example, when it reacts with concentrated nitric acid in aqueous solution to form a white precipitate of carbamide nitrate:

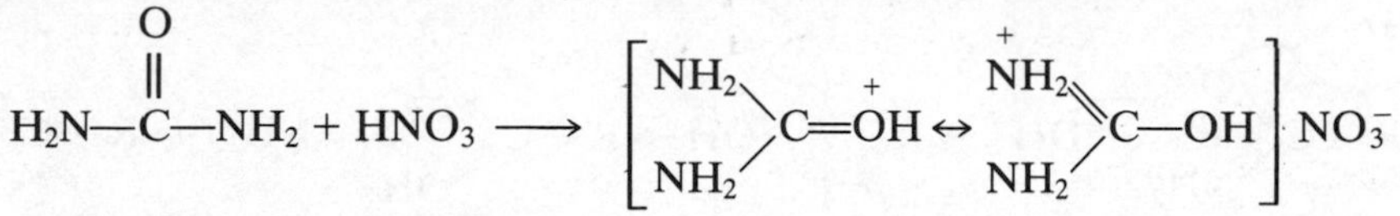

$$H_2N-\overset{\overset{\displaystyle O}{\|}}{C}-NH_2 + HNO_3 \longrightarrow \left[\begin{array}{c} NH_2 \\ NH_2 \end{array} \!\!\!\! C=\overset{+}{O}H \leftrightarrow \begin{array}{c} \overset{+}{N}H_2 \\ NH_2 \end{array} \!\!\!\! C-OH \right] NO_3^-$$

carbamide
m.p. = 133°C

The **polyamides** constitute an important class of synthetic fibres, e.g. nylon (*see* Section 20.3(i)).

10.4.1   Hydrolysis

Like other acid derivatives, acid amides can be hydrolysed. Hydrolysis is usually slow but is catalysed by acid or base. When a simple amide is treated with an alkali, then ammonia is rapidly evolved on *warming* the mixture:

$$RCONH_2 + OH^- \rightarrow RCOO^- + NH_3$$

e.g.

$$CH_3CONH_2 + OH^- \rightarrow CH_3COO^- + NH_3$$

ethanamide
m.p. = 82°C

(A substituted amide RCONHR' yields an amine R'NH$_2$ instead of ammonia.) In contrast, on treatment with alkali, ammonium salts evolve ammonia *in the cold*, without warming; thus, ammonium ethanoate can be readily distinguished from ethanamide, for example, by this test.

The hydrolysis of carbamide occurs in the soil and the ammonia obtained is then oxidized and utilized by plants. The reaction is catalysed by the enzyme *urease*:

$$H_2NCONH_2 + 2OH^- \rightarrow 2NH_3 + CO_3^{2-}$$

Nitrous acid also reacts with the NH$_2$ group of amides (cf. reaction with amines, Section 11.3.2(iv)), resulting in hydrolysis of the amide under mild conditions (i.e. room temperature). Nitrogen is evolved:

$$RCONH_2 + HNO_2 \rightarrow RCOOH + N_2 + 2H_2O$$

e.g.

$$H_2NCONH_2 + HNO_2 \rightarrow CO_2 + 2N_2 + 3H_2O$$

Alkaline hydrolysis of amides occurs via nucleophilic attack of hydroxide ion on the carbon atom of the amide group:

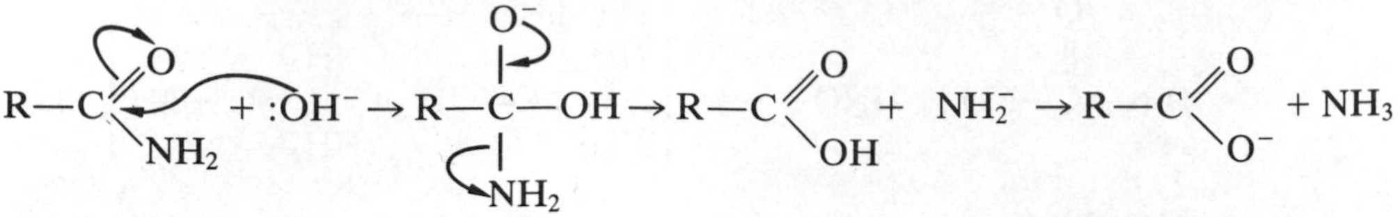

$$R-C{\overset{\displaystyle O}{\underset{\displaystyle NH_2}{\diagup}}} \; + \; :\!OH^- \to R-\overset{\displaystyle O^-}{\underset{\displaystyle NH_2}{\overset{\mid}{\underset{\mid}{C}}}}-OH \to R-C{\overset{\displaystyle O}{\underset{\displaystyle OH}{\diagup}}} \; + \; NH_2^- \to R-C{\overset{\displaystyle O}{\underset{\displaystyle O^-}{\diagup}}} \; + \; NH_3$$

10.4.2 Dehydration

Amides can be dehydrated to nitriles by distillation from phosphorus(V) oxide or from ethanoic anhydride:

$$RCONH_2 \xrightarrow{-H_2O} R-C{\equiv}N$$

e.g.

$$CH_3CONH_2 \to CH_3CN$$

ethanamide	ethanonitrile
m.p. = 82°C	b.p. = 82°C

This type of reaction has been used commercially for the manufacture of an important intermediate in the production of one type of nylon (*see* Section 20.3(i)). Hexanedioic acid is heated with excess ammonia, forming a diamide which is dehydrated in the presence of a boron phosphate catalyst:

$$HO_2CCH_2CH_2CH_2CH_2CO_2H \xrightarrow[\substack{\text{catalyst}\\350\text{--}450°C}]{NH_3} (H_2NOCCH_2CH_2CH_2CH_2CONH_2) \longrightarrow$$

$$NCCH_2CH_2CH_2CH_2CN$$

10.4.3 Reduction

Amides are reduced by lithium tetrahydridoaluminate under the usual conditions to give primary **amines**:

$$RCONH_2 \to RCH_2NH_2$$

N-substituted amides yield secondary or tertiary amines on reduction, e.g.

$$RCONHR' \to RCH_2NHR'$$

In reduction, the —NH$_2$ group is *not* displaced because it is a poor leaving group; this contrasts with the displacement of the alkoxy group during reduction of an ester (*see* Section 10.1.2):

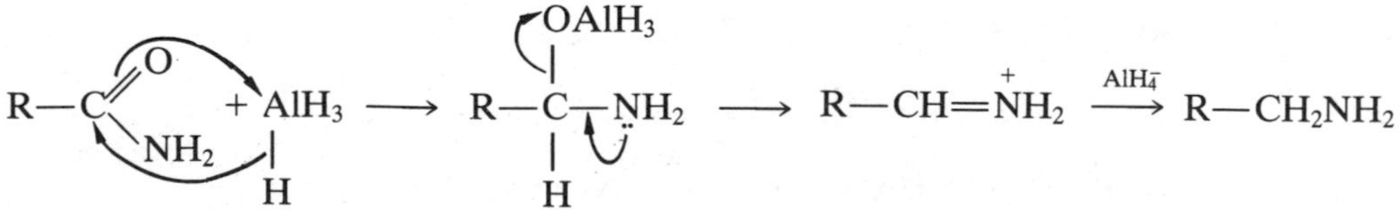

$$R-C{\overset{\displaystyle O}{\underset{\displaystyle NH_2}{\diagup}}} \; + \; AlH_3 \underset{H}{\overset{\mid}{\longrightarrow}} R-\overset{\displaystyle \overset{-}{O}AlH_3}{\underset{\overset{\mid}{H}}{\overset{\mid}{\underset{\mid}{C}}}}-NH_2 \longrightarrow R-CH{=}\overset{+}{N}H_2 \xrightarrow{AlH_4^-} R-CH_2NH_2$$

10.4.4   Hofmann rearrangement

The reaction involves the conversion of an amide into an amine, with the loss of one carbon atom, by the action of bromine and aqueous potassium hydroxide:

$$RCONH_2 + Br_2 + 4OH^- \longrightarrow RNH_2 + CO_3^{2-} + 2Br^- + 2H_2O$$

This reaction was discovered by the German chemist A. W. von Hofmann* in 1881 and is clearly a **rearrangement** reaction (*see also* Section 1.6), because the R alkyl group is initially bonded to carbon but finally becomes bonded to nitrogen. Not very many such reactions are encountered at an elementary level, although they are, in fact, very common in organic chemistry.

The reaction involves several steps. First, bromination of the amide at the nitrogen atom takes place when bromine is added to the solid amide and alkali is then added until the mixture becomes pale yellow:

$$RCONH_2 + Br_2 + OH^- \rightarrow RCONHBr + Br^- + H_2O$$

This solution is then added gradually to a hot solution of alkali. A proton is removed from the nitrogen atom by the alkali, and this step is followed by the loss of bromide ion. The resulting species rearranges to an **isocyanate**:

$$RCONHBr + OH^- \xrightarrow{70°C} R-\overset{\|}{\underset{O}{C}}-\ddot{N}: + Br^- + H_2O \longrightarrow R-N=C=O$$

isocyanate

Finally, the isocyanate is hydrolysed by the alkali to the amine:

$$R-N=C=O + H_2O \xrightarrow{OH^-} R-NH_2 + CO_2$$

(absorbed by alkali to give CO_3^{2-})

Carbamide also takes part in such a reaction; the initial product, hydrazine N_2H_4, is immediately oxidized to nitrogen. The reaction is conveniently performed using a solution of sodium chlorate(I) (*sodium hypochlorite*):

$$H_2NCONH_2 + 3OCl^- + 2OH^- \rightarrow N_2 + 3Cl^- + CO_3^{2-} + 3H_2O$$

The reaction is quantitative and provides a means of estimating the amount of carbamide in a given solution (e.g. in urine) by measurement of the volume of nitrogen evolved.

* Hofmann's name is associated with three reactions discussed in this book, namely the ammonolysis of halogenoalkanes (*see* Section 5.3.1(iv)), discovered in 1849, the rearrangement of acid amides (sometimes known as the *Hofmann degradation* of amides) and the elimination reaction of quaternary ammonium hydroxides (*see* Section 11.3.3).

Thus, the acid amides are generally much less reactive than acid chlorides or anhydrides. An important amide, industrially, is carbamide, which is manufactured by heating together carbon dioxide and excess ammonia at about 200°C and 20 MN m^{-2} pressure:

$$CO_2 + 2NH_3 \rightarrow H_2NCONH_2 + H_2O$$

It is widely used as a fertilizer and for making thermosetting plastics and resins (*see* Section 20.2.2(i)).

10.5 Chemistry of nitriles

The functional group of the homologous series of **nitriles** is the **cyano** group —CN. The nitriles are named by adding *nitrile* to the name of the alkane (substituting *o* for the final *e*) containing the same number of carbon atoms. The simplest nitrile is thus ethanonitrile CH_3CN which is a colourless liquid (b.p. = 82°C), readily soluble is water. It is widely used as a solvent. The higher members of the series are only slightly soluble in water.

In the series of **isonitriles**, the functional group is —NC, i.e. it is the nitrogen atom, not the carbon atom, which is attached to the alkyl group. Isonitriles have extremely objectionable odours and are *highly toxic*. The nitriles are very useful in organic synthesis since they readily undergo hydrolysis and reduction. The behaviour of the isonitriles is in marked contrast.

10.5.1 Hydrolysis

Nitriles are readily hydrolysed by acid or alkali, the reaction proceeding via an acid amide as intermediate. The product, after acidification if necessary, is a carboxylic acid, e.g.

$$RCN + 2H_2O \rightarrow RCO_2^-NH_4^+ (\xrightarrow{H^+} RCO_2H + NH_4^+)$$

In contrast, isonitriles are only hydrolysed by acid, not by alkali. The products are an amine and methanoic acid. In acid solution, the amine will be present as a salt:

$$RNC + 2H_2O \rightarrow RNH_2 + HCOOH$$

10.5.2 Reduction

Nitriles contain a carbon–nitrogen triple bond and hence can undergo reduction by catalytic hydrogenation of the triple bond. Alternatively, lithium tetrahydridoaluminate may be used as reducing agent.

$$R-C{\equiv}N + 2H_2 \rightarrow R-CH_2NH_2$$

An isonitrile yields a secondary, rather than primary, amine:

$$R-N{\equiv}C + 2H_2 \rightarrow R-NHCH_3$$

Summary

1 **Esters** can be represented by the general formula $RC\overset{\displaystyle O}{\underset{\displaystyle OR'}{\big<}}$, where R
and R' represent alkyl groups (not necessarily the same ones). They are
usually neutral liquids, immiscible with water. Their chemical properties
are summarized below.

(i) Hydrolysis
Esters are readily hydrolysed by acid or alkali but acid hydrolysis does
not go to completion as an equilibrium is established. Alkaline hydrolysis
is therefore more useful and yields an alcohol and, after acidification, a
carboxylic acid:

$$RCOOR' + H_2O \xrightarrow{\text{OH}^-} RCOOH + R'OH$$

(ii) Reduction
Reduction is conveniently carried out by lithium tetrahydridoaluminate
and the products are alcohols:

$$RCOOR' \xrightarrow{\text{LiAlH}_4} RCH_2OH + R'OH$$

Esters are widely used as solvents, as flavouring agents, in the manufacture
of soap and of margarine, and in the manufacture of synthetic polyester
fibres.

2 **Acid chlorides** can be represented by the general formula $RC\overset{\displaystyle O}{\underset{\displaystyle Cl}{\big<}}$ and
are usually highly reactive liquids. They undergo the following reactions.

(i) Hydrolysis
$$RCOCl + H_2O \rightarrow RCOOH + HCl$$

(ii) Reaction with alcohols
$$RCOCl + R'OH \rightarrow RCOOR' + HCl$$

This method of making esters is useful because it does not involve an
equilibrium reaction.

(iii) Reaction with ammonia and amines
$$RCOCl + NH_3 \rightarrow RCONH_2 + HCl$$
$$RCOCl + R'NH_2 \rightarrow RCONHR' + HCl$$

(iv) Reaction with salts of carboxylic acids

$$RCOCl + R'COO^- \rightarrow RCOOCOR' + Cl^-$$

(v) Reduction
This can be effected by catalytic hydrogenation or by lithium tetrahydrido-aluminate. The product is an alcohol:

$$RCOCl \rightarrow RCH_2OH$$

(vi) Friedel–Crafts reaction

$$C_6H_6 + RCOCl \xrightarrow{AlCl_3} C_6H_5COR + HCl$$

3 **Acid anhydrides** are derived in principle by the elimination of water from 2 molecules of carboxylic acid (which may be the same or different) and can be represented by the general formula

$$\begin{array}{ccc} O & & O \\ \| & & \| \\ R-C & -O- & C-R' \end{array}$$

where R and R' may, or may not, be identical alkyl groups. The chemical properties of acid anhydrides are very similar to those of acid chlorides but the anhydrides are less reactive. The reactions involve nucleophilic substitution.

(i) Hydrolysis

$$(RCO)_2O + H_2O \rightarrow 2RCOOH$$

(ii) Reaction with alcohols

$$(RCO)_2O + R'OH \rightarrow RCOOR' + RCOOH$$

(iii) Reaction with ammonia and amines

$$(RCO)_2O + NH_3 \rightarrow RCONH_2 + RCOOH \text{ (as ammonium salt)}$$

$$(RCO)_2O + R'NH_2 \rightarrow RCONHR' + RCOOH \text{ (as salt)}$$

4 **Acid amides** have the general formula $RC\begin{smallmatrix} \diagup O \\ \diagdown NH_2 \end{smallmatrix}$. N-Substituted amides

have the general formula $RC\begin{smallmatrix} \diagup O \\ \diagdown NHR' \end{smallmatrix}$ (or $RC\begin{smallmatrix} \diagup O \\ \diagdown NR'_2 \end{smallmatrix}$), where R and R'

are alkyl groups (not necessarily the same). Acid amides are usually neutral solids. They are much less reactive than acid chlorides or anhydrides.

(i) Hydrolysis

Amides evolve ammonia *on warming* with alkali (ammonium salts evolve ammonia on treatment with alkali *in the cold*):

$$RCONH_2 + OH^- \rightarrow RCOO^- + NH_3$$

(ii) Dehydration

$$RCONH_2 \xrightarrow{P_4O_{10}} R—CN$$

(iii) Reduction

$$RCONH_2 \xrightarrow{LiAlH_4} RCH_2NH_2$$

$$RCONHR' \rightarrow RCH_2NHR'$$

(iv) Hofmann rearrangement

$$RCONH_2 + Br_2 + 4OH^- \rightarrow RNH_2 + CO_3^{2-} + 2Br^- + 2H_2O$$

5 **Nitriles** have the general formula $R—C\equiv N$. **Isonitriles** have the general formula $R—N\overset{+}{\equiv}\overset{-}{C}$. Both are generally liquids. The reactions of the two functional groups are in marked contrast.

(i) Hydrolysis

$$RCN + 2H_2O \xrightarrow{H^+} RCO_2NH_4 (\xrightarrow{H^+} RCO_2H + NH_4^+)$$

$$RNC + 2H_2O \xrightarrow{H^+} RNH_2 + HCOOH$$

(ii) Reduction

$$RCN \xrightarrow{LiAlH_4} RCH_2NH_2$$

$$RNC \rightarrow RNHCH_3$$

Questions

1 (a) Show by means of equations how ethanoyl (*acetyl*) chloride, CH_3COCl, reacts with each of the following reagents and name the principal organic product in each case:
   (i) Concentrated aqueous ammonia
   (ii) Ethanol
   (iii) Phenylamine (*aniline*)
   (b) Each of the above reactions proceeds by a similar mechanism. How may the role of ammonia, ethanol or phenylamine molecules in this mechanism be described?

(c) How may phenylamine be regenerated from the product of its reaction with ethanoyl chloride?

(d) What use is sometimes made in preparative organic chemistry of the reaction of an amine with ethanoyl chloride?

JMB

2　(a) Name the reagents required to prepare ethanoyl (*acetyl*) chloride, state the conditions required and give the equation for the reaction.

(b) In each case, state the type of reaction and outline the mechanism by which ethanoyl chloride reacts with (i) water, (ii) benzene.

(c) Calculate the volume of 2M (2N) NaOH which is required to neutralize the products of the action of water on 0.05 mol ethanoyl chloride, naming the indicator you would use in the titration.

JMB

3*　(a) Consider the reaction:

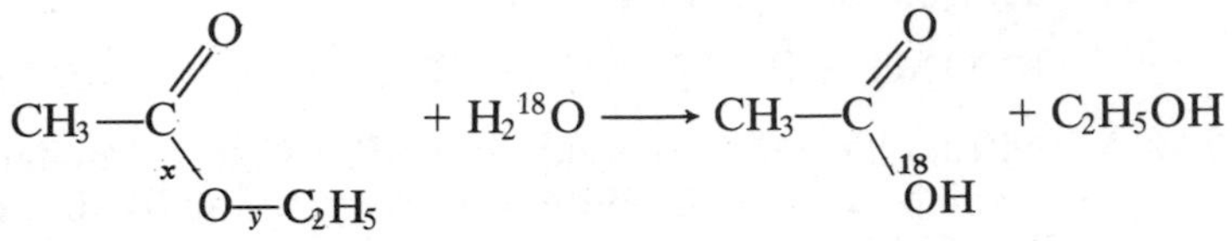

$$CH_3-C{\overset{O}{\underset{\underset{x}{O-_y-C_2H_5}}{\diagdown}}} + H_2{}^{18}O \longrightarrow CH_3-C{\overset{O}{\underset{\overset{18}{OH}}{\diagdown}}} + C_2H_5OH$$

Which bond, x or y, breaks in the ester? Explain.

(b) From aqueous solutions containing $CH_3-C{\overset{O}{\underset{\overset{18}{OH}}{\diagdown}}}$ the ethanoic (*acetic*) acid which may be isolated contains equal amounts of

$$CH_3-C{\overset{{}^{18}O}{\underset{OH}{\diagdown}}} \text{ and } CH_3-C{\overset{O}{\underset{\overset{18}{OH}}{\diagdown}}}. \text{ Suggest a reason for this.}$$

(c) Predict the position of the isotopic label in the products of the following reactions:

(i) 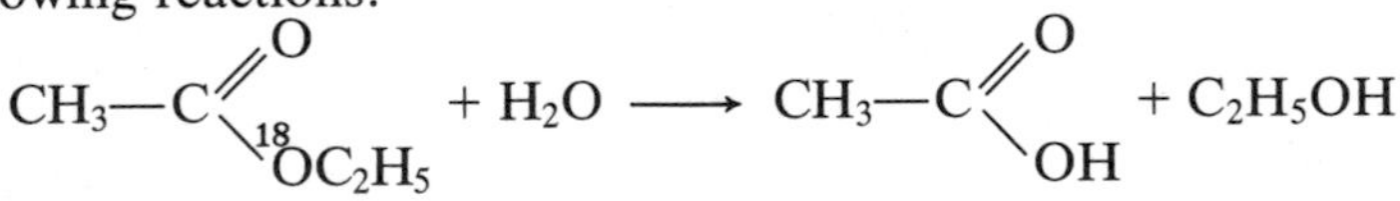 $CH_3-C{\overset{O}{\underset{\overset{18}{OC_2H_5}}{\diagdown}}} + H_2O \longrightarrow CH_3-C{\overset{O}{\underset{OH}{\diagdown}}} + C_2H_5OH$

(ii) 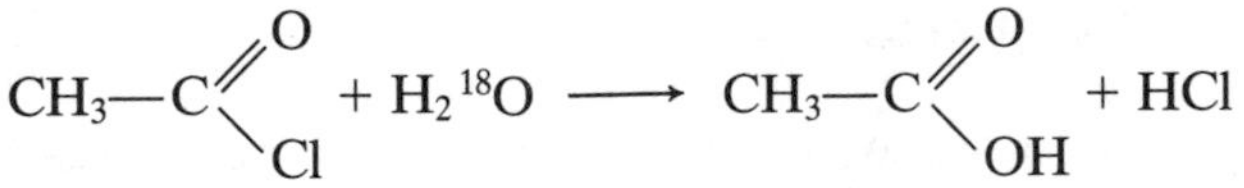 $CH_3-C{\overset{O}{\underset{Cl}{\diagdown}}} + H_2{}^{18}O \longrightarrow CH_3-C{\overset{O}{\underset{OH}{\diagdown}}} + HCl$

(d) Bearing in mind the observations referred to above, how would you attempt to synthesize $CH_3-C{\overset{O}{\underset{\overset{18}{OC_2H_5}}{\diagdown}}}$, given only a supply of ethanol (unlabelled), $H_2{}^{18}O$ and any other inorganic reagents which you require?

Oxford & Cambridge

4**   Answer *either* (a) *or* (b).
        ... (b) Compound **A**, C_4H_7NO, is a neutral liquid, stable to sodium
        hydroxide solution but easily hydrolysed by aqueous acid to give **B**,
        C_3H_9NO. On treatment with nitrous acid, **B** evolves nitrogen giving **C**,
        $C_3H_8O_2$, which, with ethanoic (acetic) anhydride, gives **D**, $C_5H_{10}O_3$.
        Catalytic hydrogenation of **A** gives a base **E**, $C_4H_{11}NO$, which does not
        evolve nitrogen when treated with nitrous acid but instead forms a water-
        insoluble liquid.
        On heating at 200°C, compound **A** forms an isomeric neutral product **F**
        which may be hydrolysed by acid or base to compound **G**, $C_4H_8O_3$.
        Suggest a structure for each of the compounds **A** to **G**.

                                                                        Cambridge

5*   The equation for a reaction by which ethyl benzoate may be prepared is

        $$C_6H_5CO_2H + C_2H_5OH \rightleftharpoons C_6H_5CO_2C_2H_5 + H_2O$$

        20 g of benzoic acid and 16 g of dry ethanol are placed in a $100\,cm^3$ round-
        bottomed flask fitted with a water-cooled reflux condenser which has a
        guard-tube containing anhydrous calcium chloride. The apparatus is in
        a fume cupboard. The flask is heated on a sand-tray so that the mixture
        boils gently. A brisk current of dry hydrogen chloride is passed into the
        reaction mixture until the boiling solution is saturated and then the flow
        is adjusted until a gentle stream of bubbles is passing through the flask.
        After refluxing for $1\frac{1}{2}$ h, the flask and its contents are cooled and poured
        into a separating funnel containing about $200\,cm^3$ of water. About $10\,cm^3$
        of tetrachloromethane are added and the mixture is shaken vigorously:
        on standing, the solution of ethyl benzoate in tetrachloromethane sep-
        arates sharply and rapidly. The ethyl benzoate layer is separated and
        transferred to another funnel. It is shaken with a dilute solution of sodium
        carbonate until all the free acid is removed. The ethyl benzoate layer is
        dried using lumps of anhydrous calcium chloride. The mixture is filtered
        into a dry $50\,cm^3$ distillation flask fitted with a 360°C thermometer and
        an air-cooled condenser. The tetrachloromethane is distilled by direct
        heating using a wire gauze, and then the ethyl benzoate, collecting the
        fraction boiling at 210–214°C.

	Ethanol	Benzoic acid	Ethyl benzoate	Tetrachloro-methane
molar mass (g mol^{-1})	46	122	150	—
density (g cm^{-3})	0.79	—	1.05	1.59
boiling point (°C)	—	—	213	77

(a) What is the most convenient way of measuring out the 16 g of ethanol?
(b) What is the principal difference between a solution of hydrogen
     chloride in ethanol and a solution of hydrogen chloride in water?
(c) What are the functions of the hydrogen chloride in this preparation?
(d) What reagents would you use to prepare dry hydrogen chloride?

(e) Calcium metal may be used for drying ethanol but not sodium metal because sodium reacts with alcohols. Write the equation for the reaction between sodium and ethanol.

(f) Draw a diagram of the apparatus described at the beginning of the preparation. (Details of the preparation of the hydrogen chloride are not required.)

(g) Ethyl benzoate is almost insoluble in water, so why are $10 \, cm^3$ of tetrachloromethane added to promote the separation of the ester from the large volume of water?

(h) How will you know when all the free acid is removed?

(i) If the final yield is 15.4 g of ethyl benzoate, what is the percentage yield?

                                                                    Nuffield

6*   Two isomers, A (boiling point 54°C) and B (boiling point 57°C), are both neutral liquids containing 48.65% carbon, 8.11% hydrogen and 43.24% oxygen (by mass in each case). Both dissolve when boiled with aqueous sodium hydroxide and a colourless liquid can be distilled from either solution. The distillate from A gives a yellow precipitate with iodine and sodium hydroxide whereas that from B shows no reaction with that particular reagent. The remaining alkaline solution obtained from A reduces potassium permanganate but that from B does not. After acidification with sulphuric acid an acidic liquid can be distilled from both solutions.
Suggest structures for A and B and explain all the reactions described above. Write *two* other structures corresponding to the same composition as A and B and indicate why you reject them as possibilities.

                                                                    Oxford

7*   Explain, giving equations and *essential* experimental details, how you would prepare a reasonably pure sample of ethanoic (*acetic*) anhydride ($CH_3COOCOCH_3$) in the laboratory, starting from glacial (anhydrous) ethanoic (*acetic*) acid.
A mixture of 10.2 g of ethanoic (*acetic*) anhydride and 4.65 g of phenyl-amine (*aniline*) ($C_6H_5NH_2$) was warmed gently until no more reaction occurred, and then shaken for some time with an excess of water. Explain the reactions involved. How much 2M sodium hydroxide solution would be needed to neutralize the aqueous mixture produced?

                                                                    Oxford

Chapter 11
Amines

11.1 Homologous series of amines

The amines may be looked upon as organic derivatives of ammonia in which one or more hydrogen atoms have been replaced by alkyl groups. Three classes of amines can be distinguished according to the number of hydrogen atoms replaced. They are, respectively, *primary, secondary* and *tertiary*:

$$CH_3—NH_2 \qquad \begin{array}{c} CH_3 \\ \diagdown \\ CH_3 \diagup \end{array} NH \qquad \begin{array}{c} CH_3 \\ \diagdown \\ CH_3—N \\ \diagup \\ CH_3 \end{array}$$

methylamine   dimethylamine   trimethylamine
b.p. = −7°C   b.p. = 7°C   b.p. = 3°C
primary   **secondary**   **tertiary**

It must be appreciated that the use of these terms in the amine series differs from that in the alcohol or halogenoalkane series (*see* Sections 6.1 and 5.1, respectively). Propan-2-ol $CH_3CH(OH)CH_3$ is a secondary alcohol, but the corresponding amine $CH_3CH(NH_2)CH_3$ is a primary amine, because only one hydrogen atom of ammonia has been replaced by an alkyl group, i.e. the amine contains the **amino** group —NH_2. A fourth class of compound analogous to ammonium salts are called **quaternary ammonium salts**, e.g. quaternary methylammonium iodide $(CH_3)_4N^+I^-$.

The common method of naming simple amines at the present time does not generally follow the pattern established so far. These names contain no root, based on the corresponding alkane, but instead just a prefix based on the alkyl group(s), and a suffix -*amine* indicating the class of compound. This system is illustrated by the examples quoted above. Isomerism is common and may arise from within the alkyl group(s) or from the different classes of amine. The names of these isomers may follow the pattern just described (the carbon atom carrying the free valence of an alkyl group must be numbered 1, hence $(CH_3)_2CH—$ is the 1-methylethyl group), or they may, in certain cases, follow a substitutive nomenclature based on the amino group —NH$_2$. The following examples should make the matter clearer:

CH$_3$CH$_2$CH$_2$NH$_2$ CH$_3$CHCH$_3$ CH$_3$CH$_2$NHCH$_3$ (CH$_3$)$_3$N

                             |

                          NH$_2$

propylamine   1-methylethylamine   N-methylethylamine   trimethylamine
(or 1-aminopropane)   (or 2-aminopropane)   b.p. = 36°C   b.p. = 3°C
b.p. = 49°C   b.p. = 33°C

Each class of amine forms a homologous series. Some physical data are provided in *Table 11.1* for primary amines. Alternatively, all of these amines could be named on a substitutive basis as 1-aminoalkanes.

Table 11.1

Name	Formula	Melting point (°C)	Boiling point (°C)	Density (g cm^{-3})
methylamine	CH$_3$NH$_2$	−93	−7	—
ethylamine	CH$_3$CH$_2$NH$_2$	−81	17	0.689
propylamine	CH$_3$CH$_2$CH$_2$NH$_2$	−83	49	0.719
butylamine	CH$_3$CH$_2$CH$_2$CH$_2$NH$_2$	−51	76	0.740
pentylamine	CH$_3$(CH$_2$)$_3$CH$_2$NH$_2$	−55	104	0.766
hexylamine	CH$_3$(CH$_2$)$_4$CH$_2$NH$_2$	−19	130	0.766
heptylamine	CH$_3$(CH$_2$)$_5$CH$_2$NH$_2$	−18	157	0.775
octylamine	CH$_3$(CH$_2$)$_6$CH$_2$NH$_2$	0	180	0.783

The three methylamines are products of the decay of fish and have characteristic fishy odours. Salt-water fish contain trimethylamine oxide (CH$_3$)$_3$$\overset{+}{N}$—$\overset{-}{O}$ in their tissues and on putrefaction trimethylamine is produced. The diamine, 1,4-diaminobutane (*putrescine*) is a product of the decomposition of proteins in flesh by bacterial action. More complex amines also occur naturally and have important physiological or pharmacological effects. *Histamine* is produced in many allergies such as hay fever and is also a component of bee and wasp stings. Other examples include the hormone *adrenaline*, the synthetic drugs *amphetamines* which stimulate the central nervous system, and the quaternary ammonium compound *acetylcholine*, the chemical transmitter of nerve impulses. The aromatic amines (*see* Chapter 14) are important intermediates in the manufacture of certain dyes.

11.2 Physical properties of amines

The lower members are gases or volatile liquids and generally possess a strong smell of ammonia; the secondary and tertiary amines have a more fishy odour but this disappears with decreasing volatility. The lower members are generally soluble in water (intermolecular bonding occurs) and react with it to some degree (*see* Section 11.3.1). In the pure liquid state there is some association of the molecules via hydrogen bonding but since nitrogen is less electronegative than oxygen, the bonds are weaker than in alcohols. Nevertheless, the boiling points of amines are higher than alkanes of corresponding relative molecular mass (*see Figure 2.1*). In the pure state, tertiary amines cannot enter into hydrogen bonding and hence their boiling points are roughly similar to non-associated liquids of similar relative molecular mass; they are also less soluble in water. The molecular association in aqueous solution, and in the pure liquid state, is shown below for a primary amine:

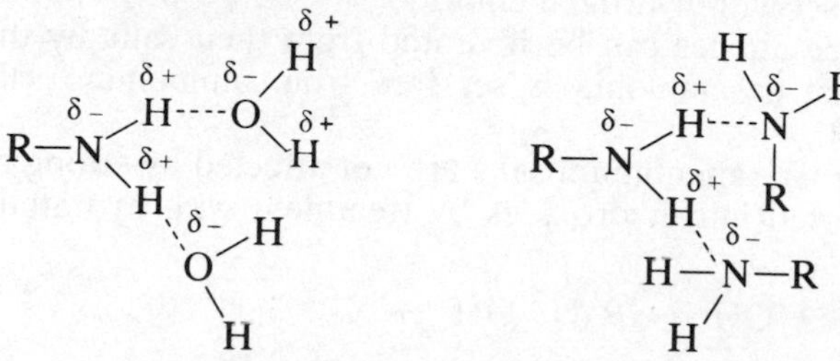

The simple volatile amines are highly flammable and should be handled with care. The ease of oxidation of amines to complex products causes discoloration of the colourless liquids on standing.

11.3 Chemical properties of amines

The lone pair of electrons on the nitrogen atom dominates the chemistry of the amines and causes them to function as Lewis bases (*see* Section 1.7) or nucleophiles.

11.3.1 Basic properties

The lower amines dissolve in water and react with it to form an alkaline solution:

$$RNH_2 + H_2O \rightleftharpoons RNH_3^+ + OH^-$$

As bases, the lower amines are slightly stronger than ammonia, a fact attributable to the slight positive inductive effect of the alkyl group(s) (*see* Section 1.4):

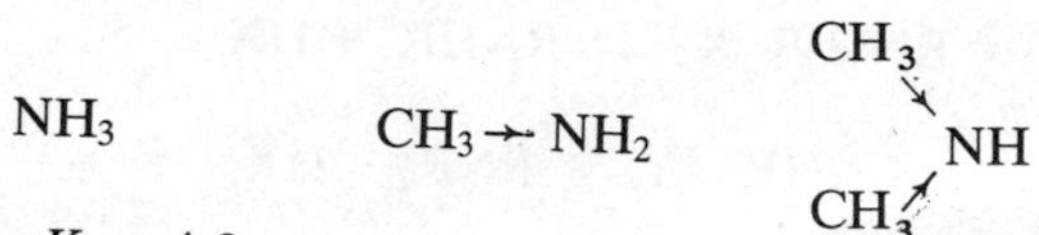

$$NH_3 \qquad CH_3 \rightarrow NH_2 \qquad \begin{array}{c} CH_3 \\ \diagdown \\ NH \\ \diagup \\ CH_3 \end{array}$$

$$pK_b = 4.8 \qquad pK_b = 3.4 \qquad pK_b = 3.2$$

Amines will also dissolve in mineral acids to form salts, which can be obtained as crystalline solids on evaporation of the water:

$$RNH_2 + HX \rightarrow RNH_3^+ X^-$$

e.g.

$$CH_3NH_2 + HCl \rightarrow CH_3NH_3^+ Cl^-$$

methylamine   methylammonium chloride
b.p. = $-7°C$ (*methylamine hydrochloride*)
                              m.p. = 225°C

The more volatile amines, like ammonia, give white fumes when brought into contact with hydrogen chloride.

The free amines can be liberated from their salts by the addition of alkali, just as ammonia is set free from ammonium salts by similar treatment.

Quaternary ammonium salts are not affected by strong bases, but can be converted into hydroxides by treatment with hydrated silver oxide, e.g.

$$R_4N^+ X^- + OH^- \rightarrow R_4N^+ OH^- + X^-$$

The reaction is more effectively carried out by using a basic ion-exchange resin whereby halide ions are replaced by hydroxide ions (*see* Section 21.2.4(v)). These quaternary ammonium hydroxides are strong bases which are completely dissociated in solution. They will absorb carbon dioxide from the atmosphere and will liberate ammonia from ammonium salts. On heating strongly, they undergo an elimination reaction (*see* Section 11.3.3), e.g.

$$(CH_3CH_2)_3\overset{+}{N}-CH_2CH_3 \xrightarrow{heat} (CH_3CH_2)_3N + CH_2{=}CH_2 + H_2O$$
$$OH^-$$

11.3.2 Nucleophilic reactions

The presence of a lone-pair of electrons on the nitrogen atom of amines permits them to act as electron-pair donors, namely nucleophiles.

(i) Reaction with halogenoalkanes
This reaction has already been discussed in Section 5.3.1(iv). Amines react with halogenoalkanes on strong heating in alcoholic solution; primary and secondary ones give a mixture of products arising from successive reactions between amine and halogenoalkane:

$$RNH_2 + R'X \rightarrow R\overset{+}{N}H_2R' X^- \rightleftharpoons RNHR' + HX$$

$$RNHR' + R'X \rightarrow R\overset{+}{N}HR_2' X^- \rightleftharpoons RNR_2' + HX$$

$$RNR_2' + R'X \rightarrow R\overset{+}{N}R_3' X^-$$

The reaction is not, therefore, very promising for synthetic purposes. The complete alkylation of an amine to the quaternary salt is, however, of importance if it is desired to carry out an elimination reaction with the amine (*see* Section 11.3.3).

(ii) Reaction with aldehydes and ketones
This reaction has been discussed in Section 8.3.3. The products are usually unstable.

(iii) Reaction with acid chlorides and anhydrides
Primary and secondary amines react readily with acid chlorides and acid anhydrides to form N-substituted amides (*see* Section 10.2.3 and Section 10.3.3):

$$RNH_2 + R'COCl \rightarrow RNHCOR' + HCl$$

$$RNH_2 + (R'CO)_2O \rightarrow RNHCOR' + R'CO_2H$$

$$R_2NH + R'COCl \rightarrow R_2NCOR' + HCl$$

Tertiary amines do not possess a hydrogen atom bonded to nitrogen and do *not* form amides with acid chlorides or anhydrides.

   This type of reaction is very useful in synthetic work because amines are very prone to oxidation but amides are not. Thus, the amine can be 'protected' temporarily against oxidation by conversion into a substituted amide, which can regenerate the free amine at a subsequent stage when hydrolysed with dilute mineral acid (*see* Section 10.4.1).

   The mechanism of nucleophilic attack in these examples follows the paths

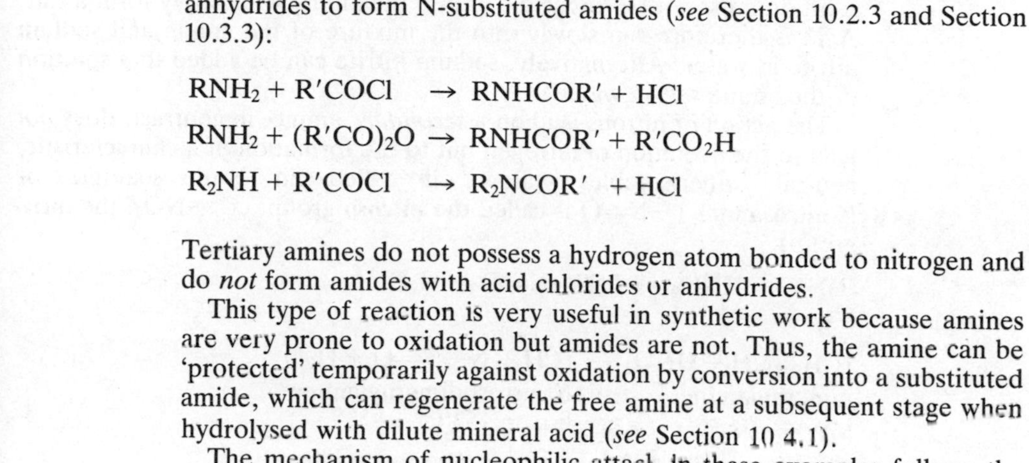

$$R\ddot{N}H_2 + R'\!-\!X \longrightarrow R\overset{+}{N}H_2R'\, X^-$$

$$R\ddot{N}H_2 + R'\!-\!C\!\!\begin{smallmatrix}O^{\delta-}\\Cl\end{smallmatrix} \longrightarrow R\overset{+}{N}H_2\!-\!\underset{R'}{\overset{O^-}{C}}\!-\!Cl \longrightarrow R\overset{+}{N}H_2COR' + Cl^- \longrightarrow RNHCOR'$$

$$+ HCl$$

$$R\ddot{N}H_2 + \begin{smallmatrix}R'\\C\!=\!O\\O\\C\!=\!O\\R'\end{smallmatrix} \longrightarrow R\overset{+}{N}H_2\!-\!\underset{R'}{\overset{O^-}{C}}\!-\!OCOR' \longrightarrow R\overset{+}{N}H_2COR' + R'COO^-$$

$$\longrightarrow RNHCOR' + R'COOH$$

(iv) Reaction with nitrous acid
Nitrous acid is thermodynamically unstable with respect to its disproportionation products and hence is always generated in the reaction

medium (*in situ*) by the addition of dilute mineral acid to sodium nitrite (*care*: poisonous). The reaction of amines with nitrous acid enables the three classes of amines to be distinguished fairly readily.

When nitrous acid reacts with a *primary* amine, nitrogen is evolved (nitrogen(II) oxide and nitrogen(IV) oxide are also formed by the spontaneous decomposition of nitrous acid; performing the reaction at low temperature helps to minimize this decomposition):

$$RNH_2 + HNO_2 \rightarrow ROH + N_2 + H_2O$$

In practice, a variety of organic products are formed and the alcohol shown in the equation above is only one of them (*see* next page). The solution must not be too acidic or else the amine will simply form a salt. Acid is therefore run slowly into the mixture of the amine and sodium nitrite in water. Alternatively, sodium nitrite can be added to a solution of the amine salt in water.

The action of nitrous acid on a *secondary* amine, in contrast, does *not* lead to the evolution of nitrogen but to the formation of a characteristic, neutral, ether-soluble, yellow oil called an N-nitrosoamine or **N-nitrosamine** (—N=O is called the **nitroso** group, cf. —NO_2, the nitro group):

$$R_2NH + HNO_2 \rightarrow R_2N—N=O + H_2O$$

e.g.

$$(CH_3)_2NH + HNO_2 \rightarrow (CH_3)_2N—N=O + H_2O$$
dimethylamine   N-nitrosodimethylamine
b.p. = 7°C   b.p. = 153°C

Once again, the solution must not be too acidic.

Little, if any, reaction is observed when a *tertiary* amine is added to nitrous acid; *no* nitrogen is evolved and no yellow oil is formed. The strongly basic amine simply dissolves in the acid solution to form a salt:

$$R_3N + HX \rightleftharpoons R_3\overset{+}{N}H\,X^-$$
(where X = Cl or NO_2)

The mechanism of the reaction of the primary and secondary amines with the nitrous acid may be envisaged as involving nucleophilic attack by the amine on the acid thus:*

$$H—O—N=O + H^+ \rightleftharpoons \underset{\overset{|}{H}}{H—\overset{+}{O}—N=O}$$

primary amine:

$$R\ddot{N}H_2 + O=N—\overset{+}{O}H_2 \longrightarrow R\overset{+}{N}H_2—N=O + H_2O \longrightarrow RNH—N=O + H_3O^+$$

* Also present in the solution may be O=N—X (where X = Cl,NO_2) or even $\overset{+}{N}O$, any of which can be attacked in a similar manner.

Further reaction occurs and an extremely unstable **diazonium** compound is produced. The diazonium compound can react in several ways to give mixed products:

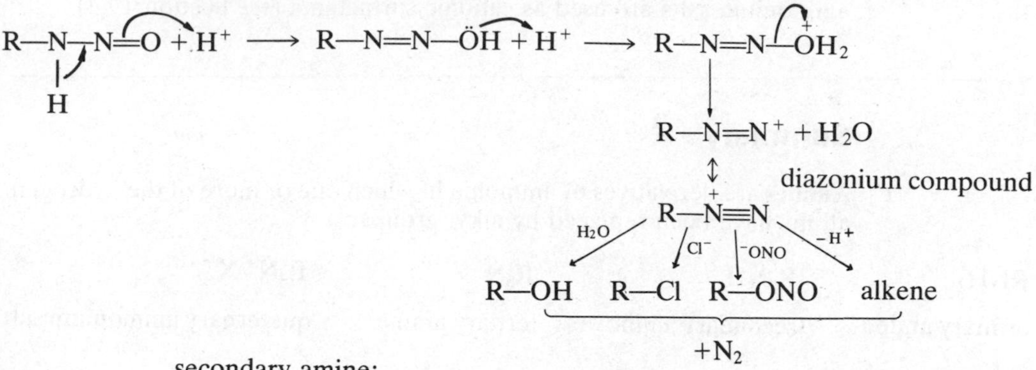

$$R—N—N{=}O + H^+ \longrightarrow R—N{=}N—\ddot{O}H + H^+ \longrightarrow R—N{=}N—\overset{+}{O}H_2$$
$$\underset{H}{|}$$

$$R—N{\equiv}N^+ + H_2O$$

$$\updownarrow \qquad \text{diazonium compound}$$

$$\overset{+}{R—N{\equiv}N}$$

$$\overset{H_2O}{\diagup} \quad \underset{Cl^-}{\Big|} \quad \underset{^-ONO}{\Big|} \quad \overset{-H^+}{\diagdown}$$

$$R—OH \quad R—Cl \quad R—ONO \quad \text{alkene}$$

$$+N_2$$

secondary amine:

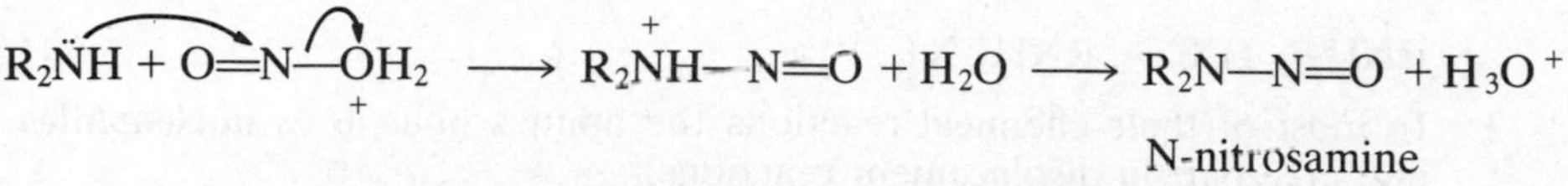

$$R_2\ddot{N}H + O{=}N—\overset{+}{O}H_2 \longrightarrow R_2\overset{+}{N}H—N{=}O + H_2O \longrightarrow R_2N—N{=}O + H_3O^+$$

N-nitrosamine

11.3.3   Elimination reactions

Amines do not undergo any direct elimination reaction but can be made to eliminate the elements of ammonia, or an amine, in a reaction discovered by the German chemist A. W. von Hofmann in 1851 (*see also* footnote to Section 10.4.4). The amine, be it primary, secondary or tertiary, is first converted into a quaternary ammonium salt, by alkylation with excess of a halogenoalkane such as iodomethane (*see* Section 11.3.2(i)):

$$RCH_2CH_2NH_2 + 3CH_3I \rightarrow RCH_2CH_2\overset{+}{N}(CH_3)_3\,I^- + 2HI$$

The salt is then converted into a quaternary ammonium hydroxide (*see* Section 11.3.1) which is heated to a moderate temperature. Elimination takes place and an alkene is formed:

$$RCH_2CH_2\overset{+}{N}(CH_3)_3\,OH^- \rightarrow RCH{=}CH_2 + (CH_3)_3N + H_2O$$

In the case of a secondary or tertiary amine, the alkyl groups attached to the nitrogen atom may not be identical and, when the quaternary ammonium hydroxide undergoes elimination, more than one alkene is produced.

11.4   Manufacture and uses of amines

Amines are usually made by the reduction of nitroalkanes or nitriles by catalytic hydrogenation, or by the reaction between a halogenoalkane and ammonia (*see* Chapter 14 for aromatic amines).

Amines are used as anti-oxidants and certain ones are important in the manufacture of nylon (*see* Section 20.3(i)). The aromatic amines are used for making certain dyes and are very useful in synthesis. Quaternary ammonium salts are used as cationic surfactants (*see* Section 19.4).

Summary

1 **Amines** are derivatives of ammonia in which one or more of the hydrogen atoms have been replaced by alkyl groups:

RNH_2 R_2NH R_3N $R_4N^+ X^-$

primary amine   secondary amine   tertiary amine   quaternary ammonium salt

2 Amines are weak organic bases and form crystalline salts with acids:

$$RNH_2 + HX \rightarrow RNH_3^+ X^-$$

3 In most of their chemical reactions the amines behave as **nucleophiles** and take part in displacement reactions:

$$RNH_2 \begin{cases} \xrightarrow{R'X} & R\overset{+}{N}H_2R' X^- \rightleftharpoons RNHR' + HX \\ \xrightarrow{R'COCl} & R'COCHR \\ \xrightarrow{(R'CO)_2O} & R'CONHR \end{cases}$$

Secondary amines react in a similar fashion but tertiary amines do *not* react with acid chlorides or anhydrides as they do not possess any hydrogen bonded to the nitrogen.

4 Each class of amine reacts characteristically with nitrous acid and the reactions form the basis of a method of distinguishing the three:

$$RNH_2 + HNO_2 \rightarrow ROH + N_2 + H_2O \text{ (plus other products)}$$
$$R_2NH + HNO_2 \rightarrow R_2N-N{=}O + H_2O$$
$$R_3N + HNO_2 \rightleftharpoons R_3\overset{+}{N}H \ NO_2^-$$

5 Quaternary ammonium hydroxides undergo **elimination** and this reaction provides a route to an alkene from an amine:

$$RCH_2CH_2NH_2 \xrightarrow[\text{(ii) Ag}_2\text{O}]{\text{(i) 3CH}_3\text{I}} RCH_2CH_2\overset{+}{N}(CH_3)_3 \ OH^- \xrightarrow{\text{heat}} RCH{=}CH_2 + (CH_3)_3N$$
$$+ H_2O$$

Questions

1   The reaction between primary amines and nitrous acid can be represented
     by the sequence

$$RNH_2 \xrightarrow[-H_2O]{NO^+} RN_2^+ \rightarrow R^+ + N_2$$

                     A   B

(a) Name the types of species represented by A and B.
(b) How do the nature of R and the reaction conditions determine whether
    or not the species A can be detected in this reaction sequence?
(c) If R is $CH_3CH_2CH_2$, give the structural formulae of **two** of the reaction
    products formed in aqueous solution.
(d) If R is C_6H_5, write equations for the reactions which take place with
    A (i) in warm dilute sulphuric acid and (ii) in aqueous potassium
    iodide. Name the principal organic product in each case.

                                                                       JMB

2*   Give the structural formula, indicating the characteristic functional group,
     of (a) a *carboxylic acid amide*, (b) a *primary aliphatic amine*, and (c) a
     primary aromatic amine.
     Describe **two** ways in which the chemical properties of amides differ from
     those of amines and **two** ways in which they resemble those of amines.
     How may your example in (a) be converted into an amine, in (b) be
     converted into an amide, and in (c) be converted into a nitrile?
                                                              Oxford & Cambridge

3*   This question concerns the following eight nitrogen-containing
     compounds:

     A NH_3 E $(C_2H_5)_2NH$
     B $C_2H_5NH_2$ F $(C_2H_5)_3N$
     C $C_2H_5CONH_2$ G $(C_2H_5)_4N^+Br^-$
     D $C_6H_5NH_2$ H $CH_3(CH_2)_2NH_2$

(a) Give the names of compounds C and F.
(b) Place A,B,C and D in order of *increasing* basic strength, the least
    basic first. Explain your reasoning.
(c) E is more basic than B. Suggest an explanation.
(d) How would you convert F into G?
(e) Describe a chemical method to distinguish between compounds B and
    H.

                                                                     London

4   Write an essay on organic amines. Your answer should include reference
     to methods of synthesis, physical and chemical properties and the use of
     amines as intermediates in the synthesis of other organic compounds.
                                                                       JMB

Chapter 12
Aromatic hydrocarbons

12.1 Introduction

The first three classes of compounds to be discussed in this book were alkane, alkene and alkyne hydrocarbons—in order of increasing degree of unsaturation. The present chapter deals with another important class of hydrocarbons which, although apparently highly unsaturated, contradict this view in many of their chemical properties.

During the nineteenth century, a variety of compounds, isolated for the first time, were found to be particularly rich in carbon and chemically less reactive than the majority hitherto encountered. Among them was benzene, a volatile hydrocarbon first isolated by Michael Faraday in 1825; the formula C_6H_6 was proposed for it in 1834. Other compounds were recognized as derivatives of benzene and because some of them had a pleasant smell they were called **aromatic** compounds. This adjective is not a very appropriate one, but it has been retained for historical reasons and is widely used. In this book its use is restricted to benzene and its derivatives, but in fact the word is far from being synonymous with benzenoid (meaning pertaining to benzene). There are many chemical systems, unrelated to benzene, which possess certain important features of benzene chemistry and which are therefore described as aromatic. Aromatic hydrocarbons related to benzene are sometimes called **arenes**. The chemical and physical properties characteristic of such aromatic systems will be made clear in the course of this chapter.

12.2 Structure of benzene

To early workers with benzene, its most puzzling feature was its relative lack of chemical reactivity, despite a high formal degree of unsaturation as indicated by the molecular formula C_6H_6. Benzene is not easily oxidized, and remains unaffected by cold alkaline solutions of manganate(VII) and by concentrated nitric acid. Nor does it easily undergo *addition* reactions, so characteristic of unsaturated compounds. This lack of reactivity was difficult to reconcile with any simple unsaturated structures which might be considered for it, e.g. $HC{\equiv}CCH_2CH_2C{\equiv}CH$ (which actually represents an entirely different compound, hexa-1,5-diyne b.p. = 86°C). Furthermore, when benzene undergoes *substitution* reactions, only one monosubstituted derivative, of general formula C_6H_5X, can be obtained, indicating that all six hydrogen atoms in the molecule of benzene are chemically equivalent, or indistinguishable, and hence similarly situated within the molecule. In contrast, three isomeric *disubstituted* derivatives can usually be obtained.

An attempt to rationalize these rather bewildering facts was made in 1865 by the German chemist August Kekulé, inspired by a famous dream in which a vision of snakes grasping their own 'tails' suggested to him a closed-ring, or cyclic, structure for benzene. Such a structure had not previously been contemplated for an organic compound. A symmetrical, cyclic structure for benzene readily accommodates the fact that the six hydrogen atoms are equivalent and, in order to retain the well-established tetravalency of carbon, Kekulé suggested that there were three carbon–carbon double bonds within the ring, occupying alternate positions:

Such a structure was not in accord with the experimental observations of the lack of reactivity of benzene. On the basis of this structure, one would anticipate that benzene would be even more reactive than an alkene possessing only one double bond. Again, while the structure accounted for the formation of only one monosubstituted benzene C_6H_5X, it was at variance with the fact that only three isomers of a disubstituted derivative $C_6H_4X_2$ were known. According to the proposed structure, *four positional isomers* should be possible, in which the two substituted atoms or groups X take up positions on carbon atoms 1 and 2, 1 and 3, 1 and 4 and 1 and 6, respectively (the relative positions 1 and 2 and 1 and 6 in this structure are not identical, because between carbon atoms 1 and 2 there is a double bond, but between 1 and 6 only a single bond). In order to surmount these difficulties, Kekulé then proposed, in 1869, an entirely novel (and as it turned out quite incorrect) structure, in which the alternate double bonds occupied no fixed positions in the ring but oscillated round it. This was depicted by writing two 'Kekulé structures' which were believed to be constantly interconverting:

There has never been any direct evidence to suggest that benzene is an equilibrium mixture of these two forms, but a clearer picture of benzene's structure emerged in the twentieth century, as a result of new experimental observations and more sophisticated theories of bonding.

Certain significant features of the benzene ring were established by the pioneering X-ray diffraction studies of benzene derivatives, made in 1929 by the English crystallographer Kathleen Lonsdale. Her work demonstrated

(1) that the six carbon atoms in the benzene molecule were indeed arranged in a ring as suggested by Kekulé,
(2) that the ring was symmetrical and planar,
(3) that the carbon–carbon bond lengths in the ring were all equal, with a value of 0.139 nm, i.e. intermediate between that for a carbon–carbon single bond (0.154 nm) and that for a carbon–carbon double bond (0.134 nm).

The existence of alternate double bonds in the ring, at variance in any case with the chemical properties of benzene and with the isomerism described previously, was thus directly disproved.

Thermodynamic evidence, too, proved valuable. The experimentally derived standard enthalpies of combustion of carbon, hydrogen and benzene permit a standard enthalpy of formation of benzene to be calculated, using Hess's law:

$$6C(s) \quad + 6O_2(g) \quad \rightarrow 6CO_2(g) \qquad \Delta H_c^{\ominus} = -(6 \times 394)\,kJ = -2364\,kJ$$
$$3H_2(g) + 3/2O_2(g) \rightarrow 3H_2O(l) \qquad \Delta H_c^{\ominus} = -(3 \times 286)\,kJ = -858\,kJ$$
$$6CO_2(g) + 3H_2O(l) \rightarrow C_6H_6(l) + 15/2O_2(g) \qquad \Delta H_c^{\ominus} = +3268\,kJ$$

Hence
$$6C(s) \quad + 3H_2(g) \quad \rightarrow C_6H_6(l) \qquad \Delta H_f^{\ominus} = +46\,kJ$$

Benzene cannot actually be synthesized from its constituent elements, but this figure for its standard enthalpy of formation may be regarded as experimentally based (i.e. based on the experimental enthalpies of combustion shown). This figure may be contrasted with a purely theoretical one, calculated on the basis of a Kekulé type of structure and of mean, or average, bond energies. The calculation utilizes Hess's law again:

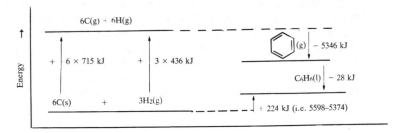

The 5346 kJ is made up of

$$6\,E(C—H) = -6 \times 413\,kJ = -2478\,kJ$$
$$3\,E(C=C) = -3 \times 610\,kJ = -1830\,kJ$$
$$3\,E(C—C) = -3 \times 346\,kJ = -1038\,kJ$$

Thus, on the basis of the Kekulé type of structure, the theoretical enthalpy of formation of benzene is $\Delta H^{\ominus} = +224\ \text{kJ mol}^{-1}$. This value is quite different from the experimentally derived value of $\Delta H^{\ominus} = +46\ \text{kJ mol}^{-1}$. However, the use of mean bond energies, and the assumptions implicit in their use, mean that too much quantitative reliance should not be placed on these figures. The calculations do suggest, however, that, even in qualitative terms only, the benzene molecule is thermodynamically more stable with respect to its constituent elements than any single, conventional structure would lead one to believe:

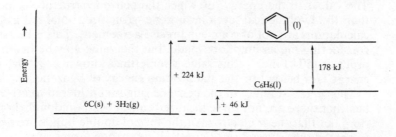

This general conclusion is borne out, too, by thermodynamic calculations based on experimental values of enthalpies of hydrogenation.

Benzene, like alkenes, will undergo catalytic hydrogenation (*see* Section 12.4.1(i)):

$$C_6H_6(l) + 3H_2(g) \rightarrow \quad \bigcirc \quad (l) \qquad \Delta H^{\ominus} = -206\ \text{kJ}$$

Now, if the hypothetical Kekulé structure ('cyclohexatriene') was correct, the standard enthalpy of hydrogenation would be expected to be about three times the value for cyclohexene, a similarly cyclic alkene containing only one carbon–carbon double bond:

$$\bigcirc (l) + H_2(g) \rightarrow \quad \bigcirc \quad (l) \qquad \Delta H^{\ominus} = -119\ \text{kJ}$$

and therefore

$$\bigcirc (l) + 3H_2(g) \rightarrow \quad \bigcirc \quad (l) \qquad \Delta H^{\ominus} \approx -357\ \text{kJ}$$

The experimental standard enthalpy of hydrogenation of benzene is thus much less than the one calculated on the basis of the 'cyclohexatriene' structure for benzene, implying again that the benzene molecule is thermodynamically more stable than the structure written for it suggests:

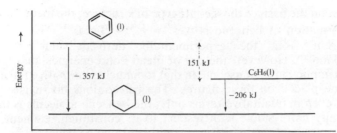

The values of the energy, by which the benzene structure is more stable than the hypothetical 'cyclohexatriene' structure, obtained in these two calculations are not of course in precise agreement. This is not surprising, considering the assumptions made, but the value may be taken as of the order of 150 kJ mol^{-1}; this value is sometimes known as the **delocalization energy** (*see* below) or the **stabilization energy** of benzene.

This 'extra' stability of the benzene ring is considered to arise from the fact that there are not three localized carbon–carbon double bonds in the ring, but that the π-electrons of the formal double bonds are **delocalized**, or spread out, symmetrically over all six carbon atoms of the ring, giving *all* the carbon–carbon bonds partial double bond character (cf. bond lengths from X-ray measurements). The situation can be viewed in orbital terms by considering the carbon atoms to be *sp^2 hybridized* (*see* Section 1.3) as in alkenes. On each carbon atom, there will also be a singly-occupied p-orbital. Overlap of these p-orbitals with their immediate neighbours gives rise to *three* new π molecular orbitals (containing a total of six electrons) of significantly lower energy (cf. thermodynamic calculations). The spatial distribution of the electrons, or the electron density, of these orbitals is such as to effectively sandwich the benzene ring between two 'hoops' of electronic charge, lying above and below the plane of the ring:

Traditional covalent bond notation is unable to express this state of affairs and so the unusually stable, cyclic, delocalized electron system, characteristic of aromatic systems, is represented by a circle ○. The structure of benzene is therefore conveniently represented by a circle lying within the hexagonal ring:

(note that the hexagonal ring on its own represents cyclohexane).

There are occasions when this representation is inadequate and it becomes necessary, at an elementary level, to make use of the concept of canonical forms (*see* Section 1.5), whereby the real structure of benzene is considered to be intermediate between, or a resonance hybrid of, the two classical Kekulé representations (canonical forms). In general, however, the other representation is satisfactory and is generally used throughout the following chapters.

12.3 Naming of benzene derivatives

The majority of benzene compounds are named as substituted derivatives, the name of the substituting group(s) prefixing the word benzene. Thus, for example, $C_6H_5NO_2$ is nitrobenzene and C_6H_5Br is bromobenzene. When there are two or more substituents in the ring, then the relative positions of the groups are indicated by numbering the carbon atoms of the ring from 1 to 6. Thus, the three possible isomers of dinitrobenzene $C_6H_4(NO_2)_2$ are described as follows:*

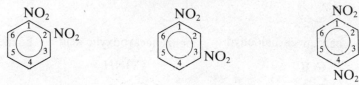

1,2-dinitrobenzene   1,3-dinitrobenzene   1,4-dinitrobenzene

When the two substituents are different, the names of the groups prefix benzene in alphabetical order, the first-named group being considered to be bonded to carbon 1. Thus:

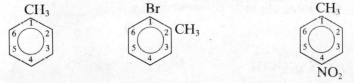

methylbenzene   bromo-2-methylbenzene   methyl-4-nitrobenzene

When the side-chain contains a functional group, then an alternative method of naming becomes possible and is sometimes to be preferred. The grouping C_6H_5-, derived from benzene, is known as the **phenyl** group (for historical reasons) and is the typical **aryl** group, a term corresponding to the alkyl group of non-aromatic, or **aliphatic**, compounds. A compound may sometimes be named *either* as a substituted benzene, *or* as a phenyl-substituted aliphatic compound. In some cases, attention is focused on

* In the past, the 2- (and the equivalent 6-), 3- (and equivalent 5-) and 4-positions in disubstituted benzenes have been called the *ortho* (*o*-), *meta* (*m*-) and *para* (*p*-) positions, respectively, so that these three dinitrobenzenes would be described as *o*-, *m*- and *p*-dinitrobenzene, respectively.

the functional group in the substituent and it is then often convenient to name the compound as a phenyl-substituted derivative. Thus, $C_6H_5CH{=}CH_2$ is named phenylethene, i.e. ethene in which one hydrogen atom has been substituted by a phenyl group. $C_6H_5NH_2$ is called phenylamine, by analogy with the common naming of aliphatic amines. On the other hand, while $C_6H_5CH_3$ is usually called methylbenzene (rather than phenylmethane), $C_6H_5CH_2OH$ is often called phenylmethanol, because interest lies in its reactions as an alcohol.

The exceptions to this substitutive nomenclature are the aldehyde, ketone and acid derivatives of benzene. The names of the corresponding aliphatic classes use suffixes -al, -one and -oic, etc., added to the root of the name of the hydrocarbon from which each is derived. This cannot be done for aromatic compounds (for example, there can be no benzeneoic acid) and so there are several specific suffixes used in conjunction with the word benzene, as illustrated below:

benzenecarbaldehyde 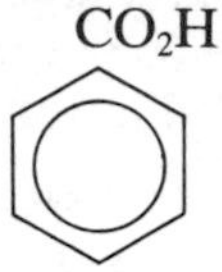   benzenecarboxylic acid    benzenecarbonitrile

benzenecarbonyl chloride   benzenecarboxamide

In the case of aromatic ketones, the compounds are named as phenyl-substituted aliphatic compounds, e.g.

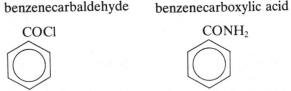

phenylethanone 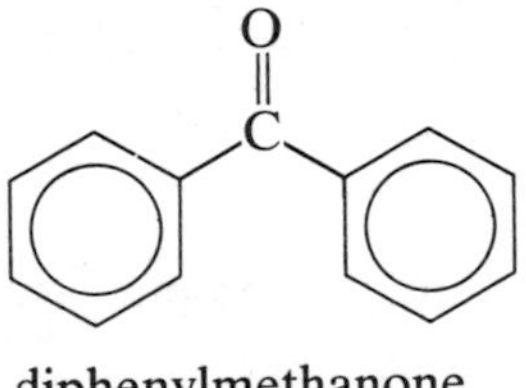   diphenylmethanone

These points will become clearer in the course of the following chapters.

12.4 Chemistry of benzene

Benzene is a colourless, flammable liquid of boiling point 80°C and melting point 5.5°C. It is immiscible with, and less dense than, water. As well as being used in the formation of chemical derivatives, it has been widely used as a solvent, but its carcinogenic (cancer-producing) nature means that there is a possible health risk, resulting from inhalation of vapour, or from skin contact, on prolonged exposure. For many simple laboratory purposes, particularly for use as a general solvent, the much less harmful methylbenzene (*see* Section 12.5) can be conveniently substituted for it.

The particularly characteristic delocalized system of π-electrons within the benzene molecule means that benzene is not easily attacked by acids, alkalis or oxidizing agents, unlike other unsaturated compounds such as alkenes. Like alkenes, however, it does undergo some **addition** reactions, although **substitution** reactions are much more characteristic of benzene's chemical behaviour.

12.4.1 Addition reactions

(i) Catalytic hydrogenation

Benzene can be hydrogenated, but not as readily as alkenes and alkynes. The addition of hydrogen is brought about, in the presence of a metal like nickel as catalyst, at high temperatures using hydrogen under considerable pressure. Even when using a very active form of platinum as catalyst, the conditions are, typically, 10 h hydrogenation at 10–15 MN m^{-2} and 100°C. These conditions illustrate the relative kinetic stability of the aromatic system as compared with a non-aromatic, but unsaturated, one. Each molecule of benzene takes up 3 molecules of hydrogen H_2 and cyclohexane is the product:

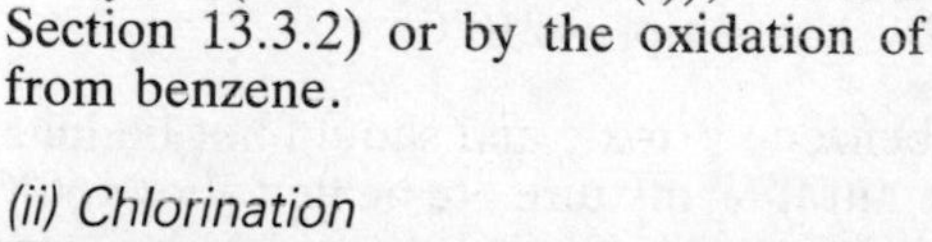

The catalytic reduction of aromatic systems in this way is important commercially. Cyclohexanol, used as an intermediate in the manufacture of nylon (*see* Section 20.3(i)), is made by the reduction of phenol (*see* Section 13.3.2) or by the oxidation of cyclohexane, which is obtained from benzene.

(ii) Chlorination

The addition of chlorine to the benzene ring requires a free-radical (*see* Section 1.6) reaction which is brought about by irradiating the mixture with ultraviolet light. Three mol of chlorine Cl_2 are taken up per mol of benzene and the solid product, hexachlorocyclohexane, is a mixture of geometrical isomers (*see* Section 16.2; there are eight possible ones for this structure):

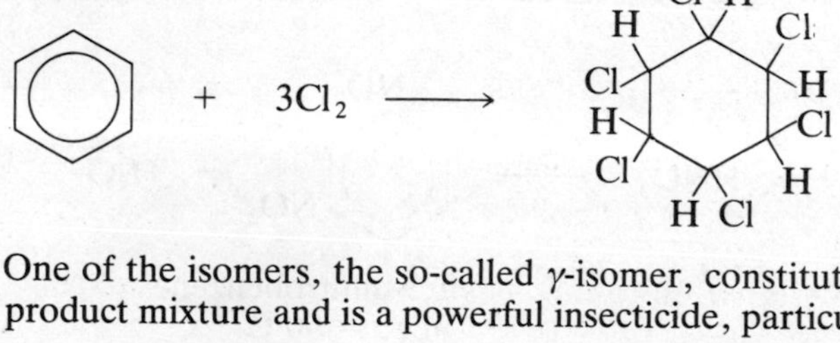

One of the isomers, the so-called γ-isomer, constitutes about 10% of the product mixture and is a powerful insecticide, particularly effective in the soil. It was one of the first commercial organochlorine insecticides and was introduced by ICI in 1943. It is marketed under various names such as Gammexane, Lindane or benzene hexachloride (BHC).

If certain catalysts are present then *substitution* by chlorine, rather than *addition*, occurs (*see* Section 12.4.2(iii)).

12.4.2   Substitution reactions

These are the most characteristic reactions of the benzene ring. Under appropriate conditions, one or more groups may be substituted for hydrogen atoms in the ring.

(i) Nitration

The substitution of one or more nitro groups into the benzene ring is effected by the use of a special nitrating mixture of concentrated nitric and sulphuric acids (*see also* Section 12.4.4). Benzene and the nitrating mixture are mixed with gentle swirling (the two are immiscible); an exothermic reaction takes place and the temperature should not be allowed to rise above about 60°C. The main product of reaction is a pale yellow liquid with a slight bitter-almond smell. It can be isolated by steam distillation (*see* Section 21.2.2(iii)), or by carefully pouring the reaction mixture into water (to remove the acid), separating, washing and drying the organic layer and then distilling it:

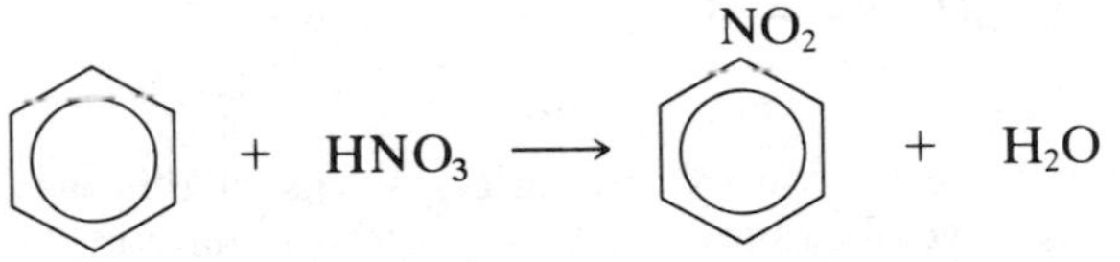

$$\text{benzene} + HNO_3 \longrightarrow \text{nitrobenzene} + H_2O$$

nitrobenzene
b.p. = 210°C

The vapour of nitrobenzene is toxic and should not be inhaled.

If benzene and the nitrating mixture are heated above 60°C for an hour or so, then a second nitro group is substituted in the ring. When the reaction mixture is poured into water, a pale yellow solid, a *dinitrobenzene*, is obtained. The presence of a nitro group in the ring makes it more difficult to introduce a second group (the rate of nitration is a million times lower than for benzene, hence the higher temperature required) and it also dictates the position of the incoming group; in this case the product is 1,3-dinitrobenzene (for explanation, *see* Section 12.4.5):

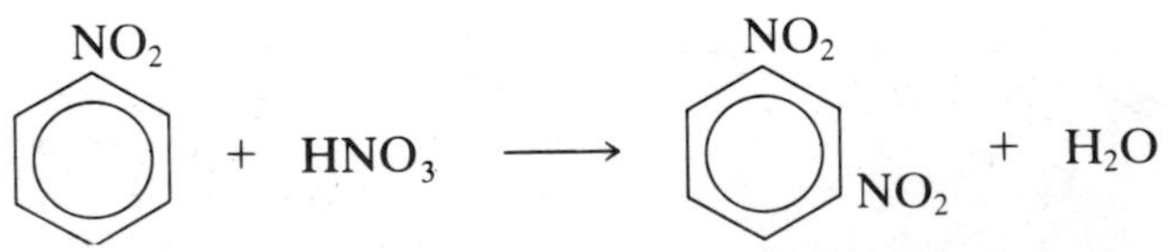

$$\text{nitrobenzene} + HNO_3 \longrightarrow \text{1,3-dinitrobenzene} + H_2O$$

1,3-dinitrobenzene
m.p. = 90°C

Under very vigorous conditions (temperature 100°C, several days), a third nitro group can be introduced into the ring:

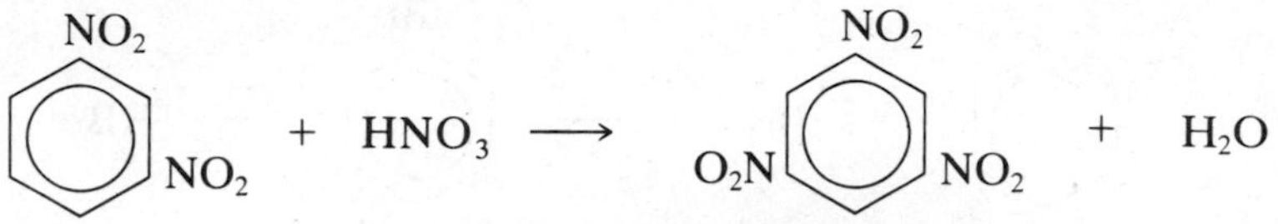

1,3,5-trinitrobenzene
m.p. = 122°C

Nitration of benzene and its derivatives is an important laboratory reaction, because the nitro group is easily reduced to an amino group —NH_2, which can be converted into a variety of ·other functional groups via diazonium salts (*see* Section 14.3). Commercially, phenylamine (*aniline*) is manufactured by the reduction of nitrobenzene (*see* Section 14.4).

(ii) Sulphonation

Under the conditions of nitration previously described, benzene is unattacked by concentrated sulphuric acid, but if it is heated under reflux for about 6 h with the concentrated acid, or preferably with 'fuming' sulphuric acid (*oleum*), **sulphonation** takes place. The **sulphonic acid** group —SO_3H is introduced into the ring (the sulphur being directly bonded to carbon) and the very soluble benzenesulphonic acid is formed:

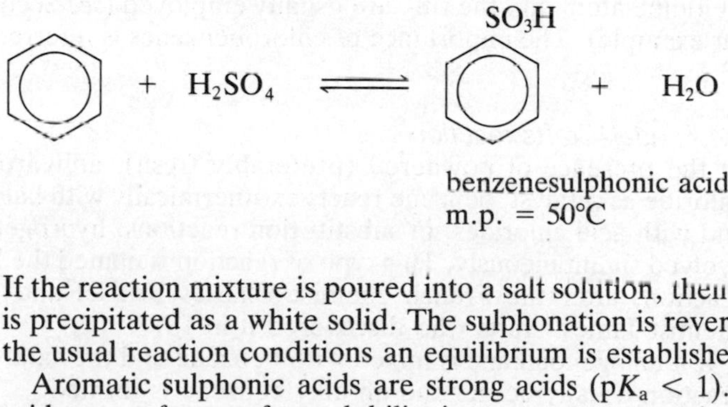

benzenesulphonic acid
m.p. = 50°C

If the reaction mixture is poured into a salt solution, then the sodium salt is precipitated as a white solid. The sulphonation is reversible and under the usual reaction conditions an equilibrium is established.

Aromatic sulphonic acids are strong acids ($pK_a < 1$). The sulphonic acid group often confers solubility in water on an organic compound (*see*, for example, detergents Section 19.4). It is also often attached to polymers in cationic exchange resins (*see* Section 21.2.4(v)).

(iii) Halogenation

The ease of addition of halogens to alkenes is in marked contrast to the reaction between halogens and benzene. As already noted (Section 12.4.1), chlorine adds on to benzene only in the presence of ultraviolet light. In the presence of certain catalysts, such as iron or iron(III) or aluminium chlorides, however, both chlorine and bromine will react with benzene to form *substitution* products. Thus, if a little catalyst is added to a red-brown solution of bromine in benzene, the colour gradually fades and white 'steamy' fumes of hydrogen bromide are given off as reaction occurs:

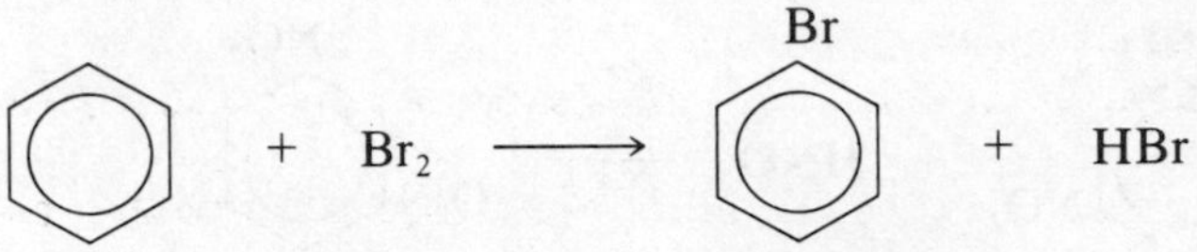

bromobenzene
b.p. = 156°C

The introduction of the second halogen atom is more difficult (i.e. the reaction is slower), but substitution can occur at both the 2- and 4- positions so that a mixture of products is obtained:

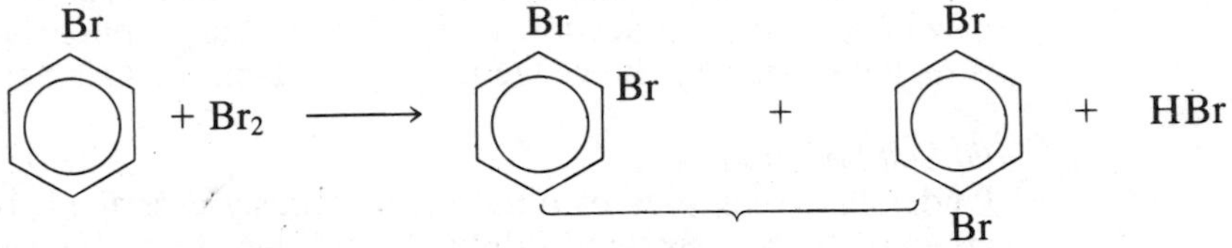

1,2-dibromobenzene   1,4-dibromobenzene
b.p. = 225°C   m.p. = 89°C

Although chlorination and bromination can be carried out quite easily, direct iodination is not very effective and indirect methods of introducing an iodine atom into the ring are usually employed (*see* Section 14.3.1(iii), for example). The importance of chlorobenzenes is referred to in Section 13.1.

(iv) Friedel–Crafts reaction

In the presence of powdered (preferably fresh), anhydrous aluminium chloride as catalyst, benzene reacts exothermically with halogenoalkanes, and with acid chlorides, in substitution reactions, hydrogen halide being evolved simultaneously. This type of reaction is named the Friedel–Crafts reaction, after the French chemist Charles Friedel and the American chemist James Crafts who discovered it in 1877.

If a halogenoalkane is mixed with benzene and the catalyst, a vigorous reaction usually occurs and an alkylbenzene is formed:

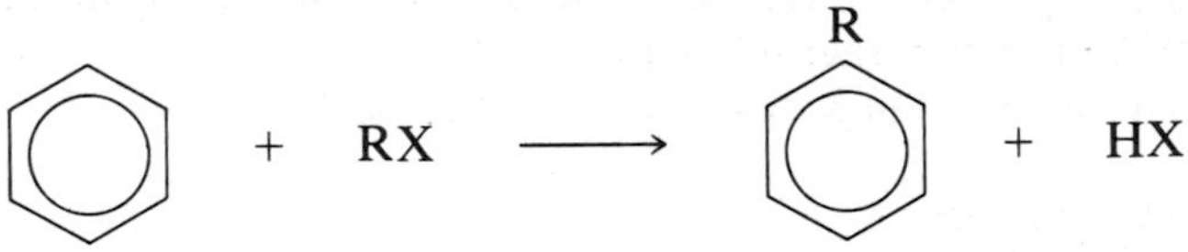

e.g.

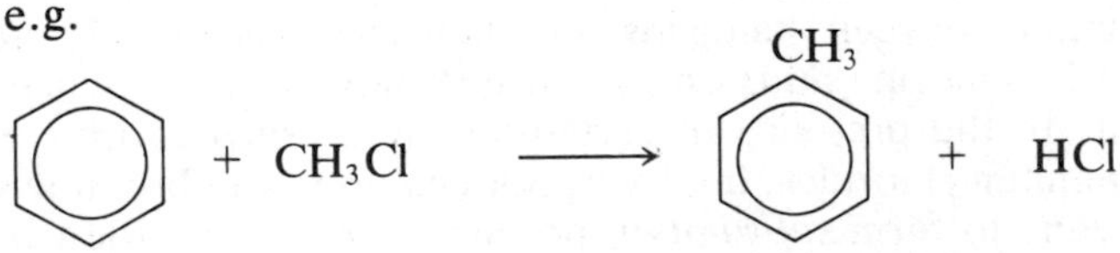

methylbenzene
b.p. = 111°C

However, the reaction suffers from the drawback that further substitution in the ring tends to occur, and also that primary alkyl groups tend to isomerize and give branched side-chains in the product (*see* Section 12.4.4(iv), for explanations).

A similar type of substitution reaction occurs when an acid chloride is used, instead of a halogenoalkane, and a ketone is the product. This reaction is not complicated by further substitution and a better yield is obtained than in the case of the halogenoalkane:

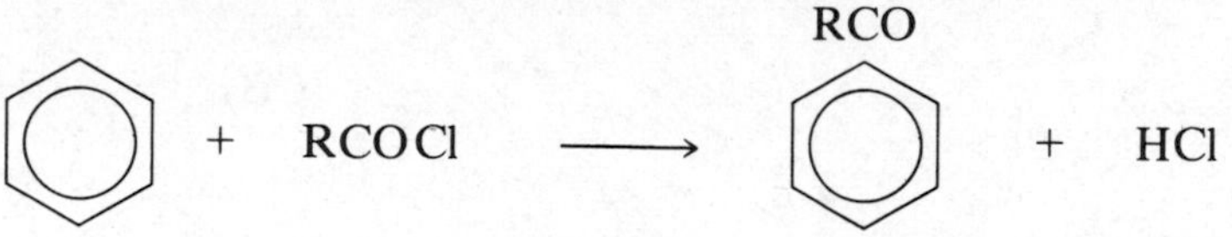

e.g.

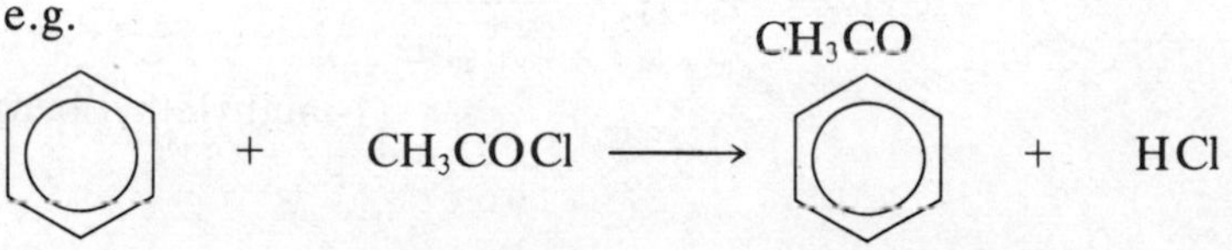

phenylethanone
b.p. = 202°C

With acid chlorides, however, excess catalyst has to be employed as some of it complexes with the ketone product (*see* Section 12.4.4(iv)).

In Friedel–Crafts reactions, benzene can be used both as reactant and solvent, but carbon disulphide may also be conveniently used. An even better solvent is 1,2-dichloroethane. Because the catalyst is soluble in this solvent, the reaction can be conducted under homogeneous conditions. In each case, the product of reaction is isolated by pouring the reaction mixture on to an ice/hydrochloric acid mixture and extracting the product.

Friedel–Crafts reactions are useful ones in both the laboratory and in industry. The important alkene phenylethene (*styrene*) which, when polymerized, forms the important thermoplastic poly(phenylethene) (*polystyrene, see* Section 20.2.1(iii)) is manufactured by the alkylation of benzene, followed by dehydrogenation:

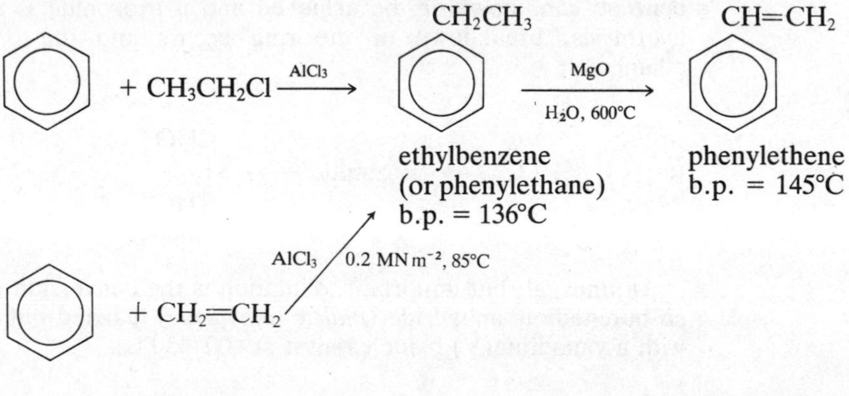

ethylbenzene
(or phenylethane)
b.p. = 136°C

phenylethene
b.p. = 145°C

Alternatively, the ethylbenzene can be made by reaction of benzene with ethene, in the presence of hydrogen fluoride, phosphoric acid or aluminium chloride as catalyst (the mechanism is very similar to the Friedel–Crafts reaction, *see* Section 12.4.4(iv)). In either case, formation of ethylbenzene is accompanied by polysubstituted products.

The reaction between benzene and an alkene is also used in the manufacture of (1-methylethyl)benzene (*cumene*), which yields phenol and propanone on oxidation (*see* Section 13.3.2):

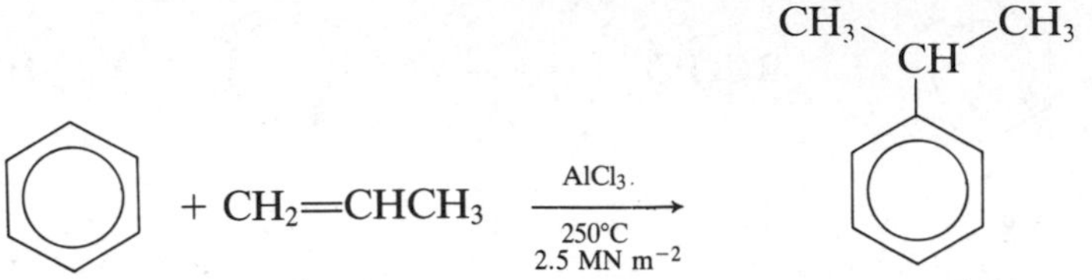

$$\text{benzene} + CH_2{=}CHCH_3 \xrightarrow[\substack{250°C \\ 2.5\ MN\ m^{-2}}]{AlCl_3} \text{(1-methylethyl)benzene}$$

(1-methylethyl)benzene
b.p. = 152°C

A similar reaction is used industrially for the manufacture of hydrocarbons such as dodecylbenzene, which is converted into biodegradable ('soft') detergents (*see* Section 19.4) by sulphonation:

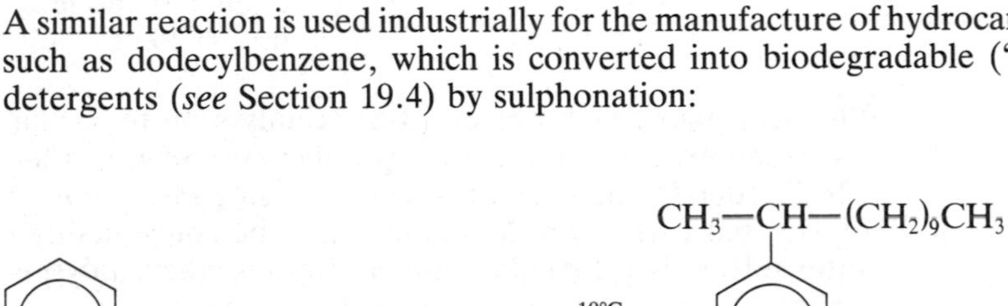

$$\text{benzene} + CH_2{=}CH(CH_2)_9CH_3 \xrightarrow[\text{HF}]{10°C} \text{[}CH_3{-}CH{-}(CH_2)_9CH_3\text{]}$$

12.4.3 Oxidation reactions

Unlike alkenes, benzene is very resistant to oxidation, even by alkaline solutions of manganate(VII) where attack is very slow. Ozonolysis of benzene can, however, be achieved and a *triozonide* is obtained. On hydrolysis, breakdown of the ring occurs and the ozonide yields ethanedial:

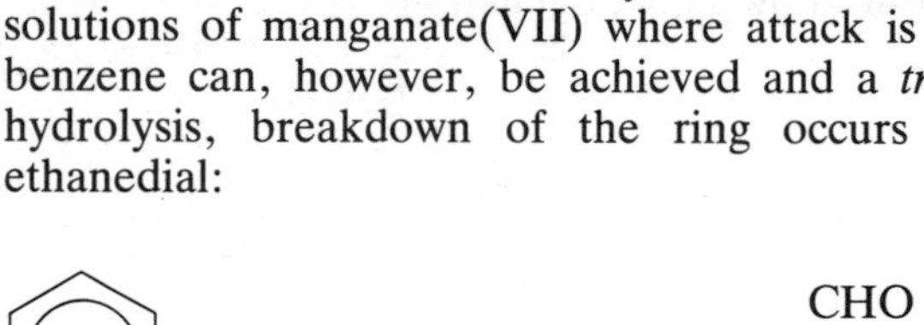

$$\text{benzene} + O_3 \longrightarrow \text{triozonide} \longrightarrow 3\begin{array}{l} CHO \\ | \\ CHO \end{array}$$

ethanedial

An unusual, but important, oxidation is the conversion of benzene to *cis*-butenedioic anhydride (*maleic anhydride*), carried out commercially with a vanadium(V) oxide catalyst at 400–450°C:

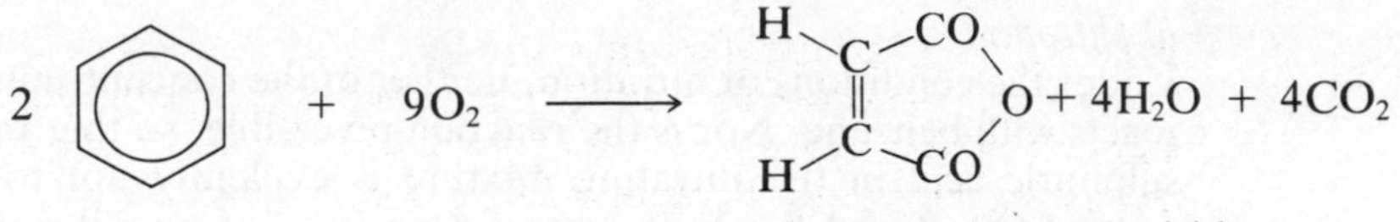

$$2 \; \bigcirc \quad + \quad 9O_2 \quad \longrightarrow \quad \overset{H}{\underset{H}{\diagdown}}\overset{C}{\underset{C}{\parallel}}\overset{CO}{\underset{CO}{\diagup}}O + 4H_2O + 4CO_2$$

cis-butenedioic anhydride
m.p. = 54°C

Like other volatile hydrocarbons, benzene is flammable and, because of its relatively high carbon content, will burn in air with a black, sooty flame. In oxygen, complete combustion to carbon dioxide and water can occur:

$$C_6H_6 + 15/2\, O_2 \rightarrow 6CO_2 + 3H_2O$$

12.4.4 Mechanism of substitution reactions

In all the substitution reactions just described, it is believed that benzene acts as a nucleophile, or source of electrons, and that the reagents all provide, under suitable reaction conditions, an electrophile as attacking species. The planar benzene ring permits easy access to its delocalized π-electrons for an incoming electrophile. Once the electrophile has become bonded to a carbon atom of the ring by an electron-pair bond, however, the distribution of π-electrons is temporarily upset. Since the electrophile is generally a positively charged species, the ring now carries a positive charge, spread out over the other five carbon atoms, and the thermodynamic stability of the aromatic ring is temporarily destroyed. It is regained by the loss of a proton from the carbon atom to which the electrophile is bonded:

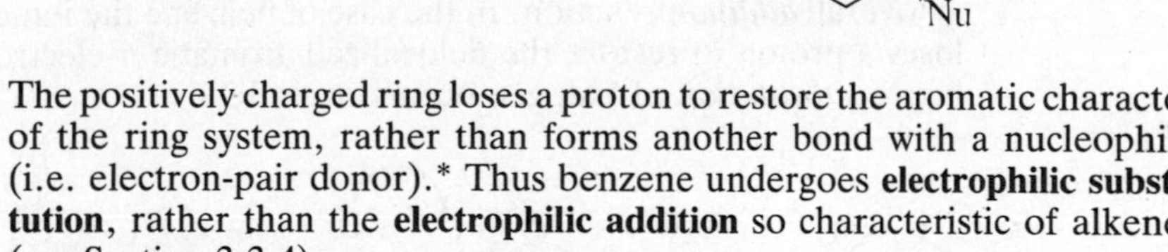

The positively charged ring loses a proton to restore the aromatic character of the ring system, rather than forms another bond with a nucleophile (i.e. electron-pair donor).* Thus benzene undergoes **electrophilic substitution**, rather than the **electrophilic addition** so characteristic of alkenes (*see* Section 3.3.4).

---

* By using bond energies, approximate values of the enthalpy change ΔH for the addition and substitution reactions may be calculated. When E = Cl, the values are about $+20\, kJ\, mol^{-1}$ and $-120\, kJ\, mol^{-1}$, respectively. Thus the substituted product is the more stable, thermodynamically, and hence is the favoured product.

✳ (i) Nitration

Under the conditions of nitration, neither of the concentrated acids alone reacts with benzene. Nor is the reaction reversible, so that the role of the sulphuric acid in the nitrating mixture is evidently not to remove the water formed and hence influence a position of equilibrium. Freezing-point depression measurements show that the addition of nitric acid to the sulphuric acid produces a freezing-point depression corresponding to *four* solute species, and not to one expected if the nitric acid dissolved unchanged. This can be accounted for in terms of the abstraction of a hydroxide ion from the nitric acid by sulphuric acid:

$$2H_2SO_4 + HNO_3 \rightleftharpoons 2HSO_4^- + H_3O^+ + NO_2^+$$

The electrophile so produced is the **nitryl** cation (or *nitronium* ion) NO_2^+. This is a well-established ion, found in salts such as $NO_2^+ClO_4^-$ and $NO_2^+BF_4^-$. The nitryl cation attacks the benzene as outlined previously (i.e. $E^+ = NO_2^+$). Thus, the function of the sulphuric acid is to generate the nitryl cation from the nitric acid.

(ii) Sulphonation

This takes place most readily when fuming sulphuric acid (*oleum*) is used, and it seems likely that the electrophile present is the dissolved sulphur trioxide, since the sulphur atom in this molecule may be considered to be electron-deficient. Thus:

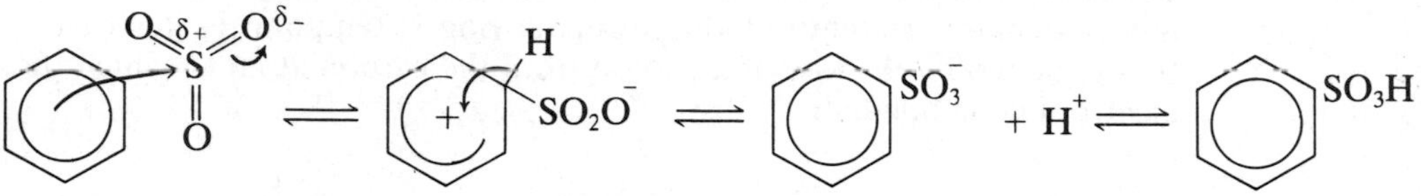

(iii) Halogenation

The reaction of a halogen with an alkene involves the initial, electrophilic attack of the halogen on the alkene to produce an intermediate which may be formulated either as a carbocation, or a bromonium-type ion (*see* Section 3.3.4). Subsequent nucleophilic attack by a halide ion results in an overall *addition* reaction. In the case of benzene the ionic intermediate loses a proton to restore the delocalized aromatic π-electron system and hence *substitution* of halogen occurs overall:

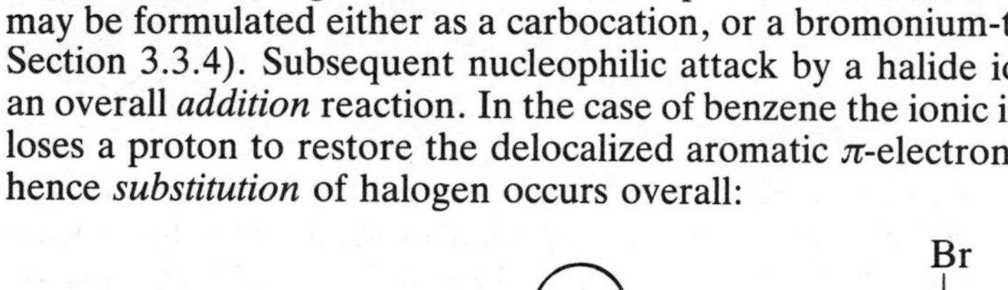

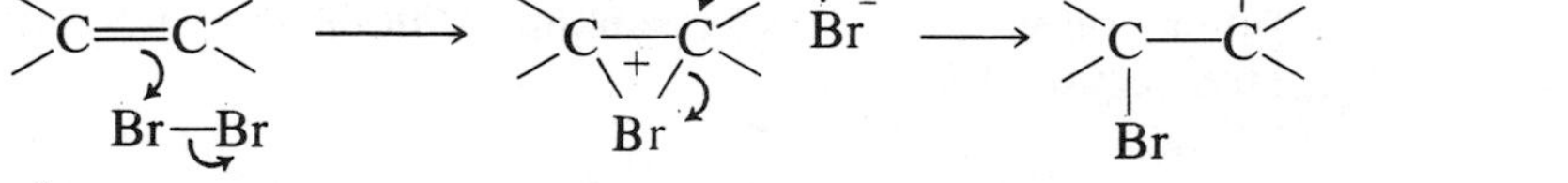

However, the benzene is a less powerful nucleophile than an alkene and a catalyst of the Lewis acid type (*see* Section 1.7) is required to promote the heterolytic (ionic) fission of the halogen molecule:

$$\overset{\delta+}{Br} \overset{\delta-}{Br} AlCl_3 \rightarrow Br^+ [AlCl_3Br]^-$$

A Lewis acid catalyst, such as aluminium chloride or ion(III) chloride (sometimes known as 'halogen carriers'), may be added directly, or else elemental iron may be added; the latter will react with the halogen to form an iron(III) halide.

Thus, the halogenation of benzene and its derivatives involves an ionic mechanism, rather than the free-radical one characteristic of the halogenation of alkanes.

(iv) Friedel–Crafts reactions

These require a Lewis acid type of catalyst, usually aluminium chloride, to generate an electrophile.

With halogenoalkanes, it is likely that a carbocation is generated as electrophile. This would explain the fact that reaction of benzene with unbranched halogenoalkanes usually leads to a mixture of products in which the alkyl substituent is branched:

$$RX + AlCl_3 \rightleftharpoons R^+ [AlCl_3X]^-$$

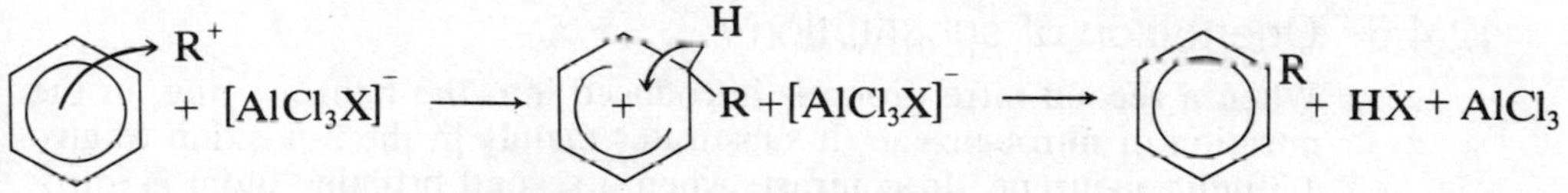

A primary carbocation would isomerize:

$$AlCl_3 + RCH_2CH_2Cl \rightleftharpoons [RCH_2CH_2]^+ [AlCl_4]^-$$

$$\Updownarrow$$

$$[R\overset{+}{C}HCH_3]$$

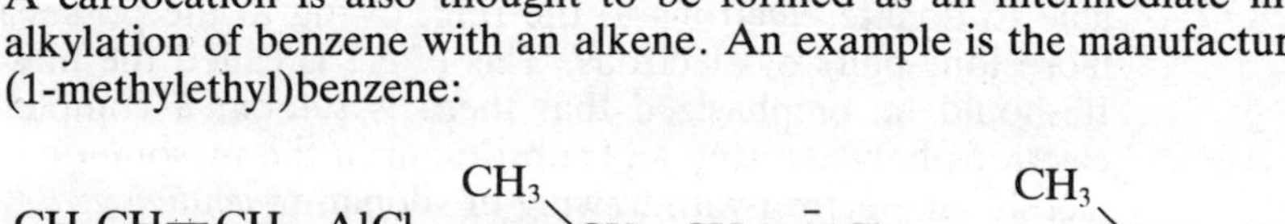

A carbocation is also thought to be formed as an intermediate in the alkylation of benzene with an alkene. An example is the manufacture of (1-methylethyl)benzene:

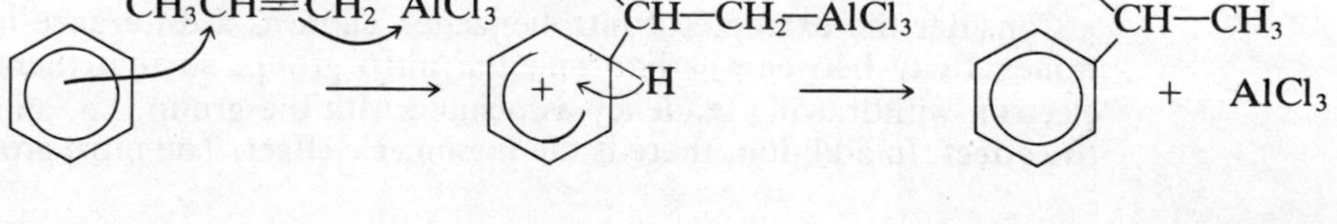

In the Friedel–Crafts **acylation** of benzene to form a ketone, an electrophilic **acylium ion** is thought to be involved, e.g.

$$RCOCl + AlCl_3 \rightleftharpoons [RCO]^+ [AlCl_4]^-$$
acylium ion

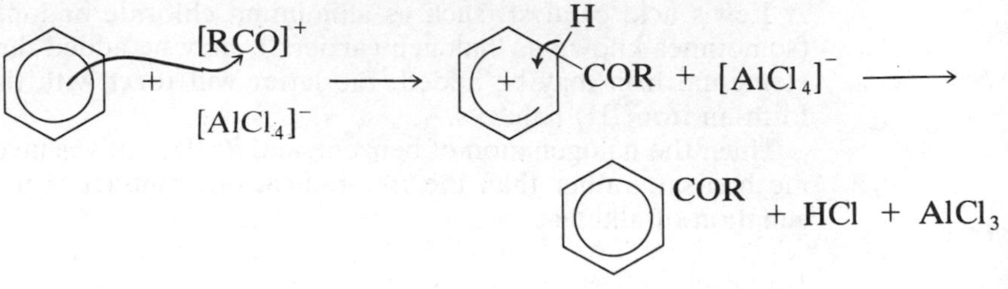

Some of the catalyst also complexes with the ketone product:

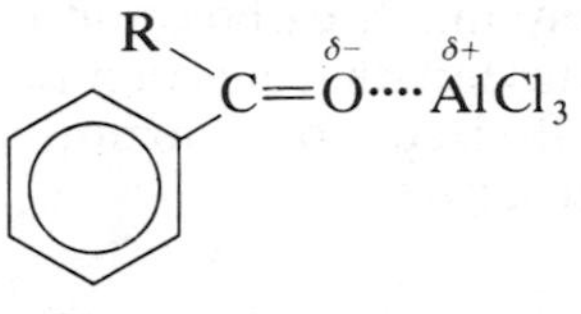

12.4.5   Orientation of substitution

When a second nitro group is introduced into the benzene ring, in the nitration of nitrobenzene, it substitutes mainly in the 3-position to give 1,3-dinitrobenzene. In contrast, when a second bromine atom is introduced into a benzene ring, in the bromination of bromobenzene, it substitutes mainly in the 2- and 4-positions (the 6-position is identical to 2) to give a mixture of 1,2- and 1,4-dibromobenzenes. It is clear, then, that the nature of an existing substituent in the ring influences, and indeed *directs*, the position of substitution of a second group. This is because it affects the relative electron densities around each of the carbon atoms of the ring. The electron density at a carbon atom in the ring may be influenced by an electron-withdrawing effect of a substituent, owing to a difference in electronegativity between the atom of the substituent to which the ring is bonded and carbon. This is an **inductive effect** (*see* Section 1.4). Alternatively, the substituent may be able to take up some of the π-electrons of the ring, via its own vacant orbitals, or it may be able to donate electrons to the ring, owing to the possession of one or more lone-pairs of electrons. This effect is called the **mesomeric effect**. It should be emphasized that there is seldom a complete transfer of electrons between ring and substituent in the mesomeric effect; there is merely an electron-withdrawing or -donating *tendency*.

Consider the example of nitrobenzene. There is a difference in electronegativity between carbon and the nitro group, so that there is an electron-withdrawing tendency associated with the group, i.e. an **inductive effect**. In addition, there is the **mesomeric effect**. The nitro group has a tendency to withdraw π-electrons from the ring, because they can be accommodated in orbitals on the nitrogen, made vacant by the transfer of π-electrons on to one of the oxygen atoms:

The electron-withdrawing tendency of the nitro group reduces the nucleophilic character of the ring and makes it more difficult to introduce a second group. The nitro group is said to **deactivate** the ring towards electrophilic substitution. However, the least reduction in electron density occurs at the 3-position, and hence a second substituting group is directed into this position. Unfortunately, owing to the limitations of the conventional representation of the benzene ring, the reason for this is not readily apparent. The situation can be comprehended to some extent by making use of the concept of canonical forms, all of which are considered to represent contributions to the real, overall structure. Thus, the following forms can be written for nitrobenzene:

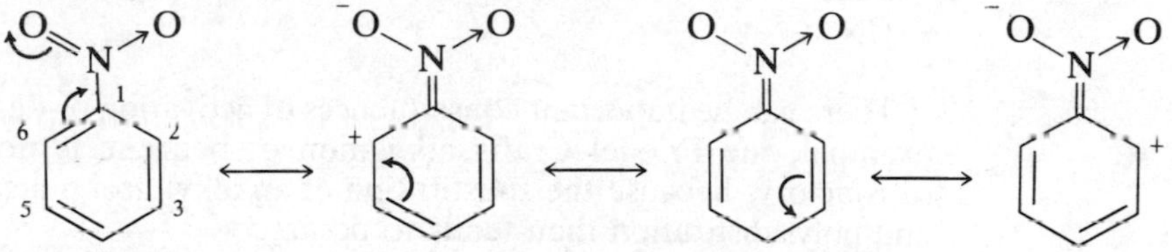

In none of these contributing structures does the formal positive charge reside on carbon atom 3 (or 5) of the ring, and it is therefore considered to be the least deactivated position with regard to electrophilic substitution. More sophisticated calculations, using molecular orbital theory, provide relative electron densities at the carbon atoms in the intermediate positive ion in electrophilic substitution by a nitro group. One such calculation gives the values shown in the formula below:

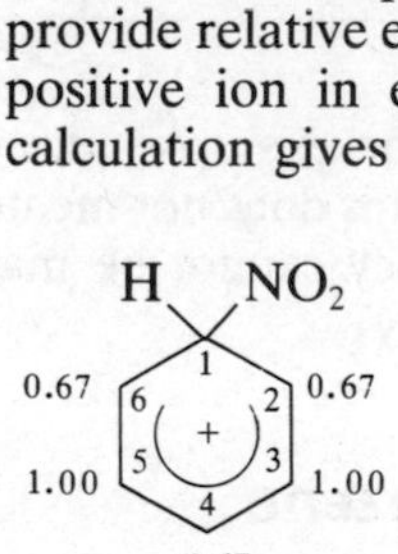

With reference to this formula, if a nitro group is to be introduced at carbon 1 during the nitration of nitrobenzene, the existing nitro group, which is electron-withdrawing, should be at carbon 3 (or 5). If methylbenzene was being nitrated, then the methyl group, which is weakly electron-donating, should be at the 2- or 4-positions, if the nitro group is to substitute at carbon 1 (*see* Section 12.5.1(iii)).

It becomes apparent, from observations on the orientation of substitution of mono-substituted benzenes, that functional groups may be classified as deactivating or activating towards electrophilic substitution in the ring. The **deactivating groups** make it more difficult to introduce a second group, and this will substitute predominantly at the 3-position, taking the first group to be at carbon 1. **Activating groups** make it easier to substitute in the ring, and generally direct an incoming group into the 2- (or 6-) and 4-positions. Rather confusingly, however, halogen atoms direct incoming groups to the 2- and 4-positions, but apparently *deactivate* the ring!

3-directing groups (deactivating)	2- and 4-directing groups (activating)
—NO$_2$	—R
—COR	—O(R or H)
—CO$_2$(R or H) (and other derivatives)	—NH$_2$
—NH$_3$	—NHCOR
—CHO	—X (halogen)—*but deactivating*
—CN	

There can be important consequences of activation or deactivation. For example, the Friedel–Crafts alkylation of benzene is not always very satisfactory, because the substitution of an alkyl group activates the ring and polysubstitution then tends to occur:

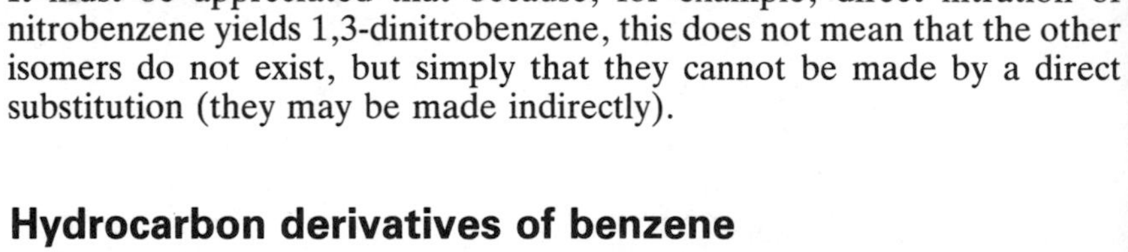

It must be appreciated that because, for example, direct nitration of nitrobenzene yields 1,3-dinitrobenzene, this does not mean that the other isomers do not exist, but simply that they cannot be made by a direct substitution (they may be made indirectly).

12.5 Hydrocarbon derivatives of benzene

The simplest member of the homologous series, of which benzene is the parent, is methylbenzene(*toluene*), a colourless, volatile liquid, immiscible with water and volatile in steam. It is much less harmful than benzene and can be conveniently substituted for it as a general solvent. Other simple hydrocarbon derivatives of benzene are shown opposite:

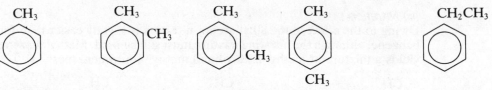

methylbenzene 1,2-dimethyl, 1,3-dimethyl, 1,4-dimethylbenzene ethylbenzene
b.p. = 111°C   b.p. = 144°C b.p. = 139°C   b.p. = 138°C   b.p. = 136°C

12.5.1   Properties of the ring

Not surprisingly, hydrocarbons such as methylbenzene show many of the properties characteristic of the benzene ring. The presence of one or more alkyl substituents or side-chains influences the position of substitution of other groups in the ring, and in general <u>activates</u> the ring towards electrophilic substitution.

(i) Hydrogenation

The ring system can be reduced, as can benzene itself, by catalytic hydrogenation. Methylbenzene, for example, yields methylcyclohexane:

$$CH_3 \qquad + 3H_2 \xrightarrow[150°C]{Ni} \qquad H_3C \quad H \qquad b.p. = 100°C$$

(ii) Halogenation

In general, the alkyl group activates the ring towards electrophilic substitution and in the presence of a catalyst or 'halogen carrier', such as aluminium chloride or iron filings, substitution by halogen takes place readily to give a mixture of isomeric products. Thus, with chlorine, methylbenzene yields a mixture of chloromethylbenzenes:

$$CH_3 \quad + \quad Cl_2 \longrightarrow \quad CH_3\text{-}Cl \quad + \quad CH_3\text{-}Cl \quad + \quad HCl$$

chloro-2-methyl   chloro-4-methyl
benzene   benzene
b.p. = 159°C   b.p. = 162°C

In the absence of such a catalyst, however, halogenation of the side-chain can be brought about (*see* Section 12.5.2(i)).

(iii) Nitration

Owing to the effect of the alkyl group, nitration is slightly easier than for benzene, although the nitrating mixture must still be used. Methylbenzene yields a mixture of methyl-2-nitro- and methyl-4-nitrobenzenes:

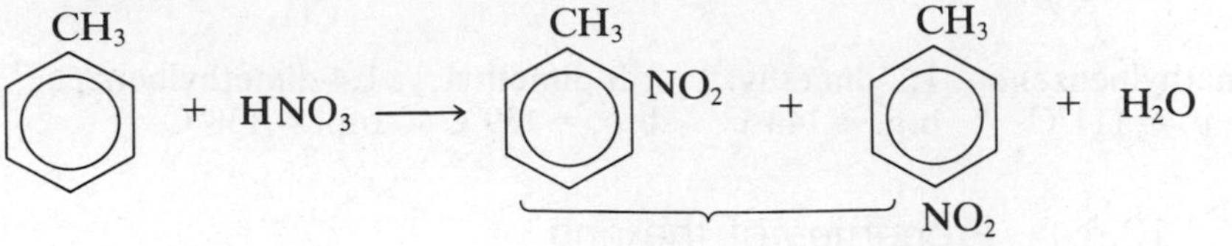

methyl-2-nitro-   methyl-4-nitro-
benzene   benzene
b.p. = 222°C   m.p. = 52°C

At higher temperatures, two, and finally three, nitro groups can be substituted in the ring. Methyl-2,4,6-trinitrobenzene is a pale yellow solid, easier to prepare than trinitrobenzene. Formerly known as TNT (*trinitrotoluene*), it is manufactured by the continuous nitration of methylbenzene with a H_2SO_4–HNO_3 mixture under vigorous conditions, and has been widely used as an explosive. It has the advantages that it is not too sensitive to shock and can be melted fairly safely:

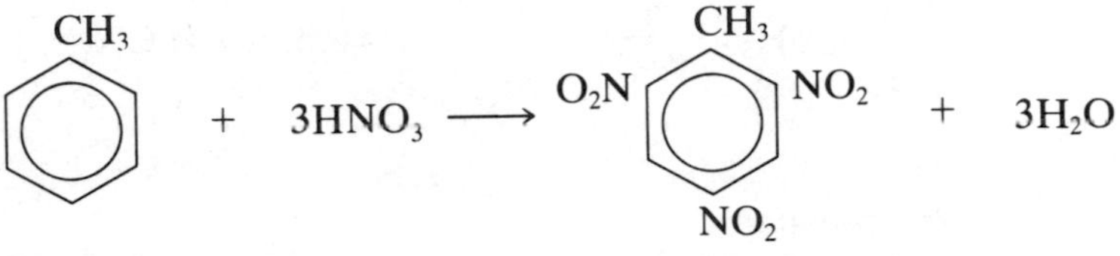

methyl-2,4,6-
trinitrobenzene
m.p. = 81°C

(iv) Sulphonation

Methylbenzene can be sulphonated with concentrated sulphuric acid, when it yields a mixture of sulphonic acids:

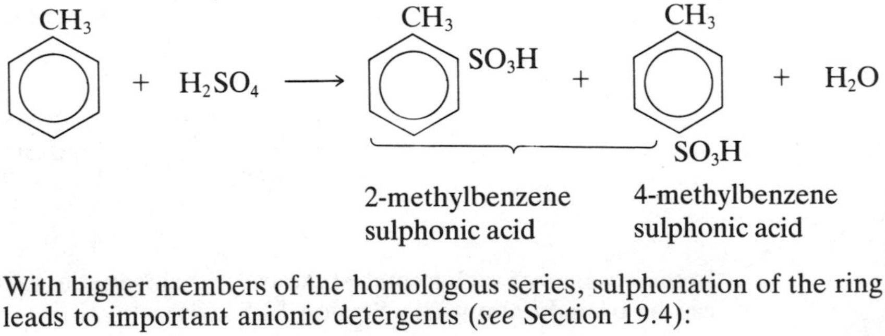

2-methylbenzene   4-methylbenzene
sulphonic acid   sulphonic acid

With higher members of the homologous series, sulphonation of the ring leads to important anionic detergents (*see* Section 19.4):

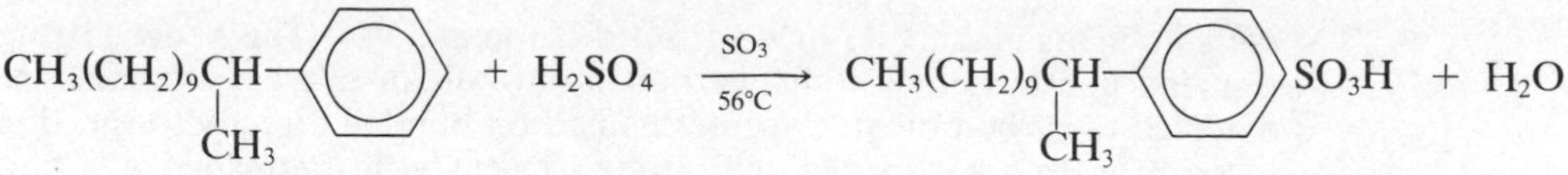

$$CH_3(CH_2)_9CH{-}\langle\bigcirc\rangle \;+\; H_2SO_4 \;\xrightarrow[56^\circ C]{SO_3}\; CH_3(CH_2)_9CH{-}\langle\bigcirc\rangle SO_3H \;+\; H_2O$$

with CH_3 substituent below the first ring and CH_3 substituent below the product ring.

(v) Friedel–Crafts reaction

These reactions take place more readily than with benzene, owing to the presence of the alkyl group which activates the ring. This can be a drawback where **alkylation** is concerned, since polysubstitution tends to take place, but, since the acyl group RCO is electron-withdrawing, **acylation** tends to give only a mono-substituted alkylbenzene. Thus:

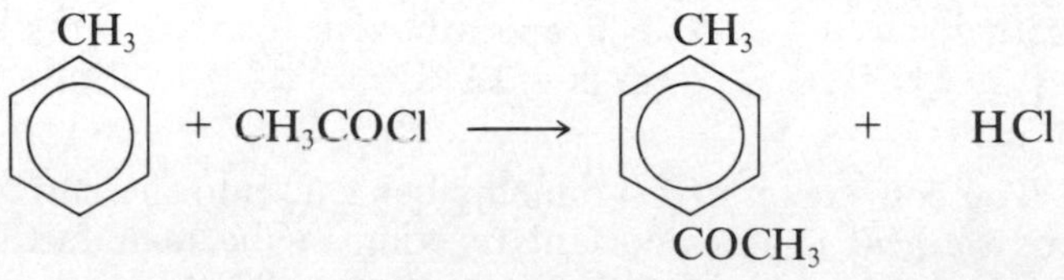

(4-methylphenyl)ethanone
b.p. = 225°C

12.5.2   Properties of the side-chain

Most of the reactions of the alkylbenzenes are reactions of the ring, but two important reactions are carried out, under the appropriate conditions, on the alkyl side-chain.

(i) Halogenation

The hydrogen atoms of the alkyl side-chain can be substituted by halogen atoms under the same reaction conditions as are used for alkanes (*see* Section 2.4.3). Thus, when chlorine is bubbled into boiling methylbenzene, exposed to ultraviolet light, substitution takes place in the methyl group. The degree of substitution can be controlled by limiting the uptake of chlorine by measuring the increase in mass of the system as chlorination proceeds:

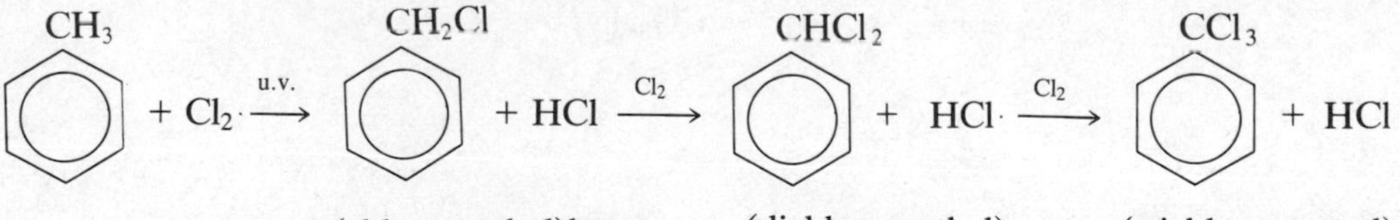

(chloromethyl)benzene	(dichloromethyl)	(trichloromethyl)
b.p. = 179°C	benzene	benzene
	b.p. = 206°C	b.p. = 221°C

(ii) Oxidation

Any alkyl group attached directly to the ring is susceptible to oxidation under vigorously oxidizing conditions, such as heating under reflux with *alkaline* manganate(VII) or *acidified* dichromate(VI). The benzene ring remains unattacked but the carbon atoms of an alkyl side-chain are oxidized to carbon dioxide, while the carbon bonded directly to the ring is oxidized to a carboxylic acid group. Thus, both methylbenzene and ethylbenzene yield benzenecarboxylic acid (*benzoic acid*) on oxidation:

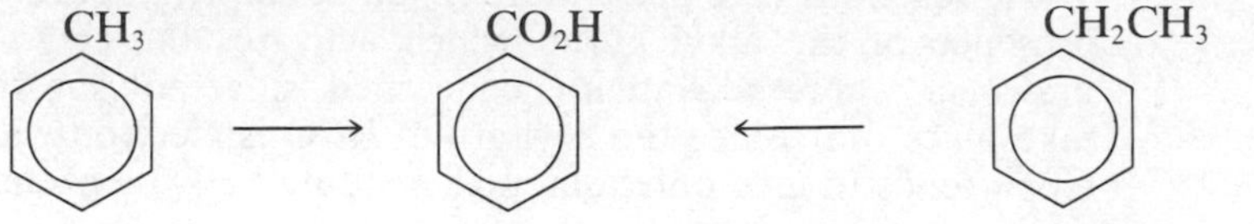

methylbenzene   benzenecarboxylic acid   ethylbenzene
b.p. = 111°C   m.p. = 122°C   b.p. = 136°C

The conversion of 1,4-dimethylbenzene into the corresponding dicarboxylic acid is an important reaction in the manufacture of polyester synthetic fibres like Terylene (*see* Section 20.3(ii)):

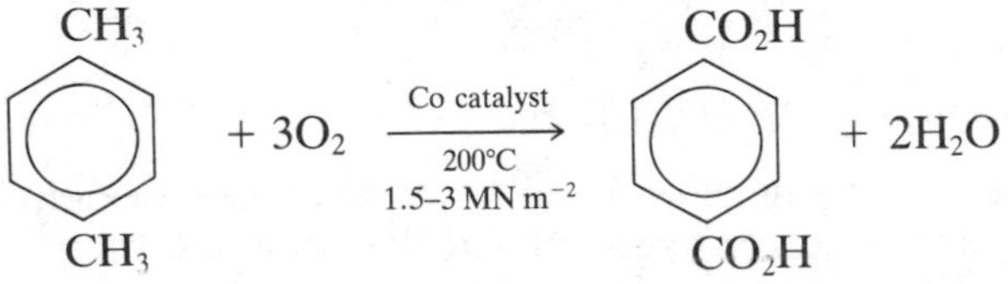

The 1,2-dimethylbenzene is also oxidized commercially into benzene-1,2-dicarboxylic anhydride (*phthalic anhydride*), which is used in the manufacture of alkyd resins and plasticizers (*see* Section 20.2.2(iv)):

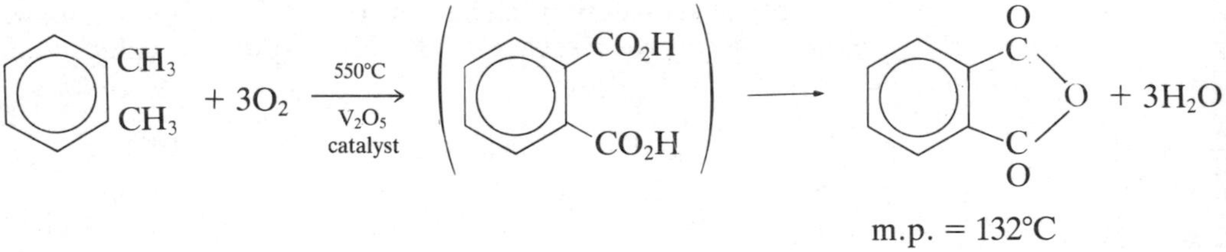

m.p. = 132°C

12.6 Manufacture and uses of aromatic hydrocarbons

Benzene is mainly obtained from the C_6–C_8 fraction of oil distillation, after it has been catalytically reformed (*see* Section 18.3.2) and alkanes and cycloalkanes removed by continuous extraction. Some is also obtained

from coal. Methylbenzene and the dimethylbenzenes (*xylenes*) are
obtained in a similar manner and, in fact, much methylbenzene is converted into benzene by **hydrodealkylation**:

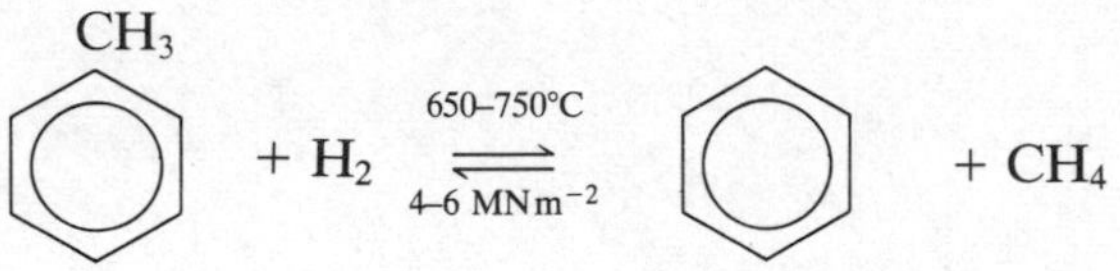

$$\text{(methylbenzene)} + H_2 \underset{4-6\ MNm^{-2}}{\overset{650-750°C}{\rightleftharpoons}} \text{(benzene)} + CH_4$$

Benzene is used to boost the octane rating of gasoline and acts as a basic
feedstock for a wide variety of manufacturing processes, some of which
are summarized below:

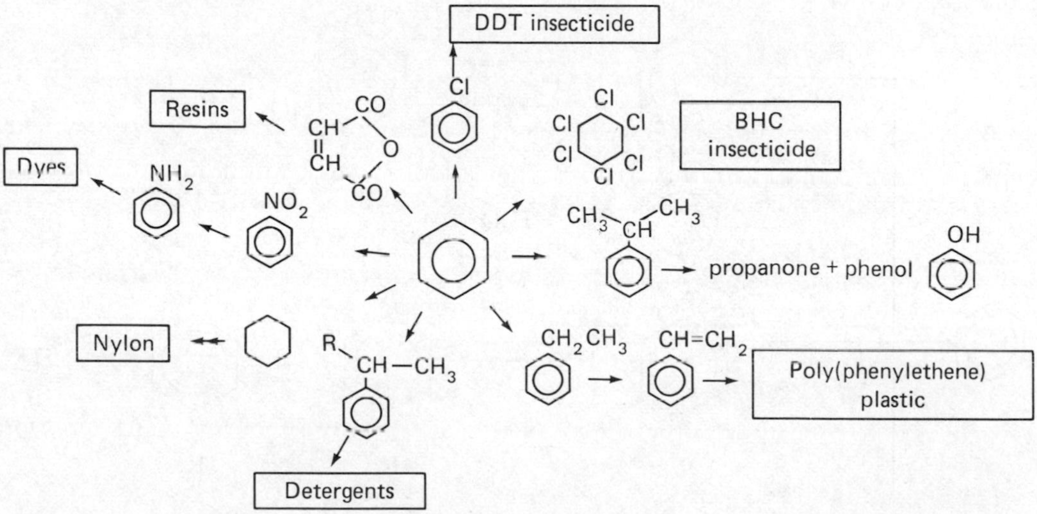

Methylbenzene, too, is used for blending with gasoline and is also a
starting material for the manufacture of explosives (for example, TNT),
polyurethanes (*see* Section 20.2.2(iii)) and the sweetener saccharin which
is 500 times as sweet as sucrose.

The dimethylbenzenes are obtained from catalytic reformate (*see* Section 18.3). The 1,4-isomer is separated by fractional crystallization and
is used as a starting material for polyester fibre manufacture (*see* Section
12.5.2(ii) and Section 20.3(ii)). The 1,2-isomer can be separated by
distillation and is used in the manufacture of alkyd resins and plasticizers
(*see* Section 12.5.2(ii) and Section 20.2.2(iv)).

Summary

1   Benzene is the simplest member of the **aromatic** class of hydrocarbons.
2   Although formally unsaturated, benzene and its homologues show little of the reactivity of alkenes, though benzene does participate in some **addition** reactions:

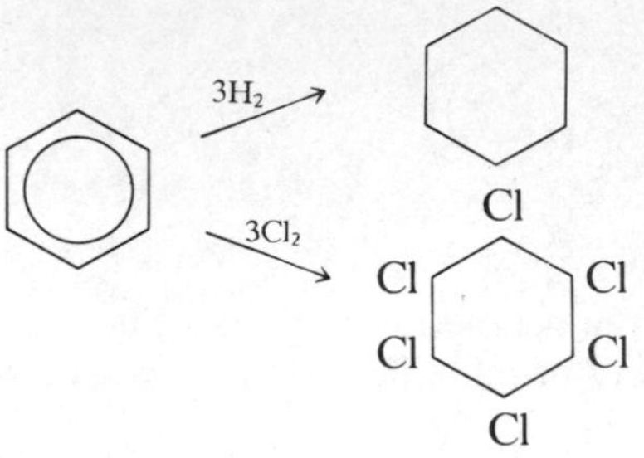

3   Much more characteristic of benzene and its homologues are the many **substitution** reactions involving electrophilic attack on the ring system:

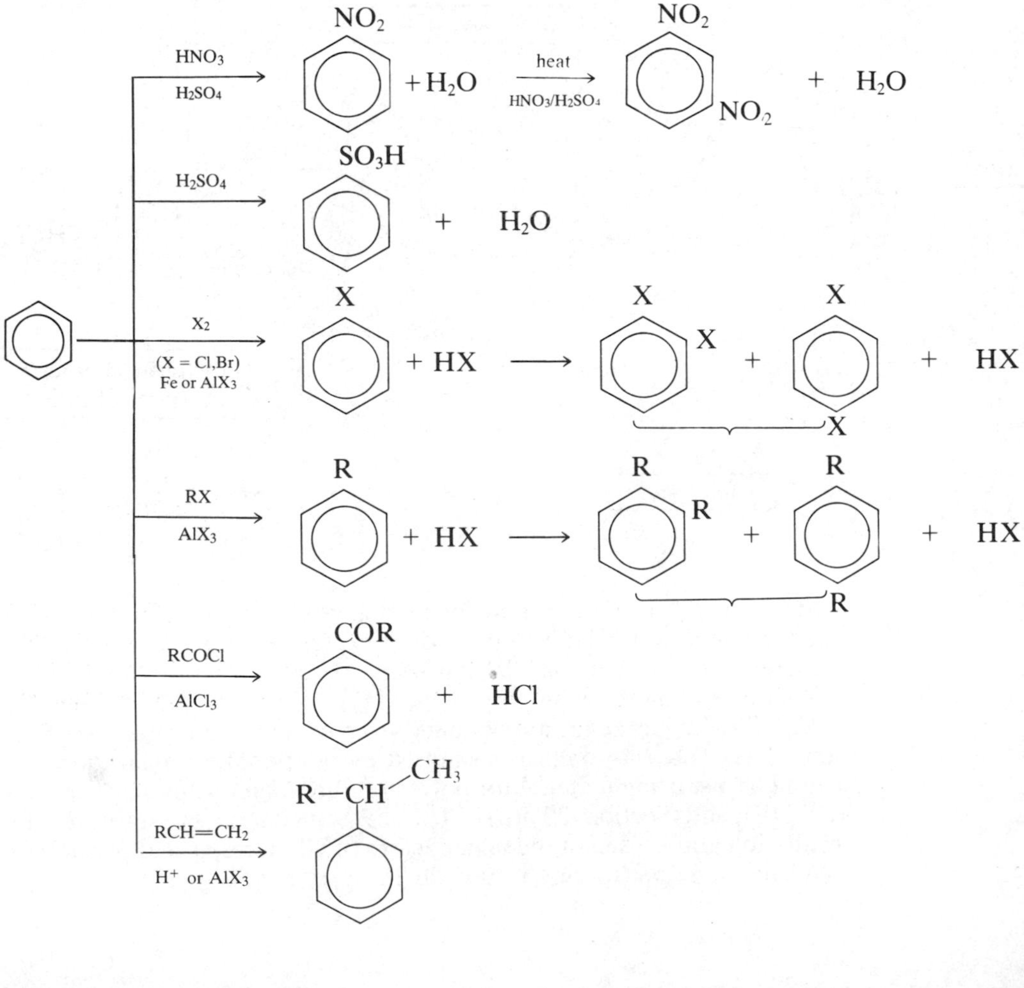

4   Alkylbenzenes undergo similar substitution reactions in the ring. Alkyl
    groups activate the ring towards electrophilic substitution and direct an
    incoming group into the 2- and 4-positions in the ring:

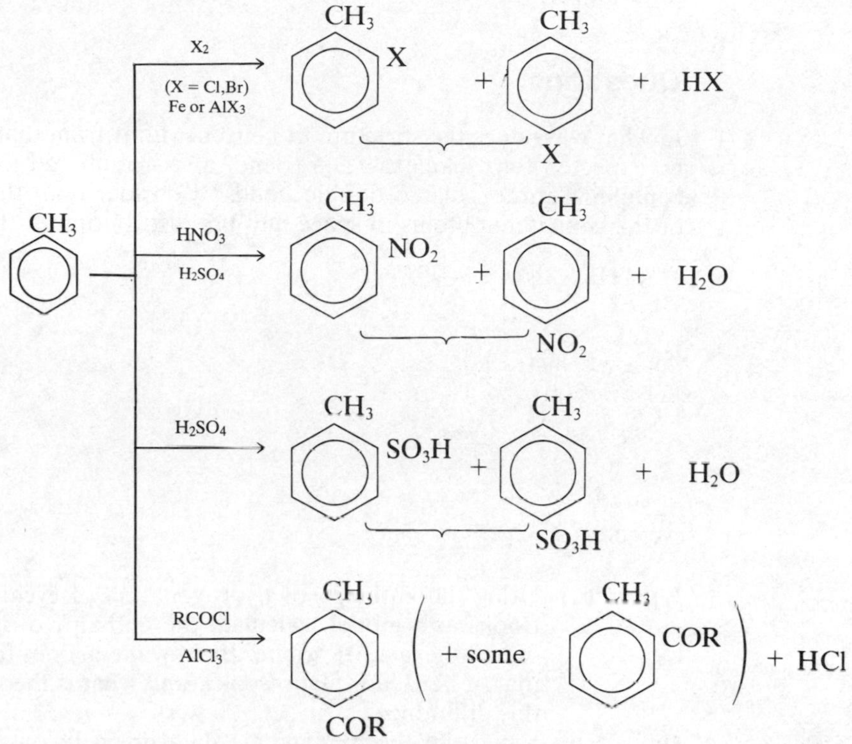

5   Alkylbenzenes can also undergo reaction at the alkyl side-chain, although
    under very different reaction conditions from those used for substitution
    in the ring, e.g.

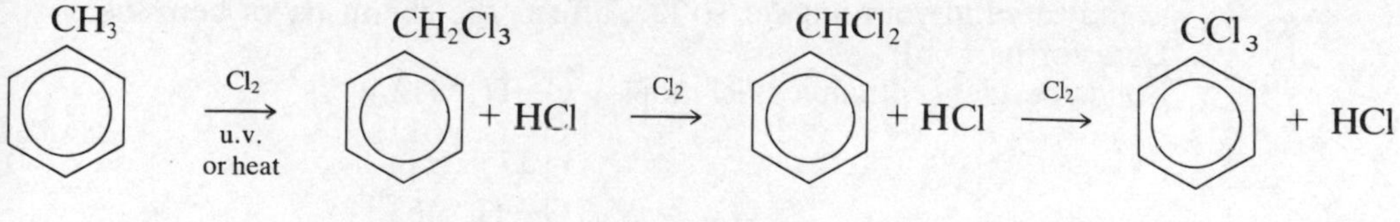

The alkyl groups are also oxidized under strongly oxidizing conditions,
e.g. heating with alkaline manganate(VII) solution. All but one of the
carbon atoms of the side-chain are lost as carbon dioxide, and the organic
product is an aromatic carboxylic acid, e.g.

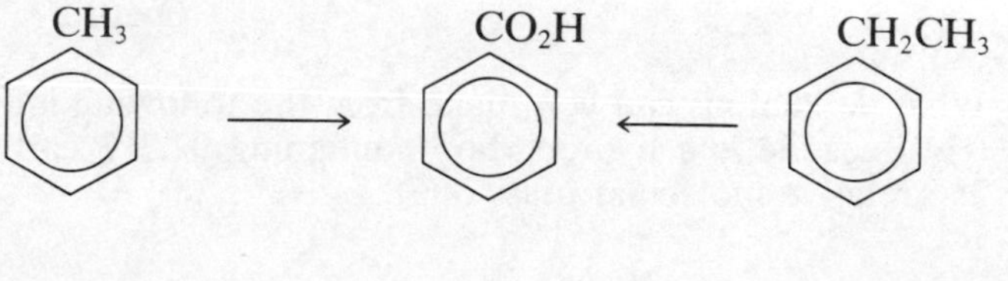

6   Benzene and the methylbenzenes are obtained from the catalytic reform-
    ing of oil fractions (to a lesser extent, they are also obtained from coal)
    and are used as feedstocks in a wide variety of organic manufacturing
    processes.

Questions

1   In what ways does the structure of benzene differ from that which would
    be expected for cyclohexa-1,3,5-triene, a six-membered ring compound
    containing three isolated double bonds? Consider both the distribution
    of the constituent atoms in space and the distribution of the electrons.

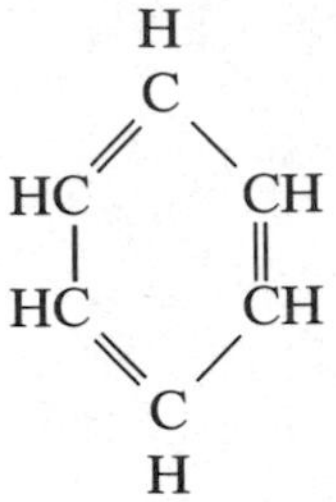

cyclohexa-1,3,5-triene

Either, (a) Using the enthalpy of hydrogenation of cyclohexene (Data
                Book) as a model, calculate the enthalpy of hydrogenation
                of cyclohexa-1,3,5-triene. By how much does this differ from
                that of benzene (Data Book), and what is the significance of
                this difference?

Or   (b) Using the thermodynamic data given below, calculate the
                enthalpy of combustion of gaseous cyclohexa-1,3,5-triene.
                By how much does this differ from that of benzene (see
                below) and what is the significance of this difference?

Give, with explanations, **two** important ways in which its special nature,
as indicated in your answer so far, affects the chemistry of benzene.
Data for part (b):

Bond energies (enthalpies)/kJ mol^{-1}
C—H	412
C=C	612
C—C	348
O—H	463
C=O	806 (in CO_2)
O=O	500 (in O_2)

Enthalpy of combustion of gaseous benzene -3314 kJ mol^{-1}
Enthalpy of vaporization of water $+41$ kJ mol^{-1}

                                                    Oxford & Cambridge

2* (a) What deductions can you make from the following information?
        (i) Benzene is a hydrocarbon containing 92.3% carbon and has a
             relative molecular mass of 78.

(ii) Under the influence of ultraviolet light, 1 mol of benzene reacts with 3 mol of chlorine without the production of hydrogen chloride.

(iii) Methylbenzene (toluene) reacts with alkaline potassium permanganate solution but benzene does not.

(iv) Benzene reacts with bromine under suitable conditions to substitute one hydrogen atom per molecule for one bromine atom. When this is done, there is only one organic product.

(v) $CH_3CH{=}CH_2(g) + H_2(g) \rightarrow CH_3CH_2CH_3(g)$ $\Delta H = -126$ kJ

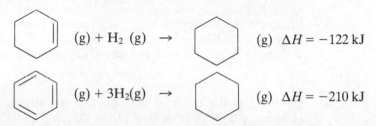

$(g) + H_2(g) \rightarrow$ $\quad\quad (g)$ $\Delta H = -122$ kJ

$(g) + 3H_2(g) \rightarrow$ $\quad\quad (g)$ $\Delta H = -210$ kJ

(b) Show, with detailed mechanistic equations, how benzene may be converted into (i) phenylethene (*styrene*),($C_6H_5CH{=}CH_2$), (ii) phenylethanone (*acetophenone*), ($C_6H_5COCH_3$).

JMB

3 The reaction between methylbenzene (*toluene*) and chlorine in the presence of ultraviolet light is said to proceed via a free-radical mechanism.
(a) State two characteristics of a free radical.
(b) Name one of the products of the reaction between methylbenzene and chlorine under these conditions and write an equation representing its formation.
(c) When methylbenzene reacts with chlorine in the dark in the presence of iron(III) chloride, the reaction proceeds by a different mechanism and different products are obtained.
   (i) How is the mechanism of the reaction under these conditions usually described?
   (ii) Name one of the organic products obtained by the reaction of chlorine with methylbenzene in the presence of iron(III) chloride and write an equation representing its formation.
(d) Give one other example from organic chemistry to illustrate how experimental conditions may affect the products of a reaction between two given reactants.

JMB

4 (a) The following reaction sequence can be used to prepare phenylethene (*styrene*) from benzene:

$$C_6H_6 \xrightarrow{\text{(i)}} C_6H_5COCH_3 \xrightarrow{\text{(ii)}} C_6H_5CH(OH)CH_3 \xrightarrow{\text{(iii)}} \text{phenylethene}$$

State the essential reagents required for each of the steps (i), (ii), and (iii).

(b) (i) Write down the structural formula of phenylethene.
    (ii) Write down the structural formula of the product you would
         expect to be formed in the ionic addition of HBr to phenylethene.
(c) Describe briefly how a specimen of poly(phenylethene) (*polystyrene*)
    can be produced from phenylethene.
(d) Draw the structure of poly(phenylethene) showing the repeating unit
    of the polymer.

JMB

5   Describe typical reactions of benzene with electrophilic reagents. Write
    a mechanism for the nitration of benzene and explain why benzene
    preferentially undergoes substitution rather than addition.
    Suggest a method of preparing the alcohol $C_6H_5CH(OH)CH_3$, using
    ethanoyl (*acetyl*) chloride and any other common reagents.

JMB

6* Describe, giving essential reagents and conditions, how methylbenzene
    (*toluene*) can be brominated (a) in the ring and (b) in the side chain.
    Write the mechanisms of the reactions involved.
    An aromatic compound $X, C_7H_6Br_2$, was converted into an alcohol
    Y, C_7H_7BrO, by heating with aqueous potassium hydroxide. The alcohol
    Y was oxidized with aqueous potassium manganate(VII); this reaction
    produced 4-bromobenzenecarboxylic (*4-bromobenzoic*) acid.
    Identify X and Y, write equations for the reactions described, and give
    the mechanism of the conversion of X into Y. Write down the structures
    of the products you would expect to be formed by the reaction of X (i)
    with ethanolic ammonia and (ii) with aqueous potassium cyanide.

JMB

7* Starting with methylbenzene (*toluene*) how could five of the following be
    synthesized in one or more steps?
    (a) bromo-4-methylbenzene   (b) (4-methylphenyl)methanol
    (c) (bromomethyl)benzene   (d) (cyanomethyl)benzene
    (e) 4-methylphenylamine   (f) benzenecarbonyl chloride

    Where possible, give the mechanisms of the reactions involved.

Imperial College

Chapter 13
Derivatives of benzene

13.1 Introduction

The purpose of this chapter is to consider the reactions and properties of the main functional groups when they are bonded directly to a benzene ring. Particular emphasis is placed on the way in which the properties of the group are influenced, or even modified, by the aromatic ring and how the functional group, in turn, may modify the properties, or reactivity, of the benzene ring. A separate chapter is devoted to the aromatic amines because of their synthetic importance.

When a functional group is separated from an aromatic ring by one or more carbon atoms, i.e. when it is in a side-chain, there is little, if any, electronic interaction and the properties of the functional group are almost invariably identical to aliphatic counterparts. Little mention is therefore made of such compounds in this chapter; their chemistry may be considered to have been discussed in previous chapters.

13.2 Aromatic halogeno compounds

Chloro- and bromo-substituted benzenes can be prepared by the direct halogenation of benzene using a halogen carrier as catalyst (see Section 12.4.2(iii)).

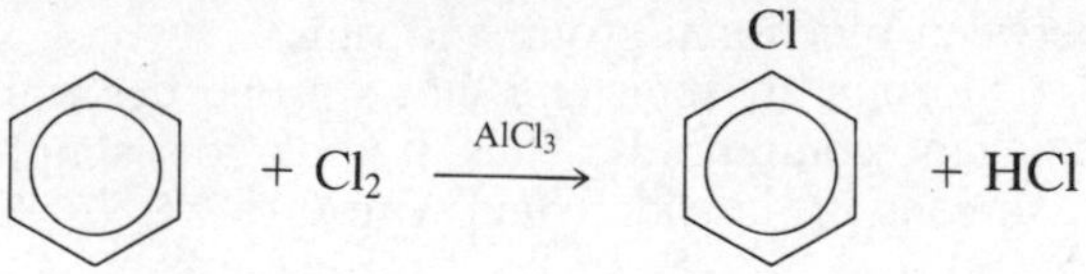

$$\text{(benzene)} + Cl_2 \xrightarrow{AlCl_3} \text{(chlorobenzene)} + HCl$$

chlorobenzene
b.p. = 132°C

Under these conditions, iodine has little action on benzene and the iodo-derivatives are usually prepared indirectly via diazonium salts (see Section 14.3.1(iii)). With alkylbenzenes, substitution in the side-chain is brought about by reacting the halogen with boiling alkylbenzene, in the presence of ultraviolet light:

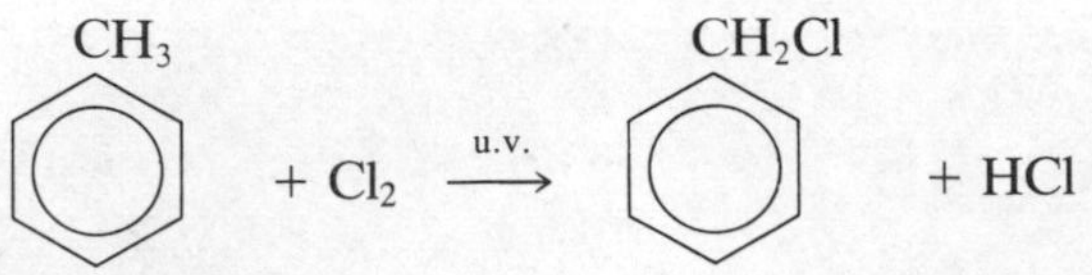

(chloromethyl)benzene
b.p. = 179°C

The reaction proceeds by way of a free-radical mechanism and occurs more easily than the photohalogenation of methane (*see* Section 2.4.3). As halogenation proceeds, the mass of the organic product increases and halogenation can be stopped at the required stage by following the uptake of halogen by weighing.

In any alkyl side-chain, halogen substitution takes place first at the carbon atom bonded directly to the aromatic ring, since an unpaired electron on such a carbon atom is stabilized by electronic interaction with the ring:

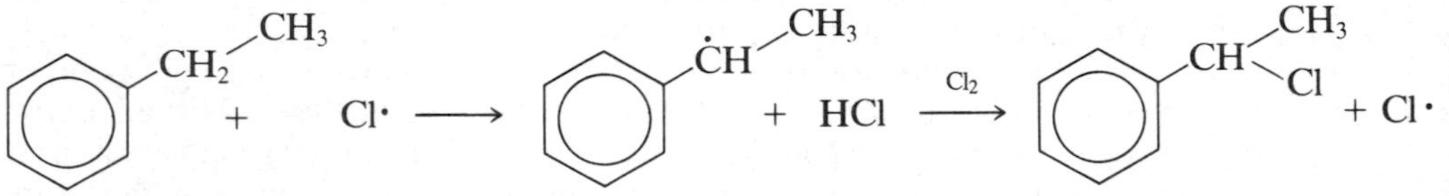

13.2.1 Chemical properties of aromatic halogeno compounds

Chlorobenzene is a colourless liquid of boiling point 132°C, which is close to the boiling point of a hydrocarbon of similar relative molecular mass. It is immiscible with water and can be steam distilled (*see* Section 21.2.2(iii)). (Chloromethyl)benzene, too, is a colourless liquid (b.p. = 179°C), and should be handled with care as it is a powerful lachrymator which can cause intense irritation to the eyes. A comparison of the properties of these two compounds well illustrates the effect of electronic interaction between functional group and ring.

In general, (chloromethyl)benzene shows properties which are typical of aliphatic halogeno compounds. Thus, it undergoes simple nucleophilic substitution reactions in the side-chain with OH^-, CN^-, OR^- and NH_3 (*see* Section 5.3.1), e.g.

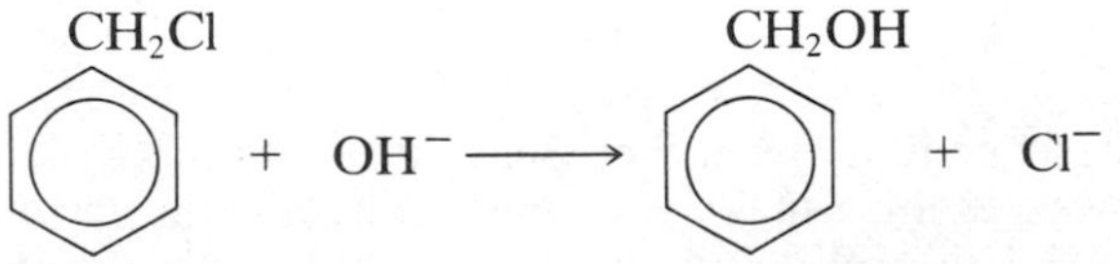

phenylmethanol
b.p. = 206°C

Dichloromethyl and trichloromethyl benzenes also undergo hydrolysis, when the products are an aldehyde and a carboxylic acid, respectively (substitution is followed by elimination):

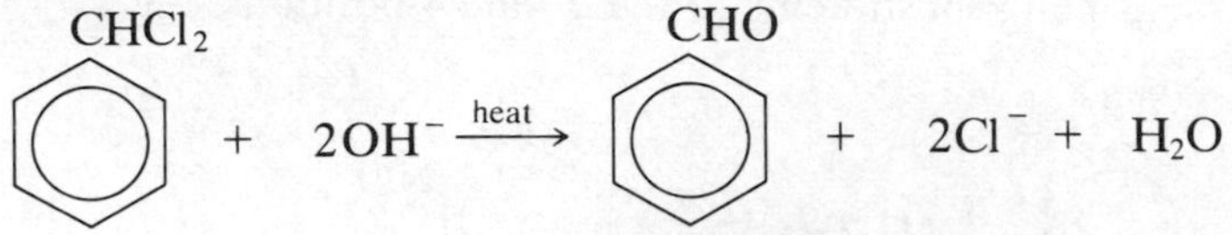

$CHCl_2$–C$_6$H$_5$ + $2OH^-$ $\xrightarrow{\text{heat}}$ CHO–C$_6$H$_5$ + $2Cl^-$ + H_2O

benzenecarbaldehyde
b.p. = 179°C

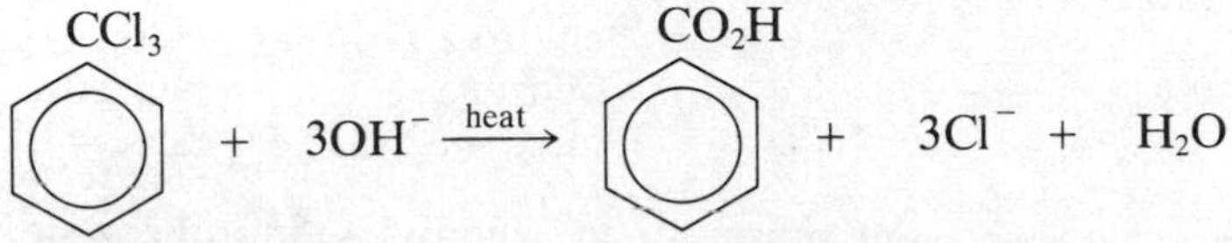

CCl_3–C$_6$H$_5$ + $3OH^-$ $\xrightarrow{\text{heat}}$ CO_2H–C$_6$H$_5$ + $3Cl^-$ + H_2O

benzenecarboxylic acid
m.p. = 122°C

In contrast, chlorobenzene is unreactive and reacts with none of these reagents under ordinary conditions. It *can* be hydrolysed but the conditions are steam at 300°C and 15 MN m^{-2}; it is unattacked by boiling water or alkali:

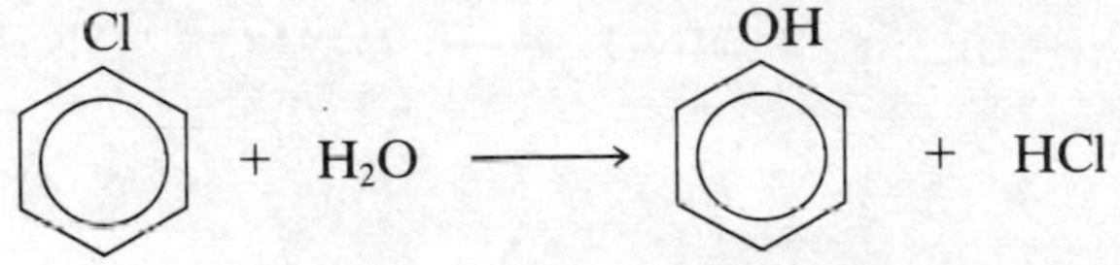

Cl–C$_6$H$_5$ + H_2O $\longrightarrow$ OH–C$_6$H$_5$ + HCl

phenol
m.p. = 41°C

Bromo- and iodo-benzenes also readily form a Grignard compound (*see* Section 5.3.4), when treated with magnesium in an ether solvent, and these derivatives are extremely useful in organic synthesis:

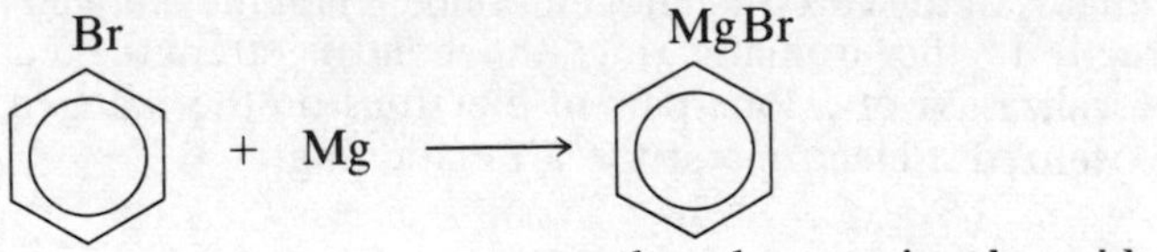

Br–C$_6$H$_5$ + Mg $\longrightarrow$ MgBr–C$_6$H$_5$

phenylmagnesium bromide

e.g.

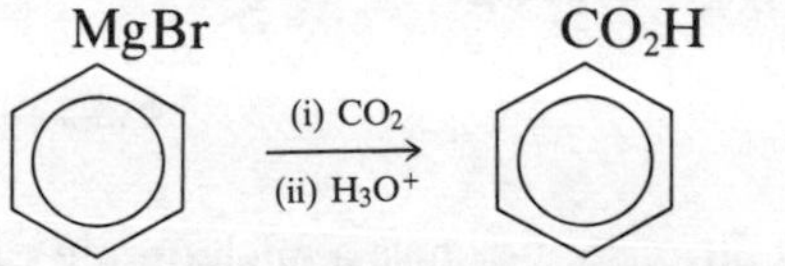

MgBr–C$_6$H$_5$ $\xrightarrow[\text{(ii) } H_3O^+]{\text{(i) } CO_2}$ CO_2H–C$_6$H$_5$

benzenecarboxylic acid
m.p. = 122°C

Halogeno benzenes are more difficult to nitrate or sulphonate, for example, than benzene because the halogen atom bonded directly to the ring deactivates the ring towards electrophilic attack, although it directs incoming substituents to the 2- and 4-positions, e.g.

$$\text{Cl} \quad + \quad HNO_3 \quad \longrightarrow \quad \text{Cl} \quad NO_2 \quad + \quad \text{Cl} \quad + \quad H_2O$$

$$\overbrace{}^{} NO_2$$

chloro-2-nitro-   chloro-4-nitro-
benzene   benzene
m.p. = 32°C   m.p. = 84°C

However, the presence, in chloro-2,4-dinitrobenzene, of two nitro groups in the 2- and 4-positions makes the chlorine atom more easily replaced than in chlorobenzene, and the compound undergoes substitution with OH⁻ and ⁻OCH₃, for example. Its reaction with hydrazine is important as the product is 2,4-dinitrophenylhydrazine, used in the identification of carbonyl compounds (*see* Section 8.3.3):

$$O_2N \quad NO_2 \quad Cl \quad + \quad H_2NNH_2 \quad \longrightarrow \quad O_2N \quad NO_2 \quad NHNH_2 \quad + \quad HCl$$

m.p. = 197°C

The corresponding fluorodinitrobenzene is used extensively in peptide chemistry to determine the N-terminal amino-acid in a peptide chain (*see* Section 17.9). Both the chloro and fluoro compounds must be handled with great care as they can act as extreme skin irritants.

The explanation for the difference in reactivity between ring substituted and side-chain substituted halogeno compounds involves the relative bond strengths of the carbon–halogen bond. When the halogen atom is bonded directly to the aromatic ring, the bond is strengthened by the partial delocalization of a lone-pair of electrons on the halogen atom with the delocalized π-electrons of the aromatic ring:

Substitution is more difficult, also, because a nucleophile cannot approach from behind the carbon atom, along a line coaxial with the carbon–halogen bond (cf. S_{N^2} mechanism, Section 5.3.2).

When, however, another carbon atom intervenes between the ring and the halogen atom, as for example in (chloromethyl)benzene, then delocalization of some of the halogen's electrons with the π-electrons of the ring becomes impossible and the carbon–halogen bond is not further strengthened. (Chloromethyl)benzene behaves similarly to aliphatic halogenoalkanes, although halogeno compounds of the type $ArCH_2X$ (where Ar = an aryl group) are actually more reactive than the RCH_2X type (where R = an alkyl group).

13.2.2   Manufacture and uses of aromatic halogeno compounds

The chlorobenzenes are of some synthetic importance. Chlorination of benzene at 40–60°C, in the presence of aluminium chloride as catalyst, will give the mono- or di-substituted products, depending on the ratio of chlorine to benzene used:

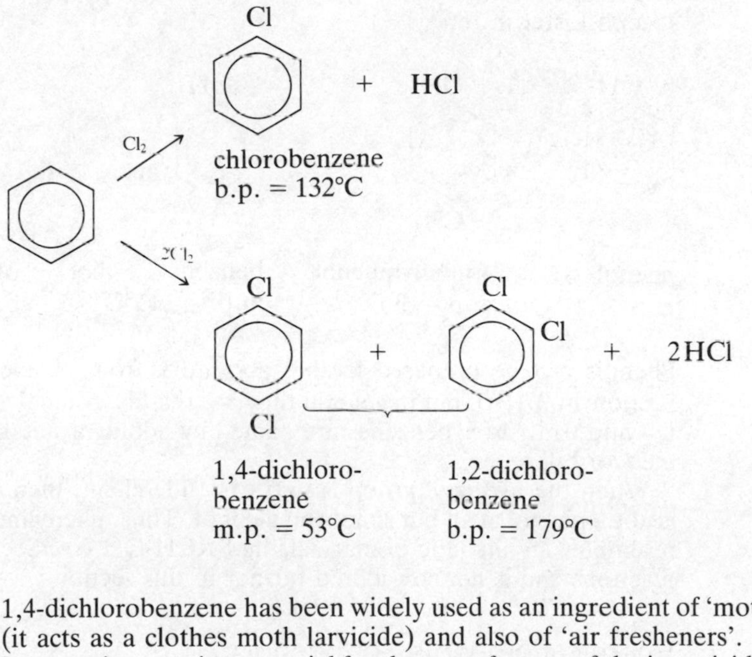

chlorobenzene
b.p. = 132°C

1,4-dichloro-
benzene
m.p. = 53°C

1,2-dichloro-
benzene
b.p. = 179°C

1,4-dichlorobenzene has been widely used as an ingredient of 'moth-balls' (it acts as a clothes moth larvicide) and also of 'air fresheners'. Chlorobenzene is a starting material for the manufacture of the insecticide DDT, which is made by the acid-catalysed condensation of chlorobenzene with trichloroethanal:

DDT
m.p. = 109°C

DDT (*dichlorodiphenyltrichloroethane*) was first made in 1874, but its insecticidal properties were not discovered until 1939 by Paul Müller of Geigy Ltd, who was awarded the Nobel Prize in 1948. It was introduced commercially in 1942 and has played an outstanding role in the eradication of malaria and typhus. Its uncontrolled use, allied to its persistence in the environment, has produced some unwanted ecological effects, but it is still important in developing countries where pests can regularly destroy up to one-third of food crops during growth and storage.

13.3 Phenols and hydroxy derivatives

Aromatic compounds in which a hydroxyl group is attached directly to the benzene ring belong to the class called **phenols**. The simplest example is phenol itself, formerly known as *carbolic acid*; it was the first antiseptic to be used in surgery, being introduced for this purpose by the surgeon Joseph Lister in 1867.

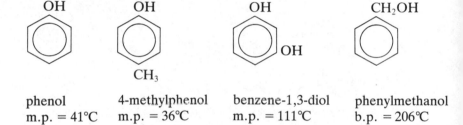

phenol   4-methylphenol   benzene-1,3-diol   phenylmethanol
m.p. = 41°C   m.p. = 36°C   m.p. = 111°C   b.p. = 206°C

Phenols can be prepared in the laboratory from diazonium salts (*see* Section 14.3.1(i)), but in general they are readily available commercially. Di- and tri-hydroxybenzenes are named by adding a suitable *-ol* suffix to the word benzene.

When the hydroxyl group is part of a side-chain, then the compound is not a phenol at all but simply an alcohol. Thus, phenylmethanol closely resembles an aliphatic primary alcohol RCH_2OH (*see* Section 6.3) in its reactions and is not considered further in this section.

13.3.1 Chemical properties of phenols

Phenol itself is a colourless, hygroscopic, crystalline solid of melting point 41°C. It gradually turns pale pink in air (probably owing to oxidation), is partly miscible with water and is readily soluble in many organic solvents. It is toxic and can cause very severe burns if left in contact with the skin for any length of time; it should therefore be handled with the greatest care. It is readily available and it is not necessary to prepare it in the laboratory.

The hydroxyl group, when directly bonded to the benzene ring, influences the reactivity of the ring and the latter, in turn, considerably modifies the properties of the hydroxyl group, compared with alcohols.

When one or more carbon atoms intervene between ring and hydroxyl group, these properties coincide with those of alcohols again (cf. phenylmethanol).

(i) Acidic properties

Phenol dissociates slightly in aqueous solution and is thus a weak acid (pK_a = 10). This contrasts sharply with the neutral character of aliphatic and aromatic alcohols (e.g. cyclohexanol pK_a = 18):

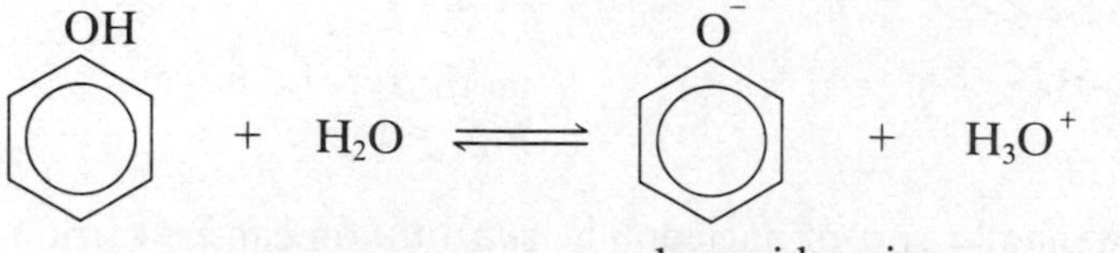

phenoxide anion

This enhanced acidity arises from stabilization of the negative charge on the phenoxide anion, which causes the equilibrium position to lie further over to the right-hand side of the equation than in the case of alcohols. The stabilization is caused by delocalization of the negative charge over all carbon atoms of the ring as well as the oxygen atom:

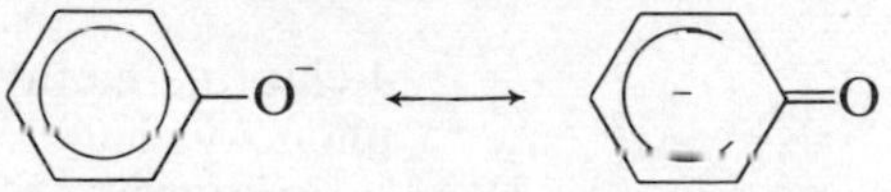

No such stabilization can occur with the alkoxide ion:

$$RCH_2OH + H_2O \rightleftharpoons RCH_2O^- + H_3O^+$$

Owing to their acidity, phenols will dissolve in aqueous alkali, no matter whether they are soluble in water or not; addition of excess acid will liberate the free phenol from its salt. However, phenols are weaker acids than carbonic acid H_2CO_3 (pK_a phenol = 10, pK_{a_1} H_2CO_3 = 6.4), and they will not liberate carbon dioxide from carbonates or hydrogencarbonates, as will carboxylic acids. Thus, such a test will readily distinguish a phenol from a carboxylic acid. A phenol may be separated from a carboxylic acid by the addition of aqueous carbonate, followed by extraction of the mixture with an organic solvent such as ethoxyethane; the acid remains in the aqueous layer as its salt, while the phenol is extracted into the organic layer.

The presence of electron-withdrawing substituents in the ring of a phenol will enhance its acidity, as the electron-withdrawal weakens the oxygen–hydrogen bond, e.g.

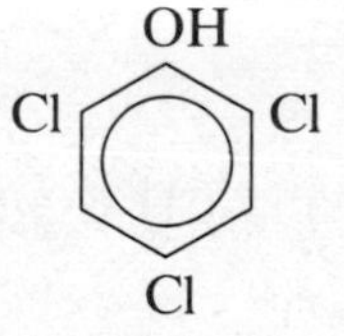

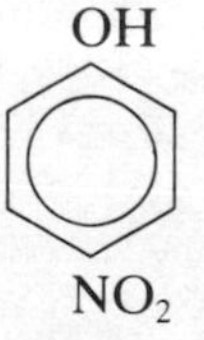

The phenoxide anion is analogous in some respects (*see* previous discussion) to the alkoxide ion ⁻OR and can be used in the Williamson ether synthesis (*see* Section 15.6.1), although the methyl and ethyl ethers are usually prepared using the dialkylsulphate in alkaline solution, e.g.

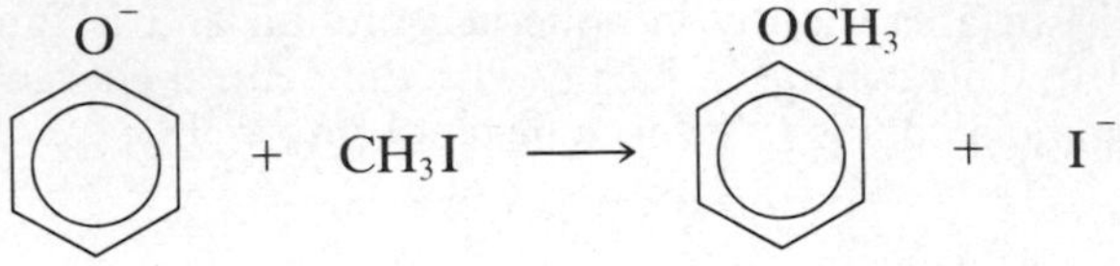

methoxybenzene
b.p. = 154°C

A similar sort of nucleophilic substitution can be carried out with halogenated carboxylic acids and some very important selective herbicides (the 'hormone' weedkillers) are manufactured in this way, e.g.

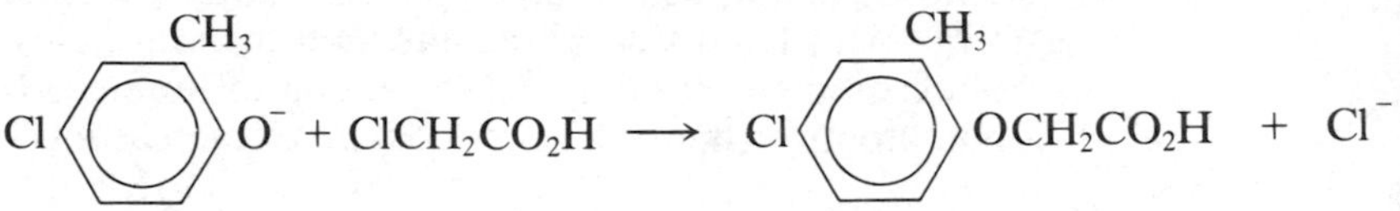

4-chloro-2-methyl
phenoxyethanoic acid
(also known as MCPA)

(ii) Ester formation

Like alcohols, phenols form esters but not under the same reaction conditions. Phenols will *not* react directly with carboxylic acids but will react with an acid chloride or acid anhydride (*see also* Section 13.5.1(iii)). The reaction is usually carried out in the presence of a base (e.g. pyridine), to remove the acid which is also formed, e.g.

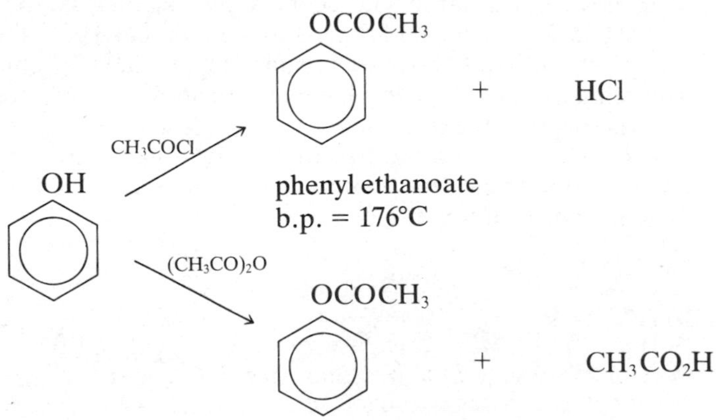

phenyl ethanoate
b.p. = 176°C

Thus, aspirin can be made as follows:

$$\underset{\substack{\text{2-hydroxybenzenecarboxylic acid} \\ \text{m.p.} = 159°C}}{\text{(benzene ring with CO}_2\text{H and OH)}} + (CH_3CO)_2O \longrightarrow \underset{\substack{\text{aspirin (2-ethanoyloxybenzenecarboxylic acid)} \\ \text{m.p.} = 136°C}}{\text{(benzene ring with CO}_2\text{H and OCOCH}_3)} + CH_3CO_2H$$

Under these reaction conditions, only the phenolic group takes part in ester formation.

An aromatic acid chloride, such as benzenecarbonyl chloride C_6H_5COCl, is considerably less reactive than an aliphatic acid chloride (*see* Section 13.5.1(iv)), and can be used in the presence of excess aqueous alkali. The alkali reacts with phenol to form the phenoxide anion, which is a better nucleophile than phenol itself, and also serves to remove hydrochloric acid formed during the reaction. Under these conditions the reaction is known as the **Schotten–Baumann reaction** (*see also* Section 14.2.2(ii)), after the German chemist Carl Schotten and a colleague, E. Baumann. The acid chloride is added to a solution of phenol in alkali and the whole shaken vigorously. The ester precipitates out:

$$\text{(benzene ring)}COCl + {}^-O\text{(benzene ring)} \longrightarrow \underset{\substack{\text{phenyl benzenecarboxylate} \\ \text{m.p.} = 69°C}}{\text{(benzene ring)}COO\text{(benzene ring)}} + Cl^-$$

(iii) Coloration with iron(III) chloride

Phenols generally give a characteristic blue or violet colour (not a precipitate), when added to a *neutralized* aqueous solution of iron(III) chloride. Solutions of iron(III) salts are acidic; a neutral solution is formed by adding ammonia to the iron(III) solution until a precipitate of iron(III) hydroxide just appears, followed by drops of the iron(III) chloride solution until the red-brown precipitate just disappears. This is a useful, simple test for phenols.

In alcohols, the hydroxyl group can be directly replaced by other groups such as halogens; this is not so with phenols. Unlike alcohols, phenols do not dehydrate and when they are oxidized they do not usually give simple products. The benzene ring influences the properties of the hydroxyl group in phenols and vice versa. The hydroxyl group has a powerfully *activating* effect on the benzene ring so that electrophilic substitution reactions of phenol are carried out much more easily than those of benzene itself. The ease of reaction is reflected in the very different, much milder, reaction conditions employed with phenol, as illustrated in the following reactions. The hydroxyl group directs incoming groups into the 2- and 4-positions.

(iv) Halogenation

No catalyst is required and halogenation takes place readily when an aqueous solution of chlorine or bromine is added to phenol (solid or in aqueous solution). Substitution occurs so readily that the *tri*halogenated product usually precipitates out. The turbidity produced is used sometimes as a test for phenol, e.g.

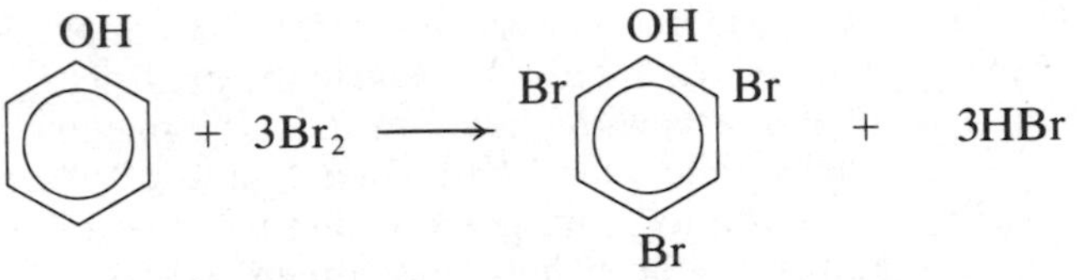

OH   OH
⬡ + 3Br$_2$ ⟶ Br⬡Br + 3HBr
                                    Br

2,4,6-tribromophenol
m.p. = 94°C

Thus, in the presence of excess phenol, the bromine solution is decolorized. An alkene also decolorizes bromine solution (*see* Section 3.3.1(ii)), but *no* precipitate is formed, so no confusion between the two need arise.

The halogenated phenols are much more acidic than phenol itself and will usually react with a carbonate (pK_a of 2,4,6-trichlorophenol is 7.6). They also possess a powerfully antiseptic action; such compounds are constituents of antiseptics like TCP and Dettol.

If 2,4,5-trichlorophenol is condensed with methanal, an important compound *hexachlorophene*, used as a mouthwash and in toothpastes, is obtained:

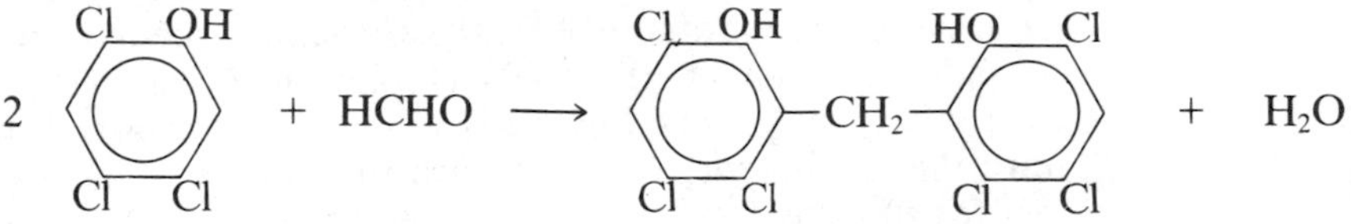

Cl OH   Cl OH   HO Cl
2 ⬡   + HCHO ⟶ ⬡—CH$_2$—⬡   + H$_2$O
Cl Cl   Cl Cl   Cl Cl

(v) Nitration

Nitration of phenol occurs about 1000 times faster than nitration of benzene, and it will proceed using dilute nitric acid, rather than the usual nitrating mixture. If the mixture is kept cool, a mixture of 2- and 4-nitrophenols is obtained:

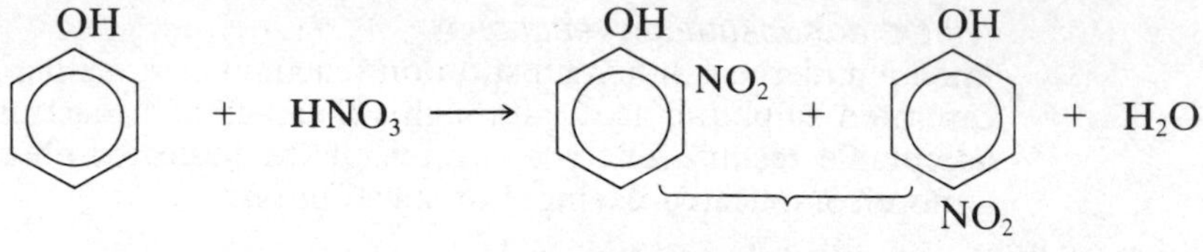

OH
⬡ + HNO₃ ⟶

OH
⬡ NO₂ +

OH
⬡ + H₂O
         NO₂

2-nitrophenol   4-nitrophenol
m.p. = 45°C   m.p. = 114°C

The 2-nitrophenol can be conveniently separated from the 4-isomer by steam distillation; the latter remains dissolved in hot water (*inter*molecular hydrogen bonding) while the 2-isomer, which exhibits *intra*molecular hydrogen bonding, is volatile in steam and distils over:

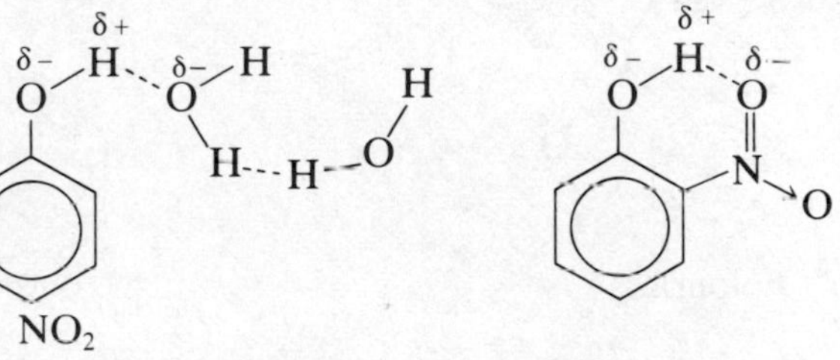

intermolecular
hydrogen bonding with water

intramolecular
hydrogen bonding

The use of dilute nitric acid as nitrating agent strongly suggests that a different mechanism is involved here, and it is thought that initial attack on the phenol is by the **nitrosyl** cation (or *nitrosonium* ion) NO^+ rather than the nitryl cation NO_2^+. Oxidation of the nitroso group to the nitro group then takes place.

If the conventional nitrating mixture of concentrated nitric and sulphuric acids is used, and the temperature is allowed to rise during the nitration, the final product is 2,4,6-trinitrophenol:

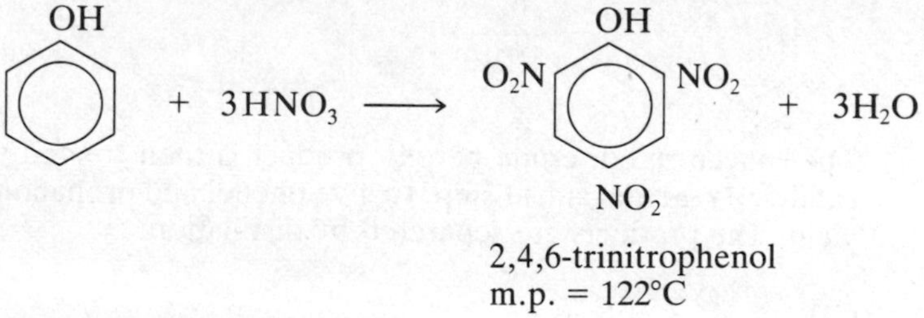

OH
⬡ + 3 HNO₃ ⟶

OH
O₂N ⬡ NO₂ + 3 H₂O
         NO₂

2,4,6-trinitrophenol
m.p. = 122°C

This is a yellow, crystalline compound, which was formerly used as a dye and is also a powerful explosive. The presence of three electron-withdrawing nitro groups in the ring enhances the acidity of the phenol very considerably; its pK_a is 0.4, and it was formerly known as *picric acid*.

(vi) Other substitution reactions

Phenol undergoes other substitution reactions, e.g. sulphonation by concentrated sulphuric acid, although Friedel–Crafts reactions do not give acceptable results. The preparation of the indicator phenolphthalein is a more complicated example of substitution:

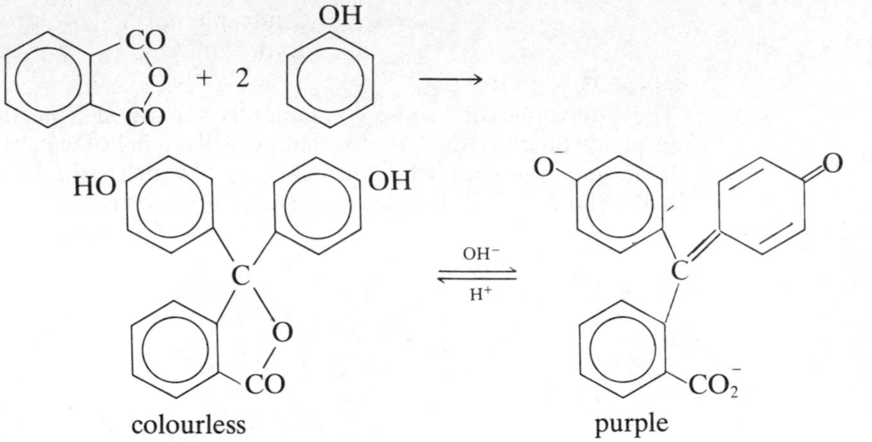

$$\text{phthalic anhydride} + 2\ \text{phenol} \longrightarrow$$

colourless $\underset{H^+}{\overset{OH^-}{\rightleftharpoons}}$ purple

13.3.2   Manufacture and uses of phenols

Phenol is now generally manufactured by the oxidation of (1-methylethyl)benzene (*see* Section 12.4.2(iv)), in a process developed in the UK by Distillers Co. Ltd. Its major advantage, and the reason for it replacing other routes, is that the co-product is propanone, which is required commercially in large quantities as a solvent. The first step of the process involves the oxidation of the hydrocarbon to a peroxo compound:

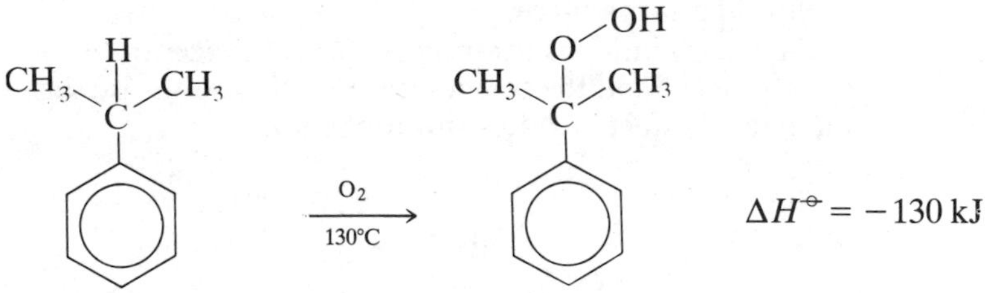

$$\xrightarrow[130°C]{O_2} \qquad \Delta H^{\ominus} = -130\ \text{kJ}$$

The concentrated, crude peroxo product is then treated with acid, in a catalysed **rearrangement** step, to give phenol and propanone in over 90% yield. The products are separated by distillation:

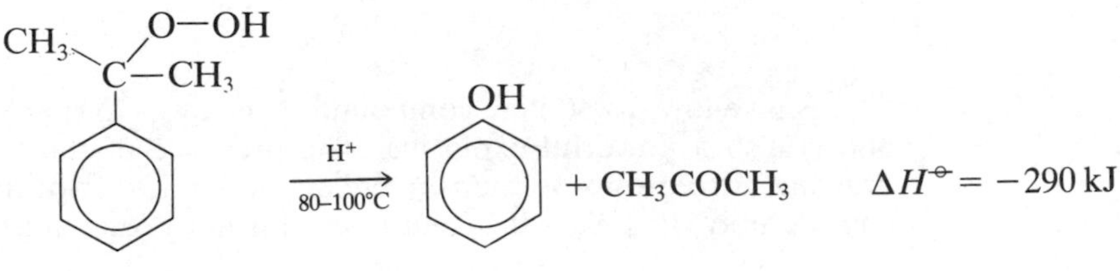

$$\xrightarrow[80-100°C]{H^+} \qquad + \ CH_3COCH_3 \qquad \Delta H^{\ominus} = -290\ \text{kJ}$$

Phenol is used extensively in the manufacture of thermosetting plastics such as Bakelite (*see* Section 20.2.2(i)), of explosives and of many pharmaceutical products. Polychlorinated phenols are used as antiseptics, fungicides and wood-preservers.

Hydrogenation of phenol to cyclohexanol is the start of one process for the manufacture of *caprolactam*, used to make nylon. This route is not as potentially dangerous as the catalytic air oxidation of the hydrocarbon cyclohexane (this was used by the Nypro Company at the time of the 1974 Flixborough explosion in the UK, but has since been replaced by the safer route):

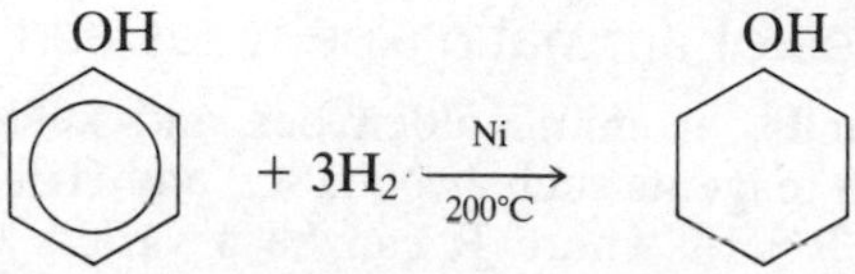

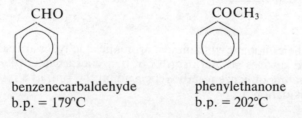

13.4 Aldehydes and ketones

The simplest aromatic aldehyde is benzenecarbaldehyde (*benzaldehyde*), a colourless oily liquid, b.p. = 179°C, with a strong smell of almonds. It is found as a component of a complex compound in the seeds of bitter almonds and in peach and cherry seeds. The simplest ketone is phenylethanone, a colourless oily liquid, b.p. = 202°C:

CHO

benzenecarbaldehyde
b.p. = 179°C

COCH₃

phenylethanone
b.p. = 202°C

Benzenecarbaldehyde can be prepared by the controlled oxidation of phenylmethanol or methylbenzene, or by the alkaline hydrolysis of (dichloromethyl)benzene:

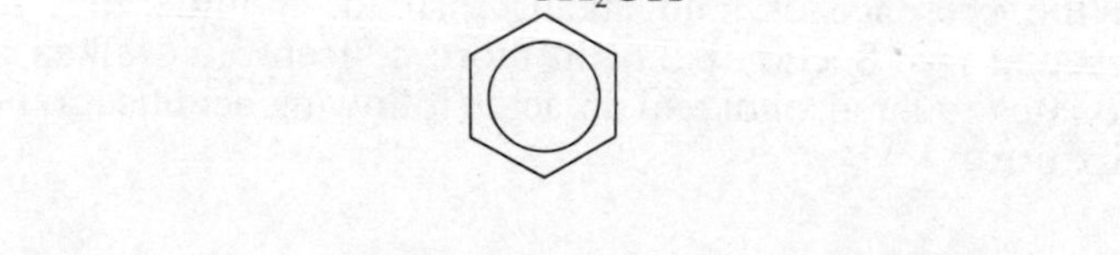

CH₃ $\xrightarrow[\text{H}_2\text{SO}_4]{\text{MnO}_2}$ CHO $\xleftarrow{\text{2OH}^-}$ CHCl₂

CH₂OH

Aromatic ketones are generally prepared by the Friedel–Crafts reaction (*see* Section 12.4.2(iv)):

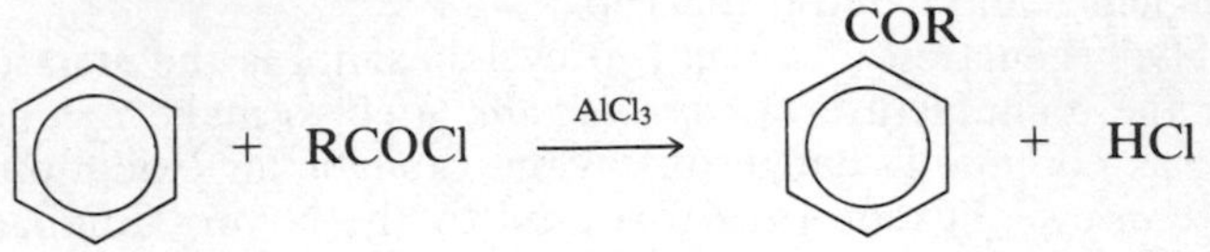

$$\text{C}_6\text{H}_6 + \text{RCOCl} \xrightarrow{\text{AlCl}_3} \text{C}_6\text{H}_5\text{COR} + \text{HCl}$$

13.4.1 Chemical properties of aromatic aldehydes and ketones

As carbonyl compounds, aromatic aldehydes and ketones will form addition products with reagents such as $NaHSO_3$ and HCN, and condensation products with RNH_2 (where R can be a variety of groups, *see* Section 8.3.3). They are also reduced by complex metal hydrides. In general, the aromatic aldehydes and ketones react a little less readily than their aliphatic counterparts, because the electron-donating ability of the benzene ring reduces the electrophilic character of the carbon atom of the carbonyl group:

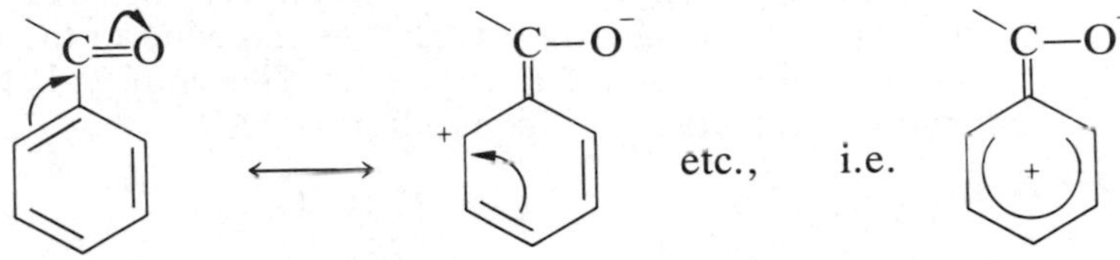

etc.,   i.e.

Like their aliphatic compounds, aromatic aldehydes are easily oxidized, while the ketones are not. Bottles of benzenecarbaldehyde often have a white coating of benzenecarboxylic acid on the rim, as a result of atmospheric oxidation:

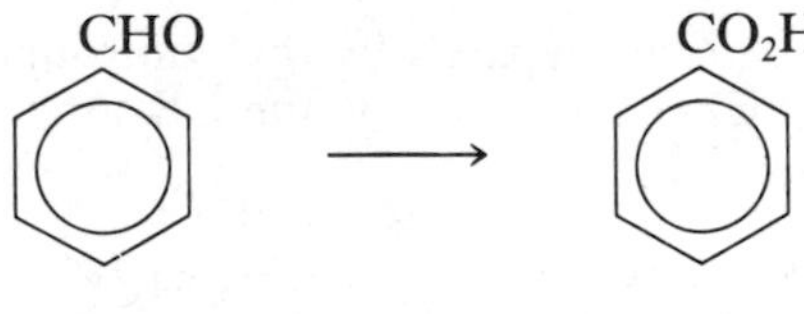

CHO $\longrightarrow$ CO_2H

Although aromatic aldehydes are reducing agents, and slowly give a positive result in Tollens's test, they have little effect on Fehling's solution.

An aromatic aldehyde like benzenecarbaldehyde differs in one important respect from aliphatic ones; it has no hydrogen atom bonded to the carbon atom bearing the aldehyde group. Consequently, unlike aliphatic ones, it does not undergo an aldol condensation on its own, or polymerize in the presence of mild alkali. Instead, it undergoes the **Cannizzaro reaction** (*see* Section 8.3.6) in more concentrated alkali and disproportionates to an alcohol and an acid (following acidification of the reaction mixture):

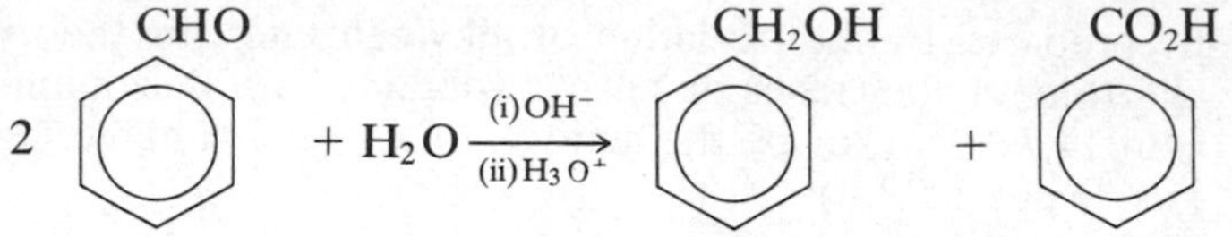

$$\text{2 } \underset{\text{CHO}}{\bigcirc} + H_2O \xrightarrow[\text{(ii)}\,H_3O^+]{\text{(i)}\,OH^-} \underset{\text{CH}_2\text{OH}}{\bigcirc} + \underset{\text{CO}_2\text{H}}{\bigcirc}$$

phenylmethanol   benzenecarboxylic acid
b.p. = 206°C   m.p. = 122°C

Benzenecarbaldehyde can be sulphonated and nitrated, as can aromatic ketones, but under oxidizing conditions the —CHO group is converted into the —COOH group. The conditions required for substitution in the ring are more vigorous than for benzene, from which it may be inferred that the —CHO group is *deactivating*. Substitution occurs at the 3-position in the ring.

When the carbonyl group is not directly bonded to the ring, the aromatic compounds resemble the aliphatic ones.

13.4.2 Manufacture and uses of aromatic aldehydes and ketones

Benzenecarbaldehyde can be manufactured by the oxidation of methylbenzene or the hydrolysis of (dichloromethyl)benzene. Phenylethanone is manufactured by the catalytic air oxidation of ethylbenzene.

These compounds have few commercial uses, although benzenecarbaldehyde is used as a food flavouring (e.g. in marzipan), as is the more complex *vanillin* (in vanilla beans).

13.5 Aromatic carboxylic acids and derivatives

The simplest acid is benzenecarboxylic acid, a white crystalline solid of melting point 122°C, which is sparingly soluble in cold water but soluble in hot water, from which it can be readily crystallized. It is soluble in organic solvents like benzene, in which it associates via hydrogen bonding:

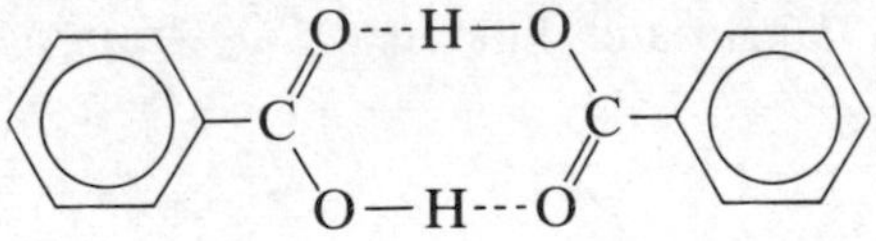

In aqueous solution, benzenecarboxylic acid is a slightly stronger acid than ethanoic acid, their respective pK_a values being 4.2 and 4.8. Electron-withdrawing groups substituted in the ring enhance the acidity of the molecule, e.g. 2-nitrobenzenecarboxylic acid, $pK_a = 2.2$. The effect of the ring on the acidity of the carboxyl group, however, falls off with increasing distance apart (although phenylethanoic acid has a similar acid strength to benzenecarboxylic acid).

Acids in which the carboxyl group is attached directly to the ring can be prepared by the oxidation of alkylbenzenes, by the hydrolysis of trichloromethylbenzenes or nitriles (obtained via diazonium salts, *see* Section 14.3.1(iii)), or by the action of CO_2 on a suitable Grignard reagent (*see* Section 5.3.4):

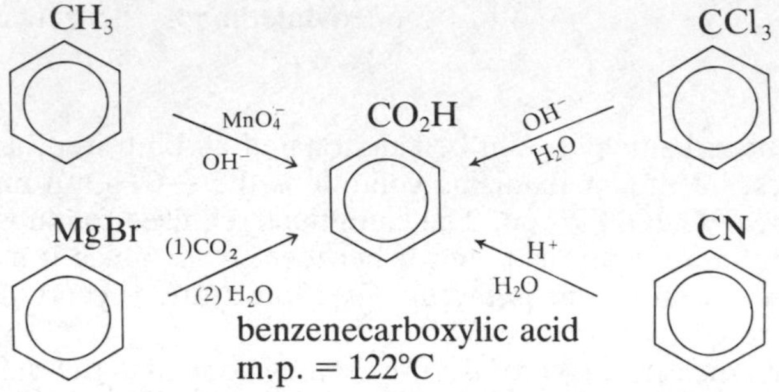

benzenecarboxylic acid
m.p. = 122°C

13.5.1 Chemical properties of aromatic carboxylic acids

The properties of the carboxyl group, unlike those of the hydroxyl group for example, are not significantly modified in aromatic compounds, whether the group is directly bonded to the benzene ring or not. One exception is the decarboxylation of aromatic carboxylic acids.

(i) Decarboxylation
When the carboxyl group is directly bonded to the benzene ring, it is fairly readily lost as CO_2 if the acid or its salt is heated with soda-lime, the process being known as **decarboxylation**, e.g.

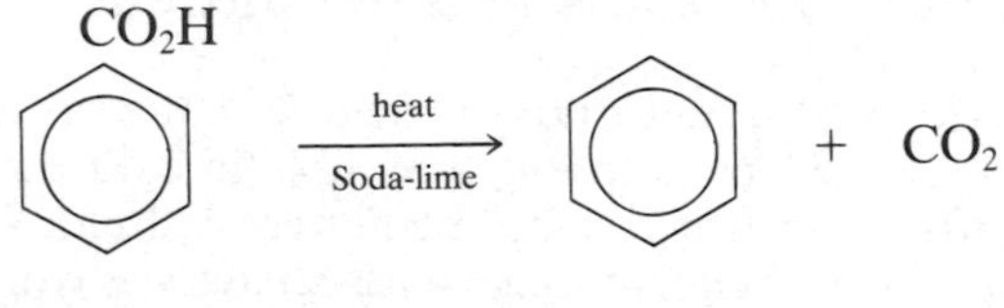

(ii) Acidic properties
Benzenecarboxylic acid, although only sparingly soluble in cold water, dissolves readily in alkali and liberates CO_2 from carbonates and hydrogencarbonates.

(iii) Ester formation
Benzenecarboxylic acid is readily esterified with an alcohol, using either concentrated sulphuric acid or dry hydrogen chloride as catalyst:

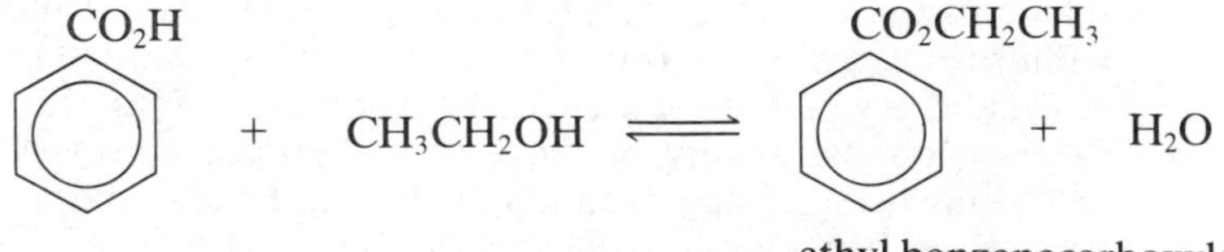

ethyl benzenecarboxylate
b.p. = 213°C

Using 2-hydroxybenzenecarboxylic acid and methanol, the strong-smelling ester known as Oil of Wintergreen is obtained (cf. the conversion of the hydroxyl group into an ester group in the preparation of aspirin, Section 13.3.1(ii)):

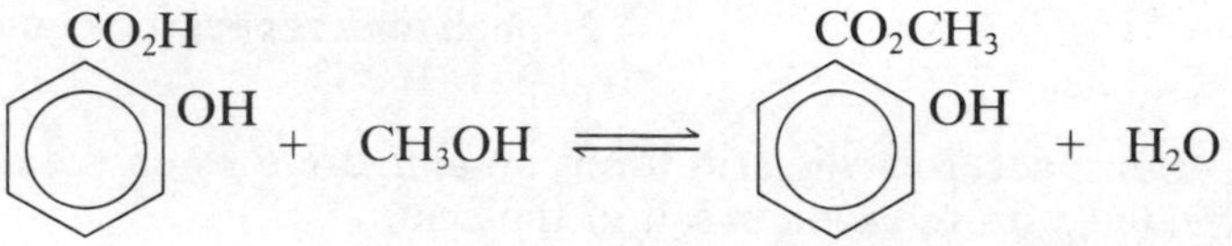

methyl 2-hydroxybenzenecarboxylate
b.p. = 223°C

(iv) Acid chloride formation
Treatment of benzenecarboxylic acid with sulphur dichloride oxide (*thionyl chloride*) gives the corresponding acid chloride, benzenecarbonyl chloride:

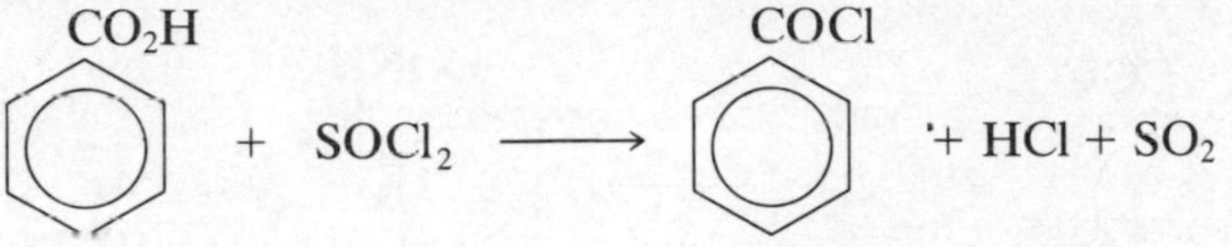

benzenecarbonyl chloride
b.p. = 197°C

It must be treated with the greatest care, since the liquid can cause burns and the vapour is pungent and a respiratory irritant. Owing to the electronic influence of the benzene ring on the carbonyl group (cf. Section 13.4.1), it is not so reactive as aliphatic acid chlorides and, since it undergoes comparatively slow hydrolysis in water, it may be used in aqueous solution. In the **Schotten–Baumann reaction** (*see* Section 13.3.1(ii)), the acid chloride reacts with either a phenol or an amine in aqueous solution. The phenol is used in an alkaline solution, since the phenoxide ion is a stronger nucleophile than the phenol itself and the alkali also prevents any insoluble benzenecarboxylic acid (formed by hydrolysis of the acid chloride) from precipitating out:

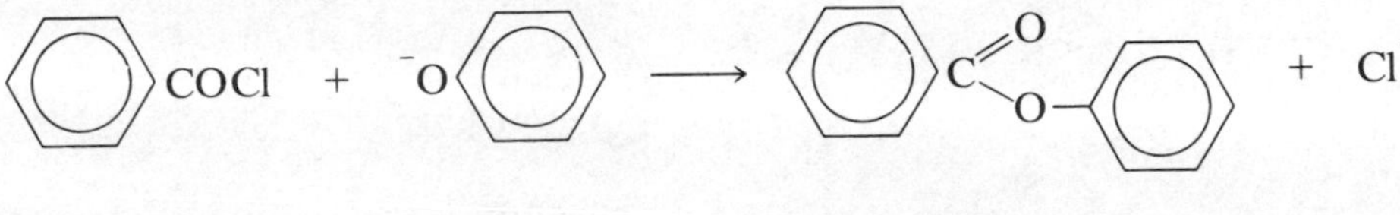

phenyl benzenecarboxylate
m.p. = 69°C

With an amine in alkaline solution, the product is an N-substituted amide:

N-phenylbenzenecarboxamide
m.p. = 162°C

Benzenecarboxylic acid forms an anhydride when the acid chloride is treated with the sodium salt of the acid:

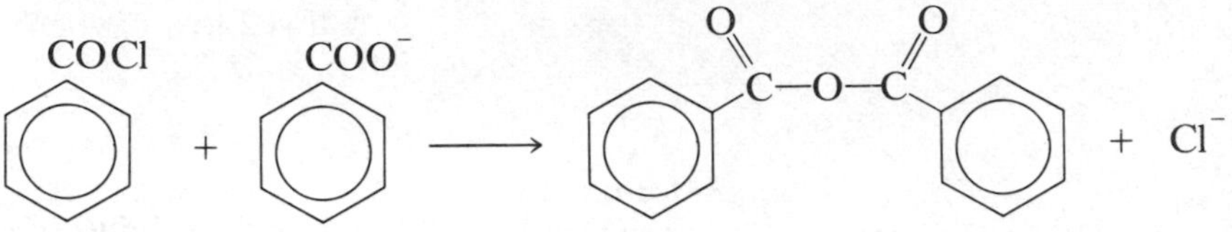

benzenecarboxylic anhydride
m.p. = 42°C

Treatment of the acid chloride with ammonia yields the amide:

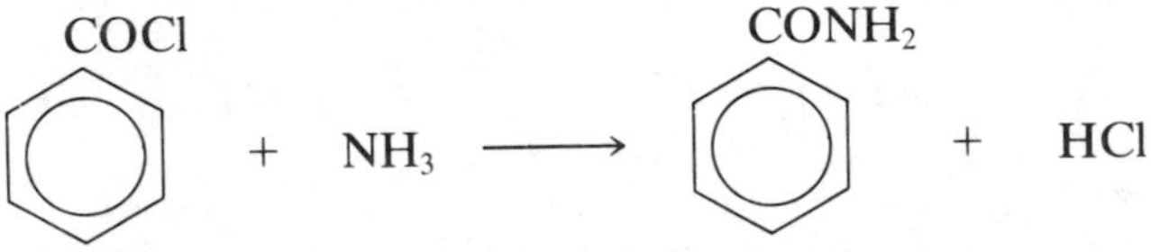

benzenecarboxamide
m.p. = 130°C

(v) Substitution reactions
With regard to electrophilic substitutions in the benzene ring, the carboxyl group of benzenecarboxylic acid is *deactivating* and directs groups to the 3-position, e.g.

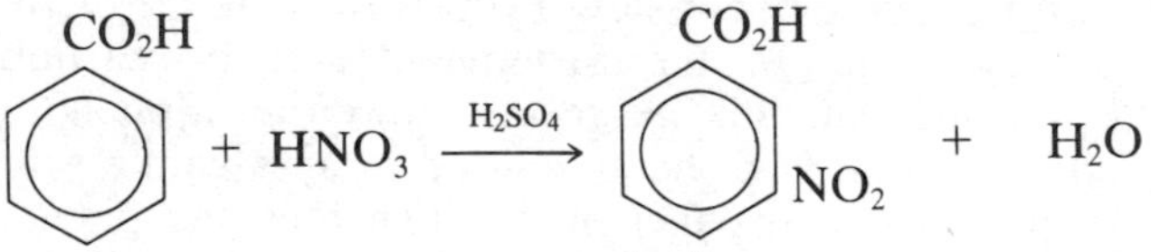

3-nitro-
benzenecarboxylic acid
m.p. = 140°C

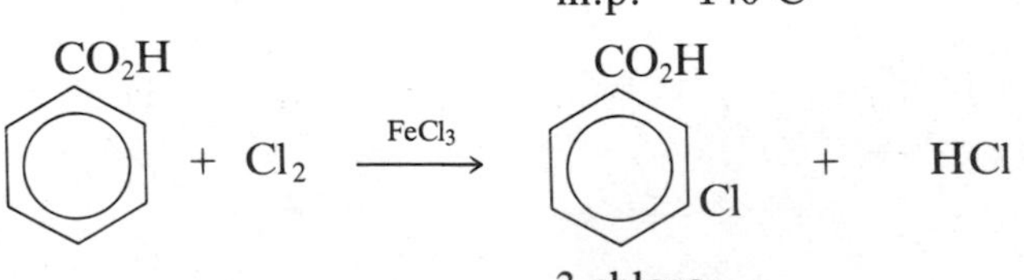

3-chloro-
benzenecarboxylic acid
m.p. = 158°C

13.5.2   Manufacture and uses of aromatic carboxylic acids

The oxidation of the corresponding alkylbenzenes affords the simple carboxylic acids shown below (*see also* Section 12.5.2(ii)):

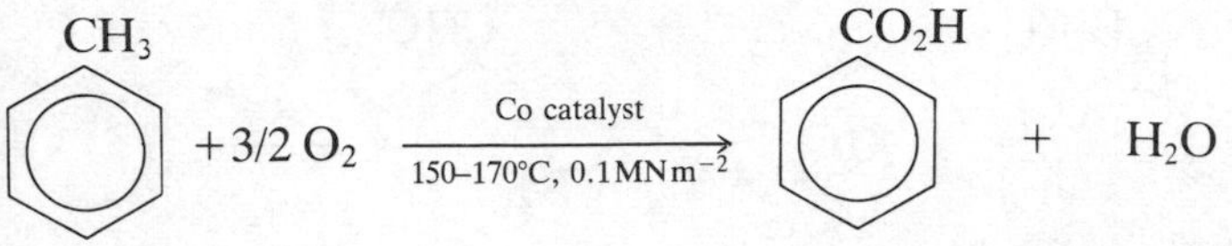

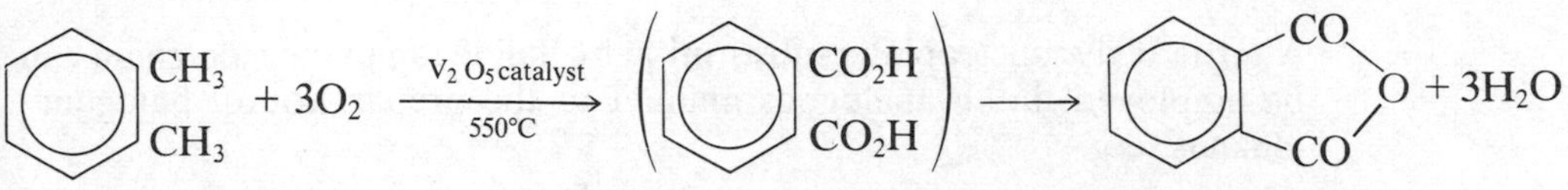

benzene-1,2-
dicarboxylic anhydride
m.p. = 132°C

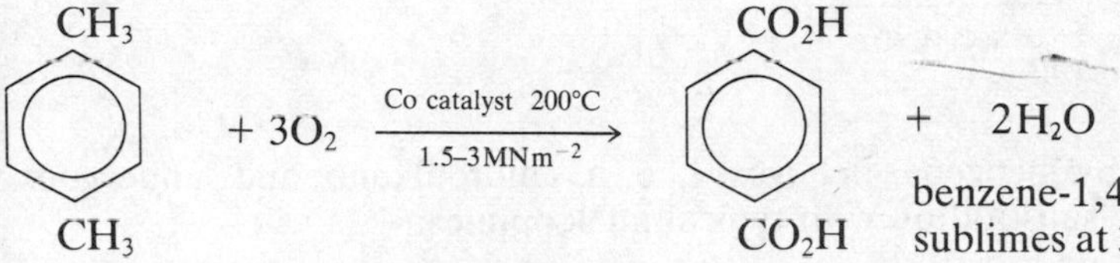

benzene-1,4-dicarboxylic acid
sublimes at 300°C

Benzenecarboxylic acid and its salts are used in the food industry as preservatives. Benzene-1,2-dicarboxylic anhydride is used as a component of alkyd resins and plasticizers (*see* Section 20.2.2(iv)). The 1,4-dicarboxylic acid is a starting material in the manufacture of polyester fibres (*see* Section 20.3(ii)). (Note that the 1,4-dicarboxylic acid cannot form an anhydride because the two carboxyl groups are too far apart.)

Summary

1   When the functional group is directly bonded to the benzene ring, its properties may be modified by electronic interaction between the two. When the functional group is separated from the ring by one or more carbon atoms, the ring has little effect on it and such compounds closely resemble their aliphatic counterparts. Both types of compound generally undergo the reactions characteristic of the benzene ring.

2   **Halogeno compounds.** Chlorination and bromination of the benzene ring are achieved with the aid of a catalyst or 'halogen carrier', e.g.

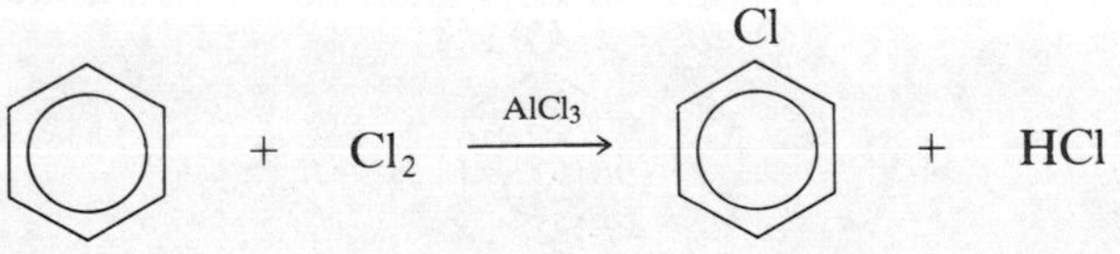

Side-chain halogenation is brought about with the boiling hydrocarbon and ultraviolet light, the reaction being a **free-radical**, rather than *ionic*, one:

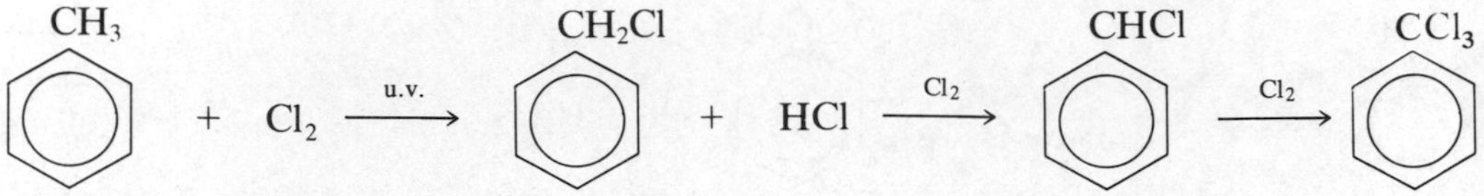

$$CH_3 \quad + \quad Cl_2 \quad \xrightarrow{u.v.} \quad CH_2Cl \quad + \quad HCl \quad \xrightarrow{Cl_2} \quad CHCl \quad \xrightarrow{Cl_2} \quad CCl_3$$

Alternatively, nucleophilic substitution by halide ion in the side-chain can be employed, in an analogous manner to the preparation of halogeno-alkanes, e.g.

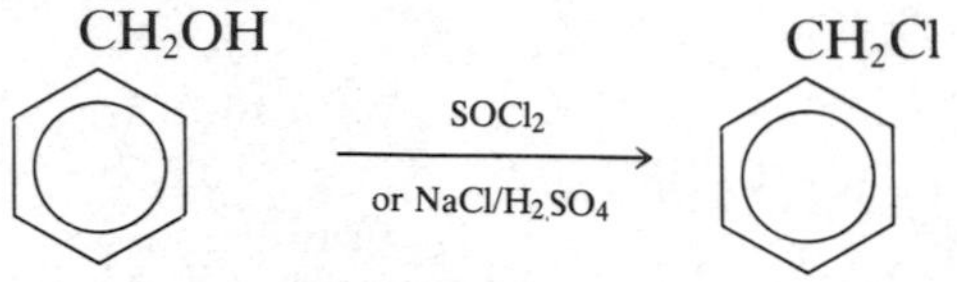

$$CH_2OH \quad \xrightarrow[\text{or NaCl/H}_2\text{SO}_4]{SOCl_2} \quad CH_2Cl$$

(Chloromethyl)benzene behaves like a chloroalkane and undergoes nucleophilic substitution with typical nucleophiles:

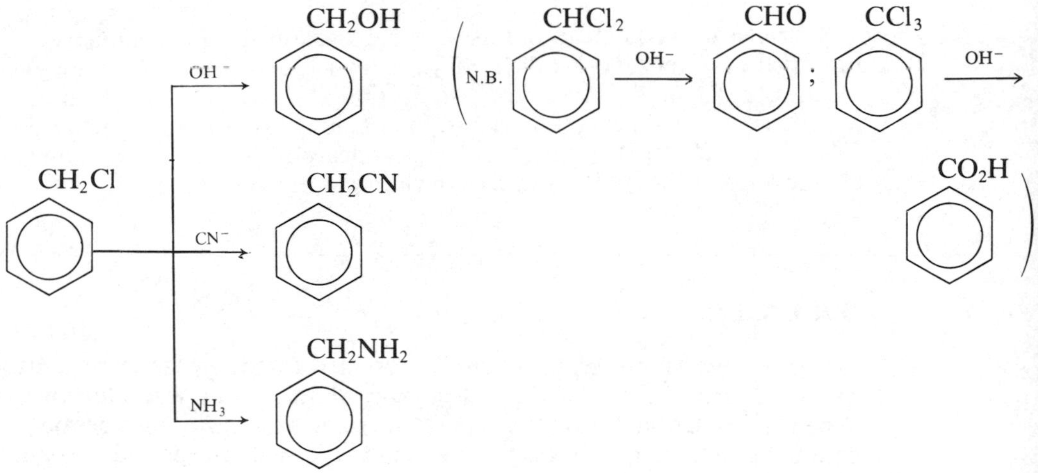

$$CH_2Cl$$

$$\xrightarrow{OH^-} CH_2OH$$

$$\left(\text{N.B.} \quad CHCl_2 \xrightarrow{OH^-} CHO \; ; \quad CCl_3 \xrightarrow{OH^-} CO_2H \right)$$

$$\xrightarrow{CN^-} CH_2CN$$

$$\xrightarrow{NH_3} CH_2NH_2$$

In contrast, chlorobenzene is unreactive and does not normally undergo these reactions. Only at 300°C and 15 MN m^{-2} will it hydrolyse:

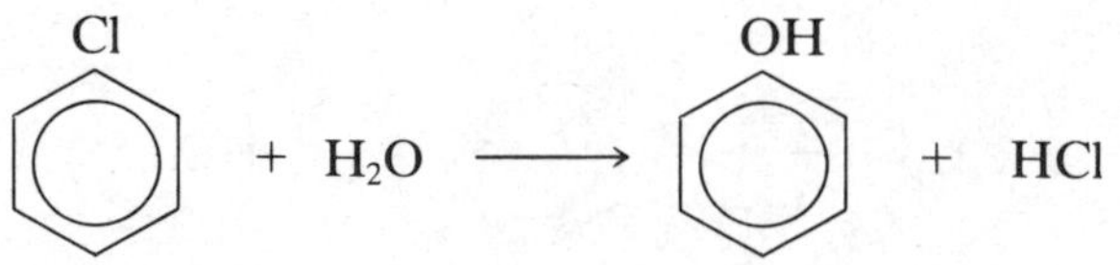

$$Cl \quad + \quad H_2O \quad \longrightarrow \quad OH \quad + \quad HCl$$

The benzene ring will enter into typical electrophilic substitution reactions in the case of both types of halogeno compound, e.g.

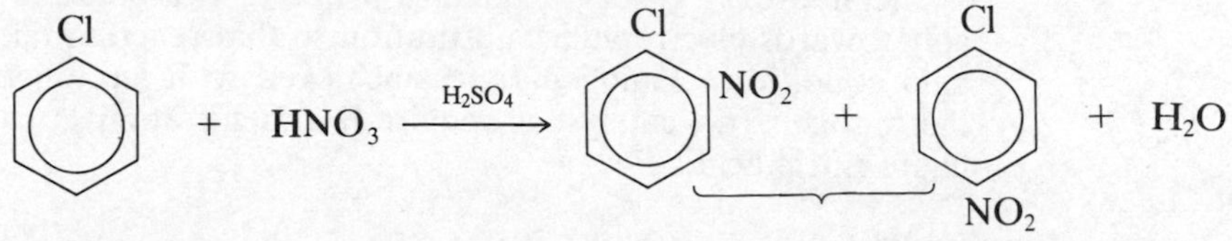

Chlorobenzene is used in the manufacture of the insecticide DDT.

3 **Phenols and hydroxy compounds.** When the hydroxyl group is part of the side-chain, the compound is an alcohol. It can be prepared in a typical manner:

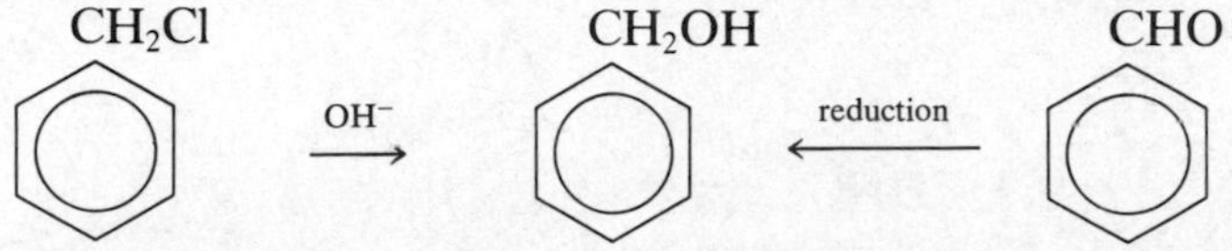

and undergoes reactions typical of an alcohol:

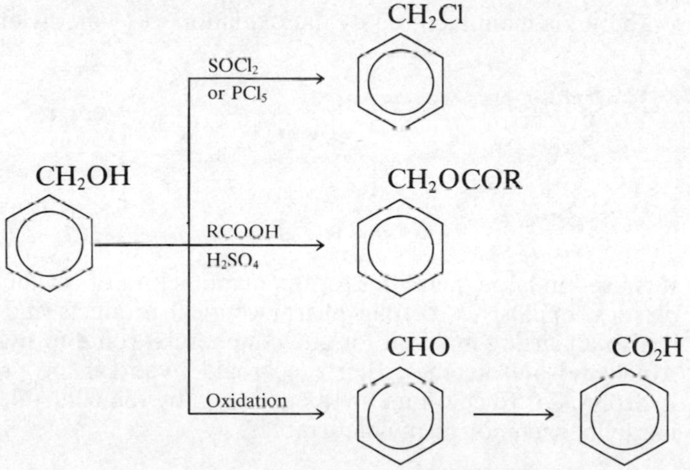

When the hydroxyl group is bonded directly to the benzene ring, the compound is a **phenol**. In contrast to alcohols, phenols are weak acids and, although they will dissolve in alkali to form salts, they are too weak to liberate CO_2 from carbonates or hydrogencarbonates.

Phenols are converted into esters by treatment with acid chlorides or anhydrides. The reaction with aromatic acid chlorides is carried out under alkaline conditions and is known as the **Schotten–Baumann reaction**, e.g.

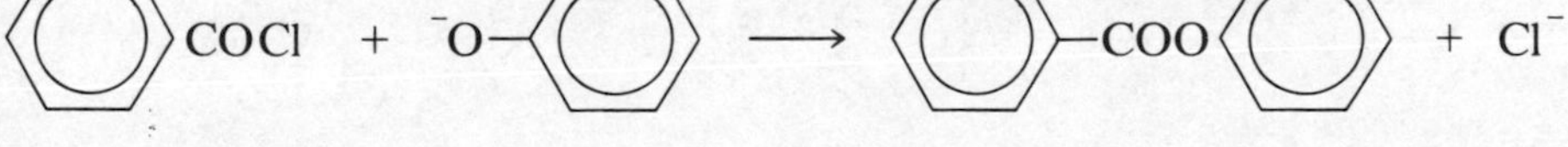

Phenols give a characteristic blue or violet coloration with neutralized iron(III) chloride solution; this is a general test for a phenol.

The hydroxyl group in phenols *activates* the benzene ring considerably towards electrophilic substitution so that reaction takes place under mild conditions. Halogenation is achieved with an aqueous solution of the halogen (no catalyst is required) and nitration is carried out with dilute nitric acid:

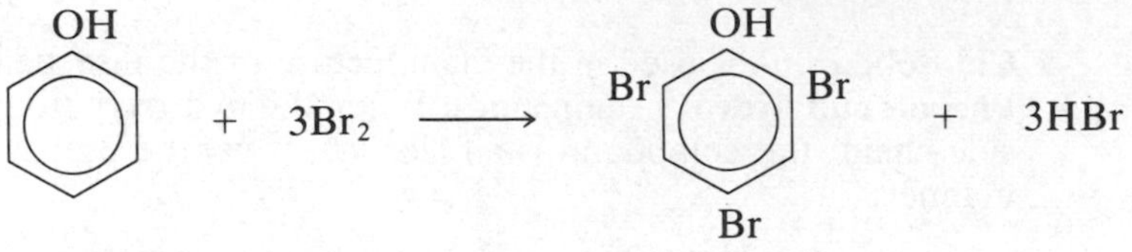

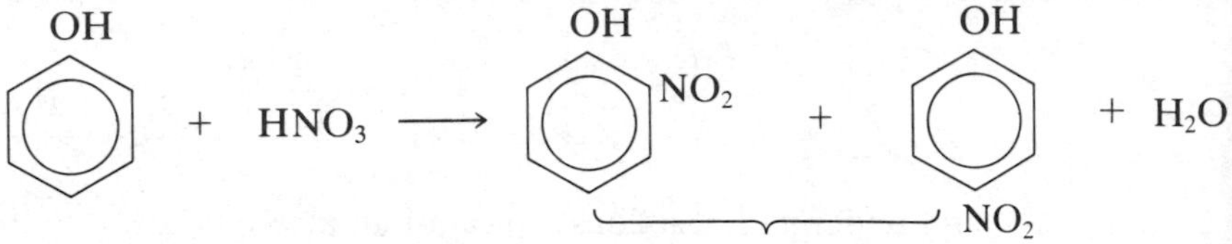

Phenol is manufactured by the oxidation of (1-methylethyl)benzene:

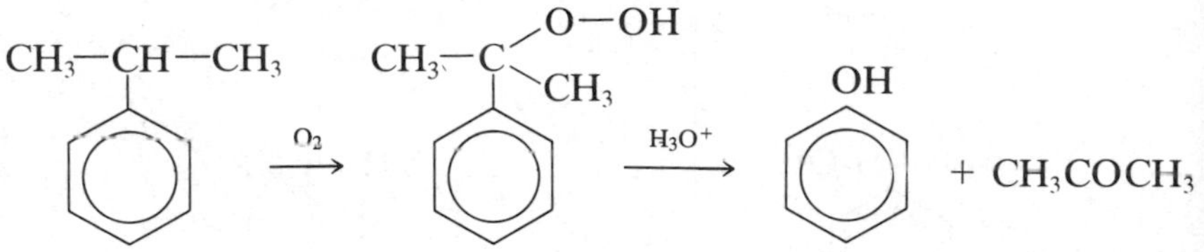

It is used in large quantities for the manufacture of certain thermosetting plastics, explosives, certain pharmaceutical products and antiseptics. It is also a starting material for one commercial route to nylon.

4 **Aldehydes and ketones.** Benzenecarbaldehyde can be prepared by the hydrolysis of (dichloromethyl)benzene or by the controlled oxidation of methylbenzene or phenylmethanol:

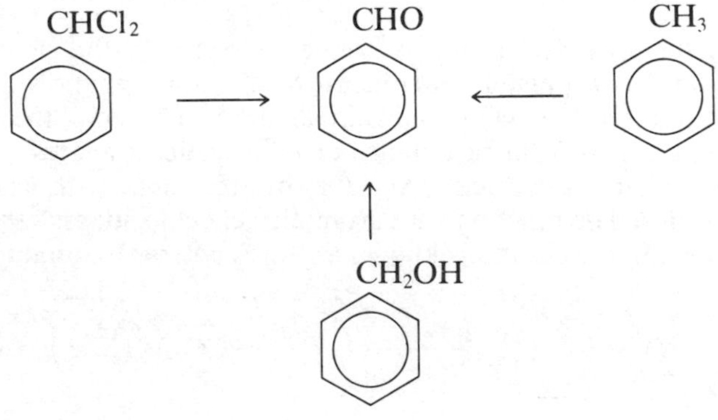

Aromatic ketones are prepared by the **Friedel–Crafts reaction**:

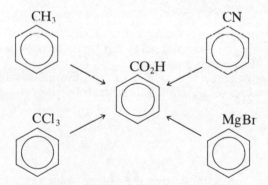

The reactions of the aromatic compounds are very similar to those of the aliphatic ones, although the aromatic ones are rather less reactive. They undergo **nucleophilic addition** reactions with $NaHSO_3$, HCN and AlH_4^-, etc., and **condensation** reactions with compounds of the type RNH_2 (where R = OH, NH_2, etc). The aldehydes are easily oxidized to carboxylic acids.

Benzenecarbaldehyde differs from those aldehydes, in which the —CHO is not directly bonded to the ring, in undergoing the **Cannizzaro reaction** in alkali:

The benzene ring in both aldehydes and ketones can be nitrated, halogenated and sulphonated, as is to be expected.

5   **Aromatic carboxylic acids.** These can be prepared by oxidation of methylbenzenes, by the hydrolysis of (trichloromethyl)benzenes or nitriles, or by carbonation of a Grignard reagent, e.g.

If the carboxyl group is bonded directly to the ring, the acid can be fairly readily **decarboxylated** by heating with soda-lime, e.g.

The acids are weak acids in aqueous solution, their strength being modified by the presence of electron-withdrawing or -donating groups substituted in the ring:

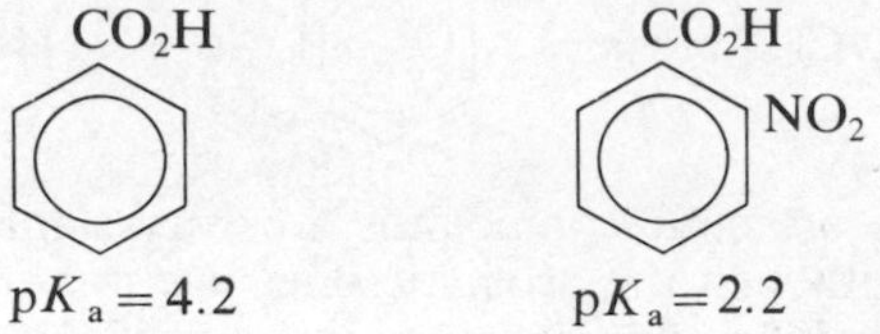

CO_2H

$pK_a = 4.2$

CO_2H NO_2

$pK_a = 2.2$

They will liberate CO_2 from carbonates and hydrogencarbonates, and hence are readily distinguishable from phenols. They can be esterified with an alcohol, using either concentrated H_2SO_4 or anhydrous HCl as catalyst:

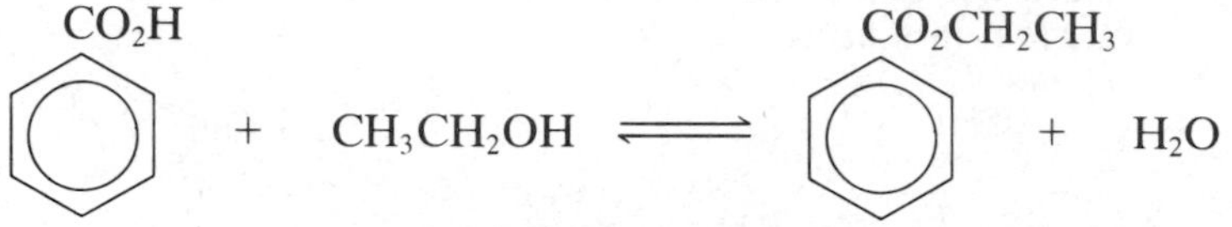

CO_2H $+$ CH_3CH_2OH $\rightleftharpoons$ $CO_2CH_2CH_3$ $+$ H_2O

The corresponding acid chlorides, made by the action of $SOCl_2$ on the acid, are less reactive than their aliphatic counterparts and can be used in aqueous solution (hydrolysis is relatively slow) when being reacted with phenols or amines in the **Schotten–Baumann reaction**:

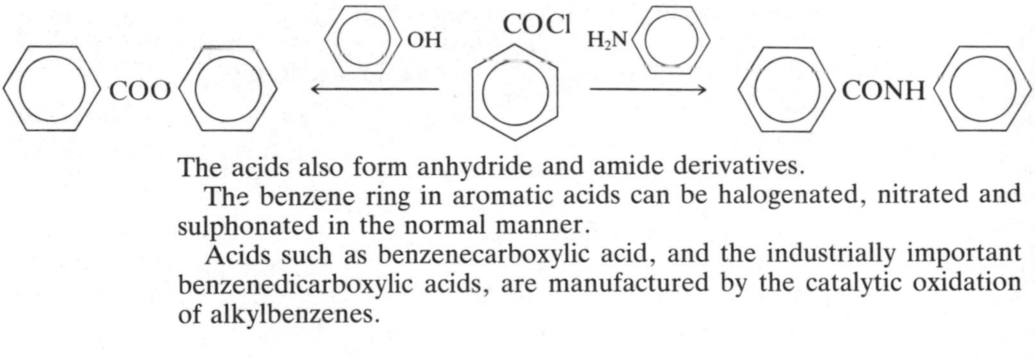

COO ← OH COCl H_2N → CONH

The acids also form anhydride and amide derivatives.

The benzene ring in aromatic acids can be halogenated, nitrated and sulphonated in the normal manner.

Acids such as benzenecarboxylic acid, and the industrially important benzenedicarboxylic acids, are manufactured by the catalytic oxidation of alkylbenzenes.

Questions

1 What do you understand by the term 'electron delocalization' (resonance)?

How may enthalpies of hydrogenation be used to support the idea of electron delocalization in the benzene molecule?

How does the electronic structure of a benzene molecule help to explain the following?

(a) Concentrated sulphuric acid is added to concentrated nitric acid in the nitration of benzene.

(b) Unlike iodomethane, iodobenzene does not react readily with aqueous sodium hydroxide.
(c) Benzenecarboxylic (*benzoic*) acid is a stronger acid than ethanoic (*acetic*) acid in aqueous solution.

JMB

2 (a) What do you observe when an acidified solution of 2,4-dinitrophenylhydrazine and a solution of an aldehyde are warmed together and then cooled?
(b) Name the mechanism of this reaction.
(c) Write equations for the reaction of benzenecarbaldehyde (*benzaldehyde*) with the following:
  (i) 2,4-dinitrophenylhydrazine;
  (ii) hydroxylamine.
(d) Benzenecarbaldehyde (*benzaldehyde*) can give two different products $A, C_7H_8N_2$, and $B, C_{14}H_{12}N_2$, with hydrazine (NH_2NH_2). Write structures for A and B.

JMB

3* Describe, with essential experimental details, a laboratory method for preparing benzenecarbonyl (*benzoyl*) chloride from benzenecarboxylic (*benzoic*) acid.
Show, by means of equations, how the acid chloride can be used to prepare (a) benzenecarboxamide (*benzamide*) and (b) phenyl benzenecarboxylate (*benzoate*).
Describe what you would observe when benzenecarbonyl chloride is warmed with an excess of aqueous sodium hydroxide, and write a mechanism for the reaction which takes place.
Explain why Terylene and nylons are degraded by sodium hydroxide whereas poly(ethene) (*polythene*) is inert under the same conditions.

JMB

4* Describe briefly how each of the following can be prepared in the laboratory, starting from methylbenzene (*toluene*) and any other organic or inorganic reagents. Give equations and outline the mechanisms of the reactions where possible.

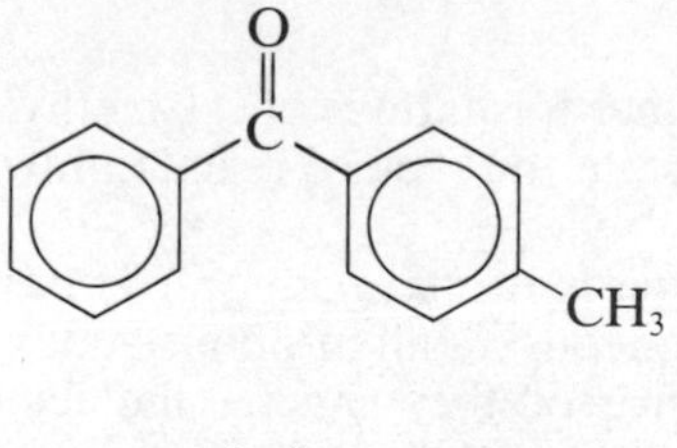

(A) 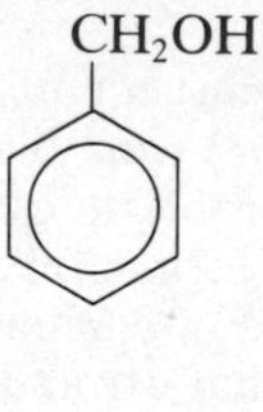   (B) 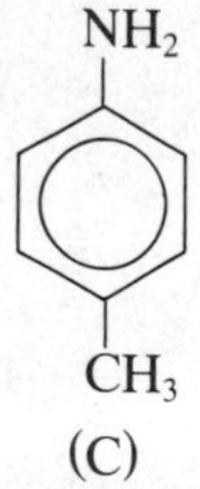   (C)

State briefly how you would carry out experiments to demonstrate each of the following:

(i) that the hydroxyl group of substance (B) is in the side chain and not
   directly attached to the ring;
(ii) that substance (C) is a primary aromatic amine.

JMB

5   Describe the characteristic reactions of the hydrocarbon benzene.
    How does the presence of the benzene ring affect the reactivity of the
    substituent group: (a) in bromobenzene, and (b) in phenol?
    How does the presence of the substituent affect the reactivity of the
    benzene ring: (a) in nitrobenzene, and (b) in methylbenzene (toluene)?
    If possible give reasons for the effects.

Imperial College

6*   Deduce the structures of the compounds **A** to **G** in the scheme below,
     giving your reasons (hv represents irradiation with light):

$$C_7H_6Cl_2 \xrightarrow[\text{Boil}]{\text{NaOH aq}} C_7H_7ClO$$

$$\textbf{A} \qquad\qquad\qquad \textbf{B}$$

$$Cl_2 \downarrow hv \qquad\qquad \downarrow CrO_3$$

$$C_7H_5Cl_3 \xrightarrow[\text{Boil}]{\text{Na}_2\text{CO}_3\text{ aq}} C_7H_5ClO$$

$$\textbf{C} \qquad\qquad\qquad \textbf{D}$$

$$Cl_2 \downarrow hv \qquad\qquad \downarrow KMnO_4$$

$$C_7H_4Cl_4 \xrightarrow[\text{Boil}]{\text{Na}_2\text{CO}_3\text{ aq}} C_7H_5ClO_2$$

$$\textbf{E} \qquad\qquad\qquad \textbf{F}$$

$$C_7H_4ClNO_4 \xleftarrow[\text{conc. HNO}_3]{\text{conc. H}_2\text{SO}_4}$$

$$\textbf{G}$$

G is a mixture of only **two** compounds.

Oxford & Cambridge

7*   (a) Write an equation for the formation of (1-methylethyl)benzene
        (*cumene*) from benzene. State the type of reaction involved and give
        an appropriate catalyst.
    (b) In the formation of cumene hydroperoxide, air is passed through a
        mixture of cumene and aqueous alkali at about 90°C.
        (i)   What structural feature of the cumene molecule permits the
              formation of the hydroperoxide?
        (ii) Why is it necessary to use cooling coils during the oxidation?
        (iii) What is the main reason for the use of multiple reactors?
    (c) Write the overall equation for the acid catalysed decomposition of
        cumene hydroperoxide. Name the **two** main products.

(d) Give the name or structural formula of **one** of the by-products formed during the oxidation or decomposition stages of the cumene process.

JMB

8* Describe in outline how **four** of the following changes might be effected:

(a) 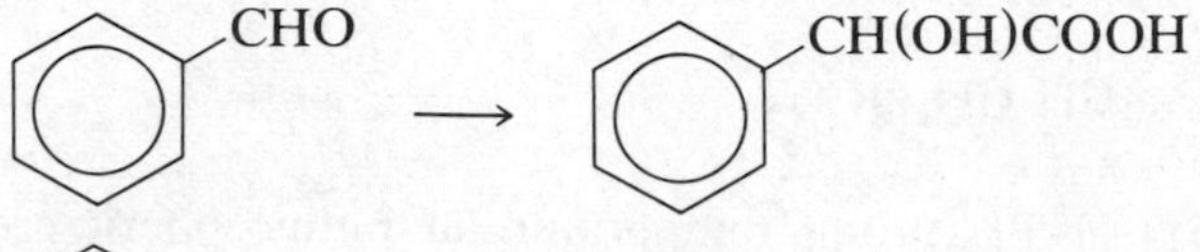 [benzene ring with CHO] $\longrightarrow$ [benzene ring with CH(OH)COOH]

(b) 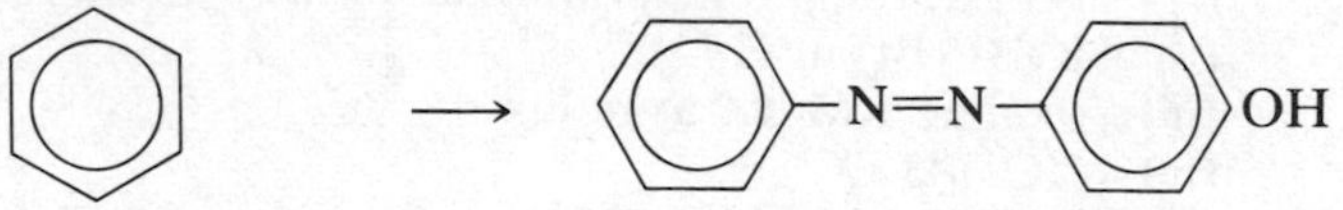 [benzene ring] $\longrightarrow$ [benzene ring]$-N=N-$[benzene ring]OH

(c) 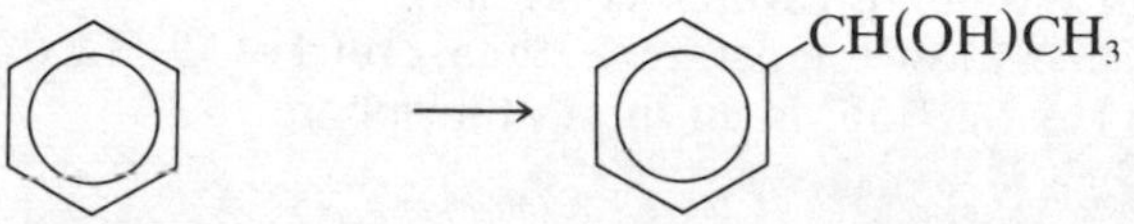 [benzene ring] $\longrightarrow$ [benzene ring with CH(OH)CH$_3$]

(d) 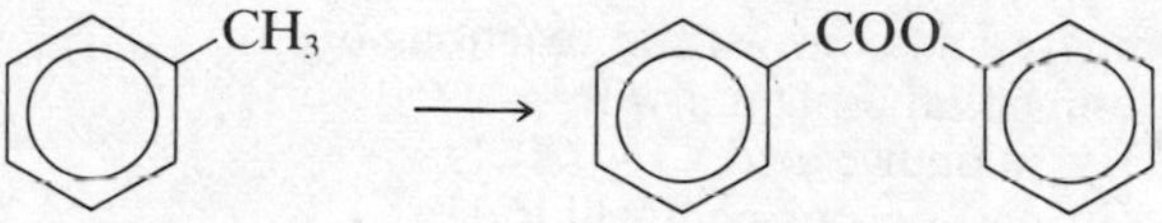 [benzene ring with CH$_3$] $\longrightarrow$ [benzene ring with COO benzene ring]

(e) 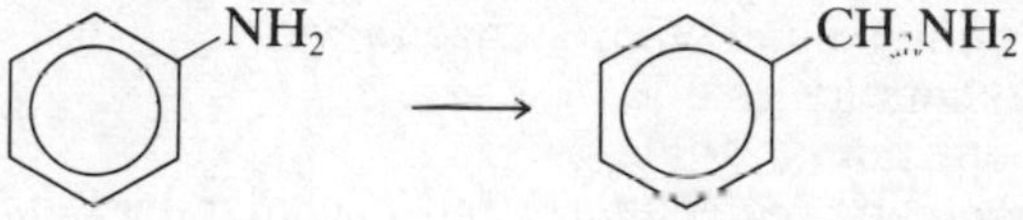 [benzene ring with NH$_2$] $\longrightarrow$ [benzene ring with CH$_2$NH$_2$]

Cambridge

9** Answer **one** part of this question.
Either (a) A neutral yellow liquid (A), $C_9H_9NO_2$, decolorized a weakly acidic solution of potassium permanganate in the cold. Addition of bromine to (A) gave (B), $C_9H_9Br_2NO_2$, which gave (D), $C_9H_7NO_2$, when boiled with potassium hydroxide in ethanol. Reaction of (D) with hydrogen and platinum at room temperature and atmospheric pressure gave the weak base (E), $C_9H_{13}N$, after five molecular equivalents of hydrogen had been consumed.
Treatment of (D) with aqueous sulphuric acid in the presence of mercury(II) salts gave (F), $C_9H_9NO_3$, which reacted with bromine in aqueous sodium hydroxide to give the acid (G), $C_8H_7NO_4$. Oxidation of (G) with boiling alkaline potassium permanganate gave the acid (H), $C_8H_5NO_6$, after acidification. When (H) was heated above 150°C the solid (J), $C_8H_3NO_5$, was produced spontaneously.
The acid (G), synthesized readily from 2-methylbenzoic acid by nitration, is *not* a 1,2,3-trisubstituted benzene derivative.
Deduce possible structures for the compounds (A), (B), (D)–(J) and account for the above reactions.
Or (b) Draw the structure (or structures) of the compounds you would expect to isolate from the reaction of the following compound with **each** of the reagents listed below. *No other writing at all is required.*

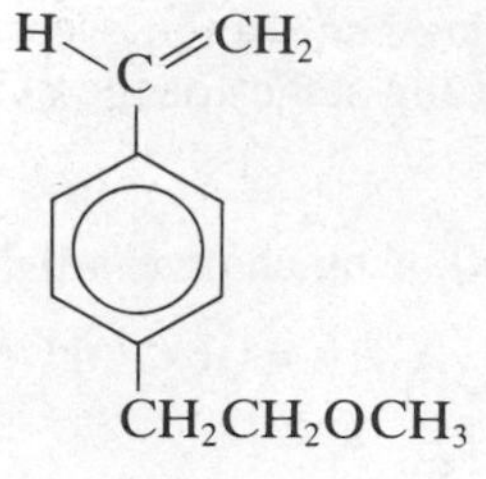

$$H\diagdown_{C}\diagup^{CH_2}$$

CH$_2$CH$_2$OCH$_3$

(i)   H$_2$/Pt at room temperature and atmospheric pressure.
(ii) HOBr (Br$_2$ in H$_2$O).
(iii) Alkaline KMnO$_4$ at reflux.
(iv) NaOH/H$_2$O
(v)   Cold dilute KMnO$_4$ solution.
(vi) O$_3$ at low temperature, then Zn/CH$_3$CO$_2$H ('ozonolysis').
(vii) Heat at 150° in an inert atmosphere.

Cambridge

10   Compare and contrast the reactions of
     (a) ethanol and phenol with
          (i) sulphuric acid,
          (ii) phosphorus pentachloride;
     (b) chloroethane and chlorobenzene with
          (i) sodium hydroxide,
          (ii) potassium cyanide;
     (c) aminoethane (*ethylamine*) and phenylamine (*aniline*) with
          (i) nitrous acid,
          (ii) ethanoyl (*acetyl*) chloride.
     Suggest, as far as possible, reasons for any differences which may exist
     in the behaviour of the compounds stated towards the same reagent.

London

11* Compound X (0.363 g) on combustion yielded carbon dioxide (0.922 g)
     and water (0.189 g). Compound X (0.725 g) was refluxed with 70% sul-
     phuric acid and the mixture made strongly alkaline. Ammonia was lib-
     erated and was distilled into 50 cm^3 of 0.1 M sulphuric acid. Subsequently,
     47.1 cm^3 of 0.085 M sodium hydroxide were required for neutralization.
     What is the empirical formula of compound X? Suggest a likely structure
     for this compound. How would your compound X behave towards:
     (a) phosphorus(V) oxide (phosphorus pentoxide);
     (b) nitrous acid (HNO$_2$);
     (c) bromine;
     (d) bromine and sodium hydroxide, followed by isolation of the product
         and treatment of this with bromine?

Oxford

Chapter 14
Aromatic amines

14.1 Introduction

The aromatic amines resemble in structure the aliphatic amines, one or more of the alkyl groups being replaced by a phenyl C_6H_5— group. Thus, in a *primary* aromatic amine, an amino group is bonded directly to the benzene ring. In *secondary* and *tertiary* amines, one and both hydrogen atoms, respectively, of the amino group are replaced by either alkyl or phenyl groups:

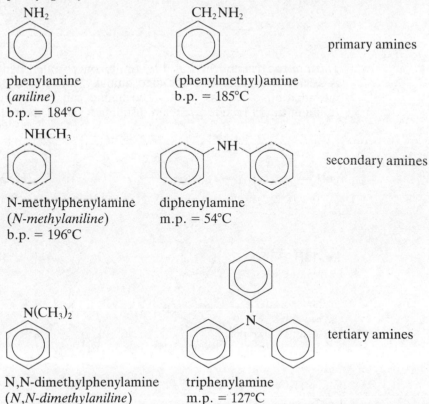

phenylamine
(*aniline*)
b.p. = 184°C

(phenylmethyl)amine
b.p. = 185°C

primary amines

N-methylphenylamine
(*N-methylaniline*)
b.p. = 196°C

diphenylamine
m.p. = 54°C

secondary amines

N,N-dimethylphenylamine
(*N,N-dimethylaniline*)
b.p. = 193°C

triphenylamine
m.p. = 127°C

tertiary amines

The secondary and tertiary aromatic amines possess similar properties to the aliphatic ones, but the presence of an aromatic ring in an amine obviously causes differences. These are most pronounced in the case of those *primary* aromatic amines where the amino group is bonded directly to the benzene ring. If the amino group is separated from the ring by one or more carbon atoms, as in (phenylmethyl)amine, then the amine behaves in a generally similar fashion to a primary aliphatic one. Again, this illustrates the way in which aromatic ring and functional group interact electronically when directly bonded together.

14.2 Preparation and chemical properties of aromatic amines

Primary aromatic amines are readily prepared by the reduction of the corresponding nitro derivative, using either dissolving metals, complex metal hydrides or catalytic hydrogenation (the aromatic ring is more resistant to reduction). In the laboratory, reduction is easily accomplished, using either tin or iron in acid solution:

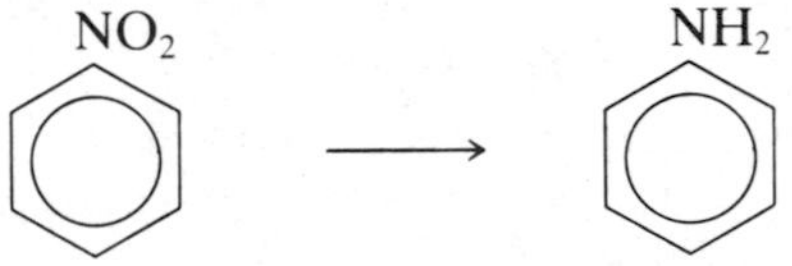

Thus, nitrobenzene is reduced by tin in concentrated hydrochloric acid. A salt of phenylamine is obtained, initially, and this is decomposed by the addition of excess alkali. The amine can be separated from the resultant tin(IV) oxide by steam distilling it out of the reaction mixture:

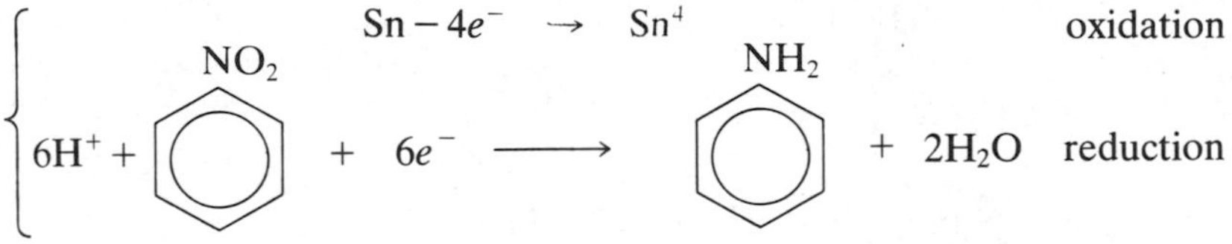

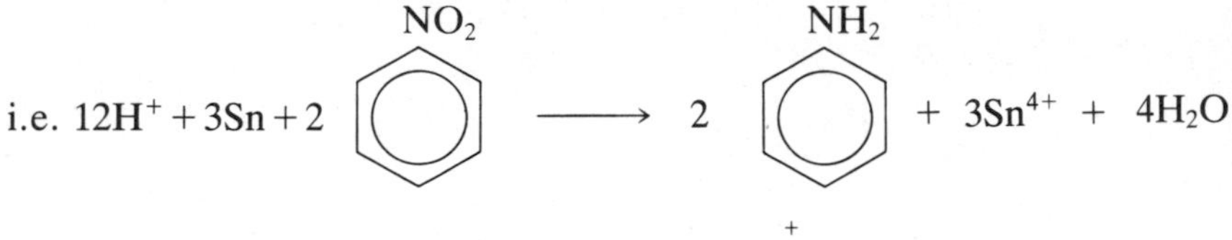

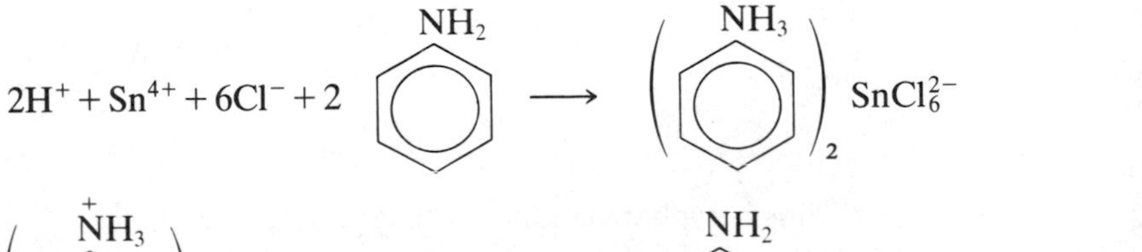

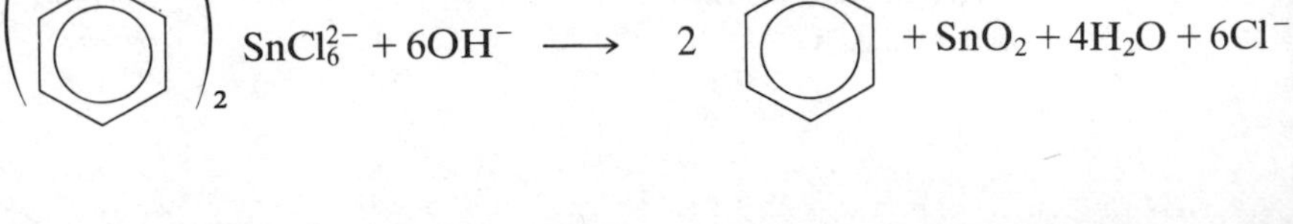

Secondary and tertiary amines with alkyl groups attached to the nitrogen can be made by reaction of a primary amine with a halogenoalkane (*see* Section 14.2.2(i)). (Phenylmethyl)amine can be made either by reduction of benzenecarbonitrile C_6H_5CN, or by heating (chloromethyl)benzene $C_6H_5CH_2Cl$ with ammonia in a nucleophilic substitution reaction.

The aromatic amines are generally high boiling-point liquids or solids at room temperature. Phenylamine (*aniline*), for example, is a colourless, oily liquid of boiling point 184°C. Like other amines, on standing in air and light it gradually darkens as a result of atmospheric oxidation. The aromatic amines are generally toxic by skin absorption, and by inhalation of vapour, and must be handled with care.

As with the aliphatic amines, much of the chemistry of aromatic ones is determined by the presence of a lone-pair of electrons on the nitrogen atom, which causes them to behave as Lewis bases and nucleophiles. The aromatic amines also display properties typical of the aromatic benzene ring.

14.2.1   Basic properties

Phenylamine is only slightly soluble in water and, although it forms an alkaline solution, it is a weaker base than ammonia ($pK_b = 4.8$):

$$C_6H_5NH_2 + H_2O \rightleftharpoons C_6H_5NH_3^+ + OH^- \qquad pK_b = 9.3$$

(Phenylmethyl)amine $C_6H_5CH_2NH_2$, in contrast, is strongly basic and resembles an aliphatic primary amine.

All the simple aromatic amines will, however, dissolve in dilute mineral acid to form salts, e.g.

$$C_6H_5NH_2 + HCl \rightarrow C_6H_5NH_3^+ Cl^-$$
$$\text{phenylammonium chloride}$$
$$(\textit{phenylamine hydrochloride})$$

This reaction allows amines to be protected against oxidation by storage in the form of their salts. The free amine can be liberated from the salt by the addition of alkali:

$$C_6H_5NH_3^+Cl^- + OH^- \rightarrow C_6H_5NH_2 + H_2O + Cl^-$$

The loss of basic character of phenylamine, compared with ammonia, is a consequence of the partial delocalization of the nitrogen's lone-pair of electrons over the benzene ring; consequently, the lone-pair is not so readily available for donation to other species. This is not the case with (phenylmethyl)amine, since a carbon atom intervenes between amino group and benzene ring; no such delocalization is possible and this amine is a stronger base than phenylamine:

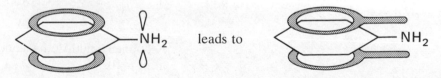

The presence of electron-withdrawing substituents in the ring reduces the basic character of the amine still further, so that 2,4-dinitrophenylamine, for example, is insoluble even in mineral acids.

14.2.2 Nucleophilic reactions

(i) Reaction with halogenoalkanes
Phenylamine reacts as a nucleophile on heating with halogenoalkanes to give first a secondary and then a tertiary amine (see also aliphatic case, Section 11.3.2(i)). Treatment of the salts formed liberates the free amines:

$$C_6H_5\ddot{N}H_2 + R-I \longrightarrow C_6H_5\overset{+}{N}H_2R\ I^- (\xrightarrow{OH^-} C_6H_5NHR)$$

$$C_6H_5\overset{+}{N}H_2R\ I^- + C_6H_5\ddot{N}H_2 \rightleftharpoons C_6H_5\ddot{N}HR + C_6H_5\overset{+}{N}H_3\ I^-$$

$$C_6H_5\ddot{N}HR + R-I \longrightarrow C_6H_5\overset{+}{N}HR_2\ I^- (\xrightarrow{OH^-} C_6H_5NR_2)$$

Sometimes quaternary salts are also formed.

(ii) Reaction with acid chlorides and anhydrides
Primary and secondary amines react with these reagents to form N-substituted amides. Tertiary amines do not react, since they do not possess a hydrogen atom bonded to the nitrogen, e.g.

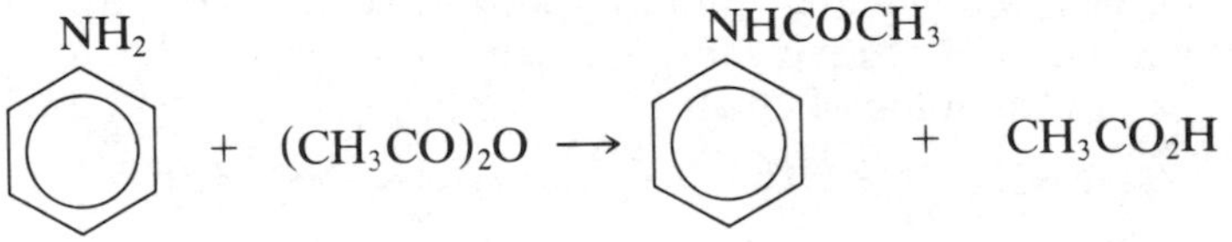

NH₂ + (CH₃CO)₂O → NHCOCH₃ + CH₃CO₂H

N-phenylethanamide
m.p. = 114°C

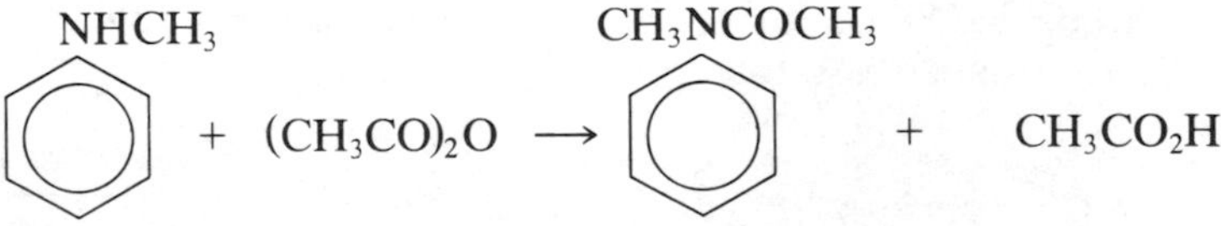

NHCH₃ + (CH₃CO)₂O → CH₃NCOCH₃ + CH₃CO₂H

N,N-phenylmethylethanamide
m.p. = 102°C

Similar compounds have important pharmaceutical properties and are widely used:

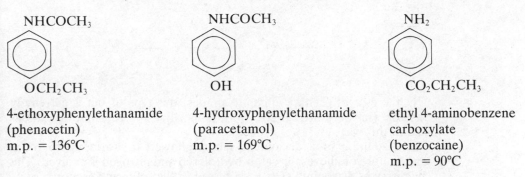

4-ethoxyphenylethanamide
(phenacetin)
m.p. = 136°C

4-hydroxyphenylethanamide
(paracetamol)
m.p. = 169°C

ethyl 4-aminobenzene
carboxylate
(benzocaine)
m.p. = 90°C

Phenacetin and paracetamol are used for treating headaches and neuralgia, and benzocaine is an important local anaesthetic.

2-Ethanoylaminobenzenecarboxylic acid, the product of ethanoylation (*acetylation*) of 2-aminobenzenecarboxylic acid, is of interest because the crystals emit light when they are fractured or rubbed together; the crystals are said to be *triboluminescent*.

When benzenecarbonyl chloride is used with an aromatic amine, in the presence of alkali, the reaction is known as the **Schotten–Baumann reaction** (*see also* Section 13.3.1(ii)).

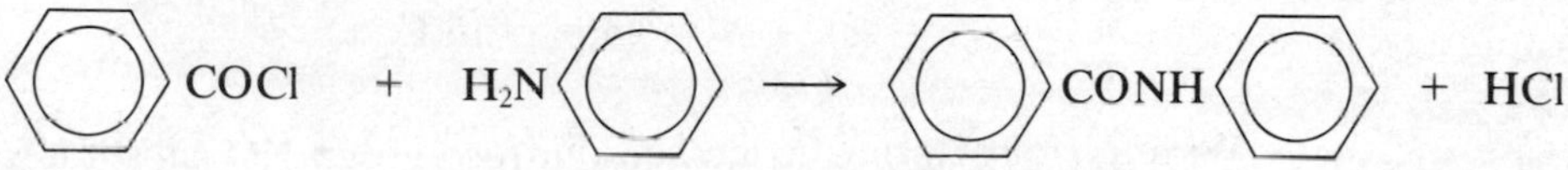

$$\text{⬡—COCl} + \text{H}_2\text{N—⬡} \longrightarrow \text{⬡—CONH—⬡} + \text{HCl}$$

N-phenylbenzenecarboxamide
m.p. = 162°C

The conversion of amines into amides is a useful way of 'protecting' the amine against oxidation (*see* Section 14.2.3).

(iii) Reaction with nitrous acid
If an aqueous solution of sodium nitrite (*care*: poisonous) is added to a solution of a *primary* aromatic amine in aqueous mineral acid at a temperature of 0–5°C, a **diazonium salt** is formed, e.g.

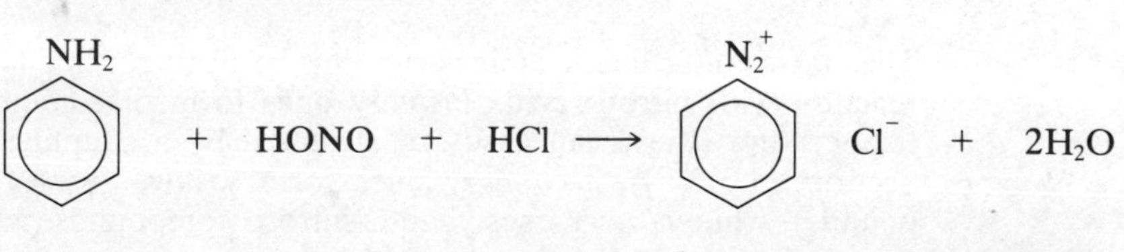

$$\text{⬡—NH}_2 + \text{HONO} + \text{HCl} \longrightarrow \text{⬡—N}_2^{+} \quad \text{Cl}^- + 2\text{H}_2\text{O}$$

benzenediazonium chloride

Several canonical forms can be written for this type of salt, showing how the positive charge is stabilized by delocalization:

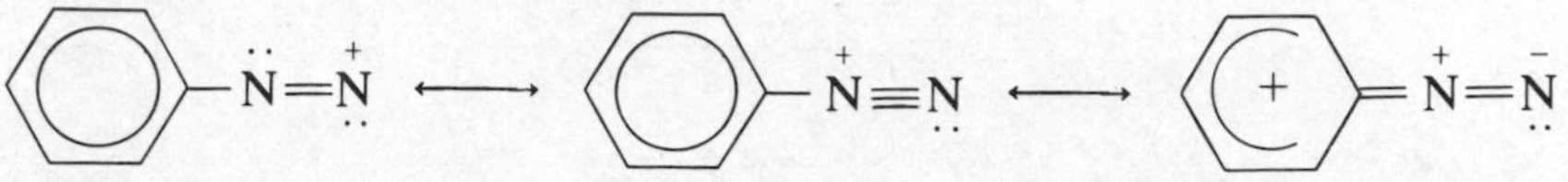

In the case of primary aliphatic amines, this type of salt is extremely unstable (*see* Section 11.3.2(iv)), as the positive charge cannot be stabilized in this way.

If a solution of a diazonium salt is allowed to warm up to room temperature or above, the salt is hydrolysed and nitrogen is evolved. The chemistry of diazonium salts is discussed in the following section.

A *secondary* aromatic amine forms an N-nitroso compound with nitrous acid and thus resembles its aliphatic counterpart, e.g.

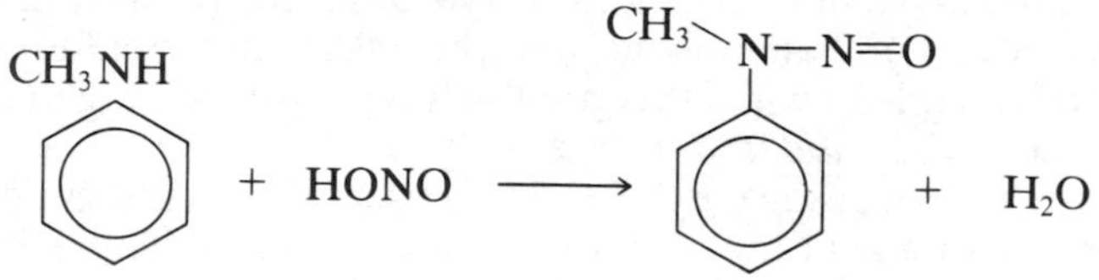

N-nitroso(phenylmethyl)amine
m.p. = 15°C

With a *tertiary* aromatic amine, a nitroso group NO substitutes in the 4-position of the benzene ring and the product may precipitate in the form of its hydrochloride, e.g.

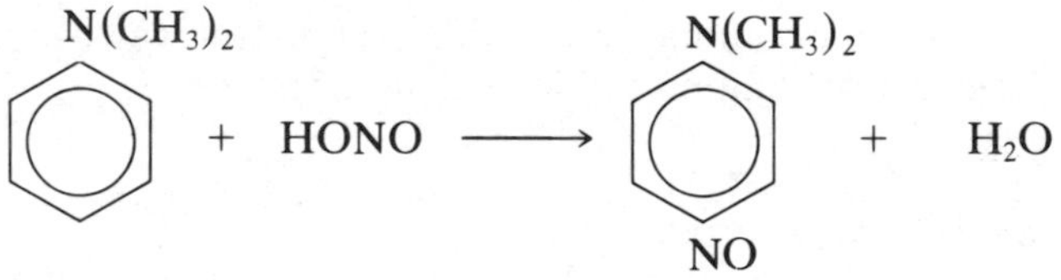

m.p. = 87°C
(m.p. of hydrochloride = 177°C)

The classes of aromatic amines can thus be distinguished readily by their reaction with nitrous acid. *Primary* ones form diazonium salts at low temperature (these can easily be detected by a coupling reaction, *see* Section 14.3.2(ii)), *secondary* ones form yellow oils (N-nitroso compounds), while *tertiary* ones yield C-nitroso compounds (which may precipitate in the form of the hydrochloride).

14.2.3 Substitution reactions of the ring

The amino group *activates* the benzene ring towards electrophilic substitution and thus halogenation, for example, takes place readily, without a catalyst, when an aqueous solution of chlorine or bromine is added to the amine (cf. phenols, Section 13.3.1(iv)):

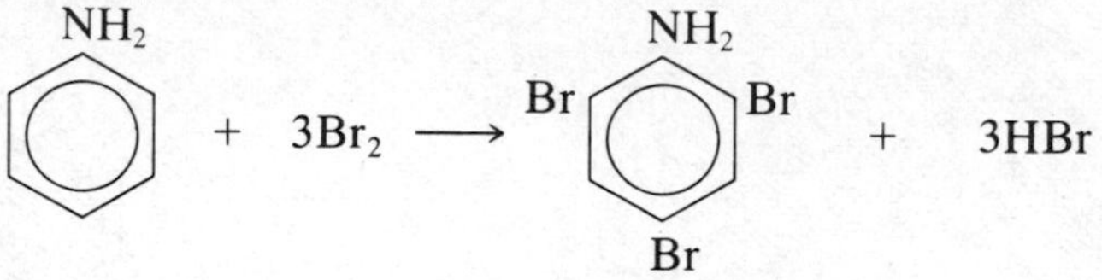

2,4,6-tribromo-
phenylamine
m.p. = 122°C

The ring is so activated that the *tri*bromo derivative is formed, and precipitates out. To achieve monobromination, for example, the amino group is converted into an amide (*see* Section 14.2.2(ii)). The ring is less activated and bromination takes place essentially just at the 4-position. The amide can be converted back into an amine by hydrolysis afterwards:

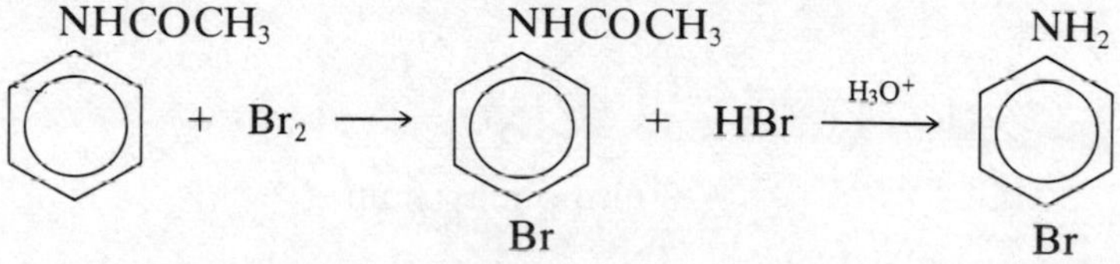

4-bromophenylethanamide   4-bromophenylamine
m.p. = 166°C   m.p. = 66°C

Sulphonation of phenylamine produces the important acid 4-amino-benzenesulphonic acid (*sulphanilic acid*) used in making certain dyes (*see* Section 14.3.2(ii)):

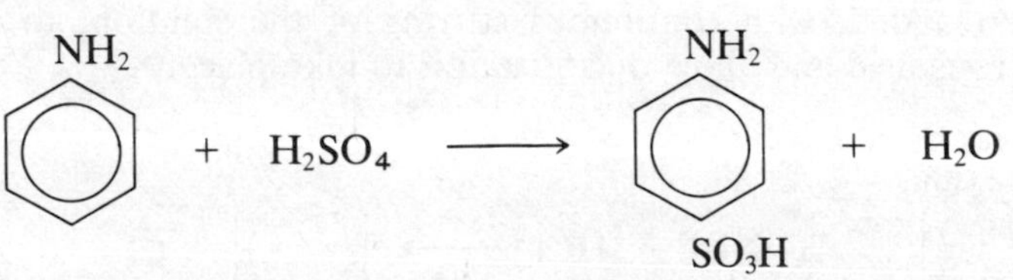

4-aminobenzenesulphonic acid

Nitration tends to cause oxidation of the amino group but it can be 'protected' by ethanoylation (*acetylation*); the ethanoyl (*acetyl*) group can be removed, following nitration, by hydrolysis:

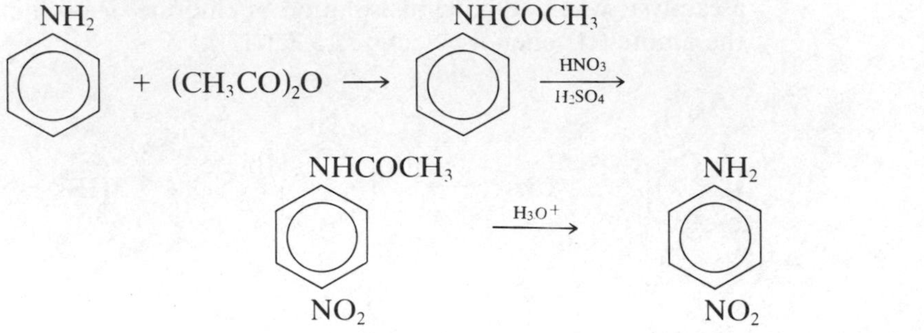

NH$_2$ NHCOCH$_3$

⬡   + (CH$_3$CO)$_2$O ⟶   ⬡ $\xrightarrow[\text{H}_2\text{SO}_4]{\text{HNO}_3}$

NHCOCH$_3$ NH$_2$

⬡ $\xrightarrow{\text{H}_3\text{O}^+}$ ⬡

NO$_2$ NO$_2$

4-nitrophenyl-   4-nitrophenylamine
ethanamide   m.p. = 147°C
m.p. = 215°C
(some 2-substitution occurs)

3-nitrophenylamine cannot be prepared by direct nitration of the amine, or derivative, but if 1,3-dinitrobenzene is treated with sodium sulphide, or polysulphide, reduction can be stopped after *one* nitro group has reacted, since the reduction of the second group is very much slower:

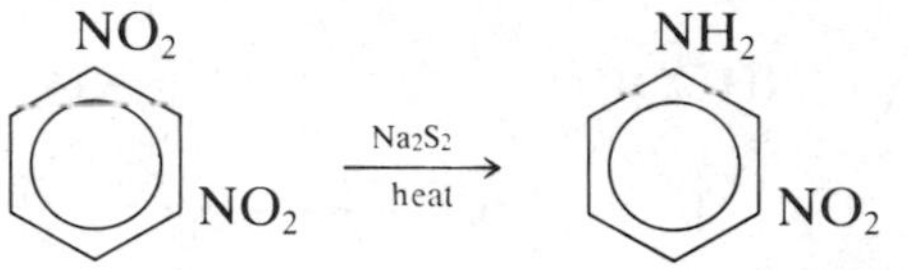

NO$_2$ NH$_2$

⬡ NO$_2$ $\xrightarrow[\text{heat}]{\text{Na}_2\text{S}_2}$ ⬡ NO$_2$

3-nitrophenylamine
m.p. = 114°C

14.3   Diazonium salts

These salts are of synthetic importance because of their reactivity. They are not easily isolated in the pure state (when dry, some are very explosive), and are usually prepared and used in aqueous solution. The preparation of a diazonium salt from a primary aromatic amine is called **diazotization**. The amine is normally dissolved in excess mineral acid and cooled to about 0–5°C. A solution of sodium nitrite (*care*: poisonous) is then added, with continuous stirring of the solution, to generate the nitrous acid and allow diazotization to take place:

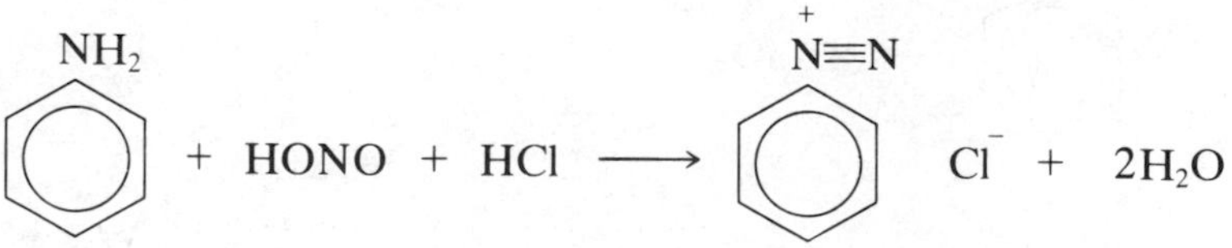

NH$_2$ $\overset{+}{N}{\equiv}N$

⬡   + HONO + HCl ⟶   ⬡   Cl$^-$ + 2H$_2$O

benzenediazonium chloride

The presence of excess mineral acid reduces the concentration of any free amine which could react with the diazonium salt (*see* Section 14.3.2(ii)). If the temperature is allowed to exceed 10°C, the diazonium salt will decompose.

Mechanistically, the reaction of a primary aromatic amine with nitrous acid is identical to that of an aliphatic amine:

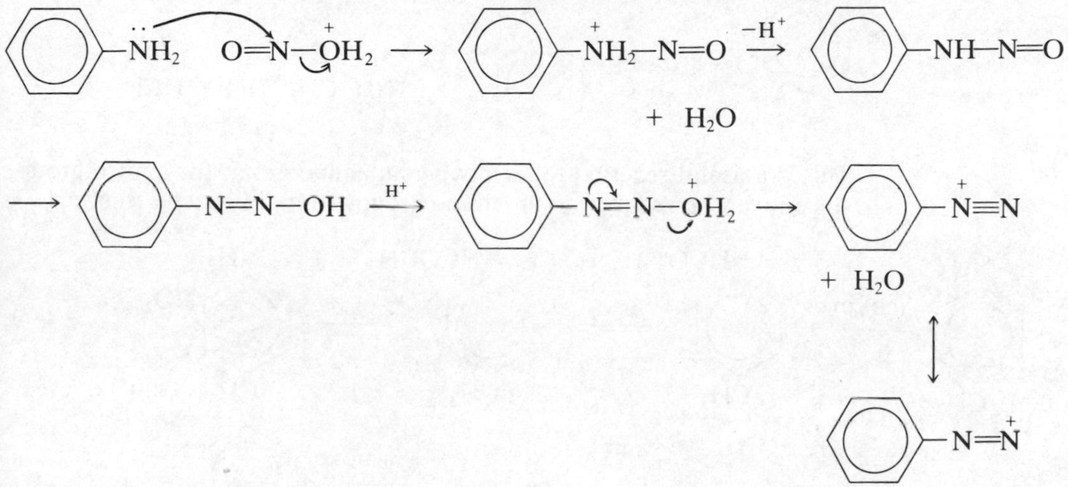

$$\text{Ph-NH}_2 \quad O{=}N{-}\overset{+}{O}H_2 \longrightarrow \text{Ph-}\overset{+}{N}H_2{-}N{=}O \xrightarrow{-H^+} \text{Ph-NH-N}{=}O$$

$$+ \; H_2O$$

$$\longrightarrow \text{Ph-N}{=}N{-}OH \xrightarrow{H^+} \text{Ph-N}{=}N{-}\overset{+}{O}H_2 \longrightarrow \text{Ph-}\overset{+}{N}{\equiv}N$$

$$+ \; H_2O$$

$$\updownarrow$$

$$\text{Ph-N}{=}\overset{+}{N}$$

The more basic the amine, the stronger its nucleophilic character and the easier it is to diazotize. Although the diazonium chloride is usually used, the hydrogensulphate, prepared using sulphuric acid as the mineral acid, is sometimes to be preferred.

The reactions of diazonium salts can conveniently be divided into two types, namely (1) those in which the nitrogen is displaced as molecular nitrogen and (2) those in which the nitrogen is retained in the product.

14.3.1   Reactions involving displacement of nitrogen

(i) Hydrolysis

If the temperature of the diazonium salt solution is allowed to rise above 10°C, the salt begins to decompose by hydrolysis. This process can be accelerated by heating the solution to about 50°C. Nitrogen is evolved and a phenol is formed, e.g.

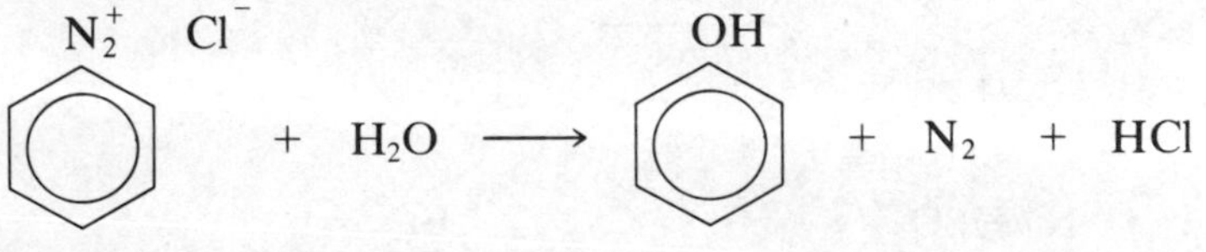

$$\text{Ph-}\overset{+}{N}_2 \; Cl^- \quad + \quad H_2O \longrightarrow \text{Ph-OH} \quad + \quad N_2 \quad + \quad HCl$$

This is a useful way of introducing a hydroxyl group into the benzene ring.

(ii) Reduction
Certain reagents, such as phosphinic acid H_3PO_2 or ethanol, reduce diazonium salts, e.g.

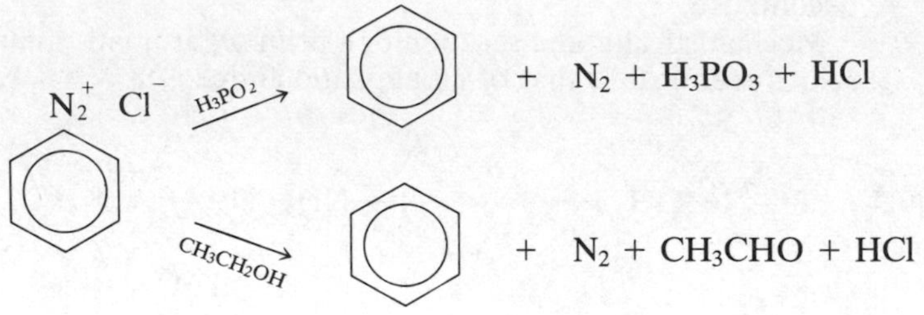

$$\underset{}{N_2^+ \ Cl^-} \xrightarrow{H_3PO_2} \bigcirc + N_2 + H_3PO_3 + HCl$$

$$\xrightarrow{CH_3CH_2OH} \bigcirc + N_2 + CH_3CHO + HCl$$

This is a useful reaction for removing an amino group (or a nitro group, since it can be reduced to an amino group) from a benzene ring, e.g.

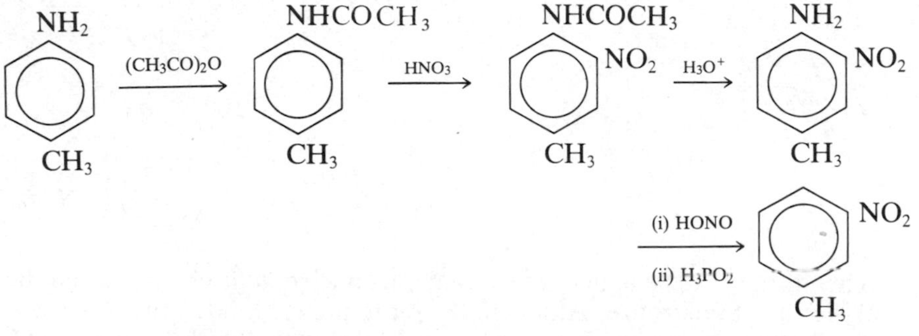

3-methylnitrobenzene cannot be made by direct nitration of methylbenzene. On the other hand, 4-methylnitrobenzene is readily available from methylbenzene and is reduced to give the starting material, 4-methylphenylamine.

(iii) Sandmeyer reactions
A way of replacing the nitrogen in a diazonium salt by chlorine, bromine or cyanide was discovered by the Swiss chemist Traugott Sandmeyer in 1884.

The reaction between a diazonium salt and chloride or bromide ion in acid solution is catalysed by the corresponding copper(I) halide at about 60°C:

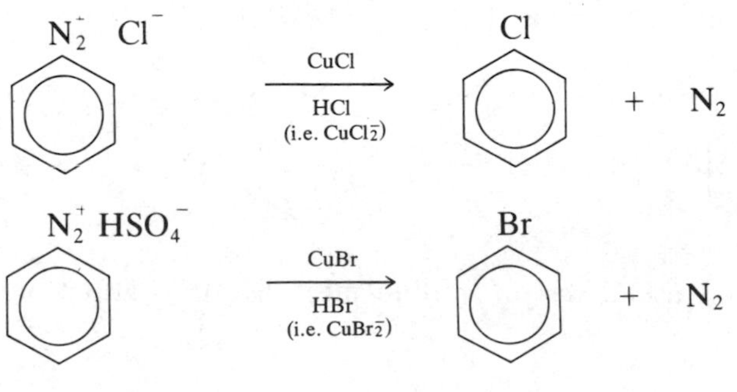

In 1890, the German chemist Ludwig Gattermann found that the formation of chlorobenzene from benzenediazonium chloride was also catalysed by powdered copper and this method is sometimes more convenient to use.

If a warm solution of copper(I) cyanide in excess potassium cyanide solution (this gives the complex $K_3Cu(CN)_4$) is employed with a diazonium salt, the nitrogen is replaced by a cyano group, e.g.

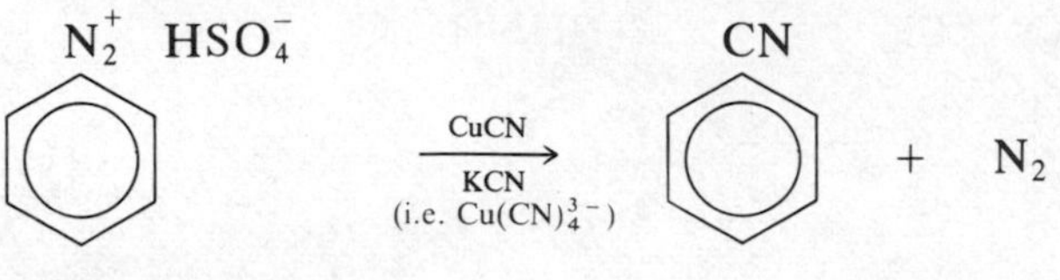

$$
\underset{\text{CuCN}}{\overset{\text{}}{}}
$$

This is a useful reaction because the cyano group cannot be introduced by nucleophilic displacement of chlorine in chlorobenzene (*see* Section 13.2.1). Such nitriles can be readily hydrolysed to a carboxylic acid, or reduced to a primary amine, e.g.

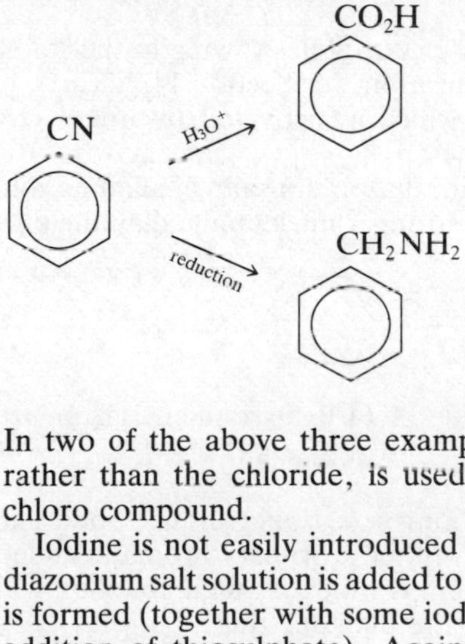

In two of the above three examples, the diazonium hydrogensulphate, rather than the chloride, is used in order to prevent formation of the chloro compound.

Iodine is not easily introduced into the benzene ring directly but, if a diazonium salt solution is added to potassium iodide solution, iodobenzene is formed (together with some iodine which can easily be removed by the addition of thiosulphate). Again, the hydrogensulphate gives a better yield than the chloride:

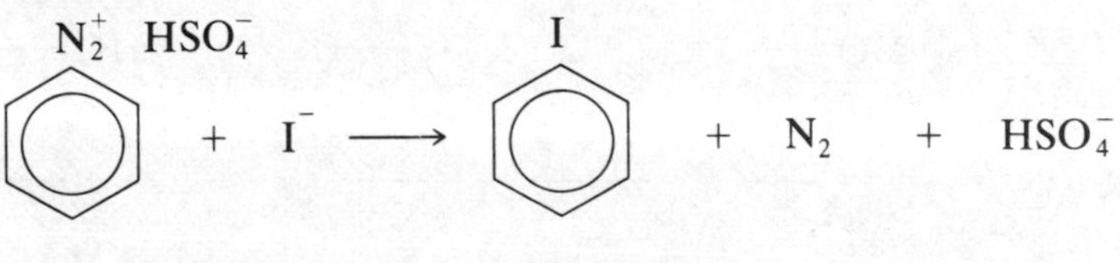

iodobenzene
b.p. = 188°C

14.3.2   Reactions involving retention of nitrogen

(i) Reduction of the diazo group

Mild reducing agents, such as $SnCl_2$, SO_2 or $NaHSO_3$, reduce the **diazo** group, $-\overset{+}{N}\equiv N$, in diazonium salts. Thus, benzenediazonium chloride is reduced to the hydrochloride of phenylhydrazine; the free phenylhydrazine can be obtained by warming with alkali:

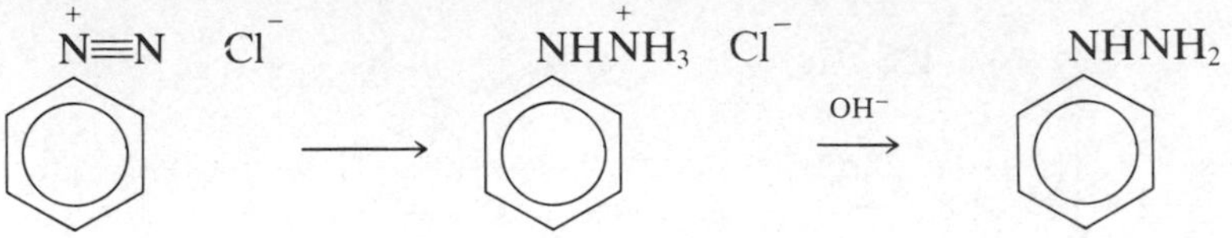

phenylhydrazine
m.p. = 19°C

Phenylhydrazine forms useful crystalline derivatives with aldehydes and ketones (*see* Section 8.3.3).

(ii) Coupling reactions

Diazonium salts are *weak* electrophiles (owing to the stabilization of the positive charge by delocalization, *see* Section 14.2.2(iii)), but will couple with aromatic compounds which are activated towards electrophilic attack, namely phenols and amines.

Phenols are coupled with diazonium salts in alkaline solution, because the anion of a phenol is a stronger nucleophile than the phenol itself, e.g.

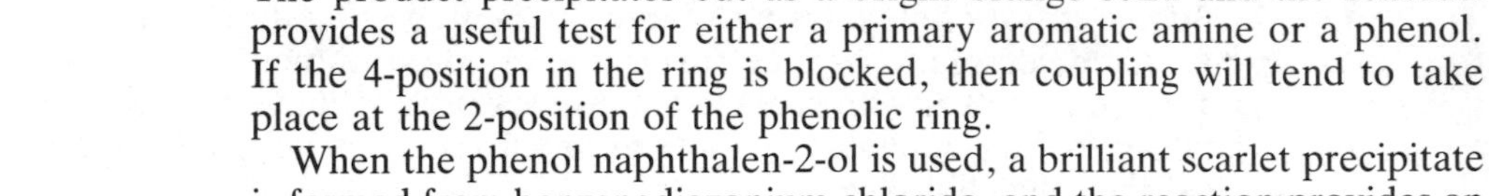

(4-hydroxyphenyl)azobenzene
m.p. = 152°C

The product precipitates out as a bright orange solid and the reaction provides a useful test for either a primary aromatic amine or a phenol. If the 4-position in the ring is blocked, then coupling will tend to take place at the 2-position of the phenolic ring.

When the phenol naphthalen-2-ol is used, a brilliant scarlet precipitate is formed from benzenediazonium chloride, and the reaction provides an even more striking test for phenylamine:

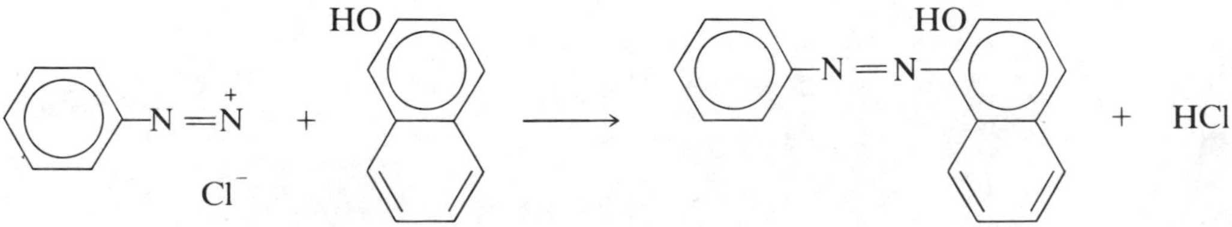

m.p. = 131°C

Diazonium salts can also be coupled with aromatic amines in weakly acidic solution, e.g.

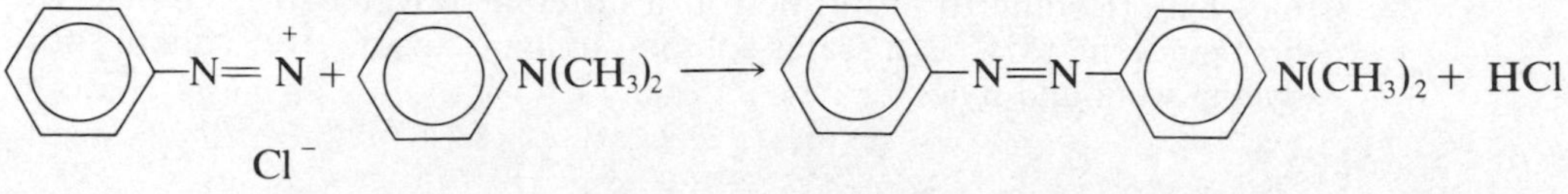

$C_6H_5-N=N^+\ Cl^- + C_6H_5N(CH_3)_2 \longrightarrow C_6H_5-N=N-C_6H_4N(CH_3)_2 + HCl$

m.p. = 117°C

In these coupling products, the presence of two aromatic rings linked by the **azo** link —N=N— imparts an orange-yellow colour to them; the structural arrangement responsible for the colour is called a **chromophore**. These compounds absorb light in the visible region of the spectrum (delocalized electrons are promoted to higher energy levels, temporarily) and hence display the colour which is complementary to that absorbed. When a coloured compound displays an affinity for a fibre, then it may be used as a dye. The simpler **azo dyes** are generally insoluble in water and are applied to a fibre such as cotton by precipitation. However, solubilizing groups, such as the sulphonic acid group, can be incorporated. Thus, the diazotization of 4-aminobenzenesulphonic acid can lead to soluble products, e.g.

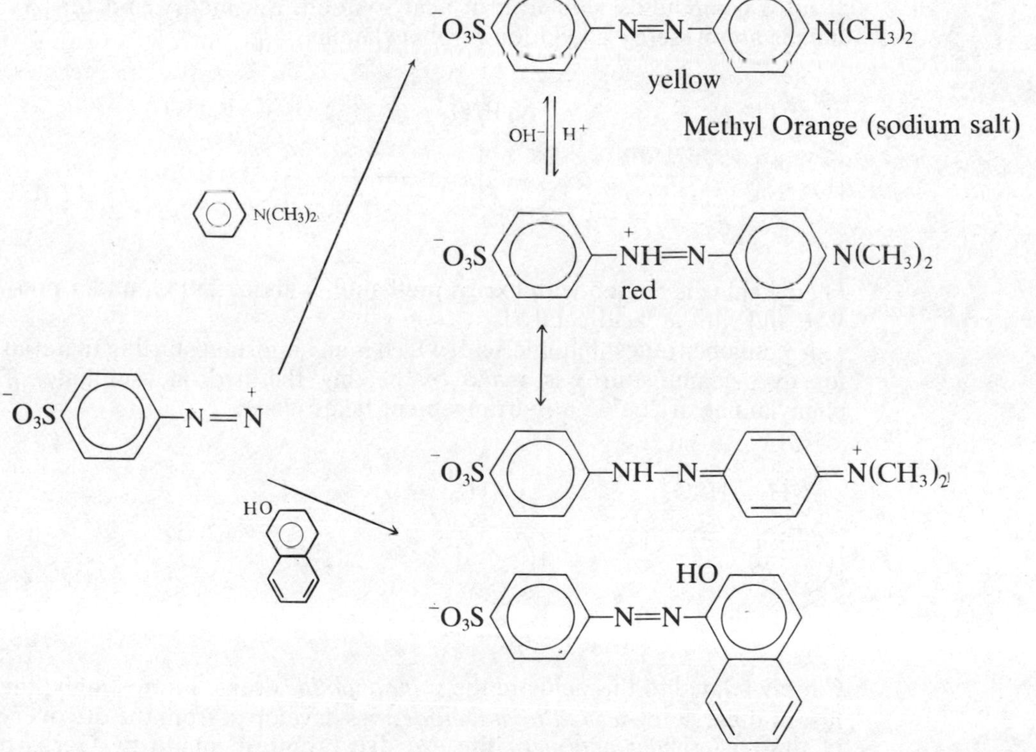

$^-O_3S-C_6H_4-N=N-C_6H_4-N(CH_3)_2$

yellow

Methyl Orange (sodium salt)

$OH^- \Updownarrow H^+$

$^-O_3S-C_6H_4-\overset{+}{N}H=N-C_6H_4-N(CH_3)_2$

red

$C_6H_4N(CH_3)_2$

$^-O_3S-C_6H_4-NH-N=C_6H_4=\overset{+}{N}(CH_3)_2$

$^-O_3S-C_6H_4-N=N$

HO-naphthalene

$^-O_3S-C_6H_4-N=N-(HO\text{-naphthalene})$

Orange II

Methyl Orange is a familiar acid-base indicator which, on protonation, displays a different colour, owing to charge delocalization over the whole molecule (leading to absorption at a different wavelength). Orange II, first made in 1876, is a water-soluble, anionic, acidic dye, suitable for dyeing wool and nylon.

14.4   Manufacture and uses of aromatic amines

Phenylamine is made by the catalytic hydrogenation of nitrobenzene (made by the nitration of benzene); the process is a continuous one and is carried out in the vapour phase. (Note that reduction of the aromatic ring requires much more vigorous conditions than are employed here.)

More specialized amines are made in batch processes by the reduction of nitro compounds with iron in acid solution. Secondary and tertiary amines are made by alkylation of phenylamine, e.g.

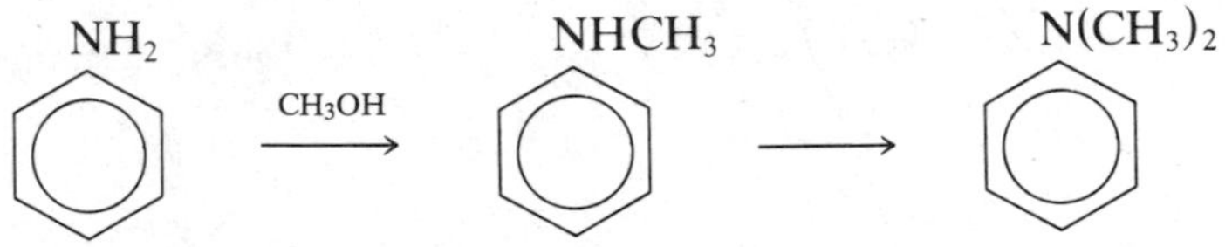

Phenylamine is heated with excess methanol at about 250°C, under pressure and with an acid catalyst.

4-Aminobenzenesulphonic acid, which is an important starting material for dyes manufacture, is made by heating the hydrogensulphate of phenylamine at 200°C; a rearrangement takes place:

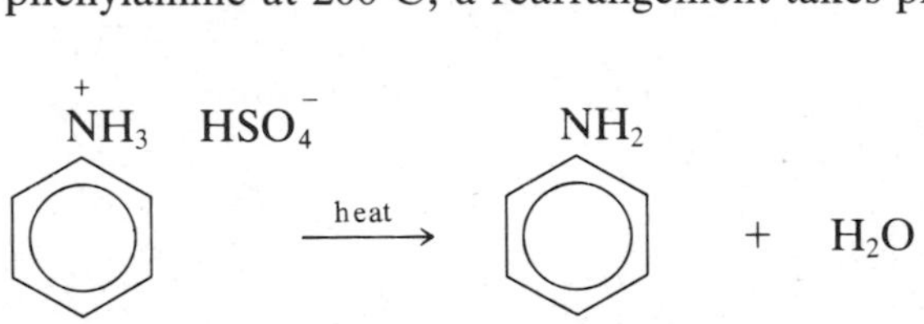

Closely related to this acid are the *sulphonamide* drugs. Interestingly, the first of these, known as *sulphanilamide*, was developed from the discovery of the bactericidal action of the azo dye Prontosil, made by Gerhard Domagk in 1935:

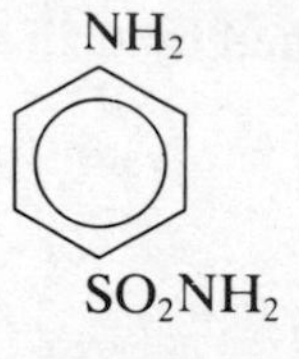

sulphanilamide 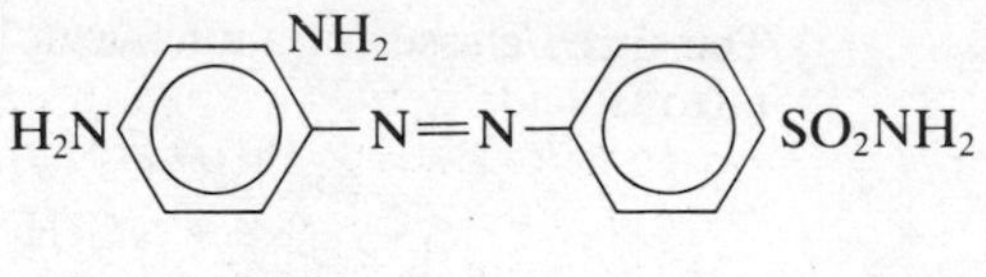   Prontosil

Phenylamine is used largely in the production of anti-oxidants for rubber and, to a lesser extent, in the manufacture of some dyes and pharmaceutical products.

Summary

1 **Aromatic amines** may possess one or more benzene rings within the molecule. Like aliphatic amines, they are classified as primary, secondary or tertiary.

2 Primary aromatic amines like phenylamine (*aniline*) are prepared by the reduction of aromatic nitro compounds. In the laboratory a dissolving metal reduction (e.g. Sn, or Fe, and concentrated HCl) is usually employed.

3 Amines in which the nitrogen is not directly bonded to the benzene ring, e.g. (phenylmethyl)amine $C_6H_5CH_2NH_2$, closely resemble the aliphatic amines. In the remaining cases, however, electronic interaction can occur between the ring and the lone-pair of electrons on the nitrogen atom. A simple aromatic amine like phenylamine shows properties characteristic of a Lewis base and of a nucleophile, as well as properties typical of the benzene ring:

$$C_6H_5NH_2 \begin{cases} \xrightarrow{HX} C_6H_5NH_3^+ X^- \quad (\xrightarrow{OH^-} C_6H_5NH_2) \\[2mm] \xrightarrow{RX} C_6H_5\overset{+}{N}H_2R\,X^- \quad (\xrightarrow{OH^-} C_6H_5NHR) \\[2mm] \text{and} \\[2mm] C_6H_5\overset{+}{N}HR_2\,X^- \quad (\xrightarrow{OH^-} C_6H_5NR_2) \\[2mm] \xrightarrow[\text{or } (RCO)_2O]{RCOCl} C_6H_5NHCOR \end{cases}$$

The three classes of amines can be distinguished by their reactions with nitrous acid:

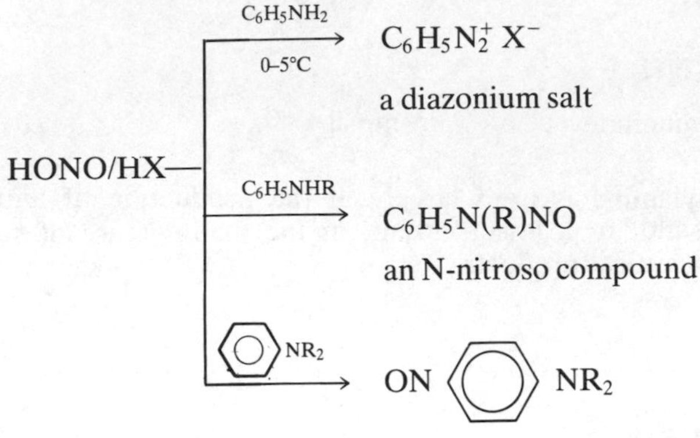

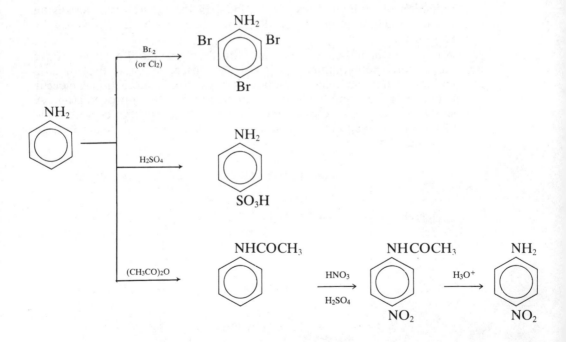

HONO/HX
$C_6H_5NH_2$
0–5°C
$C_6H_5N_2^+X^-$
a diazonium salt
C_6H_5NHR
$C_6H_5N(R)NO$
an N-nitroso compound
NR_2
ON
NR_2
a C-nitroso compound

The aromatic amines also show properties typical of the benzene ring. The amino group *activates* the ring towards electrophilic substitution:

NH_2
Br_2
(or Cl_2)
NH_2
Br
Br
Br
H_2SO_4
NH_2
SO_3H
$(CH_3CO)_2O$
$NHCOCH_3$
HNO_3
H_2SO_4
$NHCOCH_3$
NO_2
H_3O^+
NH_2
NO_2

4 *Primary* aromatic amines form **diazonium salts** on treatment with nitrous acid and mineral acid at 0–5°C. These salts are very reactive and allow a range of extremely useful synthetic transformations to take place.

(i) Displacement of nitrogen

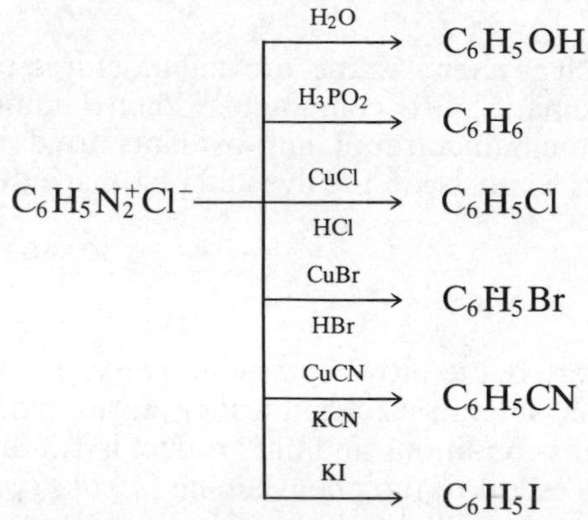

$$C_6H_5N_2^+Cl^-$$

$\xrightarrow{H_2O}$ C_6H_5OH

$\xrightarrow{H_3PO_2}$ C_6H_6

$\xrightarrow[HCl]{CuCl}$ C_6H_5Cl

$\xrightarrow[HBr]{CuBr}$ C_6H_5Br

$\xrightarrow[KCN]{CuCN}$ C_6H_5CN

$\xrightarrow{KI}$ C_6H_5I

(ii) Retention of nitrogen
Reduction of diazo group:

Coupling reactions:
with a phenol:

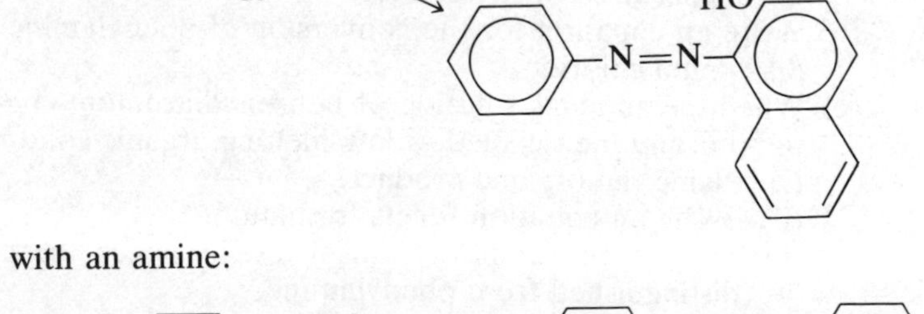

with an amine:

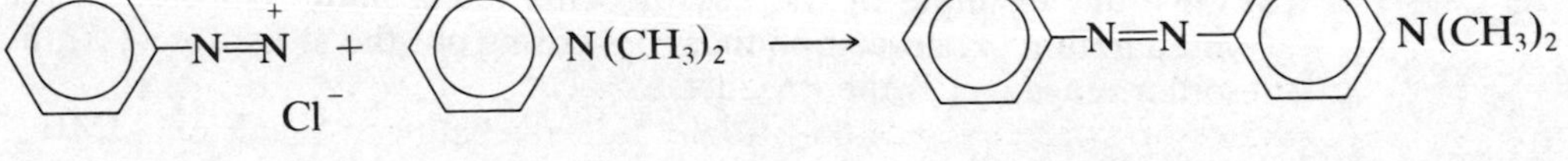

The coupling reaction with a phenol produces a bright orange-red solid and the reaction is a useful test for either a primary aromatic amine or a phenol.

5   Aromatic amines, such as phenylamine, are manufactured by the catalytic hydrogenation of aromatic nitro compounds. Phenylamine is used predominantly for the manufacture of anti-oxidants used in the rubber industry but also finds some use in the dye and pharmaceutical industries.

Questions

1   Benzene can be converted into nitrobenzene by a mixture of concentrated nitric and concentrated sulphuric acids. Further nitration of nitrobenzene requires more vigorous conditions and the product is 1,3-dinitrobenzene. If the nitrobenzene is reduced, the phenylamine (*aniline*) produced is too reactive to nitrate in a controlled fashion under normal conditions. However, after ethanoylation (*acetylation*) the resulting phenylethanamide (*acetanilide*) ($C_6H_5NHCOCH_3$) can be nitrated to give 4-nitrophenylethanamide (*4-nitroacetanilide*) mixed with a small amount of 2-nitrophenylethanamide.
(a) Comment briefly on the mechanism of the nitration of benzene.
(b) For the nitration of nitrobenzene, explain the orientation of the product and the reason for the necessary change of conditions.
(c) What is the origin of the high reactivity of phenylamine (*aniline*)?
(d) What is the electronic effect of the ethanoyl (*acetyl*) group of phenylethanamide (*acetanilide*) on the reactivity of the C_6H_5NH-moiety?
(e) Explain the orientation of the nitration products of phenylethanamide (*acetanilide*).
(f) Why is the 4-isomer the predominant product?

Imperial College

2   Nitrobenzene was reduced to phenylamine and the phenylamine converted into benzenediazonium chloride by its reaction with aqueous sodium nitrite and hydrochloric acid below 10°C.
(a) State a reagent or mixture of reagents suitable for the reduction of nitrobenzene to phenylamine.
(b) Write an equation for the conversion of phenylamine into benzenediazonium chloride.
(c) When the aqueous solution of benzenediazonium chloride is heated to 50°C and then cooled, a low-melting organic solid is formed.
(i)   Name the organic product.
(ii) Write an equation for its formation.
(iii) Suggest **one** simple chemical test by which the product can be distinguished from phenylamine.
(d) Give **one** example of the use of benzenediazonium chloride in the preparation of azo-compounds, by writing out the structure of (i) the other reagent, (ii) the product.

JMB

3   Describe in outline how phenylamine (*aniline*) may be prepared from
    (a) benzene
    (b) N-phenylethanamide (CH₃CONHC₆H₅).
    What is the effect upon phenylamine of the following reagents? Give
    equations and the structures of the products.
    (c) aqueous hydrogen chloride
    (d) ethanoyl (*acetyl*) chloride
    (e) cold nitrous acid
    (f) iodoethane (*ethyl iodide*).
    Briefly describe **two** ways in which phenylamine (b.p. 184°C) could be
    distinguished from phenylmethylamine (C₆H₅CH₂NH₂) (b.p. 184°C).

                                                             Oxford & Cambridge

4   Addition of sodium nitrite solution to an ice-cold solution of phenylamine
    (*aniline*) in hydrochloric acid yields a solution of benzenediazonium chlor-
    ide. What happens when sodium nitrite is added to a cold solution of
    1-aminopropane in hydrochloric acid? How do you account for this?
    Starting from benzenediazonium chloride solution, how would you obtain
    (a) chlorobenzene, (b) an azo dyestuff, and (c) benzene?
    How may 1,3,5-tribromobenzene be obtained from suitable starting
    materials?

                                                             Oxford & Cambridge

5* A solid organic compound C₇H₇NO₂ is thought to be either

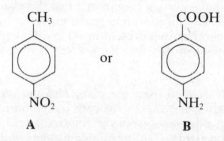

    A   **B**

    (a) Predict which of the two compounds has the higher melting point and
        briefly explain your prediction.
    (b) Suggest one simple chemical test-tube reaction which would distin-
        guish between **A** and **B**, giving the appropriate equation.
    (c) Show, by giving the appropriate reagents, how **A** can be converted
        into **B** via compound **X**,C₇H₅NO₄.
    (d) The pK_a value of benzenecarboxylic (*benzoic*) acid is 4.17 and that
        of **X** is 3.43. Explain qualitatively the reason for the difference in the
        pK_a values.

                                                                          JMB

6   Explain the following observations:
    (a) Phenylamine (*aniline*) is a weaker base than methylamine.
    (b) Ethylamine is very soluble in water.

(c) Dimethylamine on reaction with nitrous acid yields a neutral compound ($C_2H_6N_2O$).

(d) The reaction between 1-aminobutane ($CH_3CH_2CH_2CH_2NH_2$) and iodomethane (*methyl iodide*) leads to the formation of several products.

(e) Phenylamine (*aniline*) reacts readily with an aqueous solution of bromine to form 2,4,6-tribromophenylamine (*tribromoaniline*).

Cambridge

7   Methyl Orange

$$(CH_3)_2N - \!\!\bigcirc\!\! - N\!=\!N - \!\!\bigcirc\!\! - SO_3Na$$

is prepared by diazotizing 4-aminobenzenesulphonic (*sulphanilic*) acid followed by coupling the diazosulphonate with N,N-dimethylphenylamine (*dimethylaniline*).

(a) Outline the industrial preparation of (i) 4-aminobenzenesulphonic acid and (ii) N,N-dimethylphenylamine, starting from phenylamine (*aniline*). In each case, write appropriate equations and indicate the reagents and reaction conditions.

(b) Diazotization of 4-aminobenzenesulphonic acid can be carried out successfully only under certain conditions. List the essential requirements for a satisfactory diazotization reaction.

(c) State the chromophoric group present in Methyl Orange and explain briefly why the colour of this dye changes with variation in pH.

JMB

8**   A heart stimulant H, isolated from the plant *Halostachys caspica*, had the molecular formula $C_9H_{13}NO$. On vigorous oxidation with potassium permanganate, H gave benzenecarboxylic (*benzoic*) acid together with minor unidentified fragments. Careful treatment of H with nitrous acid gave a yellow nitroso-compound J ($C_9H_{12}N_2O$) which again gave benzenecarboxylic (*benzoic*) acid on vigorous oxidation. Oxidation of H with sodium iodate(VII) (*periodate*) however gave benzenecarbaldehyde (*benzaldehyde*) as the major fragment. Reaction of H with excess ethanoic (*acetic*) anhydride in pyridine gave a diethanoyl (*diacetyl*) compound K ($C_{13}H_{17}NO_3$) but careful ethanoylation (*acetylation*) with one equivalent of ethanoic (*acetic*) anhydride gave a mono ethanoyl (*acetyl*) derivative L ($C_{11}H_{15}NO_2$) which could be shown spectroscopically to be an amide. Mild oxidation of this gave a ketone M ($C_{11}H_{13}NO_2$).

Deduce the structures of H, J, K, L and M.

Imperial College

Chapter 15
Preparative methods and organic synthesis

Previous chapters have described the characteristic reactions of a variety of organic compounds and these reactions provide the means by which organic compounds may be synthesized from simpler compounds by transformation of functional groups. In contemplating a synthetic scheme, it is useful to think about reaction *types* and these are emphasized in the summaries of preparative methods which follow.

The outcome of an organic reaction may depend not only on the chemical reactants, but on the *reaction conditions* employed. For example, when warmed with a dilute, aqueous alcoholic solution of potassium hydroxide, a halogenoalkane like bromoethane undergoes a **substitution** reaction to form an alcohol but, when it is heated under reflux with a concentrated, alcoholic solution of potassium hydroxide, an **elimination** reaction takes place and an alkene is formed:

$$CH_3CH_2Br + OH^- \nearrow \text{\scriptsize dil.aq warm} \quad CH_3CH_2OH + Br^-$$
$$\searrow \text{\scriptsize hot conc. alcoholic} \quad CH_2{=}CH_2 + H_2O + Br^-$$

When an alkylbenzene, like methylbenzene, reacts with chlorine in the presence of a 'halogen carrier' such as aluminium chloride, a ring-substituted product is formed whereas the same compound, when treated at its boiling point with chlorine in the presence of ultraviolet light, forms a product in which substitution has occurred in the side-chain:

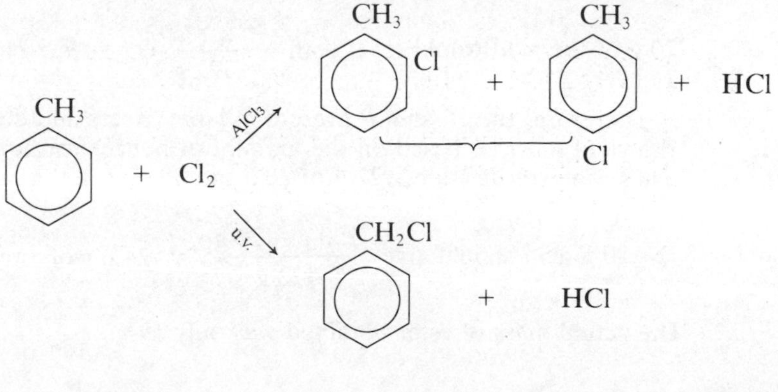

An alcohol, such as ethanol, when heated with concentrated sulphuric acid, may undergo three kinds of reaction, depending on the temperature and whether one reagent is in excess or not:

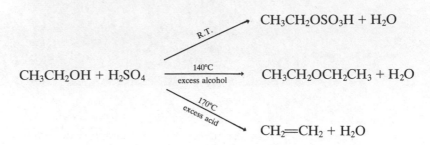

$$CH_3CH_2OH + H_2SO_4$$

$$\xrightarrow{R.T.} CH_3CH_2OSO_3H + H_2O$$

$$\xrightarrow[\text{excess alcohol}]{140°C} CH_3CH_2OCH_2CH_3 + H_2O$$

$$\xrightarrow[\text{excess acid}]{170°C} CH_2{=}CH_2 + H_2O$$

These examples underline how vital it is to know what *reaction conditions* are needed for a reaction, as well as what reagents. It should also be appreciated that, under a given set of conditions, one reaction tends to *predominate* over others rather than take place exclusively. Partly because of this, yields in organic reactions are seldom 100% and, in considering different reactions for a synthetic scheme, attention must be paid to the sort of yields which are usual for the reactions (*see also* later note on synthetic schemes, p. 270). The method of calculating the yield of an organic product is set out below.

Example. A mixture of 20 g benzenecarboxylic acid and 20 g pure ethanol was heated under reflux while a stream of dry hydrogen chloride was bubbled through for about 1½ h. The ethyl benzenecarboxylate was extracted from the cooled solution and 19 g of the pure ester were obtained on distillation. What is the yield of ester as a percentage of the theoretical amount which should have been produced?

The relevant equation is

$$C_6H_5COOH + CH_3CH_2OH \rightarrow C_6H_5COOCH_2CH_3 + H_2O$$

1 mol reacts with 1 mol to give 1 mol

$\Rightarrow$ 122 g reacts with 46 g to give 150 g

20 g of the acid would react with $\dfrac{46 \text{ g} \times 20 \text{ g}}{122 \text{ g}} = 7.5$ g of ethanol

Therefore, the ethanol is present in large excess and the calculation of the yield must be based on the amount of benzenecarboxylic acid used. Thus, on esterification, 122 g of acid should give 150 g of ester:

$\Rightarrow$ 20 g acid should give $\dfrac{150 \text{ g} \times 20 \text{ g}}{122 \text{ g}} = 24.6$ g of ester.

The actual mass of ester obtained was only 19 g.

The percentage of the theoretical amount which should have been produced

$$= \frac{19 \text{ g}}{24.6 \text{ g}} \times 100 = 77\%$$

Yield = 77%

Below are summarized methods of preparation of different classes of organic compounds. The lists are not meant to be exhaustive but should serve for revision of the reactions already encountered in previous chapters.

15.1 Alkanes

It is not often that an alkane has to be synthesized in the laboratory, unless it is a higher member of the homologous series. If the starting material is saturated, then a substitution reaction will be required, whereas if the starting material is unsaturated, then an addition reaction will be required.

15.1.1 Decomposition of Grignard reagents

$$RX + Mg \rightarrow RMgX \xrightarrow{H_2O} RH + Mg(OH)X$$

Reagents: a halogenoalkane and magnesium turnings.
Conditions: the reagents are warmed in *dry* ethoxyethane. The Grignard
                  reagent is decomposed cautiously with water.
For example:

$$CH_3(CH_2)_4CH_2Br + Mg \rightarrow CH_3(CH_2)_4CH_2MgBr \xrightarrow{H_2O} CH_3(CH_2)_4CH_3 + Mg(OH)Br$$
b.p. = 155°C   b.p. = 69°C   Yield: 30%

15.1.2 Reduction of halogenoalkanes

$$RX + H^- \rightarrow RH + X^-$$
         (from a complex
         metal hydride)

Reagents: a halogenoalkane and lithium tetrahydridoaluminate.
Conditions: the reagents are warmed in dry ethoxyethane.
For example:

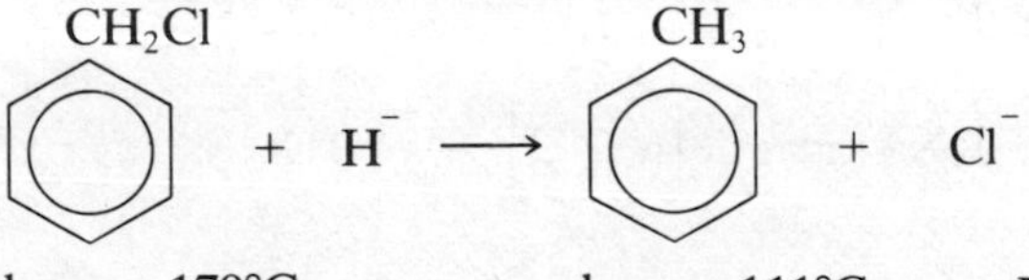

b.p. = 179°C   b.p. = 111°C   Yield: 98%

15.1.3 Catalytic hydrogenation

$$\ce{>C=C< + H_2 \longrightarrow >CH-CH<}$$

or

$$\ce{-C\equiv C- + 2H_2 \longrightarrow -CH_2-CH_2-}$$

Reagents: an alkene or alkyne (this is possibly more useful, *see* Section
15.3.2), and hydrogen.
Conditions: the mixture is warmed with Ni, Pt or Pd catalyst until uptake
of hydrogen is complete.

The reaction can be applied to $\ce{>C=C<}$ or $\ce{-C\equiv C-}$ bonds in compounds other than alkanes or alkynes, of course.
For example:

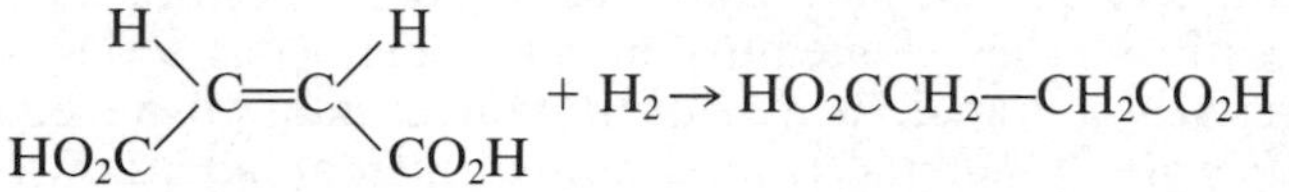

$$\underset{\text{m.p.} = 143°\text{C}}{\ce{\underset{HO_2C}{\overset{H}{>}}C=C\underset{CO_2H}{\overset{H}{<}}}} + \ce{H_2} \rightarrow \underset{\text{m.p.} = 185°\text{C} \qquad \text{Yield} = 98\%}{\ce{HO_2CCH_2-CH_2CO_2H}}$$

15.1.4 Decarboxylation

$$\ce{RCO_2H \rightarrow RH + CO_2}$$

Reagents: a carboxylic acid and 'soda-lime' (calcium oxide which has
been 'slaked' with aqueous sodium hydroxide).
Conditions: the mixture is heated strongly.
This reaction is of use for removing carboxyl groups from aromatic
compounds.
For example:

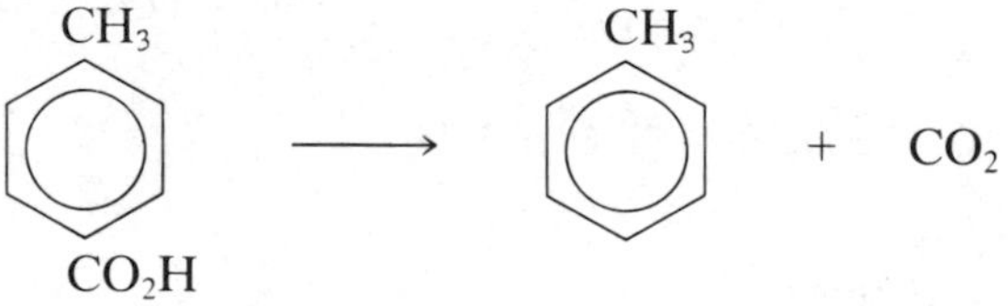

(benzene ring with CH$_3$ and CO$_2$H) $\longrightarrow$ (benzene ring with CH$_3$) + CO$_2$

m.p. = 182°C   b.p. = 111°C   Yield: 92%

15.1.5 Friedel–Crafts alkylation

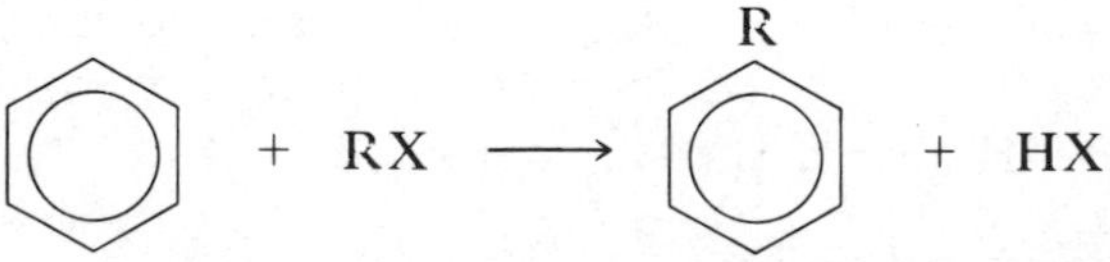

(benzene ring) + RX $\longrightarrow$ (benzene ring with R) + HX

Reagents: an aromatic compound and a halogenoalkane.

Conditions: the mixture is warmed in a solvent with aluminium, or iron(III), chloride catalyst.

With straight-chain halogenoalkanes, rearrangement of the chain to a branched-chain occurs (*see* below). Alkylation can also be carried out using an alkene and BF_3.

For example:

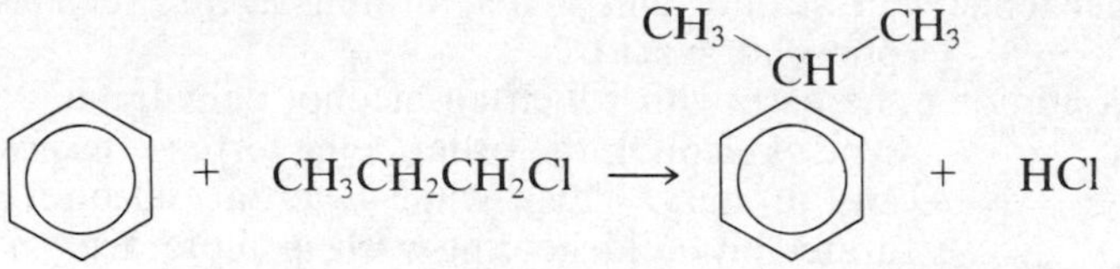

$$\text{C}_6\text{H}_6 + CH_3CH_2CH_2Cl \longrightarrow \text{C}_6\text{H}_5\text{CH(CH}_3)_2 + HCl$$

b.p. = 80°C   b.p. = 46°C   b.p. = 152°C   Yield: 78%

15.1.6 Kolbe electrolytic synthesis

$$2RCO_2^- - 2e^- \rightarrow R_2 + 2CO_2$$

Reagents: a salt of a carboxylic acid.

Conditions: the electrolysis of an aqueous methanolic solution of the salt, using a platinum anode, is carried out.

For example:

$$2CH_3(CH_2)_{12}COO^- - 2e^- \rightarrow CH_3(CH_2)_{24}CH_3 + 2CO_2$$
$$\text{m.p.} = 58°C \qquad \text{Yield: 30\%}$$

15.2 Alkenes

These may be synthesized from a saturated compound, in which case an elimination reaction is required, or from a more unsaturated compound, in which case an addition reaction is required.

15.2.1 Catalytic hydrogenation

$$-C\equiv C- + H_2 \rightarrow -CH=CH-$$

Reagents: an alkyne and hydrogen.

Conditions: the mixture is heated with a Lindlar catalyst (deactivated palladium) so that reduction is stopped at the alkene stage. Generally, the *cis* isomer of the alkene predominates.

For example:

$$CH_3C\equiv CCH_2CH_3 + H_2 \rightarrow CH_3CH\overset{cis}{=}CHCH_2CH_3$$
b.p. = 56°C   b.p. = 38°C   Yield: 55%

15.2.2   Dehydration of alcohols

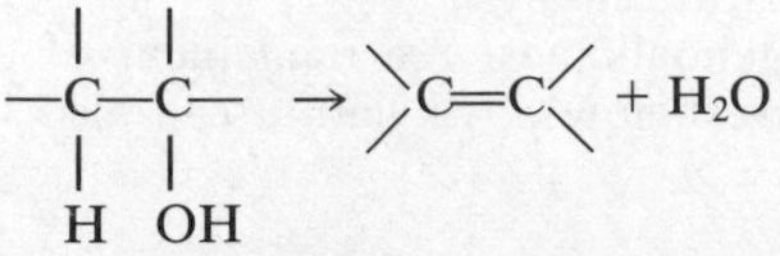

$$-\overset{|}{\underset{H}{C}}-\overset{|}{\underset{OH}{C}}- \rightarrow \; \overset{}{\underset{}{>}}C{=}C\overset{}{\underset{}{<}} + H_2O$$

Reagents: an alcohol and a dehydrating agent, such as sulphuric or phosphoric acid.

Conditions: the ease with which an alcohol dehydrates depends on the type of alcohol, the order being tertiary (easiest), secondary, and primary. Thus, while a tertiary alcohol may be dehydrated by mild heating with a dilute acid, a secondary or primary alcohol will generally require heating with concentrated acid.

For example:

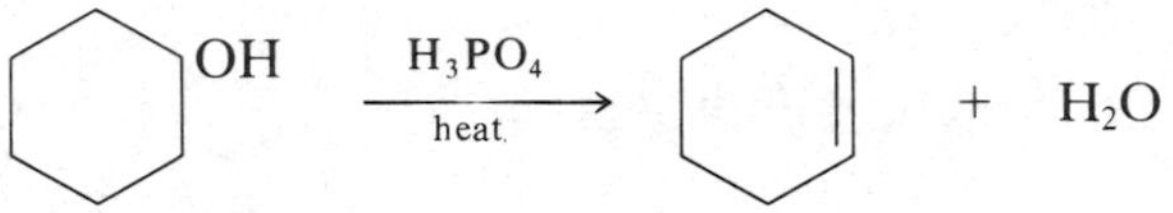

b.p. = 161°C   b.p. = 83°C   Yield: 73%

15.2.3   Dehydrohalogenation

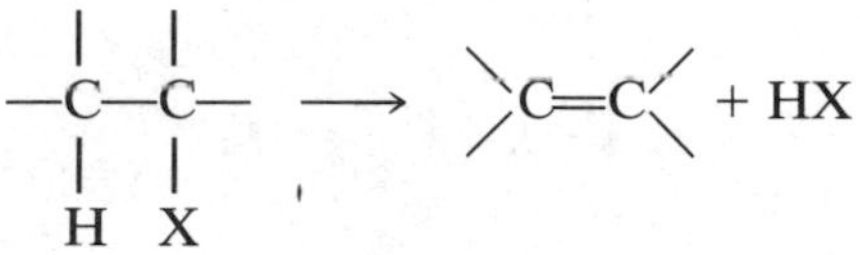

$$-\overset{|}{\underset{H}{C}}-\overset{|}{\underset{X}{C}}- \longrightarrow \; \overset{}{\underset{}{>}}C{=}C\overset{}{\underset{}{<}} + HX$$

Reagents: a halogenoalkane and a concentrated methanolic solution of potassium hydroxide.

Conditions: the two reagents are heated under reflux.

For example:

$$CH_3CH_2CHCH_2CH_3 \;\rightarrow\; CH_3CH{=}CHCH_2CH_3 + HBr$$
$$\underset{Br}{|}$$

b.p. = 119°C   b.p. = 36°C   Yield: 90%

In the case of an unsymmetrical halogenoalkane, the most highly substituted alkene will generally predominate in the reaction products.

15.2.4   Debromination

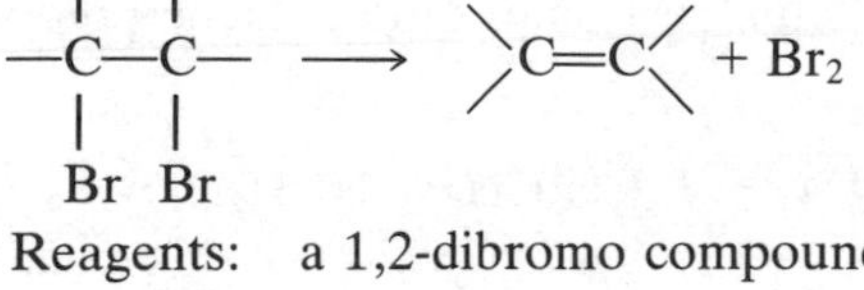

$$-\overset{|}{\underset{Br}{C}}-\overset{|}{\underset{Br}{C}}- \longrightarrow \; \overset{}{\underset{}{>}}C{=}C\overset{}{\underset{}{<}} + Br_2$$

Reagents: a 1,2-dibromo compound and zinc dust.

Conditions: the two reagents are heated in 95% ethanolic solution.
For example:

$$CH_2—CHCH_2C(CH_3)_3 \longrightarrow CH_2=CHCH_2C(CH_3)_3$$
$$\quad | \quad |$$
$$\quad Br \quad Br$$

b.p. = 78°C/1–2 kN m^{-2} b.p. = 72°C   Yield: 91%

15.3 Alkynes

If the triple bond is to be generated, then an elimination reaction will be
required. Otherwise, a substitution reaction may be employed in which
the triple bond is already present in one of the reagents.

15.3.1 Dehydrohalogenation

$$\quad | \quad |$$
$$CH—CH \longrightarrow —C\equiv C— + 2HX$$
$$\quad | \quad |$$
$$\quad X \quad X$$

Reagents: a 1,2-dihalogeno compound and a concentrated alcoholic
           solution of potassium hydroxide.
Conditions: the reagents are heated under reflux.
For example:

$$CH_3—CH—CH—CH_2CH_3 \longrightarrow CH_3—C\equiv C—CH_2CH_3 + 2HBr$$
$$\qquad | \quad |$$
$$\qquad Br \quad Br$$

b.p. = 178°C   b.p. = 55°C   Yield: 56%

15.3.2 Alkylation of alkynes

$$RX + {}^-C\equiv CH \rightarrow R—C\equiv CH + X^-; \quad also\ RC\equiv C^- + R'X \rightarrow RC\equiv CR' + X^-$$

Reagents: a halogenoalkane and the salt of a 1-alkyne.
Conditions: the 1-alkyne is passed into liquid ammonia (kept at
           <−34°C), containing sodium amide NaNH$_2$, and the
           halogenoalkane then added.
For example:

$$CH_3CH_2CH_2Br + {}^-C\equiv CH \rightarrow CH_3CH_2CH_2C\equiv CH + Br^-$$

b.p. = 72°C   b.p. = 40°C   Yield: 85%

15.4 Halogeno compounds

In general, either a substitution or addition reaction is required here. With aromatic compounds substitution may occur in the ring or the side-chain, depending on the reaction conditions.

15.4.1 Photochemical halogenation

$$RH + Cl_2 \xrightarrow{\text{u.v.}} RCl + HCl, \text{etc.}$$

This is used industrially but not normally in the laboratory, since the substitution is relatively unselective. It can be used, however, for substituting chlorine in the side-chain of an alkylbenzene, e.g.

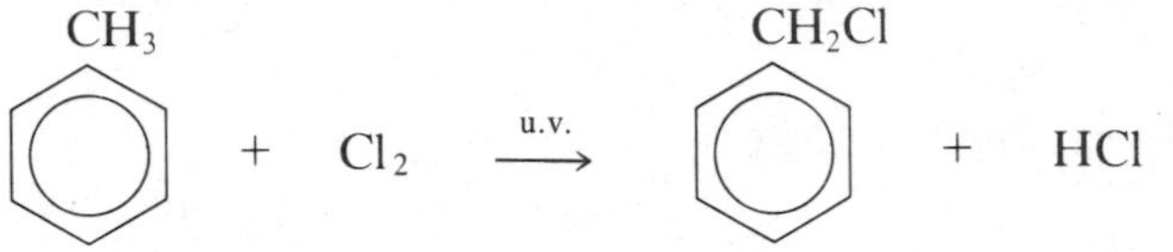

CH$_3$ +   Cl$_2$ $\xrightarrow{\text{u.v.}}$ CH$_2$Cl   +   HCl

b.p. = 111°C   b.p. = 179°C   Yield: 73%

15.4.2 Substitution of alcohols

$$-\overset{|}{\underset{|}{C}}-OH + X^- \longrightarrow -\overset{|}{\underset{|}{C}}-X + OH^-$$

Reagents: these vary, depending on the halogen to be substituted and the class of the alcohol. The corresponding hydrogen halide is generally satisfactory for chloro and bromo compounds, but a mixture of red phosphorus and iodine is often used for iodo compounds. The order of reactivity of alcohols is: tertiary > secondary > primary.

Conditions: the reagents are heated under reflux.

For example:

$$CH_3CH_2CH_2CH_2OH + HBr \longrightarrow CH_3CH_2CH_2CH_2Br + H_2O$$
b.p. = 118°C   b.p. = 101°C   Yield: 95%

15.4.3 Addition to alkenes

$$\underset{/}{\overset{\backslash}{C}}=\underset{\backslash}{\overset{/}{C}} + X_2 \longrightarrow -\overset{|}{\underset{|}{C}}-\overset{|}{\underset{|}{C}}- \\ X X$$

Reagents: the halogen and alkene.

Conditions: the reagents are mixed at room temperature. The reaction
is often carried out in a solvent, such as tetrachloromethane,
to moderate the vigour of the reaction.

For example:

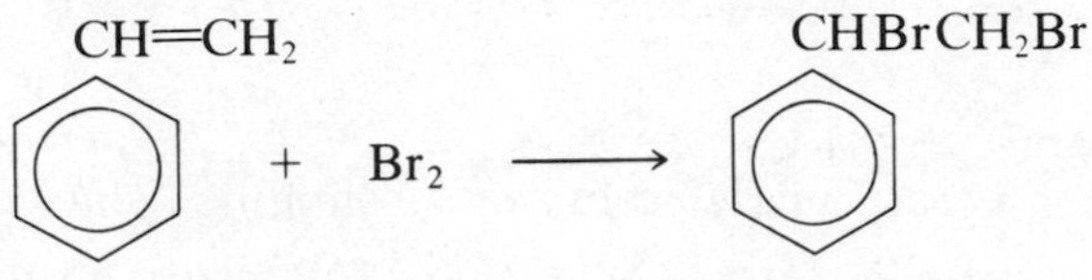

b.p. = 146°C   m.p. = 74°C   Yield: 96%

15.4.4 Hydrohalogenation

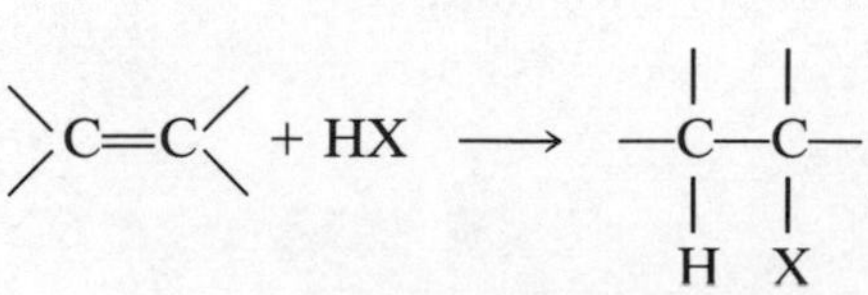

$$\text{>C=C<} + HX \longrightarrow -\overset{|}{\underset{|}{C}}-\overset{|}{\underset{|}{C}}- \atop \quad\; H\;\; X$$

Reagents: an alkene and a hydrogen halide.
Conditions: hydrogen chloride HCl can be warmed with the alkene;
            otherwise concentrated solutions of HBr, or HI, in ethanoic
            acid are mixed with the alkene.

For example:

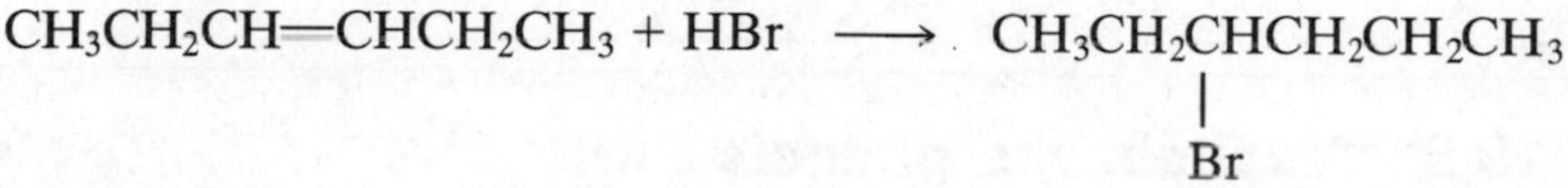

$$CH_3CH_2CH=CHCH_2CH_3 + HBr \longrightarrow CH_3CH_2CHCH_2CH_2CH_3$$
$$\qquad\qquad\qquad\qquad\qquad\qquad\qquad\qquad | \atop Br$$

b.p. = 67°C   b.p. = 142°C   Yield: 76%

(If an unsymmetrical alkene is used, the predominant product may be
predicted by the application of Markovnikov's rule, *see* Section 3.3.1(iii).)

15.4.5 Substitution in benzene ring

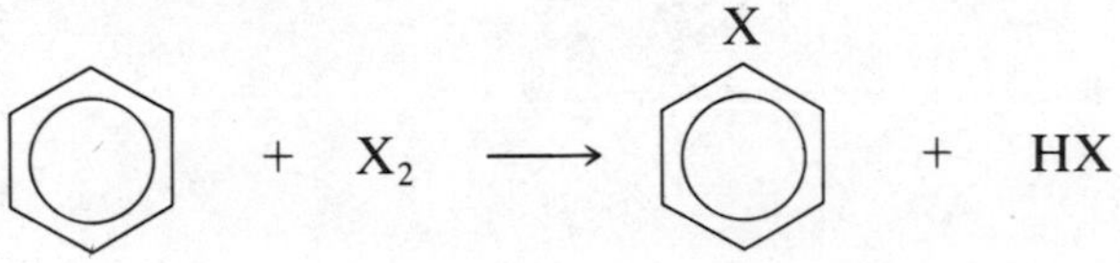

Reagents: an aromatic compound and a halogen (not iodine).
Conditions: the reagents are warmed, in the presence of fresh, *dry* alu-
            minium chloride, iron(III) chloride or iron filings, as 'halogen
            carrier'. Excess of the hydrocarbon is used as a solvent.

15.4.6   Substitution of diazonium salts

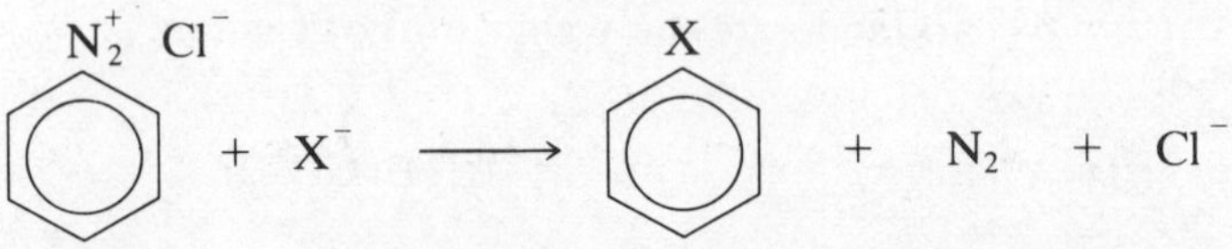

This reaction is particularly useful for substituting iodine into the ring.

Reagents: a diazonium salt and either CuCl/HCl, CuBr/HBr (**Sand-meyer reaction**) in aqueous solution, or just an aqueous solution of potassium iodide.

Conditions: the reagents are mixed at a temperature lower than 10°C (to prevent decomposition of the diazonium salt), and allowed to warm to room temperature.

For example:

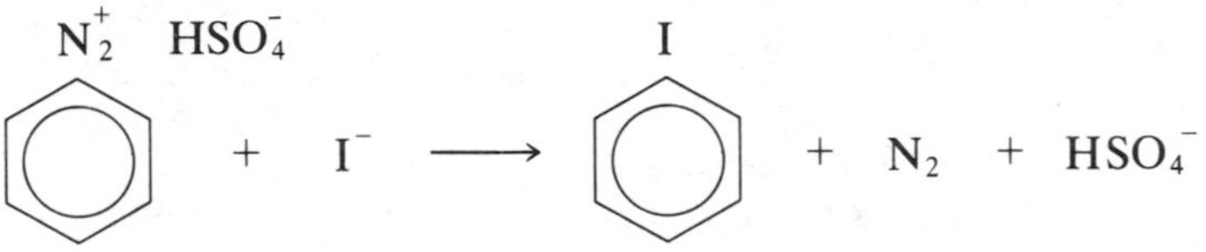

$$\text{b.p.} = 188°C \qquad \text{Yield: } 76\%$$

15.5   Alcohols and phenols

Substitution of another group by OH, the addition of hydrogen atoms across a $\diagdown$C=O bond, and the Grignard reaction, are the commonest reactions for producing alcohols.

15.5.1   Hydrolysis of halogeno compounds

$$-\overset{|}{\underset{|}{C}}-X + OH^- \longrightarrow -\overset{|}{\underset{|}{C}}-OH + X^-$$

Generally, the reverse of this reaction is more useful (*see* Section 15.4.2).

Reagents: a halogeno compound and potassium hydroxide.

Conditions: the reagents are heated under reflux in an aqueous, ethanolic solution. (The order of reactivity of the halogeno compounds is iodo > bromo > chloro).

15.5.2   Hydrolysis of esters

$$RCOOR' + OH^- \rightarrow RCOO^- + R'OH$$

Reagents: an ester and alkali.
Conditions: the reagents are heated under reflux in aqueous, ethanolic
            solution.

This reaction is really only useful when the ester is obtained from
natural sources and the alcohol component is an unusual one.

15.5.3   Reduction of carbonyl compounds

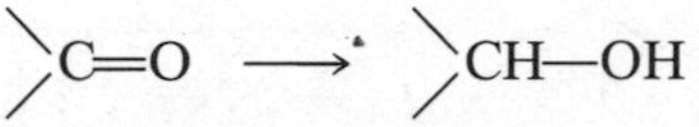

$$\begin{array}{c}\diagdown\\\diagup\end{array}C{=}O \longrightarrow^{\cdot} \begin{array}{c}\diagdown\\\diagup\end{array}CH{-}OH$$

Reagents: a complex metal hydride, such as lithium tetrahydridoalu-
             minate or sodium tetrahydridoborate, and an aldehyde or
             ketone (or even an ester).
Conditions: the reagents are mixed—in the case of LiAlH$_4$, in *dry* ethoxy-
             ethane; in the case of NaBH$_4$, methanol may be used. The
             complex formed is decomposed by the very cautious addition
             of water.
For example:

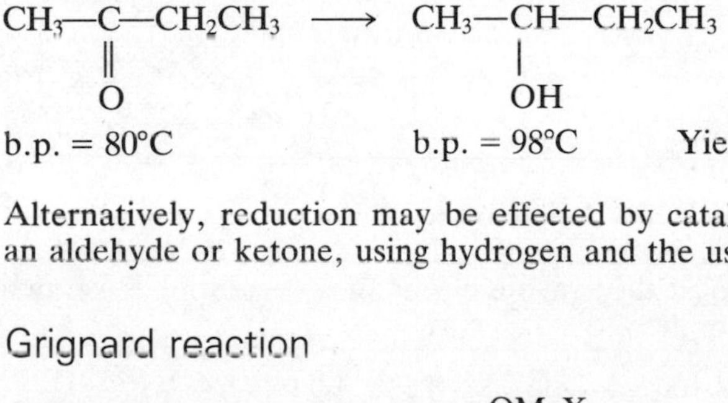

$$CH_3{-}\underset{\underset{O}{\|}}{C}{-}CH_2CH_3 \longrightarrow CH_3{-}\underset{\underset{OH}{|}}{CH}{-}CH_2CH_3$$

b.p. = 80°C   b.p. = 98°C   Yield: 87%

Alternatively, reduction may be effected by catalytic hydrogenation of
an aldehyde or ketone, using hydrogen and the usual catalyst.

15.5.4   Grignard reaction

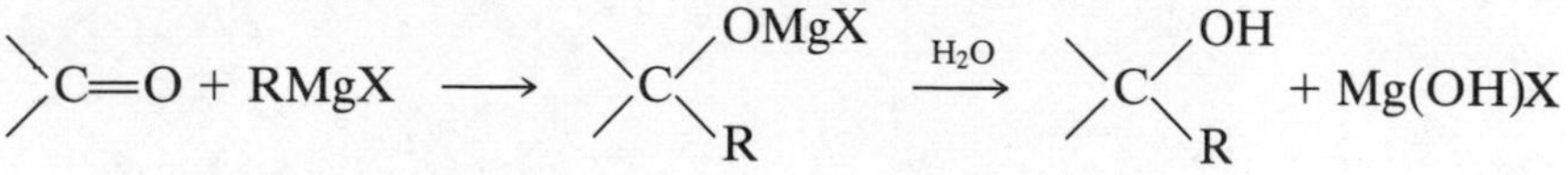

$$\begin{array}{c}\diagdown\\\diagup\end{array}C{=}O + RMgX \longrightarrow \begin{array}{c}\diagdown\\\diagup\end{array}C\begin{array}{c}\diagup OMgX\\\diagdown R\end{array} \xrightarrow{H_2O} \begin{array}{c}\diagdown\\\diagup\end{array}C\begin{array}{c}\diagup OH\\\diagdown R\end{array} + Mg(OH)X$$

Reagents: an aldehyde or ketone, and a Grignard reagent.
Conditions: the reagents are mixed in *dry* ethoxyethane at room tem-
             perature. Reaction is usually vigorous. The organometallic
             complex is decomposed cautiously with water.
For example:

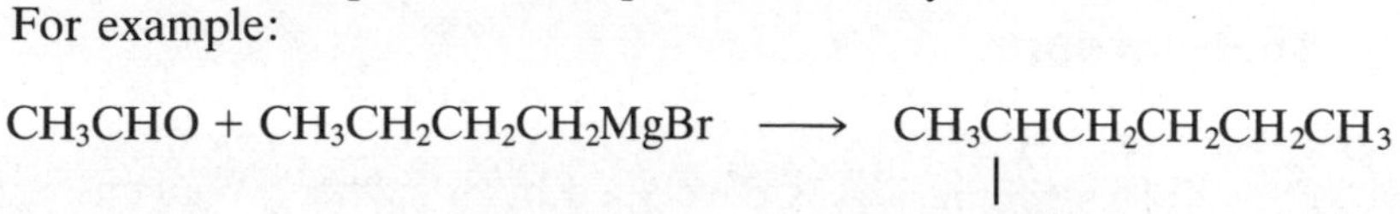

$$CH_3CHO + CH_3CH_2CH_2CH_2MgBr \longrightarrow \underset{\underset{OH}{|}}{CH_3CHCH_2CH_2CH_2CH_3}$$

b.p. = 21°C   b.p. = 136°C   Yield: 66%

15.5.5 Hydration of alkenes

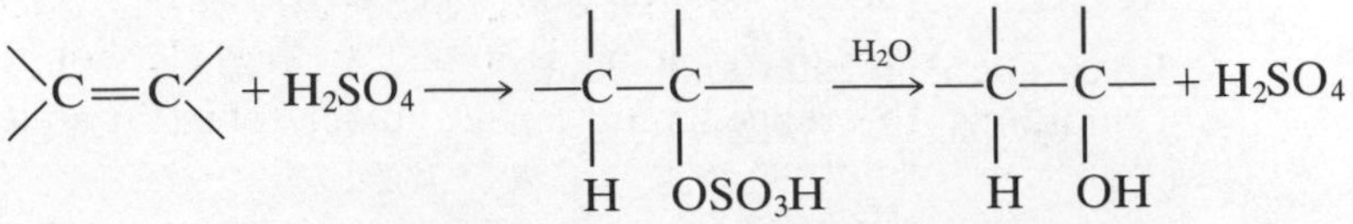

$$\text{\textbackslash}C=C\text{/} + H_2SO_4 \longrightarrow \underset{\substack{| \\ H \quad OSO_3H}}{-\overset{|}{C}-\overset{|}{C}-} \xrightarrow{H_2O} \underset{\substack{| \\ H \quad OH}}{-\overset{|}{C}-\overset{|}{C}-} + H_2SO_4$$

Reagents: an alkene and concentrated sulphuric acid.
Conditions: the alkene is mixed with the conc. acid at room temperature with cooling, if necessary. The mixture is then stirred with a large excess of water.
For example:

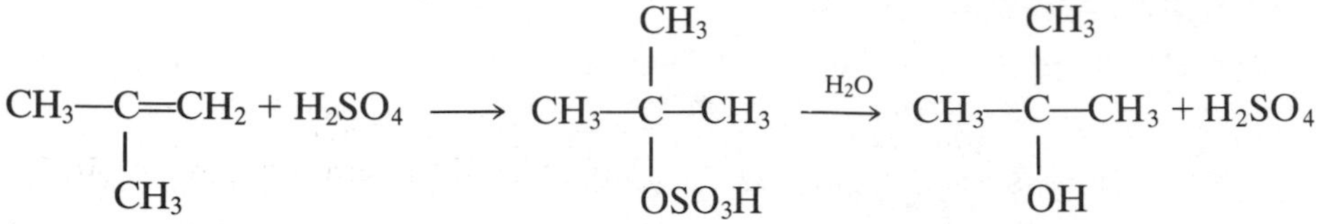

$$CH_3-\underset{\substack{| \\ CH_3}}{C}=CH_2 + H_2SO_4 \longrightarrow CH_3-\underset{\substack{| \\ OSO_3H}}{\overset{\substack{CH_3 \\ |}}{C}}-CH_3 \xrightarrow{H_2O} CH_3-\underset{\substack{| \\ OH}}{\overset{\substack{CH_3 \\ |}}{C}}-CH_3 + H_2SO_4$$

b.p. = −7°C   b.p. = 82°C   Yield: 40%

15.5.6 Hydrolysis of diazonium salts

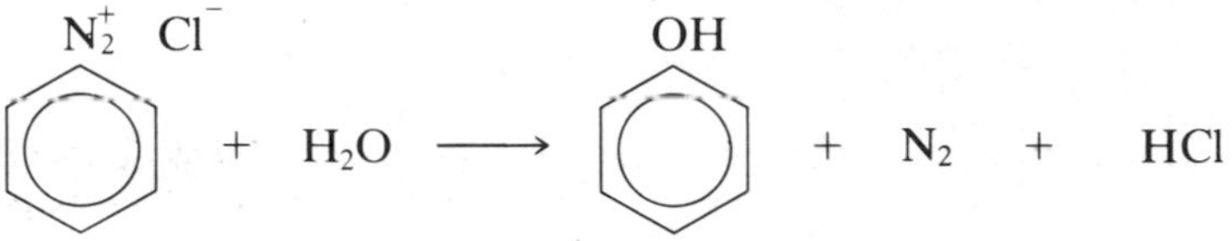

$$\underset{}{\overset{N_2^+ \ Cl^-}{\bigcirc}} + H_2O \longrightarrow \underset{}{\overset{OH}{\bigcirc}} + N_2 + HCl$$

This reaction is an important method of making phenols.
Reagents: a diazonium salt.
Conditions: the aqueous diazonium salt solution is warmed to about 60°C.
For example:

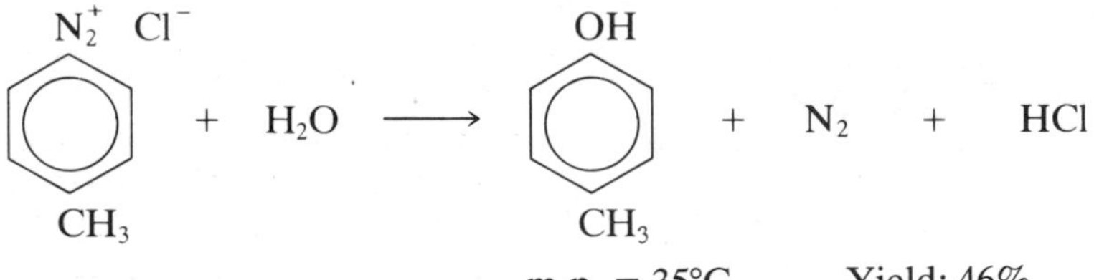

$$\underset{CH_3}{\overset{N_2^+ \ Cl^-}{\bigcirc}} + H_2O \longrightarrow \underset{CH_3}{\overset{OH}{\bigcirc}} + N_2 + HCl$$

m.p. = 35°C   Yield: 46%

15.6 Ethers

15.6.1 Williamson ether synthesis

This synthesis was developed by Alexander Williamson in 1852 and is particularly useful for preparing unsymmetrical ethers:

$$RX + {}^-OR' \rightarrow R-O-R' + X^-$$

Reagents: a halogenoalkane and an alkoxide.
Conditions: the reagents are heated under reflux, in the alcohol from
which the alkoxide is derived (very often a calculated amount
of sodium metal is added to the alcohol to prepare the
alkoxide).
For example:

$CH_3I + {}^-OCH_2CH_2CH_2CH_3 \rightarrow CH_3{-}O{-}CH_2CH_2CH_2CH_3 + I^-$
b.p. = 42°C   b.p. = 70°C   Yield: 71%

15.6.2   Dehydration of alcohols

This method is limited to the preparation of symmetrical ethers and is
only applicable to relatively simple compounds:

$2ROH \rightarrow ROR + H_2O$

Reagents: an alcohol and concentrated sulphuric acid.
Conditions: the two reagents are heated until the ether begins to distil.
For example:

$2CH_3CH_2CH_2CH_2OH \rightarrow CH_3CH_2CH_2CH_2OCH_2CH_2CH_2CH_3 + H_2O$
b.p. = 118°C   b.p. = 144°C   Yield: 60%

15.7   Aldehydes and ketones

These are generally prepared by oxidation of hydroxy compounds but
aromatic ketones can be prepared by the **Friedel–Crafts reaction**.

15.7.1   Oxidation

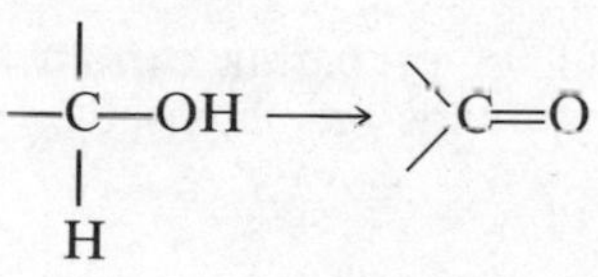

$$\underset{\underset{H}{|}}{\overset{|}{-}}C{-}OH \longrightarrow \ \ \diagdown\!\!\diagup C{=}O$$

Reagents: a primary alcohol (for an aldehyde), or a secondary alcohol
(for a ketone), and an acidified solution of either potassium
dichromate(VI) or potassium manganate(VII).
Conditions: the alcohol is heated with the oxidizing agent. If an aldehyde
is being prepared, then it is immediately distilled from the
reaction mixture, as it is formed, to prevent any further
oxidation to a carboxylic acid.
For example:

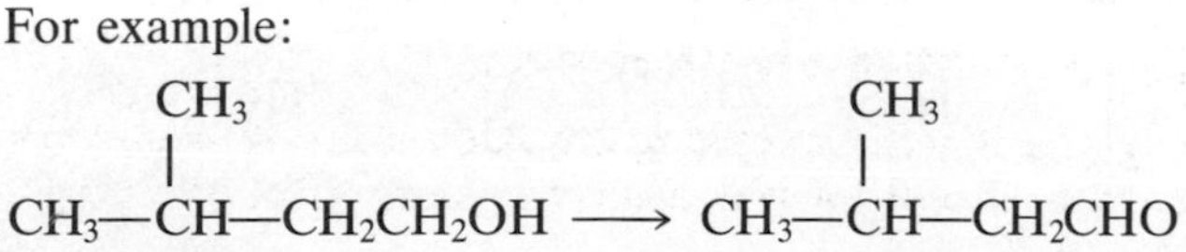

$$CH_3{-}\underset{\underset{}{}}{\overset{\overset{CH_3}{|}}{C}}H{-}CH_2CH_2OH \longrightarrow CH_3{-}\overset{\overset{CH_3}{|}}{C}H{-}CH_2CHO$$
b.p. = 129°C   b.p. = 93°C   Yield: 60%

b.p. = 161°C   b.p. = 155°C   Yield: 85%

Industrially, dehydrogenation is usually used.

15.7.2 Oxidation of alkenes

This is occasionally useful, if the necessary alkene is readily available:

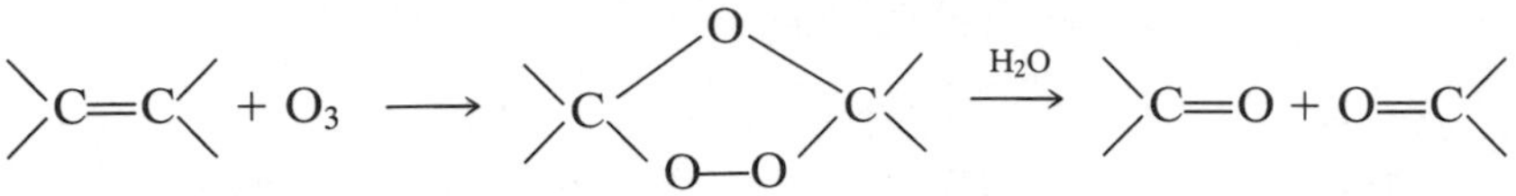

Reagents: an alkene and trioxygen (*ozone*).
Conditions: the alkene and trioxygen are reacted at low temperature and
             the resulting ozonide decomposed with water and zinc dust
             (to prevent oxidation of an aldehyde to a carboxylic acid by
             the H_2O_2 formed).

For example:

$$CH_2=C-CH_2CH_2CH_2CH_3 \longrightarrow CH_3COCH_2CH_2CH_2CH_3 \, (+HCHO)$$
$$\qquad | \qquad \qquad \qquad$$
$$\quad CH_3$$

b.p. = 92°C   b.p. = 128°C   Yield: 60%

15.7.3 Hydrolysis of 1,1-dihalogeno compounds

This reaction is sometimes useful for preparing certain aldehydes, but
generally the 1,1-dihalogeno compounds are themselves prepared from
aldehydes by heating with PCl_5:

$$-CHCl_2 + H_2O \xrightarrow{\;OH^-\;} -CHO + 2HCl$$

Reagents: a 1,1-dihalogeno compound and aqueous alkali.
Conditions: the reagents are warmed.
For example:

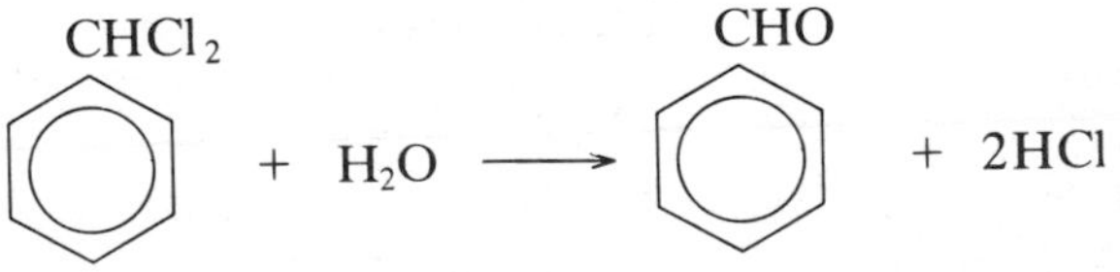

b.p. = 206°C   b.p. = 179°C   Yield: 84%

15.7.4 Friedel–Crafts acylation

This is a standard method of preparing aromatic ketones:

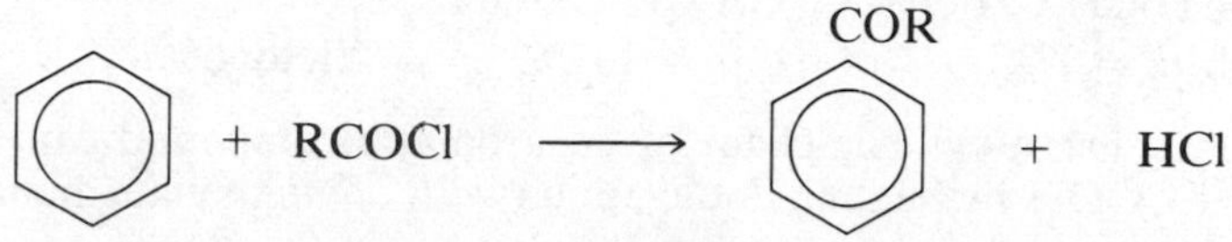

Reagents: an aromatic compound and an acid chloride.
Conditions: the acid chloride is added to the aromatic compound (either
            in excess of the latter, as solvent, or in a solvent such as 1,2-
            dichloroethane) with fresh, powdered aluminium chloride as
            catalyst. The mixture is warmed if necessary.
For example:

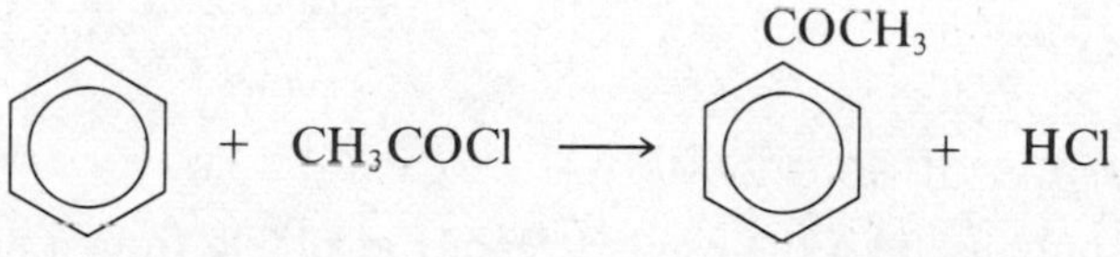

b.p. = 80°C   b.p. = 202°C   Yield: 61%

15.7.5 Hydration of alkynes

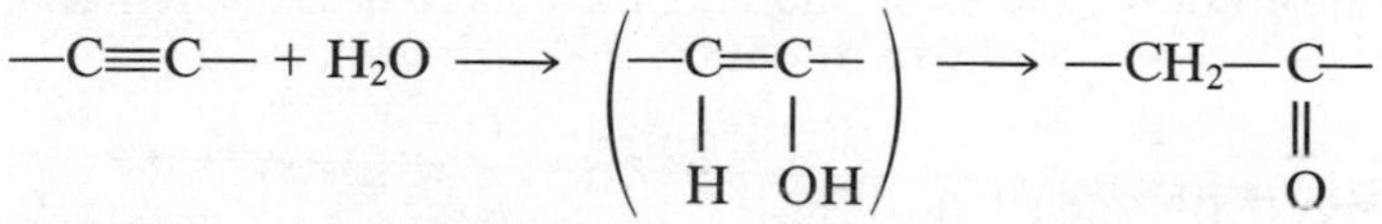

$$-C\equiv C- + H_2O \longrightarrow \left(\begin{matrix} -C=C- \\ | \quad | \\ H \quad OH \end{matrix}\right) \longrightarrow -CH_2-\underset{\underset{O}{\|}}{C}-$$

Reagents: an alkyne and dilute sulphuric acid.
Conditions: the alkyne is passed through hot, dilute sulphuric acid, con-
            taining mercury(II) sulphate as catalyst.
For example:

$$CH_3(CH_2)_3C\equiv C(CH_2)_3CH_3 + H_2O \longrightarrow CH_3(CH_2)_3\underset{\underset{O}{\|}}{C}CH_2(CH_2)_3CH_3$$

b.p. = 178°C   b.p. = 208°C   Yield: 80%

15.8 Carboxylic acids

15.8.1 Oxidation of alcohols and aldehydes

Primary alcohols can be oxidized directly, or via the corresponding aldehydes, to carboxylic acids:

$$-CH_2OH \rightarrow -CHO \rightarrow -COOH$$

Reagents: an alcohol or aldehyde, and an acidified solution of either
            potassium dichromate(VI) or potassium manganate(VII).

Conditions: the two reagents are heated under reflux.
For example:

$CH_3CH_2CH_2OH \rightarrow CH_3CH_2COOH$
b.p. = 97°C   b.p. = 141°C   Yield: 65%

N.B. The alkyl side-chain of an aromatic compound can be oxidized to
—COOH by heating under reflux with *alkaline* manganate(VII), e.g.

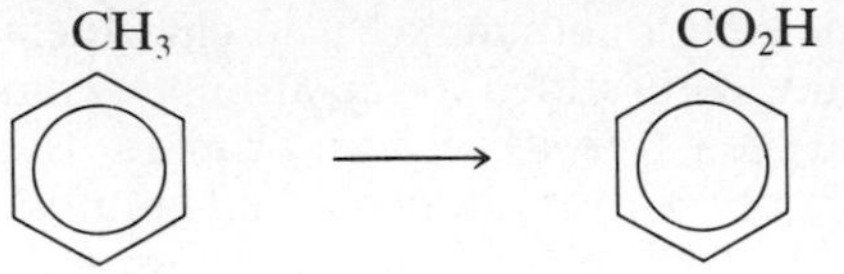

CH₃   CO₂H

b.p. = 111°C   m.p. = 122°C   Yield: 49%

15.8.2 Hydrolysis of esters

This is only useful if the ester is readily available from a natural source;
for example, triglycerides in vegetable oils:

—COOR + OH⁻   →   —COO⁻ + ROH

Reagents: an ester and aqueous alkali.
Conditions: the ester and alkali are heated under reflux. The solution is
           acidified to obtain the carboxylic acid.

15.8.3 Hydrolysis of nitriles

$$—C{\equiv}N + H_2O \longrightarrow —\underset{\underset{O}{\|}}{C}—NH_2 \longrightarrow —COOH + NH_3$$

Reagents: a nitrile and either sulphuric acid or aqueous alkali.
Conditions: the mixture is heated under reflux.
For example:

CH₂CN   CH₂CO₂H

+   2H₂O   ⟶   +   2NH₃

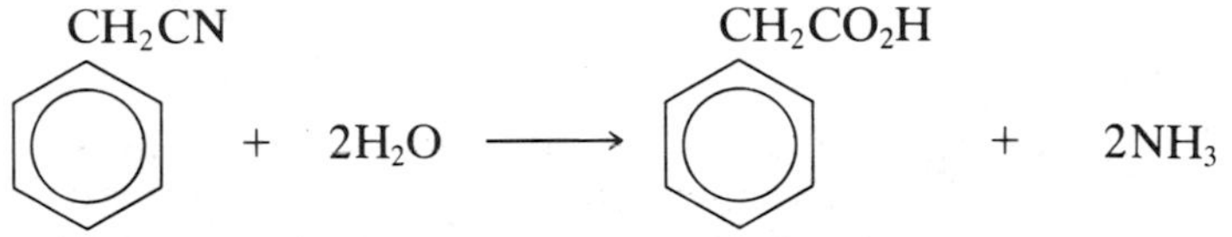

b.p. = 109°C/2 kN m⁻²   m.p. = 76°C   Yield: 78%

15.8.4 Carbonation of Grignard reagents

$$RMgX + CO_2 \rightarrow RCO_2MgX \xrightarrow{H_2O} RCOOH + Mg(OH)X$$

Reagents: a Grignard reagent and carbon dioxide.
Conditions: either solid carbon dioxide is added to the Grignard reagent in dry ether solution, or else *dry* carbon dioxide gas is bubbled into the solution. The organomagnesium complex is carefully decomposed with water.

For example:

$$CH_3CH_2CH_2CH_2MgBr + CO_2 \rightarrow CH_3CH_2CH_2CH_2COOH$$
$$\text{b.p.} = 187°C \qquad \text{Yield: } 80\%$$

15.8.5   Oxidation of alkenes

The oxidation of an alkene with trioxygen (*ozone*) is occasionally useful, as the example below shows:

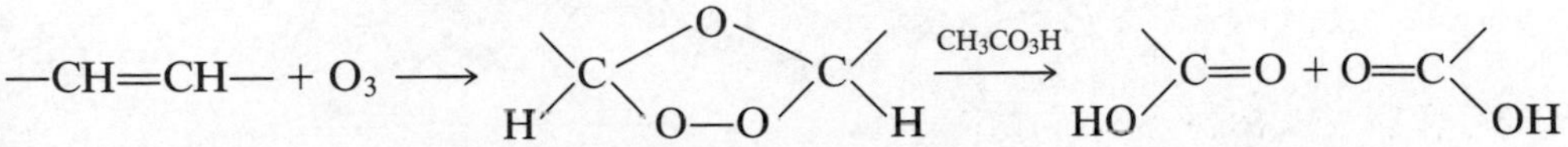

Reagents: an alkene and trioxygen (*ozone*).
Conditions: the alkene and trioxygen are mixed at low temperature. The ozonide is decomposed to carboxylic acids by the addition of peroxoethanoic acid CH_3CO_3H in ethanoic acid.

For example:

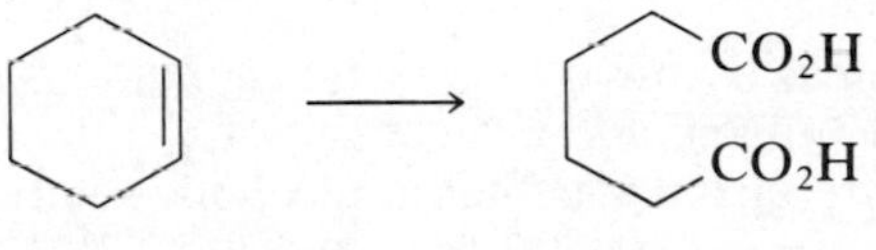

b.p. = 83°C   m.p. = 152°C   Yield: 60%

15.9   Esters

15.9.1   Esterification of acids with alcohols

$$RCOOH + R'OH \rightleftharpoons RCOOR' + H_2O$$

Reagents: an acid and an alcohol.
Conditions: the two reagents (one in excess) are heated in the presence of a small amount of concentrated sulphuric acid as catalyst and the ester distilled from the mixture. Alternatively, *dry* hydrogen chloride, as acid catalyst, may be bubbled through the mixture under reflux.

For example:

$$CH_3COOH + CH_3CH_2CH_2CH_2OH \rightleftharpoons CH_3COOCH_2CH_2CH_2CH_3 + H_2O$$
b.p. = 118°C   b.p. = 118°C   b.p. = 125°C   Yield: 69%

15.9.2   Acylation of alcohols

$$\text{ROH} + \text{R'COCl} \rightarrow \text{ROCOR'} + \text{HCl}$$
$$\text{or} \qquad\qquad\qquad \text{or}$$
$$(\text{R'CO})_2\text{O} \qquad\qquad \text{R'COOH}$$

Reagents: an alcohol and either an acid chloride or anhydride.
Conditions: the acid chloride or anhydride is added carefully to the
            alcohol, sometimes in the presence of pyridine as base and
            solvent. The reaction with the acid chloride is very vigorous
            and the anhydride is therefore sometimes to be preferred.
This method is also applicable to phenols.
For example:

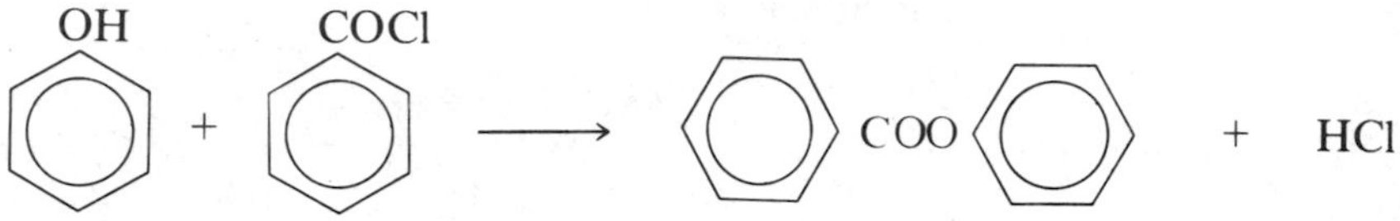

m.p. = 41°C   b.p. = 197°C   m.p. = 69°C   Yield: 76%

15.9.3   Reaction of halogenoalkanes with a silver salt of an acid

$$\text{RX} + \text{R'COOAg} \rightarrow \text{R'COOR} + \text{AgX}$$

This method is sometimes useful as the reaction conditions are very mild.
Reagents: a halogenoalkane and a silver salt of a carboxylic acid.
Conditions: the two are mixed together in alcoholic solution.

15.10   Acid derivatives

There is usually only one commonly used method of preparing acid
chlorides, anhydrides, amides and nitriles and these have each been
discussed in the appropriate section of Chapter 10.

15.11   Amines

15.11.1   Reduction of nitro compounds

$$-\text{NO}_2 \rightarrow -\text{NH}_2$$

This method can be used with nitroalkanes but is much more important
in preparing aromatic amines.
Reagents: the corresponding nitro compound and iron, or tin, and
            concentrated hydrochloric acid. (Catalytic hydrogenation or
            complex metal hydrides, like LiAlH₄, may also be used to
            effect the reduction.)

Conditions: the reagents are heated together.
For example:

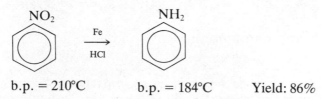

b.p. = 210°C   b.p. = 184°C   Yield: 86%

15.11.2 Reduction of nitriles

$-C\equiv N \rightarrow -CH_2NH_2$

Reagents: a nitrile and a reducing agent (either a complex metal hydride,
            or H_2 and a catalyst).
Conditions: the nitrile is added to a solution of the hydride in *dry* ethoxy-
            ethane.(Normal conditions for catalytic hydrogenation are
            employed.)
For example:

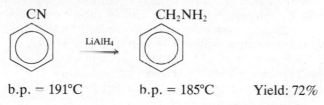

b.p. = 191°C   b.p. = 185°C   Yield: 72%

15.11.3 Reduction of amides

$-CONH_2 \rightarrow -CH_2NH_2$

Reagents: an acid amide and lithium tetrahydridoaluminate.
Conditions: the amide is added to a solution of the hydride in *dry* ethoxy-
            ethane and the complex decomposed with water.
For example:

$CH_3CON(CH_2CH_3)_2 \rightarrow (CH_3CH_2)_3N$
b.p. = 185°C   b.p. = 89°C   Yield: 50%

15.11.4 Ammonolysis of halogenoalkanes

$RX + NH_3 \rightarrow RNH_2 + HX$

This is generally a poor method as other products are formed as well.
Reagents: a halogenoalkane and ammonia.
Conditions: the reagents are heated in alcoholic solution in a sealed tube
            (a dangerous operation without special equipment).

For example:

$CH_3CH_2CH_2CH_2Br + NH_3 \rightarrow CH_3CH_2CH_2CH_2NH_2 + HBr$
b.p. = 101°C   b.p. = 76°C   Yield: 47%

15.11.5 Hofmann rearrangement of amides

$R{-}CONH_2 \rightarrow R{-}NH_2$

Reagents: an acid amide, sodium hydroxide and bromine (*care!*).
Conditions: bromine is added to a solution of the amide, followed by the
                  addition of sodium hydroxide. The mixture is then boiled
                  and the amine distilled from the mixture.

For example:

$CH_3CONH_2 \rightarrow CH_3NH_2$
m.p. = 82°C   b.p. = −7°C   Yield: 78%

15.12 Organic synthesis

Much research in organic chemistry is devoted to the preparation of new
compounds and to the development of better methods of making known
compounds. Organic synthesis is widely used to confirm structures of
newly discovered compounds (*see* Section 22.5) and often leads to the
development of new synthetic methods. The synthesis of complex molecules such as vitamin B_{12} is both intellectually challenging and practically
demanding, but it is in the course of such work that the frontiers of
organic chemistry can be extended and important new discoveries made:

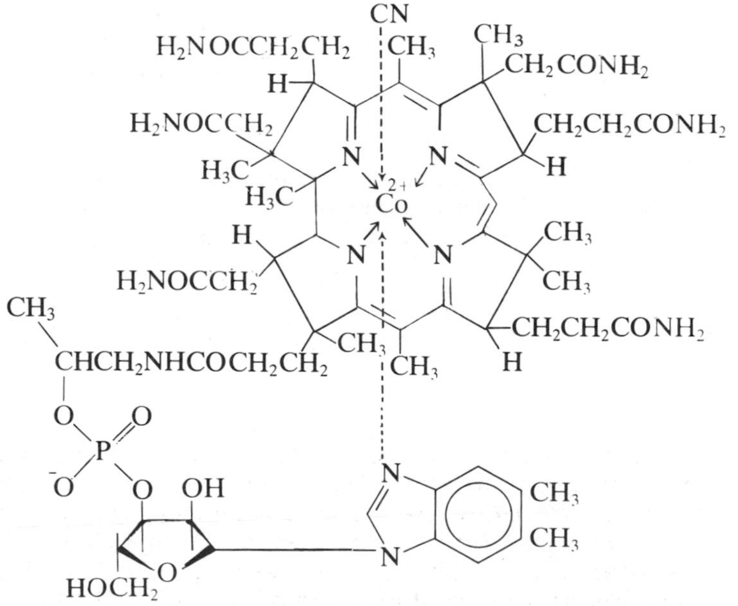

Vitamin B_{12} $C_{63}H_{90}O_{14}N_{14}PCo$

The devising of a synthetic route to a compound involves the organizing of a sequence of organic reactions, or transformations, arranged so as to lead to the desired product. Although this is a paper exercise initially, the actual practical completion of the synthesis seldom goes completely according to plan. Unanticipated difficulties are often encountered; yields may be disappointingly low or purification of intermediates along the synthetic pathway may prove very difficult. For these, and other, reasons it is important to have a choice of general synthetic methods available, so that alternative synthetic schemes can be drawn up. This is why the general preparative methods outlined in the first part of this chapter are so important in basic organic synthesis and should be thoroughly understood.

As illustration, consider the laboratory synthesis of 2-hydroxypropanoic acid (*lactic acid*) from the simple, readily available starting material, ethanol CH_3CH_2OH. Two possible synthetic schemes are shown below:

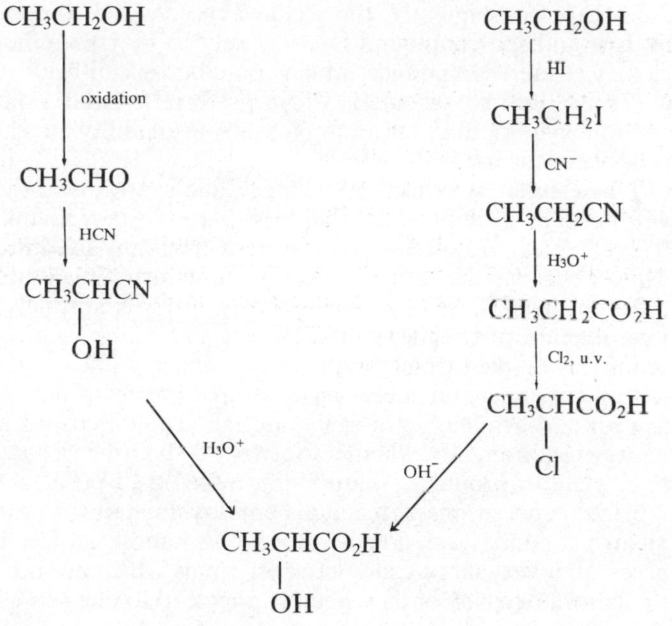

CH_3CH_2OH

$\downarrow$ oxidation

CH_3CHO

$\downarrow$ HCN

CH_3CHCN
|
OH

CH_3CH_2OH

$\downarrow$ HI

CH_3CH_2I

$\downarrow$ CN^-

CH_3CH_2CN

$\downarrow$ H_3O^+

$CH_3CH_2CO_2H$

$\downarrow$ Cl_2, u.v.

CH_3CHCO_2H
|
Cl

H_3O^+ $\searrow$ OH^- $\swarrow$

CH_3CHCO_2H
|
OH

2-hydroxypropanoic acid

The starting material contains only two carbon atoms per molecule, while the compound to be synthesized contains three carbon atoms per molecule. It is clear then that, at some point along the synthetic route, an extra carbon atom must be introduced into a molecule. This is most readily achieved by the use of cyanide ion, particularly since the cyano group can be hydrolysed to a carboxyl group. The cyano group can be introduced, either by a nucleophilic addition reaction with a carbonyl compound, or else by a nucleophilic substitution reaction with a halogenoalkane, as shown.

Which synthetic scheme is to be preferred, in principle? The first scheme contains only three steps, while the alternative scheme involves five. Suppose that every step in both schemes afforded a 70% yield (this would be very acceptable). Then, in a three-step synthesis the overall yield, based on the starting material, would be only 34%; in a five-step one it would be as low as 17%. Thus the three-step synthesis is to be preferred on grounds of efficiency. More detailed examination of the individual steps in the schemes shows that there is a rather weak link in the second scheme. The conversion of the propanoic acid to 2-chloropropanoic acid is brought about by passing chlorine into the boiling acid in the presence of ultraviolet light. However, the substitution of hydrogen by chlorine is not very selective and not only is there a chance of obtaining the 2,2-*di*substituted propanoic acid as a contaminant, but also a chance of substitution occurring at the methyl group as well (*see* Section 9.3.3(ii)). Consequently, the yield of the desired compound is likely to be low and the compound is also likely to be contaminated with other closely related compounds, which could be very difficult to separate from it. Thus, the first scheme is to be preferred. Other considerations can become paramount; for example, the availability or cheapness of the proposed starting material.

This chapter is concerned with organic synthesis in the laboratory. It will be clear from previous chapters that industrial manufacturing routes to a chemical are often quite different from any used in the laboratory. This is because the concerns of a manufacturing plant are usually rather different from those of a laboratory. Industry is normally involved in the manufacture of chemicals on a large scale, using, where possible, cheap readily available raw materials. In devising a manufacturing route, chemists and chemical engineers have to take into consideration such factors as cost and availability of raw materials, capital cost of necessary plant, energy requirements, whether the process should be a batch or continuous one, yield of product, pollution and effluent disposal. Since economy is a prime concern, many reactions employ a catalyst so that they proceed at an economic rate. Furthermore, the handling of solids, liquids and gases on a very large scale raises problems which are not encountered in the laboratory; any of these factors may lead to one route being preferred to another.

Questions

1*   The following scheme outlines three different pathways by which compound C may be prepared:

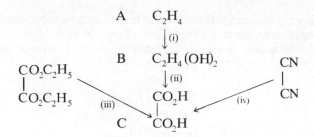

(a) Name the compounds A, B and C.

(b) Write equations and state the necessary conditions for the reactions shown at (i), (ii) and (iii).

(c) (i) Indicate possible intermediate compound(s) which might be found at stage (iv).

(ii) Suggest possible conditions for stage (iv), when cyanogen $\left(\begin{matrix} CN \\ | \\ CN \end{matrix}\right)$

is converted to C.

(d) In a quantitative check on the reaction (ii), it was found that 45 g of the crystalline product of $C, (CO_2H)_2.2H_2O$, was obtained from 30 g of B. What was the percentage yield?

London

2*   The following reaction sequence relates to the synthesis of a compound, **D**, from another substance **A**, which has the molecular formula C_3H_5N.

Compound A
   C_3H_5N
      ↓   reduction with Na in ethanol
          or $LiAlH_4$ in ether
Compound B
      ↓   reaction with $NaNO_2$ and dilute HCl
          in the cold
Compound C
      ↓   reaction with CH_3COOH and concentrated H_2SO_4
Compound D
   $C_5H_{10}O_2$

(a) Give the structural formulae of compounds **A**, **B** and **C**.

(b) (i) Name compound **D**.

   (ii) To what class of organic compounds does **D** belong?

   (iii) Give the structural formula of another compound which also belongs to this class of compounds and is isomeric with **D**.

(c) Write an equation to show what happens when **D** is boiled with an excess of aqueous sodium hydroxide.

JMB

3*   The following scheme outlines two different pathways by which lactic acid
     may be prepared from ethene. The scheme is incomplete in that some
     of the intermediate products are not given and certain reagents and
     conditions are not stated.

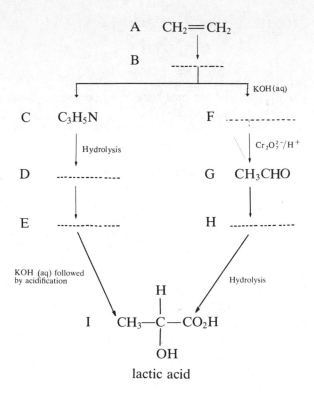

lactic acid

(a) Write down the systematic name of lactic acid.
(b) Compound B contains carbon, hydrogen and chlorine only. Its relative
    molecular mass is approximately 65.

    (i) State the formula of B.
    (ii) Outline how you would determine its relative molecular mass.

(c) Give the name and structural formula of compound C, of molecular
    formula C_3H_5N.
(d) Give the structural formulae of compounds D, E, F and H.
(e) Which of the compounds A to I possess asymmetric carbon atoms?
    Give structural formulae and indicate the asymmetric carbon atoms
    by an asterisk.
(f) Briefly explain why under these actual experimental conditions
    none of the compounds identified by you in (e) shows optical
    activity.

                                                            London

4* Using propanoic acid ($CH_3CH_2CO_2H$) as your sole source of carbon, how would you synthesize **four** of the following?
(i) $CH_3CH_2CO_2CH_2CH_2CH_3$
(ii) $CH_3CHBrCH_2Br$
(iii) $CH_3CH_2CONHCH_2CH_3$
(iv) $CH_3CH_2CHBrCH_2CH_2CH_3$
(v) CH_3COCH_3

Cambridge

5* Suggest how **five** of the following conversions may be carried out:
(a) $CH_3CH_2OH \rightarrow CH_3CH(OH)CO_2H$
(b) $CH_3CO_2H \rightarrow CHCl_3$
(c) $(CH_3)_2CHCH_2COCH_3 \rightarrow (CH_3)_2CHCH_2NHCOCH_3$
(d)

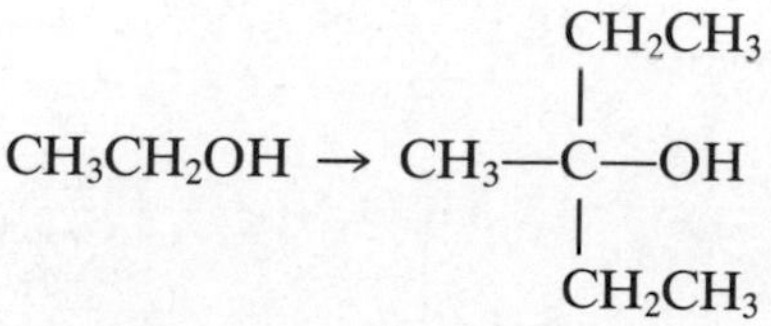

$$CH_3CH_2OH \rightarrow CH_3\overset{\displaystyle CH_2CH_3}{\underset{\displaystyle CH_2CH_3}{\overset{|}{\underset{|}{C}}}}-OH$$

(e)

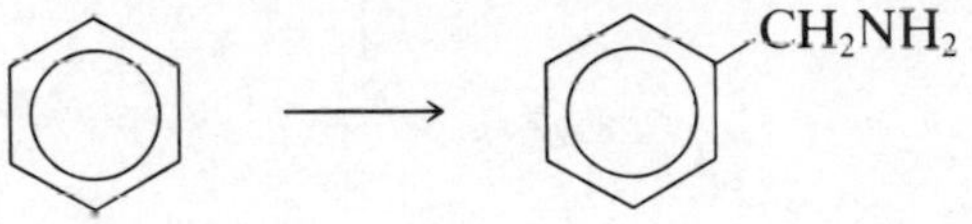

(f)

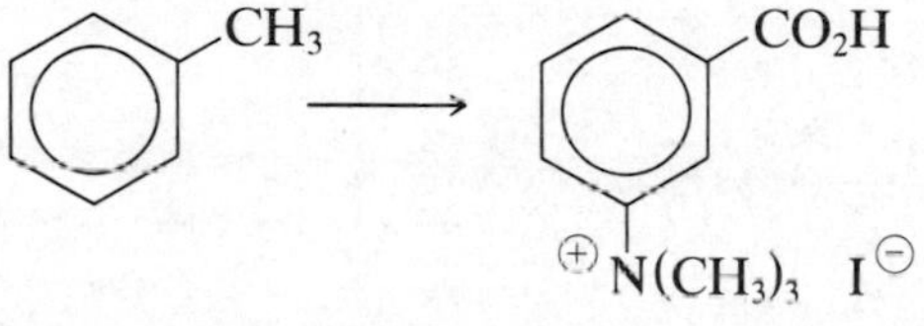

Cambridge

6* How would you carry out the following transformations? State essential conditions and reagents.

$CH_2{=}CH(CH_2)_4CO_2C_2H_5$

(a) $\longrightarrow$ $CH_2{=}CH(CH_2)_4CH_2OH$
(b) $\longrightarrow$ $CH_3(CH_2)_5CH_2OH$
(c) $\longrightarrow$ $CH_2{=}CH(CH_2)_4CONHCH_3$
(d) $\longrightarrow$ $HO_2C(CH_2)_4CO_2H$
(e) $\longrightarrow$ $CH_3CHBr(CH_2)_4CO_2C_2H_5$
(f) $\longrightarrow$ $OCH(CH_2)_4CO_2C_2H_5$

Indicate the mechanisms of reactions (c) and (e).

Oxford & Cambridge

7* Suggest methods by which the following transformations could be carried out. For each stage, indicate the reagents required and the conditions necessary.
(a) $CH_3CH_2CH_2Cl \rightarrow CH_3CH_2CH_2Br$
(b) $CH_2{=}CH_2 \rightarrow CH_3CO_2H$
(c) $C_6H_5CH{=}CH_2 \rightarrow C_6H_5CH(NH_2)CH_2NH_2$
(d) $C_6H_5COCH_3 \rightarrow C_6H_5COCl$
Describe how you would obtain (e) phenylamine (*aniline*) from a mixture of it with benzene, and (f) benzene, from a mixture of it with phenol.

London

8* The compounds A to G are represented by the given structural or molecular formulae and may be converted into each other as indicated by the reagents shown.

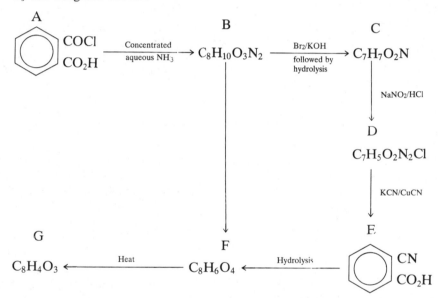

(a) (i) Are the substances B, C, D, F and G all aliphatic, all aromatic, or are there some of each?
      (ii) Briefly give your reasons.
(b) Give the structural formulae of (i) C, (ii) D, (iii) G.
(c) Describe briefly how B could be directly converted to F.
(d) (i) Two of the substances A to G are salts. Which two are these?
      (ii) Do any of the compounds shown exhibit geometrical isomerism? If so, which?
      (iii) What particular feature of the structure of F is confirmed by the fact that it readily gives G on heating?
(e) Describe with essential practical details how you would carry out the conversion of C into D in the laboratory.

London

9* Show how the following syntheses may be carried out:
    (a) Propan-2-ol from 2-bromopropane.
    (b) Propan-2-ol from propene without the intermediate formation of the
        halogenoalkane.
    In each case give the reagents, conditions and detailed mechanistic
    equation.
    Discuss in detail the factors which determine the yield of propan-2-ol, in
    each case including reference to any possible impurities.
    Which of these methods is most likely to be used for the commercial
    production of this alcohol? Give reasons for your choice.
    Extend these arguments to a consideration of the commercial production
    of 2-methylpropan-2-ol from the appropriate halogenoalkane and from
    the alkene.

                                                                          JMB

10 A large proportion of the phenol manufactured is made from benzene
    via (1-methylethyl)benzene (*cumene*).
    Describe **two** methods by which phenol may be prepared from benzene
    in the laboratory and discuss the reasons why these methods are less
    suitable than the cumene method for large-scale commercial manufacture.
    Outline the commercial importance of phenol.

                                                                          JMB

11* Methanol can be produced industrially by a variety of processes. For
    centuries it has been obtained from the destructive distillation of wood.
    However, by 1961 less than 1% of the total was produced in this way.
    In the 1920s the production of methanol from carbon monoxide and
    hydrogen was introduced. More recently methane from natural gas has
    been used as the source of carbon.
    Discuss the economic and social factors which have led to these changes
    in the manufacture of methanol.

                                                                          JMB

Chapter 16
Stereoisomerism

16.1 Isomerism

Isomerism may be defined as the existence of two or more compounds with the same molecular formula, but having one or more different physical or chemical properties. The reasons for the extensive **structural isomerism** to be found in organic chemistry have already been mentioned in Chapter 1, as has the necessity of writing structural, rather than molecular, formulae. **Structural isomerism—the existence of two or more compounds with the same molecular composition but with different structures**—arises from the different ways in which a given number of atoms can be linked together within a molecule. The more atoms there are, the more structural isomers there are likely to be. Structural isomerism may occur not only between members of different homologous series, in which case both physical and chemical properties will be different (e.g. methoxymethane CH_3OCH_3 and ethanol CH_3CH_2OH), but also between members of the same series, where the main differences may be in physical properties only (e.g. butane $CH_3CH_2CH_2CH_3$ and 2-methylpropane $CH_3CH(CH_3)CH_3$). Structural isomerism also arises from differences in the relative positions of substitution in the benzene ring, as exemplified by 1,2- and 1,4-dichlorobenzene.

In general, it is not easy to convert one structural isomer into another, although in the phenomenon known as **tautomerism** two structural isomers are in dynamic equilibrium. In such a case, the isomers are known as **tautomers** and they often differ only in the position of a hydrogen atom which is transferred from one site to another in the equilibrium process. One such example is the keto–enol tautomerism displayed by aldehydes and ketones (*see* Section 8.3.6). The ethyl ester of 3-oxobutanoic acid affords another interesting example:

$$CH_3-\overset{\overset{\displaystyle O}{\|}}{C}-CH_2-COOCH_2CH_3 \quad \rightleftharpoons \quad CH_3-\overset{\overset{\displaystyle OH}{|}}{C}=CH-COOCH_2CH_3$$

keto form   **enol** form

The enol form shows properties characteristic of unsaturated compounds and decolorizes bromine water and potassium manganate(VII), for example. It also forms a purple coloration, characteristic of enols and phenols (see Section 13.3.1(iii)), when iron(III) chloride solution is added to it. If such an addition is followed by the addition of a little bromine water, the purple colour disappears at first. As the orange colour of bromine gradually fades, owing to the reaction between the enol form and bromine, the purple colour is restored, showing that fresh enol is being produced as the equilibrium is re-established.

16.2 Stereoisomerism: geometrical isomerism

Stereoisomerism arises, not from the different ways in which atoms within a molecule can be linked together, but rather from the different positions taken up *in space* by the same atoms or groups similarly joined together. Two kinds of stereoisomerism can be distinguished—geometrical isomerism and optical isomerism.

Geometrical isomerism may arise in compounds which contain double bonds or cyclic structures. A double bond restricts free rotation of atoms or groups about the bond (see Section 1.3) and therefore usually fixes in space the positions of those atoms or groups linked to it. This type of isomerism is common in all but the simplest alkenes. Thus, there are four isomers of C_4H_8. Cyclobutane, but-1-ene and but-2-ene are clearly structural isomers but but-2-ene can exist as two geometrical isomers. In *cis*-but-2-ene, the two methyl groups lie on the same side of the planar molecule, whereas in the *trans* form they lie on opposite sides. The two isomers cannot freely interconvert because no free rotation of the groups about the $\diagup\!\!\!\!\diagdown C = C \diagdown\!\!\!\!\diagup$ bond is possible.

cis-but-2-ene   *trans*-but-2-ene

Since the chemical properties of but-2-ene are those of an alkene, the properties of these two isomers are almost identical but their physical properties differ, e.g. the boiling point of the *cis* isomer is 4°C, that of the *trans* isomer is 1°C. Sometimes geometrical isomerism *can* lead to differences in chemical behaviour, even in simple molecules. Butenedioic acid $HO_2CCH{=}CHCO_2H$ has two geometrical isomers; the *cis* isomer is sometimes known as *maleic acid* and the *trans* isomer as *fumaric* acid. Both isomers can be reduced to butanedioic acid. However, if maleic acid is heated it readily forms an acid anhydride (cf. Section 9.6(iii)), whereas fumaric acid does not readily do so. The reason for this difference lies in the spatial positions of the two carboxylic acid groups. In the *trans*

278   Stereoisomerism

isomer the two groups are too far apart to permit the ready elimination
of water to form an anhydride:

m.p. = 143°C   m.p. = 54°C   m.p. = 300°C

However, on very strong heating the *trans* isomer does form the anhydride
because the thermal energy supplied is sufficient to break the carbon–
carbon double bond temporarily and allow free rotation about the σ-
bond. Generally speaking, the *trans* form of such a pair of geometrical
isomers is, thermodynamically, the more stable, since the larger groups
are the maximum distance apart and electronic interactions are minimized.
When they contain polar bonds, geometrical isomers can often be distinguished by their dipole moments (*see* Section 1.4), as illustrated by the
following example:

net dipole moment   zero dipole moment

Geometrical isomerism also arises when carbon forms a double bond
with nitrogen, as in the oximes of an aldehyde. Here, the equivalent of
the *cis* isomer is usually designated the *syn* isomer and the other is
designated the *anti* isomer, e.g.

       syn   *anti*
m.p. = 35°C   m.p. = 125°C

Geometrical isomerism, arising from double bonds, is very widespread
in fats and oils (*see* Section 19.1). It may exert a profound influence on
the properties of a material, an excellent example being afforded by a
comparison of natural rubber with gutta-percha (*see* Section 20.4).
    Restricted rotation in a molecule may also come about, not because
of the presence of multiple bonds, but because of a cyclic structure. Thus,
for 1,2-dichlorocyclohexane (which could be made by the addition of

chlorine to cyclohexene), two geometrical structures are possible, depending on whether the two chlorine atoms are on the same side, or on opposite sides, of the molecule:

The case of hexachlorocyclohexane $C_6H_6Cl_6$, which can be made by the ultraviolet irradiation of a mixture of benzene and chlorine (*see* Section 12.4.1(ii)), is particularly interesting. There are eight geometrical isomers, one of which also shows optical isomerism (*see* Section 16.3), giving a total of nine stereoisomers in all. Only one of these isomers, referred to as the γ-isomer, shows insecticidal activity but although it is the active ingredient of certain insecticides (*see* Section 12.4.1(ii)), it is not separated from its other stereoisomers:

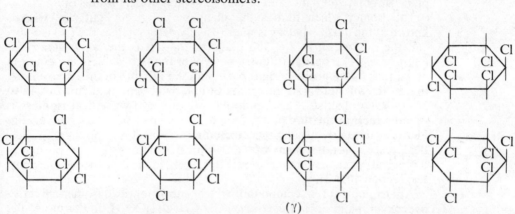

In fact, these cyclohexane ring structures are not planar, because the carbon atom has its four bonds almost tetrahedrally disposed in the normal fashion. There are two possible shapes, or **conformations**, which the ring can assume and these are referred to as *chair* and *boat* forms, the chair form being, thermodynamically, the more stable. The forms are interconvertible, however, at most temperatures:

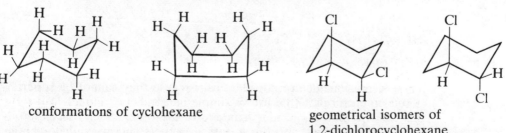

conformations of cyclohexane

geometrical isomers of
1,2-dichlorocyclohexane
(not interconvertible) in their
more stable conformations

The stereoisomerism of such ring or cyclic structures is very important in the chemistry of sugars and other carbohydrates (*see* Section 17.2).

16.3 Stereoisomerism: optical isomerism

16.3.1 Optical isomers

When light is passed through certain materials, such as the familiar Polaroid, it is said to become plane-polarized, i.e. the light can be considered to be electromagnetic radiation which is vibrating in one plane only. If such light is viewed through a long tube, as in a *polarimeter*, then a second piece of the material can be arranged as an eye-piece in such a way that there is no transmission of light at all. If a solution of a substance is then placed in the tube, it may be found necessary to rotate the Polaroid eye-piece, either clockwise or anti-clockwise, in order to prevent once more any transmission of light. A substance which causes this effect is said to be **optically active**, i.e. it has the ability to rotate the plane of polarized light, and is said to exhibit **optical isomerism**.

The isomer which rotates the plane of polarized light clockwise is known as the *dextrorotatory* isomer, usually designated as the (+) isomer, and the one which rotates the plane to the left is the *laevorotatory* or (−) isomer. The optical rotatory powers of the two isomers are equal in magnitude but opposite in sign. An equimolar mixture of the two isomers, therefore, will not rotate the plane of polarized light at all and is said to be **optically inactive**. This equimolar mixture of two optical isomers is called a **racemic mixture** or a **racemate**. Such a pair of optical isomers are identical in nearly all respects (including chemical and physical properties), except for the signs of their optical rotations (the properties in biological systems, however, may differ owing to the influence of enzymes, *see* Section 16.3.3).

An example of a compound which shows optical isomerism is 2-hydroxypropanoic acid (*lactic acid*) $CH_3CH(OH)CO_2H$. In this molecule, there are four *different* atoms or groups attached to a central carbon atom. The actual arrangement of these atoms or groups in space about the central atom is termed the **configuration** of that atom and it can be seen from the diagram below that two such arrangements are possible, which can be distinguished from one another:

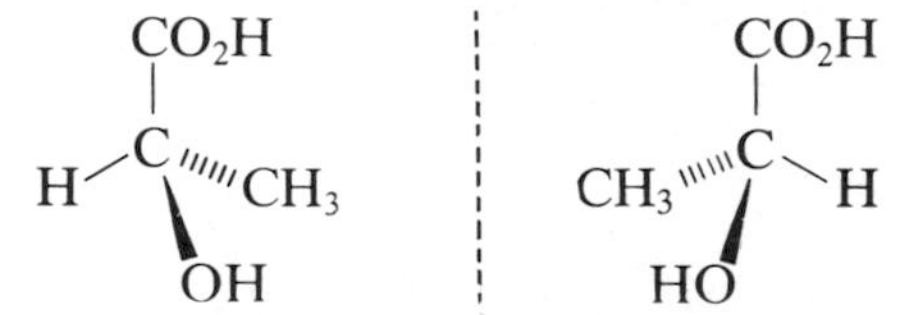

These arrangements are not identical because they cannot be superimposed on each other (the use of simple ball-and-stick models will show this clearly); one is the mirror image of the other. Such *non-superimposable* mirror-image forms are optical isomers and are called **enantiomers**. One of them will rotate the plane of polarized light clockwise and one anti-clockwise, but which one will do which cannot be determined simply by inspection of the configurations. In the first instance, the absolute configuration of an optical isomer must be determined by an X-ray diffraction study so that the precise positions of the atoms or groups in space can be located. In 1951, the Dutch chemist Johannes Bijvoet succeeded in determining the absolute configuration of (+) 2,3-dihydroxypropanal, using special X-ray diffraction techniques on a related compound. By convention, the absolute configuration of the (+) isomer is designated D- and that of the (−) isomer L-:

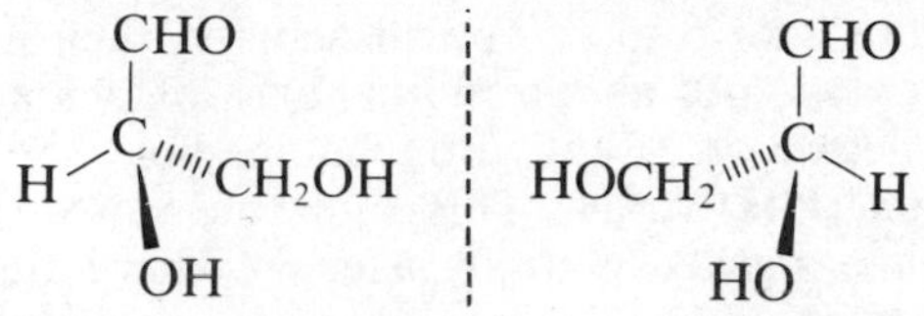

CHO   CHO
| |
H—C""""CH₂OH   HOCH₂""""C—H
| |
OH   HO

D-(+)   L-(−)

The absolute configuration of an optical isomer cannot be inferred from its sign of optical rotation, but may be determined by deriving the isomer from one whose absolute configuration is known, using chemical reactions whose stereochemical courses are known. Thus, an optical isomer of *lactic acid* $CH_3CH(OH)CO_2H$ could, in principle, be made from D-(+)-2,3-dihydroxypropanal, by first converting the —CH_2OH group to —CH_2Cl and then reducing it to —CH_3, and by oxidizing the —CHO group to —CO_2H. Since none of these reactions involves the central carbon atom, the absolute configuration of the isomer produced must be the same as that of the starting material, i.e. D-. The sign of the optical rotation of this isomer must then be determined experimentally. The results are as shown below:

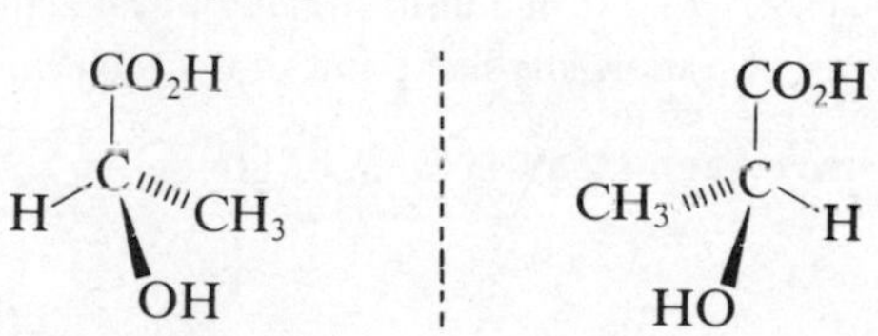

CO₂H   CO₂H
| |
H—C""""CH₃   CH₃""""C—H
| |
OH   HO

D-(+) isomer   L-(−) isomer

In the case of the optical isomers of the amino-acid alanine $H_2NCH(CH_3)CO_2H$, the (+) isomer, in contrast to the above, has the L- absolute configuration:

CO₂H   CO₂H
| |
H—C""""CH₃   CH₃""""C—H
| |
NH₂   H₂N

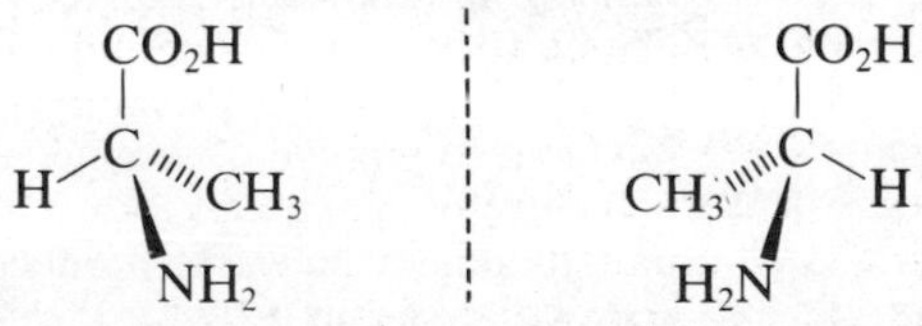

D-(−) isomer   L-(+) isomer

From a study of the optical isomers discussed so far, it should now be clear that if two or more atoms or groups about the central carbon atom of such a molecule are identical, then the mirror-image forms *will* be superimposable and such a compound cannot, therefore, display optical isomerism. The necessary condition for a molecule to display optical isomerism, or optical activity, is that it should be **dissymmetric**, i.e. not superimposable on its mirror-image. In simple organic molecules, this dissymmetry results from the presence of an **asymmetric**, or **chiral** (from the Greek *cheir*, a hand), carbon atom, i.e. one with four different atoms or groups directly bonded to it.* In more complicated cases, one can say that, if the molecule is to display optical activity, then it must possess neither a centre, a plane nor an alternating axis, of symmetry.†

Consider the molecule of 2,3-dihydroxybutanedioic acid (*tartaric acid*) $HO_2CCH(OH)—CH(OH)CO_2H$. This molecule possesses two asymmetric carbon atoms, marked with an asterisk. There are three optical isomers of this acid:

$(+)$ $(-)$ meso

* Optical activity is not confined to carbon compounds but is found in inorganic compounds also. For example, wherever an element M is bound tetrahedrally to four different elements or groups Mabcd, then optical activity is possible. M may be another element in Group 4, such as silicon or lead, or it may be an element in a complex ion, e.g. $Ni(abcd)^{2+}$. Though 6-coordinate, an ion like $[Co(NH_2CH_2CH_2NH_2)_2Cl_2]^+$ can also display optical activity.

† A *centre of symmetry* in a molecule is a point in the molecule such that any line drawn from an atom to this point, and then extended an equal distance beyond it, will arrive at an identical atom or group, e.g.

A *plane of symmetry* in a molecule simply divides the molecule into halves which are mirror-images of each other (*see* meso-tartaric acid, for example).

An *n-fold alternating axis of symmetry* in a molecule is an axis in the molecule, about which the molecule can be rotated by 360/n° and then reflected in a plane perpendicular to this axis to give an identical structure. A one-fold alternating axis of symmetry is equivalent to a plane of symmetry and a twofold alternating axis equivalent to a centre of symmetry.

If a molecule can be rotated about an axis by 360/n° to produce an identical structure, then the axis is described as an *n-fold axis of symmetry* (*not* an alternating one, note).

Two of these isomers constitute a pair of enantiomers and are optically active. The third isomer, while still possessing two asymmetric carbon atoms in its molecule, possesses a plane of symmetry and is consequently optically inactive; its mirror-image form *is* superimposable. This optically inactive form is said to be **internally compensated** (one may consider that the optical rotation of one asymmetric centre is annulled by that of the other) and is known as the **meso** form. Another optically inactive form of this acid would be a racemic mixture, i.e. an equimolar mixture of the (+) and (−) isomers, although it is capable of separation into its components (see resolution, below).

In principle, a molecule containing n asymmetric centres can give rise to a maximum of 2^n optical isomers. In the case of *tartaric acid*, where $n = 2$ and four isomers would therefore be expected, two of these isomers are the mirror-image forms of the *meso* isomer and are identical, so that there are only three optical isomers in practice. Where there is no plane of symmetry possible, however, as in the addition product of chloric(I) acid and but-2-ene $CH_3\overset{*}{C}H(OH)\overset{*}{C}HClCH_3$, then the maximum possible number of optical isomers may be obtained:

(1)   enantiomers (2)   (3)   enantiomers (4)

Optical isomers which are not enantiomers (mirror-images), such as the first and third or second and fourth of the above, are known as **diastereoisomers** and, unlike enantiomers, they may show slight differences in physical properties such as melting point or boiling point, density, solubility, etc. These differences are made use of in the separation of enantiomers.

16.3.2 Separation of optical isomers

This can clearly present difficulties. If the optical isomers are diastereoisomers, then they will show slight differences in physical properties and may be separated by techniques such as chromatography, fractional crystallization or fractional distillation (*see* Section 21.2). If, on the other hand, they are enantiomers, then they differ only in the signs of their optical rotation and their separation, called **resolution**, is much more difficult. In fact, they will first have to be converted into diastereoisomers which are capable of being separated by various means. Suppose a racemic mixture of *lactic acid* was carefully esterified with an optically active alcohol, say the (+) isomer. The resulting esters would possess two asymmetric centres, one of which (the one derived from the alcohol) would have the same absolute configuration in both isomers, i.e. the isomers produced would be diastereoisomers and not enantiomers:

$$\begin{matrix} R(+)\text{-CO}_2H \\ R(-)\text{-CO}_2H \end{matrix} + R'(+)\text{-OH} \begin{matrix} \longrightarrow R(+)\text{-COO}\!-\!R'(+) & \xrightarrow{\ H_2O\ } & R(+)\text{-CO}_2H \\ & \text{separate} \\ \longrightarrow R(-)\text{-COO}\!-\!R'(+) & \xrightarrow{\ H_2O\ } & R(-)\text{-CO}_2H \end{matrix}$$

racemic mixture   diastereoisomers

These diastereoisomers could be separated by physical methods and the separated esters carefully hydrolysed back to the acid. The racemic mixture would thus have been separated.

16.3.3   Biological significance of asymmetry

Although there are many fascinating examples of asymmetry in nature, it is at the molecular level that it is of most significance. The naturally occurring amino-acids, which are the molecular units from which proteins (and hence enzymes) are constructed, are (with one exception) all optically active and have the L-configuration (*see* diagram of alanine, p. 281). The corresponding D-enantiomer is not found in nature. This seems to indicate a fundamental asymmetry concerned with the evolution of molecules of biological importance (*see* Section 17.12).

Because of the control of many synthetic and metabolic reactions in living organisms by enzymes, optical isomerism is very important. For instance, only the (+) isomer of glucose is absorbed in the human body, the (−) isomer being excreted unchanged. In the case of optically active drugs, very often only one of the isomers is pharmacologically active, although the patient may be supplied with a racemic mixture if it is uneconomic to try to resolve it. In such a case, it is thought that the drug must interact with an asymmetric site in the body so that only one configuration will provide an accurate 'fit'. Optically active molecules are synthesized biologically with the aid of optically active enzymes, which can distinguish between (+) and (−) isomers. When a compound capable of showing optical isomerism is synthesized in the laboratory, however, a racemic mixture is almost invariably obtained.

Suppose *lactic acid* was to be synthesized in a two-step sequence from ethanal:

$$CH_3CHO + HCN \rightarrow CH_3CH(OH)CN \xrightarrow{\ H_3O^+\ } CH_3CH(OH)CO_2H$$

In the nucleophilic attack of cyanide ion on the ethanal, the cyanide ion could attack the carbon atom from either side of the double bond and, since there is nothing to cause a preferred direction of attack, it will proceed from both sides to give an equimolar mixture of the (+) and (−) isomers. This mixture must therefore yield racemic *lactic acid* on hydrolysis:

$$CH_3 \xrightarrow{} C \xrightarrow{O} \begin{array}{c} CN^- \\ \\ CN^- \end{array}$$

Reaction scheme showing:

$$\xrightarrow{H^+} \quad CH_3 - \overset{H}{\underset{OH}{C}}\text{''''}CN \quad \xrightarrow{H_3O^+} \quad CH_3 - \overset{H}{\underset{(+)\ \ OH}{C}}\text{''''}CO_2H$$

$$\xrightarrow{H^+} \quad CH_3 - \overset{H}{\underset{CN}{C}}\text{''''}OH \quad \xrightarrow{H_3O^+} \quad CH_3 - \overset{H}{\underset{(-)\ \ CO_2H}{C}}\text{''''}OH$$

} racemic mixture

16.4 Stereochemistry and mechanism

A great deal of evidence for some of the mechanisms discussed at various points in this book has come from a detailed study of the stereochemical features of the reactions. Measurements of stereochemical properties on the macro scale often provide a surprising amount of information about events occurring at the molecular level. The following examples have been chosen to complement previous descriptions of two *types* of reactions, namely substitution and addition reactions.

16.4.1 Substitution reactions

Nucleophilic substitution reactions of halogenoalkanes have already been described (*see* Section 5.3) and two mechanisms outlined (*see* Section 5.3.2). In the simplest mechanism, the S_{N^2} mechanism, a single step only is involved. The nucleophile attacks the electrophilic carbon from the opposite side to the halogen atom and, as the new bond forms, the carbon–halogen bond breaks and the halogen atom is displaced as halide ion. Suppose that the carbon atom bearing the halogen atom was an asymmetric one, as in a halogenoalkane of the type $RR'\overset{*}{C}HX$. Then, if one optical isomer was used, the configuration of this carbon atom would be **inverted** if the nucleophilic substitution followed the S_{N^2} mechanism:

$$Y\overset{..}{:}\quad \overset{R'}{\underset{R}{\underset{(+)}{\overset{\delta+\ \ \delta-}{H\text{''''}C-X}}}} \longrightarrow \quad \overset{R'}{\underset{H\ R}{\overset{\delta-\ \ \delta+\ \ \delta-}{Y\cdots C\cdots X}}} \longrightarrow \quad \overset{R'}{\underset{R}{Y-C\text{''''}H}} + \overset{-}{:}X$$

could be $(+)$ or $(-)$ in principle

However, the product is a different compound from the reactant so that such an inversion of configuration would not necessarily lead to a change in optical rotation. Starting with a $(+)$ optical isomer, the configuration might be inverted and yet the new optical isomer also have a positive sign of rotation (as has been previously mentioned, there is no correlation between sign of rotation and absolute configuration). Thus, it would not be possible to infer anything very useful about the stereochemical course of the reaction by simply recording the optical rotation of the product. Ideally, one would like to be able to carry out a substitution reaction on one optical isomer of the halogenoalkane without leading to a new product, so that any change in optical rotation could immediately be correlated with a **retention** or an **inversion** of the configuration.

Although this may seem contradictory, it *is* possible to carry out such a substitution without producing a new compound. The substitution is an *isotopic* one. Suppose an iodoalkane in solution was treated with a solution containing radioactive iodide ions. These ions are nucleophiles and begin to substitute in the iodoalkane. The rate at which such substitution occurs can be measured, by recording the rate at which radioactive iodide is taken up into the organic material. Now, if attack occurs randomly from either side of the carbon atom, then equal amounts of the $(+)$ and $(-)$ isomers will be formed and so a racemic mixture will gradually form at the same initial rate at which reaction occurs. If, however, attack is exclusively from the opposite side to the halogen atom, as in the S_{N^2} mechanism, then when half the molecules have undergone radioactive substitution there will be equal numbers of the $(+)$ and $(-)$ isomers present and hence again a racemic mixture will be formed.

The key observation, made in experiments carried out in 1935, is that *the initial rate of racemization is exactly twice the initial rate of incorporation of the radioactive iodine, i.e. the rate of substitution*. This fact is consistent only with an S_{N^2} mechanism and has been shown to be the case with a variety of halogenoalkanes (it is thought to be the case with most primary and many secondary ones, although for practical purposes one cannot make a primary one —CH_2X asymmetric at the relevant centre):

when *half* the molecules have reacted, the mixture is completely racemized

when *all* the molecules have reacted, the mixture is completely racemized

Thus, when a halogenoalkane undergoes nucleophilic substitution by an S_{N^2} mechanism, the reaction is accompanied by an **inversion** of configuration; an optically active starting material will yield an optically active product and *not* a racemic mixture.

The same general comments apply to such an investigation of an S_{N^1} mechanism. If a halogenoalkane first forms a planar carbocation, then it seems probable that there is equal likelihood of it undergoing nucleophilic attack from either side. Consequently, the product should be formed as a racemic mixture, but in this case the initial rate of racemization should be equal to the rate of substitution:

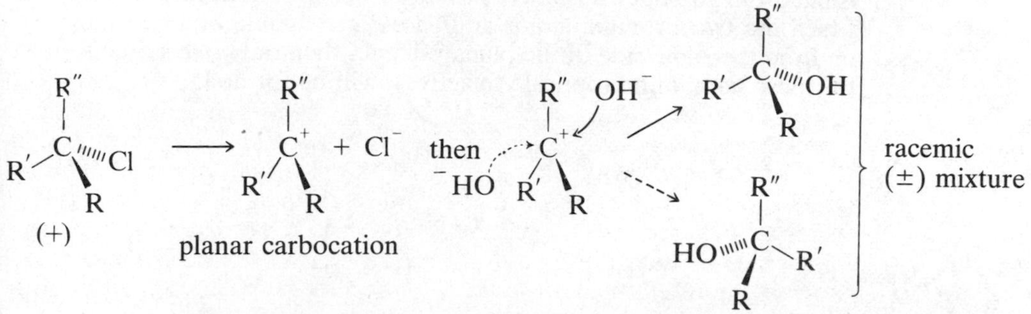

In some cases, however, some **inversion** of configuration is observed in an optically active product. This observation can be explained by assuming that the halide ion produced on initial ionization actually shields the carbocation on one side from nucleophilic attack, because it does not diffuse away rapidly enough and hence there is a preferred direction of attack leading to some inversion of configuration:

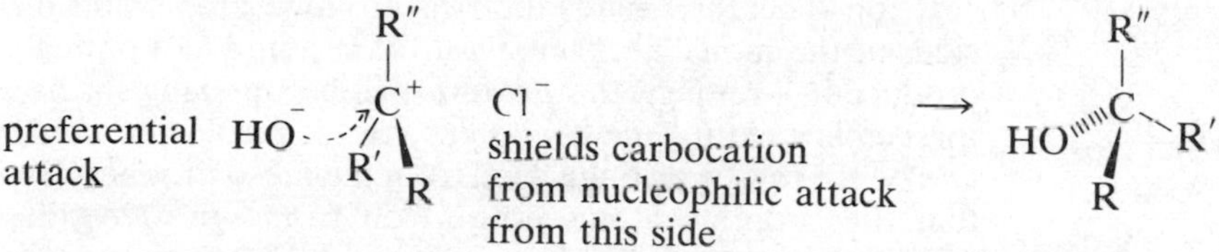

Thus, the product of an S_{N^1} substitution of a halogenoalkane would consist of a racemic ($\pm$) mixture plus some optical isomer with an inverted configuration.

16.4.2 Addition reactions

The mechanism of the electrophilic addition of halogens to alkenes has already been outlined (*see* Section 3.3.4). The stereochemical features of the reaction are outlined here as evidence for preferring the bromonium ion intermediate rather than the carbocation one:

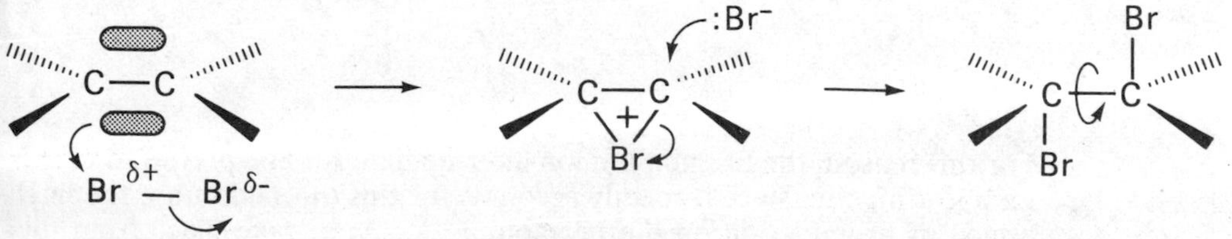

Once the two bromine atoms have added across the double bond, free rotation can occur about the carbon–carbon single bond and it is not possible to distinguish, in the product, the direction of addition of the two atoms, unless the alkene is carefully chosen. Suppose the *cis* geometrical isomer of but-2-ene was brominated. The product would then possess two asymmetric centres in the molecule and could show optical isomerism. The optical isomers produced will depend on the direction in which the two bromine atoms add across the double bond. If they add on from the same side of the planar alkene, then only one stereoisomer, the meso compound (optically inactive), will be formed:

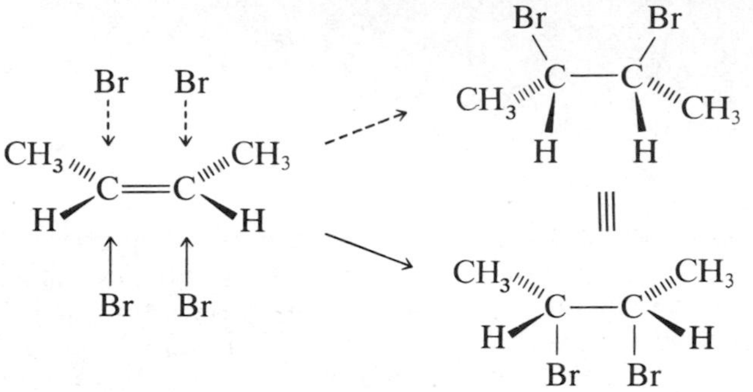

meso-compound

If, on the other hand, the two bromine atoms add on from opposite sides of the molecule, then a pair of $(+)$ and $(-)$ optical isomers will be produced. Although the mixture will be optically inactive, because it is a racemic mixture, it will be capable of resolution into its enantiomers and hence can be distinguished from the meso product. Experiment shows that the bromine atoms do add on from *opposite* sides, i.e. that the addition is in the *trans* sense:

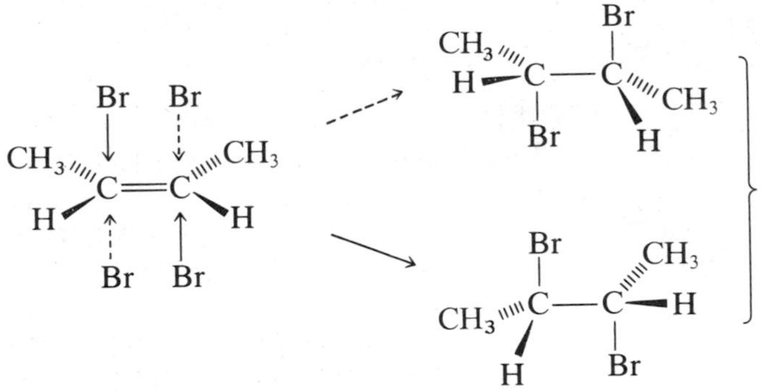

enantiomers

For this reason, the bromonium ion intermediate is to be preferred to the carbocation one, since it readily accounts for this *trans* addition; the final attack of bromide ion on the bromonium ion *must* take place from the opposite side because the bulky $\overset{+}{Br}$ would shield one side of the inter-mediate. In the case of the carbocation, one would imagine that free rotation about the carbon–carbon single bond would mean that no dis-tinction could be made in the product between the directions of attack. It has been argued that perhaps *trans* attack by bromide ion could take place before there was time for free rotation, but this argument seems less convincing.

Questions

1   The following are terms often associated with different types of isomerism found in organic chemistry:
(a) chain isomerism,
(b) positional isomerism,
(c) functional group isomerism,
(d) geometrical isomerism, and
(e) optical isomerism.
Briefly explain what is meant by each of these terms, illustrating your answer by an appropriate pair of isomers in each case.
For *two* of the pairs of isomers you have chosen, suggest how you could distinguish between the pair members.

London

2* (a) What do you understand by the terms *enantiomer (enantiomorph)*, *racemic mixture, geometrical isomer,* and *structural isomer*? Use spe-cific compounds to illustrate each term.
(b) Explain briefly what prevents the *geometrical* isomers which you have given in (a) from interconverting.
(c) Draw diagrams of each stereoisomer which may exist for each of the following structures. Indicate clearly what type(s) of isomerism is/are involved.
$C_6H_5CH{=}CHCOOH$
$C_6H_5CH_2CH(NH_2)COOH$
$C_6H_5CH{=}CHCOOCH(CH_3)C_2H_5$

Oxford & Cambridge

3   (a) Give **two** factors which may restrict rotation about carbon–carbon bonds, and indicate the consequences which this restriction may lead to in the field of isomerism. Illustrate your answer with specific compounds.
(b) What is meant by the term *chiral molecule* and to what extent is the term confined to organic chemistry? Illustrate your answer with dia-grams of specific molecules.
(c) Illustrate the importance of stereochemistry as an approach to under-standing organic reaction mechanisms.

Oxford & Cambridge

4* Write down the structures of **three** isomers for each of **two** of the following formulae, and show how the isomers could be distinguished from each other:
(i) C_4H_8O
(ii) $C_4H_{10}O$
(iii) C_5H_8
(iv) $C_4H_8O_2$

Cambridge

5* How could you distinguish chemically between **four** of these pairs of isomers?

(i) 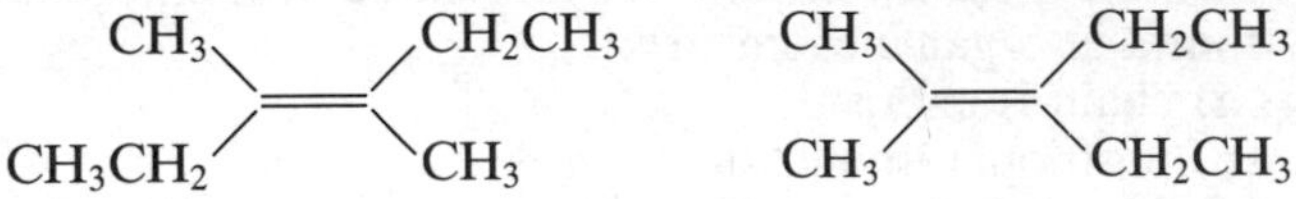
$$CH_3, \quad CH_2CH_3 \qquad\qquad CH_3, \quad CH_2CH_3$$
$$CH_3CH_2 \diagup \quad \diagdown CH_3 \qquad\qquad CH_3 \diagup \quad \diagdown CH_2CH_3$$

(ii) 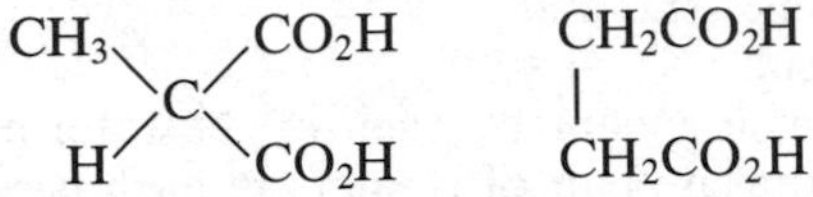
$$CH_3, \quad CO_2H \qquad\qquad CH_2CO_2H$$
$$\quad\quad C \qquad\qquad\qquad\qquad |$$
$$H \diagup \quad \diagdown CO_2H \qquad\qquad CH_2CO_2H$$

(iii) 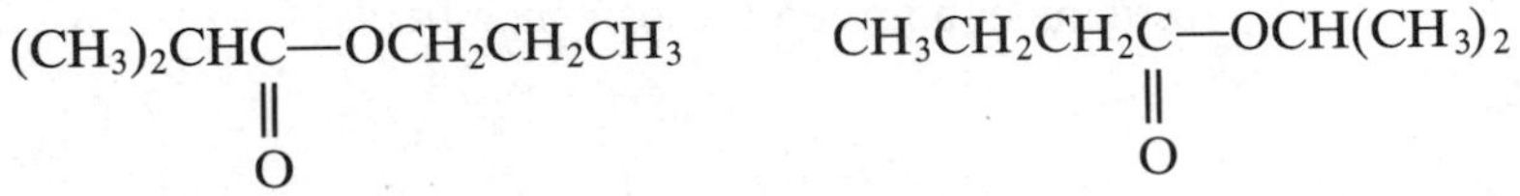 $(CH_3)_2CHC—OCH_2CH_2CH_3$ $CH_3CH_2CH_2C—OCH(CH_3)_2$
$\qquad\qquad\qquad\;\; \| \qquad\qquad\qquad\qquad\qquad\qquad\qquad\qquad \|$
$\qquad\qquad\qquad\;\; O \qquad\qquad\qquad\qquad\qquad\qquad\qquad\qquad O$

(iv) $(CH_3)_2CHNHCH_3$ $CH_3CH_2NHCH_2CH_3$

(v)

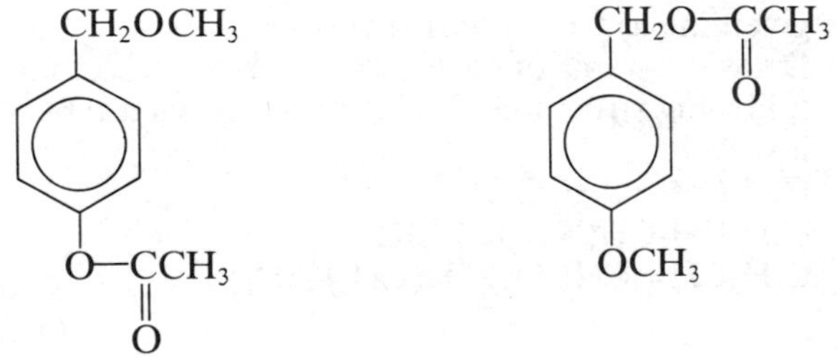

CH_2OCH_3

$O—CCH_3$
$\quad\quad\; \|$
$\quad\quad\; O$

$CH_2O—CCH_3$
$\qquad\qquad \|$
$\qquad\qquad O$

OCH_3

(vi)

CH_2Cl

CH_3

Cl

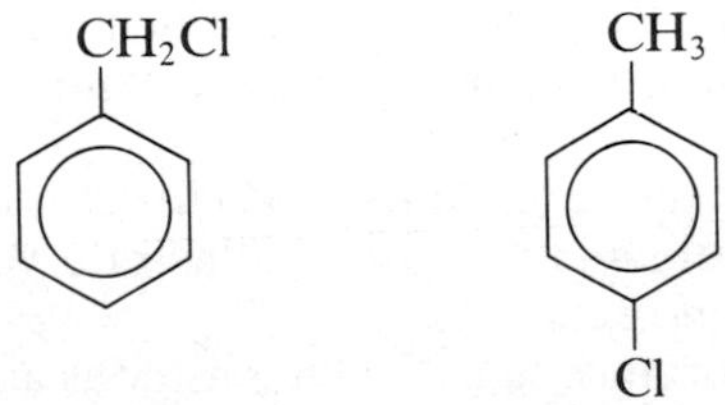

Cambridge

6*   Using the compounds shown below as your examples, explain the meaning of the terms *racemic* and *meso*:

$CH_3CHClCHClCH_3$ $CH_3CHClCH_2CH_3$

Assign possible structures to compounds **A**, **B** and **C** below, giving your reasoning. The only one of these compounds which can be resolved is **B**.

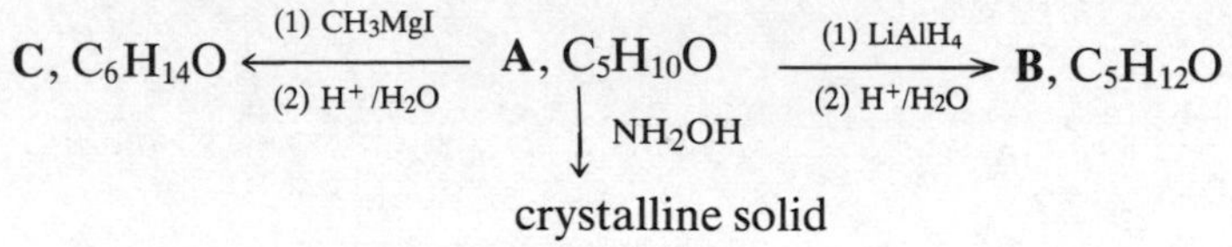

$$C, C_6H_{14}O \xleftarrow[\text{(2) } H^+/H_2O]{\text{(1) } CH_3MgI} A, C_5H_{10}O \xrightarrow[\text{(2) } H^+/H_2O]{\text{(1) } LiAlH_4} B, C_5H_{12}O$$

with **A** reacting via NH_2OH to give

crystalline solid

Oxford

7*   Explain the following:
(a) The amino acid alanine, $CH_3CH(NH_2)COOH$, is used by living cells in the synthesis of proteins. When chemically prepared alanine is fed to cells, only half of it is used for protein synthesis, but if alanine from meat extract is fed all of it is used.
(b) The optical activity of a solution containing (+)-2-iodooctane and a catalytic amount of sodium iodide in propanone (acetone) slowly decreases to zero.
(c) Two isomeric oximes are formed when butan-2-one ($CH_3COCH_2CH_3$) is treated with hydroxylamine (NH_2OH).
(d) Glycerol, $HOCH_2CH(OH)CH_2OH$, forms two monoethanoates (acetates), one of which can be resolved into optically active forms (enantiomers).

Cambridge

Chapter 17
Biological molecules: carbohydrates and proteins

Carbohydrates

17.1 Introduction

Carbohydrates contain the elements carbon, hydrogen and oxygen only and can be given a general formula $C_xH_{2y}O_y$. They take their name from the former, mistaken notion that they were hydrates of carbon (as in $C_x(H_2O)_y$). They are of great importance biologically and are widely distributed in nature, particularly in the plant kingdom, where they are produced in green plants by the process of **photosynthesis**:

$$nCO_2 + nH_2O \xrightarrow[\text{chlorophyll}]{\text{light}} C_nH_{2n}O_n + nO_2$$

The carbohydrate cellulose is the chief constituent of the structural material of many plants. Starch, another carbohydrate, is found in different parts of many plants, particularly as a food reserve stored in the roots and stems and also in many seeds. Both cellulose and starch are extremely high relative molecular mass compounds and are biological polymers. Simpler carbohydrates include the sugars fructose and glucose, which impart the sweet taste to many fruits and are also present in nectar, and therefore honey.

Although carbohydrates constitute one of the most important foods of animals, providing heat and energy, they are not present to the same extent in the animal as in the plant body. Glycogen, a carbohydrate similar to starch, is a food reserve of many animals, being stored in the liver. Lactose, a carbohydrate never found in the vegetable kingdom, is present in the milk of mammals.

There are also important materials which are made from carbohydrates on an industrial scale. These include paper, rayon, gums and celluloid.

The better-known carbohydrates are classified into three groups:

(1) *Monosaccharides.* These are isomers of $C_6H_{12}O_6$ and include sugars such as glucose and fructose.
(2) *Disaccharides.* These are isomers of $C_{12}H_{22}O_{11}$ and include the sugars sucrose, maltose and lactose.
(3) *Polysaccharides.* These are biological polymers of very high relative molecular mass and include cellulose, starch and glycogen.

17.2   Monosaccharides

Monosaccharides are polyhydroxy aldehydes or ketones, usually referred to as aldoses and ketoses, respectively. Since they share the molecular formula $C_6H_{12}O_6$, i.e. C_6, they are also described as hexoses. The most familiar of these is glucose, a white, sweet-tasting, crystalline solid which is soluble in water but almost insoluble in pure ethanol. Glucose is the main form in which carbohydrate is carried through the blood stream. On the basis of the chemical reactions outlined below, and other evidence, glucose can be assigned the structural formula

$$\overset{6}{HOCH_2}\overset{5}{CH(OH)}\overset{4}{CH(OH)}\overset{3}{CH(OH)}\overset{2}{CH(OH)}\overset{1}{CHO}$$

which shows the presence of one primary alcohol group, four secondary alcohol groups and one aldehyde group. Furthermore, there are four asymmetric carbon atoms in the molecule and hence a possible 16 (2^4) optical isomers (*see* Section 16.3), all of which have been identified. Naturally occurring glucose has the following configuration (free rotation can occur about the bonds, as shown, of course):

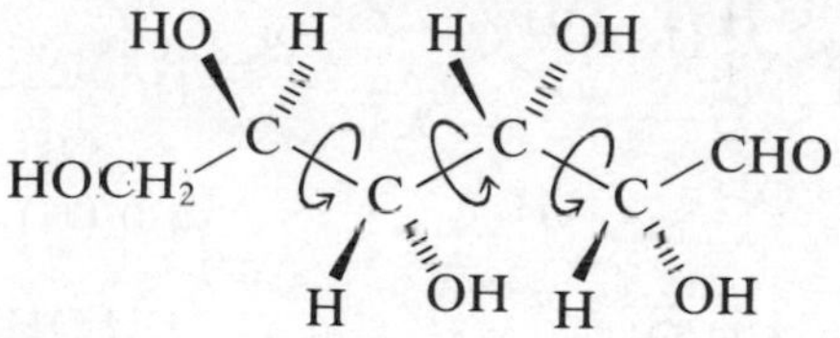

However, other evidence suggests that in solution this straight-chain form of glucose is in equilibrium with a 6-membered ring structure, formed by an internal condensation of the hydroxyl group on carbon atom 5 with the aldehyde group to form a hemiacetal (*see* Section 8.3.1(iii)):

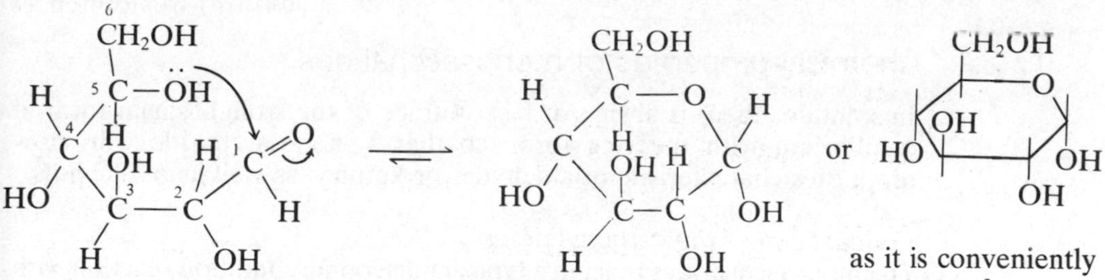

or HO

as it is conveniently represented

Hemiacetal forms predominate (over 99%) in solution and are known as the **pyranose** forms.* In these forms, a new asymmetric centre is

* The structure ⬡ is sometimes known as tetrahydropyran.

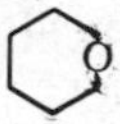

created at carbon atom 1, and hence there are two possible configurations; when the hydroxyl group is pointing upwards the form is known as the β-form, and when pointing down, the α-form. The absolute configuration of C_5 in natural glucose is identical to that of D-2,3-dihydroxypropanal (*see* Section 16.3) so that this ring form of glucose is sometimes known as **D-glucopyranose**. Solid glucose, as ordinarily prepared, is α-D-(+)-glucopyranose although it can also be prepared in the β-D-(+) form. In solution, glucose exists as an equilibrium mixture of the two (α and β) cyclic forms with a trace of the open-chain form. Hence, a glucose solution will still show some properties characteristic of an aldehyde (*see* Section 17.2.1).

Another monosaccharide is fructose, which has a sweeter taste than glucose and is more soluble in water. In its straight-chain form, fructose is a ketone (and hence a **ketose**) but in solution it forms a pyranose ring structure. In sucrose, however, it is present in a five-membered furanose ring form (*see* Section 17.3):

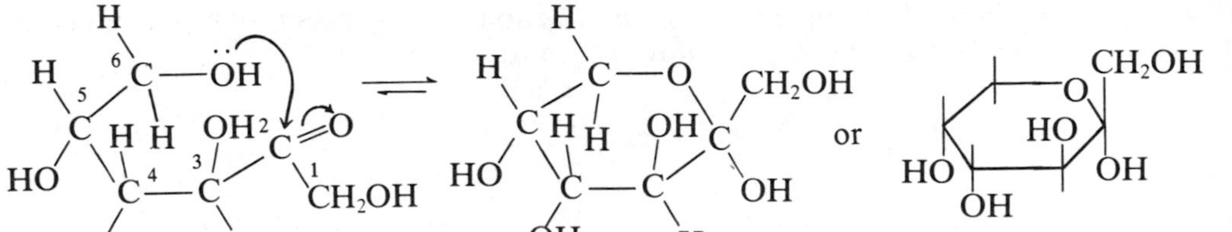

α-D-(−)-fructopyranose

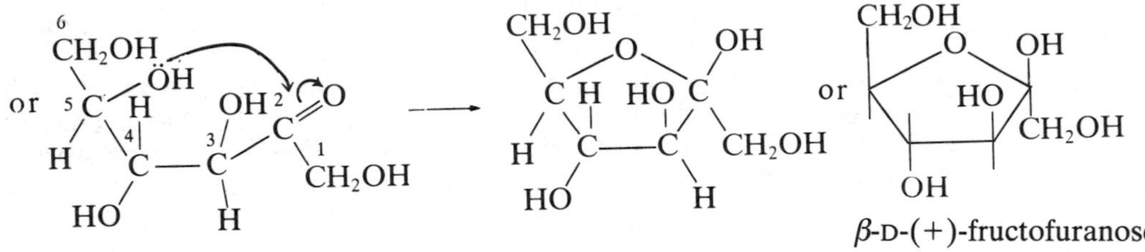

β-D-(+)-fructofuranose

17.2.1 Chemical properties of monosaccharides

In solution, there is always at least a trace of the straight-chain form in equilibrium with the ring form, so that a monosaccharide will show properties characteristic of aldehydes or ketones as well as of alcohols.

(i) Reactions of the carbonyl group
The monosaccharides react in a typical nucleophilic addition reaction with hydrocyanic acid HCN; the product may be hydrolysed to a carboxylic acid (*see* Section 8.3.1(i)):

$$HOCH_2CH(OH)CH(OH)CH(OH)CH(OH)C{\overset{O}{\underset{H}{\diagdown}}} + HCN$$

$$\longrightarrow HOCH_2CH(OH)CH(OH)CH(OH)CH(OH)CH{\overset{OH}{\underset{CN}{\diagdown}}}$$

They do not, however, form addition products with sodium hydrogen-sulphite $NaHSO_3$. The carbonyl group can also be reduced, either by 'dissolving metal' reduction (e.g. Na/Hg and ethanol, *see* Section 8.3.5(ii)), or by complex metal hydride:

$$HOCH_2CH(OH)CH(OH)CH(OH)CH(OH)C{\overset{O}{\underset{H}{<}}}$$

$$\longrightarrow HOCH_2CH(OH)CH(OH)CH(OH)CH(OH)CH_2OH$$

Monosaccharides also enter into condensation reactions. With hydroxylamine the reaction follows the normal course for aldehydes or ketones:

$$HOCH_2CH(OH)CH(OH)CH(OH)CH(OH)C{\overset{O}{\underset{H}{<}}} + NH_2OH$$

$$\longrightarrow HOCH_2CH(OH)CH(OH)CH(OH)CH(OH)CH{=}NOH + H_2O$$

but with phenylhydrazine, the reaction is more complicated and oxidation of a neighbouring hydroxyl group occurs:

$$HOCH_2CH(OH)CH(OH)CH(OH)CH(OH)C{\overset{O}{\underset{H}{<}}} + 3C_6H_5NHNH_2$$

$$\longrightarrow HOCH_2CH(OH)CH(OH)CH(OH){-}\underset{\underset{NNHC_6H_5}{\|}}{C}{-}CH{=}NNHC_6H_5$$

$$+ C_6H_5NH_2 + NH_3 + 2H_2O$$

The product is known as an **osazone** and is historically important, as it was used by the 1902 Nobel Prize winner Emil Fischer in his pioneer work on carbohydrate structures. For example, significantly both glucose and fructose form identical osazones, showing that the structures of the two sugars differ only at C_1 and C_2.

(ii) Reactions of the hydroxyl group

The simple sugars are freely soluble in water, owing to their ability to form hydrogen bonds as a result of their hydroxyl groups. These latter groups behave chemically as they do in alcohols. Thus, if glucose is treated with a mixture of ethanoic anhydride and sodium ethanoate, five ethanoyl groups are introduced into the molecule, demonstrating that there are five hydroxyl groups (including that of the hemiacetal group) which can be esterified:

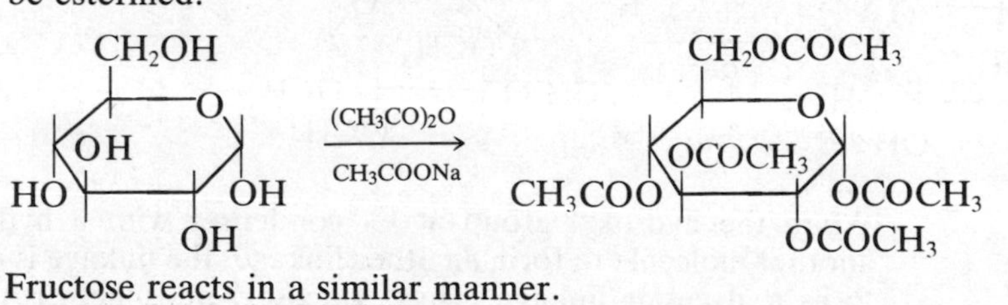

Fructose reacts in a similar manner.

The C_1 hydroxyl group, i.e. the hemiacetal hydroxyl group, is the most reactive of the hydroxyl groups in hexoses and is responsible for linking two monosaccharide units together in disaccharides (see Section 17.3).

Phosphate esters of sugars are extremely important in carbohydrate metabolism. The nucleotide adenosine triphosphate, ATP, consists of a triphosphate of the sugar ribose (a pentose monosaccharide) attached to the base adenine (the nucleoside adenine-ribose is called adenosine):

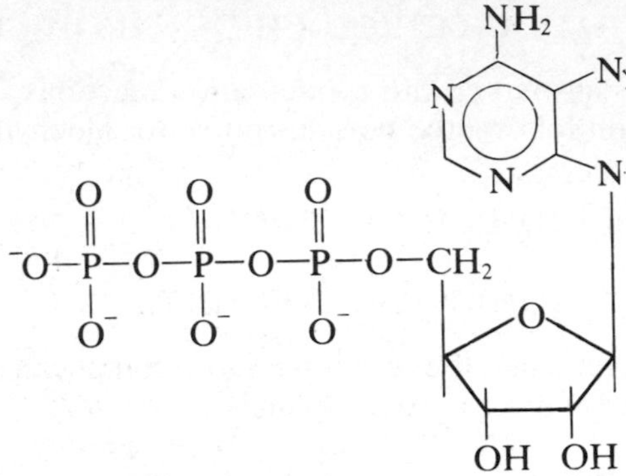

It is vitally important in transferring energy in biological systems. When energy is released in a chemical reaction (e.g. respiration), any which is not to be utilized immediately is used to form ATP from adenosine diphosphate, ADP:

ADP + phosphate + energy → ATP

An example of such a reaction is the anaerobic oxidation of glucose to *lactic acid* (see also paragraph (iii)):

$C_6H_{12}O_6$ + phosphate + 2ADP → $2\,CH_3CH(OH)COOH$ + 2ATP

This 'stored energy' can be released for use subsequently, by hydrolysis of ATP to ADP.

Hydroxyl groups can also be converted into *ethers*, by treatment of the sugar with an alkaline solution of dimethyl sulphate $(CH_3)_2SO_4$ (see also Section 13.3.1(i)). However, the ether linkage at C_1 can be easily hydrolysed by dilute hydrochloric acid, while the other hydroxyl groups remain 'protected' as ethers. These reactions were used originally to determine the ring size of sugars:

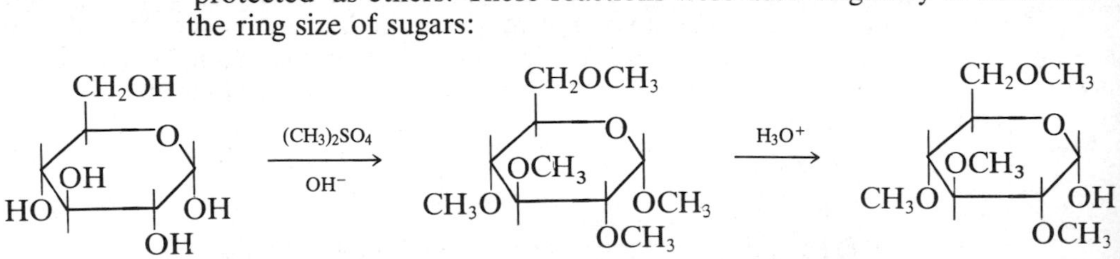

When the hydroxyl group at C_1 condenses with a hydroxyl group of another molecule to form an ether linkage, the linkage is usually referred to as a **glycoside** linkage. Thus, glucose-1-phosphate is a glycoside. An

important group of heart stimulants extracted from *digitalis* (foxglove) includes the glycoside digitoxin.

(iii) Oxidation

Monosaccharides can be burned in oxygen to carbon dioxide and water, when a considerable amount of energy is released as heat. Thus, for glucose

$$C_6H_{12}O_6 + 6O_2 \rightarrow 6CO_2 + 6H_2O \qquad \Delta H^{\ominus} = -2850\,kJ$$

This reaction is the reverse of photosynthesis which is endothermic and which utilizes energy from the sun. In the body, the oxidation of carbohydrates such as glucose takes place in a very carefully controlled fashion, involving up to twenty steps. The energy so released is used in muscular activity and cell synthesis, as well as in maintaining normal body temperature. The so-called 'calorific value' of carbohydrate is about $16\,kJ\,g^{-1}$, compared with a figure of 16–$20\,kJ\,g^{-1}$ for protein. In the absence of oxygen, glucose is metabolized either to ethanol (as in yeast cells), or else to 2-hydroxypropanoic acid (*lactic acid*) (as in muscle). When insufficient oxygen is present, glucose is converted into *lactic acid,* which is the cause of sudden stiffness in the muscles during a bout of intensive activity or exertion:

Starch (plants), Glycogen (animals) → Glucose $C_6H_{12}O_6$

→ Ethanol $2CH_3CH_2OH + 2CO_2$
→ Lactic acid $2CH_3CH(OH)CO_2H$ ($6O_2$)
→ $6CO_2 + 6H_2O$

Aldoses will reduce both Fehling's solution (or Benedict's solution, which contains *citrate* ions, rather than the tartrate ions to be found in Fehling's solution) and Tollens's solution, because they contain an aldehyde group in their open-chain form (*see* Section 8.3.7(i) and 8.3.7(ii)). Sugars like glucose are therefore **reducing sugars**. However, fructose will give a positive test with Fehling's solution although it is a ketose. The reason for this is that it is an α-hydroxyketone (a hydroxyl group is bonded to a carbon atom, the α-carbon, immediately adjacent to the carbonyl group) which can isomerize via the 'enediol' form (*see also* keto–enol tautomerism, Section 8.3.6) to an aldehyde:

$$CH_2OH - C=O \rightleftharpoons CH-OH \parallel C-OH \rightleftharpoons CHO - CH-OH$$

Fructose is therefore also classified as a reducing sugar.

However, ketoses like fructose can be distinguished from aldoses like glucose. Aldoses are oxidized by an alkaline solution of bromine, i.e. by bromate(I), whereas ketoses are not. Thus, glucose is oxidized to gluconic acid:

$$HOCH_2CH(OH)CH(OH)CH(OH)CH(OH)C\overset{\displaystyle O}{\underset{\displaystyle H}{<}}$$

$$\longrightarrow HOCH_2CH(OH)CH(OH)CH(OH)CH(OH)C\overset{\displaystyle O}{\underset{\displaystyle OH}{<}}$$

Glucose, in the urine for example, can be conveniently identified using small sticks ('Clinistix') containing the enzymes glucose oxidase and peroxidase and a reduced, colourless form of a dye. In the presence of the oxidase, glucose is oxidized to the cyclic anhydride (lactone) of the above acid; hydrogen peroxide is released which, in the presence of peroxidase, immediately oxidizes the dye on the stick to a blue form. Such tests are regularly performed to check the level of glucose excreted by diabetics.

17.3 Disaccharides

The disaccharides have the molecular formula $C_{12}H_{22}O_{11}$ and consist of two monosaccharide units (which may, or may not, be the same) linked together by a condensation reaction between hydroxyl groups to form a glycoside linkage.

The best known, and most widely used, disaccharide is sucrose (ordinary 'sugar'), which is obtained from sugar cane and sugar beet. The raw cane sugar is brown and the dark-coloured material obtained during the decoloration and crystallization of the raw sugar is sold as brown sugar. The mother liquor from the first purification process is known as *molasses* and can be used for fermentation as well as for making treacle, rum, etc.

Owing to its importance as a food, sucrose is produced in larger quantities than any other pure chemical. Pure sucrose is a colourless crystalline solid, readily soluble in water but only slightly soluble in ethanol. It melts at 160°C and solidifies to a transparent mass called *barley sugar*. If heated to 200°C, some water is lost to give a brownish substance known as *caramel*, which is used to colour gravies, rum, etc.

All disaccharides undergo acid or enzymic hydrolysis to monosaccharides. Sucrose yields glucose and fructose on hydrolysis and contains these two sugars joined together by an α-1,β-2 glycoside linkage:

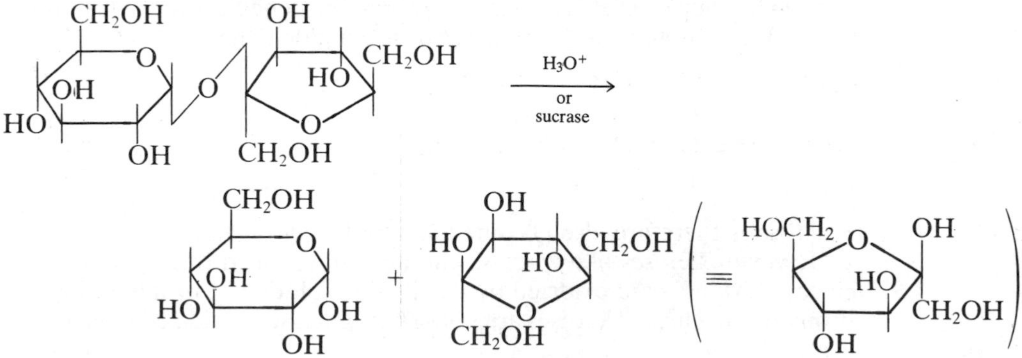

It should be noted that in sucrose the fructose is present as a five-membered **furanose*** ring rather than the six-membered **pyranose** form characteristic of the free sugar.

Natural sucrose is (+) but, as hydrolysis proceeds, the optical rotation of the solution first falls until the solution becomes optically inactive. Then, (−) rotation increases until hydrolysis is complete. The reason for this 'inversion' of sucrose is that the (−) optical rotatory power of the fructose is greater than the (+) optical rotatory power of the glucose, so that the final equimolar mixture of the two monosaccharides is (−). The mixture is often sold as 'invert sugar'. The formation of some 'invert sugar' during jam-making is important, as it prevents subsequent crystallization of sugar (the sugar acts as a preservative) from the jam during storage.

Other disaccharides include *maltose*, an intermediate in the alcoholic fermentation of starch, *cellobiose*, a hydrolysis product of cellulose which is similar to maltose except that the two glucose units are β-1,4-linked rather than α-1,4-linked, and *lactose*, the principal sugar found in the milk of mammals. They all yield monosaccharides on acid or enzymic hydrolysis:

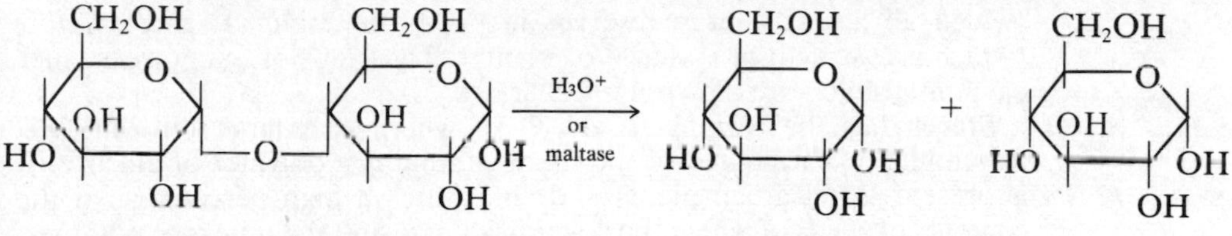

maltose   glucose

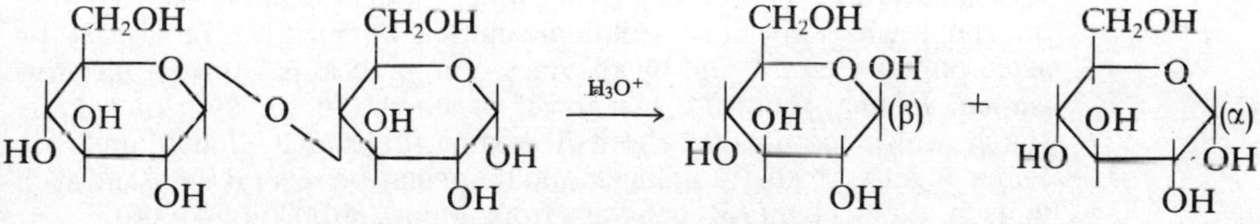

cellobiose   glucose

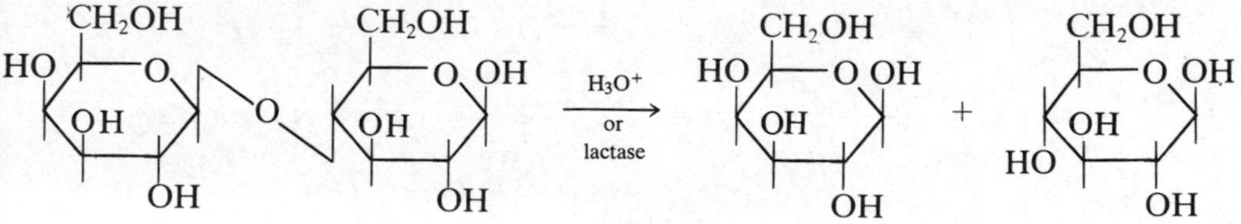

lactose   galactose   glucose

* The structure 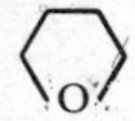 is known as tetrahydrofuran.

Both maltose and lactose are **reducing sugars**, because they both possess potential aldehyde groups at C_1 (cf. open-chain form, Section 17.2). Sucrose, on the other hand, does *not* contain a potential aldehyde group (since the link is 1,2 not 1,4) and hence is classified as a **non-reducing sugar**. Normally, it has no effect on Fehling's solution (reaction with Tollens's reagent is not reliable); if, however, sucrose is hydrolysed and the products are warmed with Fehling's solution, then reduction does occur, because of the reducing power of the glucose and fructose formed.

17.4 Polysaccharides

These are polymers of very high relative molecular mass (which in most cases are not known with certainty). Their empirical formula is $C_6H_{10}O_5$ but the relative molecular masses may vary from about 30 000 to 14 million. In contrast to the mono- and di-saccharides, the polysaccharides are amorphous and insoluble in water and organic solvents in general. Their purification is therefore difficult. Generally, they can only be brought into solution by hydrolysis to mono- and di-saccharides.

The best known polysaccharides include *starch, inulin* and *glycogen* (which all act as energy reserves in plants or animals), and *cellulose* (which is a structural material of plants). The gums of many trees, such as 'gum arabic', are also polysaccharides.

Starch has the formula $(C_6H_{10}O_5)_n$, where n is large but variable, depending on the source of the material, and is a polymer of glucose. It is present in all green plants and constitutes a high percentage of the contents of seeds of wheat, barley, maize, rice and the tubers of potatoes. It is found in the plant cells as tiny granules, the shape and size of which vary in different plant species. In hot water these granules burst, releasing the starch which goes into colloidal solution where it can be utilized or acted on more easily, e.g. by enzymes. The process is known as **gelatinization.** In fact, there are two forms of starch. In *amylose*, or soluble starch, which accounts for about 20% of natural starch, glucose units are joined together by α-1,4-linkages and there may be several thousand such units in a long chain (M_r can vary from about 160 000 to 700 000):

On acid hydrolysis, all the 1,4-linkages are cleaved and α-D-(+)-gluco-pyranose (glucose) is produced. In contrast, on enzymic hydrolysis (for example, with *diastase* or with *salivary amylase*, present in saliva), only alternate 1,4-linkages are split so that the product is the disaccharide maltose.

In *amylopectin*, the other component of starch, there are occasional 1,6-linkages between glucose units so that a branched-chain polymer is the result (with M_r over 100000, and up to about 1 million):

From these structures, it is obvious that starch will not reduce Fehling's solution. It does, however, give a very characteristic blue-black colour with a few drops of iodine solution; the colour fades on warming but reappears on cooling.

The enzymic breakdown of starch is of the utmost importance in the baking of bread. When the wheat flour is mixed with water to make dough, enzymes in the flour hydrolyse the amylose and amylopectin to maltose. When the dough is heated, starch also gelatinizes and the colloidal starch solution is then more easily broken down by enzyme action. (At the same time the elastic material, *gluten*, produced from the proteins gliadin and glutenin during the kneading of the dough, coagulates.) During the leavening of the bread by yeast, the sugar produced is fermented to ethanol and carbon dioxide (*see also* Section 17.5); the ethanol is driven off and the gas causes the bread to rise (in highly refined flour the amylases have been destroyed and sugar has to be added to the bread).

Glycogen is an important energy store in the body and is found mainly in liver and muscle. It has a similar structure to amylopectin but there is more frequent short-chain branching and the relative molecular mass is of the order of several million. Muscle glycogen is broken down under anaerobic conditions (i.e. in absence of oxygen) to *lactic acid* (*see* Section 17.2.1(iii)) but liver glycogen is broken down into glucose which, being water soluble, can be transported through the blood stream. Glycogen gives a red-brown colour with a few drops of iodine solution and is non-reducing.

Inulin, which is found in tubers of dahlias, etc., gives, on hydrolysis, β-(D)-fructose and is thus a polymer of this monosaccharide.

Cellulose consists of long chains of β-1,4-linked glucose units and has a relative molecular mass of the order of one million:

Cellulose forms fibres in which there are both crystalline regions, in which hydrogen bonding plays an important role, and amorphous regions which allow water to be absorbed very readily, owing to the availability of free hydroxyl groups. Cellulose itself is quite insoluble in water but will dissolve in ammoniacal solutions of copper(II) salts (*Schweitzer's reagent*) or in a mixture of sodium hydroxide and carbon disulphide (*see* Section 17.6).

In the human body, there are no enzymes which can act on cellulose and hence man is unable to digest it. However, ruminants like cows contain bacteria in their stomachs which readily catalyse the breakdown of cellulose in grass into cellobiose, and finally into butanoic acid, methane and carbon dioxide.

17.5 Alcoholic fermentations

The fermentation of carbohydrates can yield a variety of products, one of the most important being ethanol. Carbohydrates such as starch and cellulose are broken down to ethanol in a series of steps which are catalysed by enzymes but which do not require oxygen.

In the brewing of beer, barley is allowed to germinate in order to produce the enzyme *diastase* in the grain. After drying, the partly germinated grain, called *malt,* is hydrolysed at 60–65°C with the aid of the diastase. The starch is broken down into the disaccharide maltose:

$$(C_6H_{10}O_5)_n + n/2\ H_2O \rightarrow n/2C_{12}H_{22}O_{11}$$

The resulting liquid, called *wort,* is boiled with hops to impart the characteristic bitter flavour (the hops also act as a preservative) and the cooled solution then treated at 18°C with yeast. Two enzyme-catalysed reactions take place. First, the maltose is hydrolysed to glucose with the aid of the enzyme maltase:

$$C_{12}H_{22}O_{11} + H_2O \rightarrow 2C_6H_{12}O_6$$

The glucose so produced is then converted into ethanol and carbon dioxide with the aid of the enzyme zymase, also present in the yeast:

$$C_6H_{12}O_6 \rightarrow 2C_2H_5OH + 2CO_2$$

The final alcohol content of the beer is about 4% by volume.

In the production of wines by fermentation the grape juice, containing the sugars glucose ('grape sugar') and fructose, is fermented with the aid of micro-organisms already present on the grape skin, so that the addition of yeast is unnecessary. As well as ethanol, acids, esters and ketones are also formed in small amounts and may be responsible for characteristic flavours and 'noses'. If the carbon dioxide, liberated during the fermentation, is prevented from escaping, then a natural sparkling wine is produced, the final alcohol content being of the order of 12% by volume. The presence of other micro-organisms such as *acetobacter*, however, can catalyse the oxidation of ethanol to ethanoic acid ('wine to vinegar'), a problem which was originally studied and solved by the French scientist Louis Pasteur.

In alcoholic fermentations, the enzymes become deactivated in solutions containing more than about 15% ethanol, so that spirits and strong wines with a higher alcohol content do not turn sour when contaminated in this way. The high alcohol content of wines like port and sherry (fortified wines) is a result of the direct addition of more ethanol, whereas that of spirits is brought about by distillation.

Whisky is made by the alcoholic fermentation of starch in rye and barley, in a process similar to the brewing of beer. The cheaper grain whisky (as opposed to malt whisky) is made from maize with the addition of a little malted barley. The starch is converted into fermentable sugars which are allowed to ferment for a longer period than for beer. The solution is then fractionally distilled, using either a continuous (equilibrium) still or, for malt whisky, a pot still. Pot stills are always made out of copper which is believed to remove traces of unpleasant sulphur containing compounds during the distillation. Even the shape of the still is thought to affect the final composition of the distillate. If the spirit is allowed to mature in oakwood (old sherry casks impart a golden colour to the spirit) then changes occur in the composition of the whisky, as volatile components can diffuse slowly through the wood and escape. Even so, the final composition of whisky is immensely complex; over 256 organic compounds have been identified.

In the past, the alcoholic concentration of drinks in the UK has been expressed by an antiquated system of 'degrees of proof' (often, perhaps alarmingly, confused with percentage alcohol). Since 1979, however, the alcohol concentration has been expressed by the more comprehensible percentage content by volume, in line with other European countries. Thus, spirits contain about 40% ethanol, fortified wines 15–22%, table wines 8–15% and beer about 4%. 'Polish spirit' heads the list at an incredible 80%.

17.6 Cellulosic fibres

Cellulose occurs in the form of fibres, as in cotton seed hairs, which are utilized directly in cotton textiles. However, fibres much longer than those occurring naturally can be made from cellulose. Synthetic fibres from cellulose are called **rayon** and, in two of the three methods used in producing rayon, the cellulose fibres are simply regenerated in a different form. In 1857, the Swiss chemist Schweitzer discovered that cellulose will dissolve in an ammoniacal solution of copper(II) hydroxide. In 1890, it was discovered that when the solution is forced through tiny jets into an acid bath, the cellulose is regenerated in the form of long filaments. An alternative method of regeneration was discovered by the British chemists Charles Cross and Edward Bevan in 1891, when they found that cellulose would dissolve in a solution of sodium hydroxide and carbon disulphide to form, on ageing, a thick, golden-yellow, viscous liquid called **viscose**. This dithiocarbonate is hydrolysed by acid to regenerate the cellulose, the process being known as the **viscose process** and the commercially important product as **viscose rayon**:

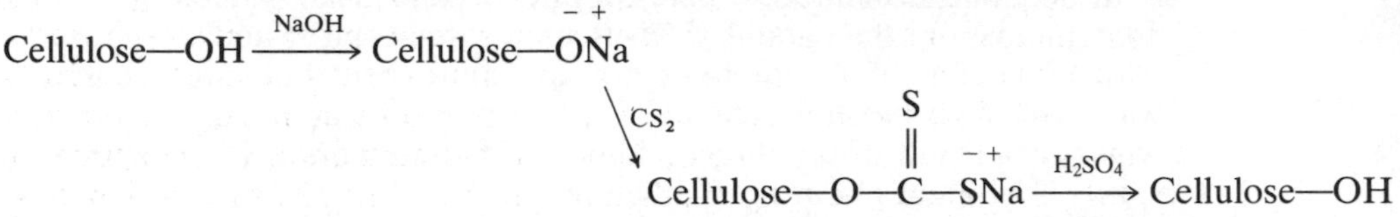

$$\text{Cellulose—OH} \xrightarrow{\text{NaOH}} \text{Cellulose—O}^-\text{Na}^+$$

$$\Bigg\downarrow {\scriptstyle CS_2}$$

$$\text{Cellulose—O—}\overset{\displaystyle S}{\overset{\displaystyle \|}{C}}\text{—S}^-\text{Na}^+ \xrightarrow{H_2SO_4} \text{Cellulose—OH}$$

When sheets of the material are produced, *cellophane* is the result (*glycerol* is added as a plasticizer to make it flexible).

Cross and Bevan also discovered that cellulose reacts with ethanoic acid and ethanoic anhydride, in the presence of sulphuric acid, to produce the ester cellulose triethanoate (Tricel), which is precipitated out of solution by the addition of dilute ethanoic acid:

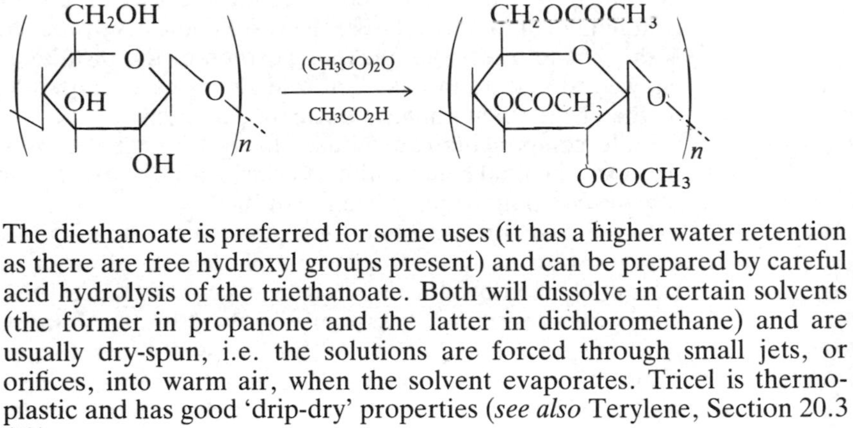

The diethanoate is preferred for some uses (it has a higher water retention as there are free hydroxyl groups present) and can be prepared by careful acid hydrolysis of the triethanoate. Both will dissolve in certain solvents (the former in propanone and the latter in dichloromethane) and are usually dry-spun, i.e. the solutions are forced through small jets, or orifices, into warm air, when the solvent evaporates. Tricel is thermoplastic and has good 'drip-dry' properties (*see also* Terylene, Section 20.3 (ii)).

Proteins

17.7　Introduction

Proteins, unlike carbohydrates, contain nitrogen as well as carbon, hydrogen and oxygen; sulphur is also usually present in small amounts.

Simple proteins can be roughly divided into two classes. The **fibrous proteins** are insoluble in water and are analogous to cellulose in plants, acting as structural material in animals. They include fibroin (in silk), collagen (in skin, cartilage and tendon bone) and keratin (in hair, nails and wool). The **globular proteins** have some degree of solubility in water, or in acids and bases, although the 'solutions' produced are really colloidal suspensions and not true solutions. Egg albumin, casein (in milk) and the proteins of blood plasma are examples of this class.

The more complex *conjugated* proteins, which contain protein combined with non-proteinaceous material, such as the blood protein haemoglobin which carries oxygen through the blood, are not considered here.

The proteins are biological polymers of immense complexity in which the monomer is an α-amino-acid. There are about 26 such amino-acids which are found in proteins and they are condensed together in long polyamide, or **polypeptide**, chains. When only a small number of units is present, the compound is sometimes referred to as a peptide (or polypeptide) rather than a protein, but the distinction is an imprecise, rather arbitrary one:

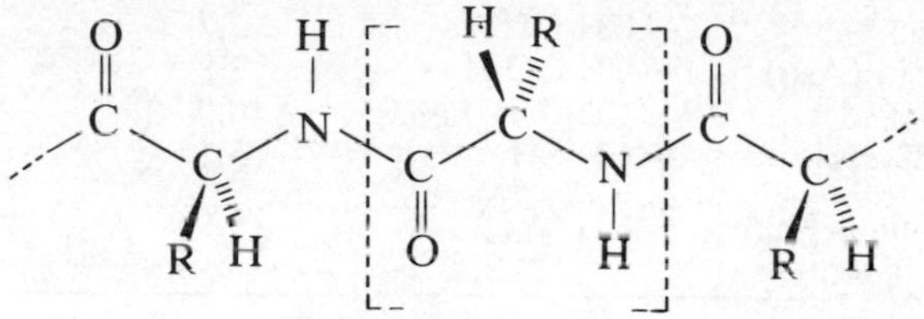

amino-acid unit

Some data on proteins are given in *Table 17.1*.

Table 17.1

Protein	No. of amino-acid units	Rel.mol.mass
insulin	51	5 700
lysozyme	119	14 400
myoglobin	153	17 000
trypsin	180	23 800

17.8 Chemistry of amino-acids

α-Amino-acids contain both an amino group —NH_2 and a carboxyl group —CO_2H attached to the same carbon atom and, with the exception of glycine $H_2NCH_2CO_2H$, this carbon atom is asymmetric so that the amino-acids can exist in enantiomeric forms. All the amino-acids which have been isolated from protein have proved to have the L-configuration (although their optical rotation may be $(+)$ or $(-)$, *see* Section 16.3.1). About 26 α-amino-acids have been identified in protein and many of these can be synthesized by the human body. Eight, however, cannot be synthesized by the human body and have to be taken in via proteins in the diet; they are termed the **essential amino-acids**. The α-amino-acids can be represented by the general formula $H_2NCHRCO_2H$, where the nature of the side-chain R can vary considerably, from a simple methyl group in alanine to a more complex cyclic structure in histidine. Although not proper chemical symbols, the first three letters of the name of an amino-acid are frequently used as short-hand, particularly in protein structures. Data on some selected, simple amino-acids are given in *Table 17.2.*

Table 17.2

Name	R group	Isoelectric point*
glycine (Gly)	—H	6.0
alanine (Ala)	—CH$_3$	6.0
serine (Ser)	—CH$_2$OH	5.7
aspartic acid (Asp)	—CH$_2$CO$_2$H	2.8
lysine (Lys)	—CH$_2$(CH$_2$)$_3$NH$_2$	10.0
cysteine (CysH)	—CH$_2$SH	5.1
phenylalanine (Phe)	—CH$_2$⟨◯⟩	5.5

* *See* following section.

17.8.1   Amphoteric nature of amino-acids

Most of the amino-acids are white, crystalline solids which are soluble in water. Since they contain both amino and carboxyl groups they can, under suitable conditions, behave as either acids or bases, i.e. display amphoteric behaviour:

$$H_2NCHRCOOH + H_3O^+ \rightleftharpoons H_3\overset{+}{N}CHRCOOH + H_2O$$
$$H_2NCHRCOOH + OH^- \rightleftharpoons H_2NCHRCOO^- + H_2O$$

At some intermediate pH value, a dipolar, or **zwitterion**, form is produced, where the hydrogen ion from the carboxyl group is transferred, not to the solvent (water) but to the basic amino group within the molecule:

$$H_2NCHRCOO^- \underset{OH^-}{\rightleftharpoons} H_3\overset{+}{N}CHRCOO^- \overset{H^+}{\rightleftharpoons} H_3\overset{+}{N}CHRCOOH$$

The pH value at this point is called the **isoelectric point**, because at it the dipolar form has no net charge. The actual value of the isoelectric point varies, depending on the nature of the R side-chain in the amino-acid; when the side-chain is neutral the value is about 6 but, for example, it is 2.8 for aspartic acid and 10 for lysine. Great use of this is made in a method of separating mixtures of amino-acids (usually derived from the hydrolysis of proteins), called **electrophoresis**. A thin strip of filter paper soaked in buffer solution has a drop of the mixture spotted on the centre and a voltage is then applied across the paper by means of electrodes. Depending on the pH of the buffer solution, some amino-acids will be in anionic form (if the pH is greater than their isoelectric point) and will therefore migrate towards the anode, some will be in cationic form (if the pH is less than their isoelectric point) and will migrate towards the cathode, while some may not migrate at all (if the pH is equal to their isoelectric point). Following the electrophoresis, the positions of the amino-acids on the paper can be located by spraying with a suitable reagent (ninhydrin, *see* Section 21.2.4(iii)), and compared with those of known standards run under the same conditions.

In the solid state, the amino-acids exist in their dipolar, or zwitterion, form and consequently have unusually high melting points for organic compounds (e.g. Gly m.p. $\approx 240°C$) and are insoluble in non-polar solvents.

17.8.2 Reactions of the amino group

(i) As outlined above, amino-acids can behave as bases by reacting with acids to form salts, e.g.

$$H_2NCH_2COOH + HCl \rightarrow H_3\overset{+}{N}CH_2COOH\ Cl^-$$

(ii) The amino group reacts with nitrous acid in a similar manner to a primary aliphatic amine, e.g.

$$H_2NCH_2CO_2H + HONO \rightarrow HOCH_2COOH + N_2 + H_2O$$

Since nitrogen is evolved quantitatively, the amino group can be estimated in this way in analysis.

(iii) Amino-acids take part in a nucleophilic substitution reaction with the aromatic compound 1-fluoro-2,4-dinitrobenzene in alkaline solution, e.g.

$$O_2N\!\!\left\langle\bigcirc\right\rangle\!\!\overset{NO_2}{F} + H_2NCH_2COOH \longrightarrow O_2N\!\!\left\langle\bigcirc\right\rangle\!\!\overset{NO_2}{NHCH_2COOH} + HF$$

The product has a characteristic yellow colour and the reaction is of great importance in determining the N-terminal amino-acid in a peptide chain (*see* Section 17.9).

(iv) Amino-acids will react with acid chlorides or anhydrides in a manner similar to amines, e.g.

$$H_2NCH_2COOH + RCOCl \rightarrow RCONHCH_2COOH + HCl$$
$$\text{(or } (RCO)_2O) \qquad\qquad\qquad\qquad \text{(or RCOOH)}$$

17.8.3   Reactions of the carboxyl group

(i) Amino-acids can be esterified by the Fischer–Speier method, e.g.

$$H_2NCH_2COOH + CH_3CH_2OH \xrightarrow{HCl} \overset{+}{H_3}NCH_2COOCH_2CH_3 + Cl^- + H_2O$$

(ii) The carboxyl group of one amino-acid molecule can be condensed with the amino group of another molecule and so on, until a polyamide or peptide chain is built up:

$$\ldots + H_2NCHRCOOH + H_2NCHR'COOH + \ldots$$
$$\longrightarrow \ldots -HNCHRCONH\,CHR'CO- \ldots$$

A carboxylic acid is usually condensed with an amine in the laboratory via the acid chloride (*see* Section 11.3.2(iii)), but in the case of amino-acids, a wide range of special reagents and methods has been developed to overcome the many problems associated with peptide synthesis. (Note that, in the case of the polyamide fibre nylon, the condensation is brought about via a salt, *see* Section 20.3(i)).

Proteins and peptides, as polyamides, give the **biuret** test. When a little dilute sodium hydroxide solution is added to a compound containing the —CONH— linkage, followed by a drop of very dilute copper(II) sulphate solution, a deep pink (to purple) coloration appears. Biuret itself is formed when carbamide is heated strongly above its melting point and the residue allowed to solidify:

$$H_2NCONH_2 + H_2NCONH_2 \xrightarrow{heat} H_2NCONHCONH_2 + NH_3$$
$$\text{biuret}$$

17.9   Structure of proteins

Proteins are organic polymers of α-amino-acids with extremely high relative molecular masses. They are difficult to purify, and hence to characterize, and the techniques used include chromatography, electrophoresis and dialysis. In dialysis, the crude protein is placed inside a permeable membrane such as cellophane, or visking tubing, surrounded by water. Small molecules or particles, such as those of salts and sugars, diffuse slowly through the membrane into the water, while the much larger protein molecules remain behind (cf. 'molecular sieves', *see* Section 2.3). Estimates of the relative molecular masses of proteins may be derived from certain colligative property measurements, such as osmotic pressure, from turbidity (light scattering) measurements and from ultracentrifugation (where the relative molecular mass determines the rate of sedimentation of the protein in an ultracentrifuge).

In proteins the amino-acids are condensed together via —CONH— (peptide) linkages and their linear sequence constitutes what is called the **primary structure** of the protein. Since the acids in the sequence are usually represented by symbols based on the first three letters of their names, it is important to appreciate that the convention used is that such a symbol represents the molecule with the amino group on the left and the carboxyl group on the right. Thus, in a dipeptide consisting of the amino-acids glycine and alanine, there are *two* possible sequences:

$$CH_3$$
$$|$$
Gly.Ala   i.e. $H_2NCH_2CONHCHCOOH$

and CH_3
$$|$$
Ala.Gly   i.e. $H_2NCHCONHCH_2COOH$

In the case of proteins and larger peptides, the molecular chain itself is not linear but is arranged in the form of a coiled helix. The precise conformation of the helix is determined by the amino-acids present and the nature of their side-chains; hydrogen bonding and electronic interactions between functional groups play an important part in establishing this **secondary structure**. Even the helix, however, is usually twisted, coiled and folded in space to give the overall molecular shape, or **tertiary structure**, which is also maintained by intermolecular forces and is often influenced by the presence of water within the protein structure (*see Figure 17.1*).

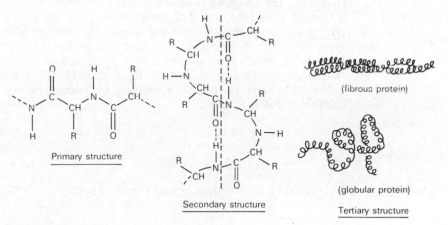

Primary structure

Secondary structure

(fibrous protein)

(globular protein)

Tertiary structure

Figure 17.1 Structural organization of proteins

In fibrous proteins, such as keratin, the helical chains are obviously twisted round each other in a thread-like manner, whereas the globular proteins, like egg or blood serum albumin, consist of heavily folded helical chains.

The actual three-dimensional organization of protein structure is of critical importance in effecting the catalytic action of enzymes (*see* Section

17.10.2). In the process of **denaturation** of a protein, all or part of the tertiary structure is lost and the secondary structure is also affected. This is because intermolecular forces, but not the covalent bonds of the primary structure, are affected. Denaturation can be brought about by the action of heat, as when albumin precipitates from heated egg-white, and by acids and bases (pH), as when casein is precipitated on acidification of milk. Certain enzymes can also denature proteins.

The **determination of protein structure** involves the identification of the sequence of amino-acids in the molecular chain(s) and the three-dimensional conformation of the molecule as a whole. The former generally depends on a combination of total and selective hydrolyses, performed on samples of the protein. If a protein or peptide is boiled with concentrated hydrochloric acid for several hours, the peptide, or amide, bonds are cleaved and the individual amino-acids are released. These are usually identified by the use of two-way paper chromatography (*see* Section 21.2.4(iii)) or ion-exchange chromatography (*see* Section 21.2.4(v)). The relative amounts are also determined. These are established routinely by a fully automated 'amino-acid analyser' which permits protein hydrolysis, separation of amino-acids in the hydrolysate, and identification of the relative amounts present by reacting them with *ninhydrin* in solution; with amino-acids, this reagent generally gives a blue colour, the intensity of which can be correlated with the concentration of the amino-acid in the solution.

The next step is to identify the two amino-acids at the end of a molecular chain, i.e. the so-called *N-terminal* and *C-terminal* acids. If the material is treated with 1-fluoro-2,4-dinitrobenzene (FDNB) under alkaline conditions at about 45°C for several hours, the acid with the free amino group, i.e. the N-terminal acid, reacts with this reagent (*see* Section 17.8.2(iii)). When the material is then subjected to total hydrolysis, the N-terminal acid is liberated in the form of its FDNB derivative which is yellow. This can be readily identified by paper chromatography (*see* Section 21.2.4(iii)), e.g.

The C-terminal acid can be identified by heating the material with hydrazine N_2H_4, when all the peptide bonds are broken and all the amino-acids *except* the C-terminal one are released as their hydrazides. The C-terminal acid can be identified by paper chromatography again. Alternatively, enzymatic hydrolysis may sometimes be used; enzymes like the carboxypeptidases often catalyse the release of the C-terminal acid only.

Total hydrolysis of a protein or peptide allows the total amino-acid content to be determined, together with the molar ratios in which the amino-acids are present. In order to determine the sequence of the amino-acids, i.e. the primary structure, the material has to be submitted to selective partial hydrolyses. These will give rise to a number of smaller peptides which can be separated and analysed for their own amino-acid sequence. Once the molecular fragments have been identified, they have to be reassembled to provide the total amino-acid sequence. Suppose, in the following hypothetical example, a tripeptide yielded the amino-acids glycine, alanine and serine in an equimolar ratio. The possible structures would be:

Gly.Ala.Ser   Ala.Gly.Ser   Ser.Ala.Gly
Gly.Ser.Ala   Ala.Ser.Gly   Ser.Gly.Ala

Suppose the N-terminal acid was found to be glycine. Then the only structures possible would be:

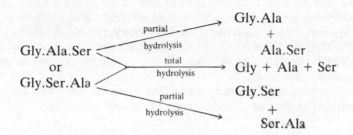

If the C-terminal acid was also identified (this is not always so easy), then there would be only one possible structure. In general, a study of the fragments of the partial hydrolysis of a peptide allows an unambiguous sequence to be determined.

This account is, of course, simplified and there are many, very difficult problems to be overcome in determining the primary sequence of a protein. It took the chemist Frederick Sanger about 10 years to determine the primary structure of the peptide hormone insulin, which contains 51 amino-acid units in the form of two peptide chains cross-linked by two sulphur atoms (a disulphide bridge).

The determination of the complete tertiary structure of a protein is a daunting task which involves the application of specialized X-ray diffraction techniques to crystalline material. It was first achieved by the chemist John Kendrew, in 1957, for the protein myoglobin ($M_r = 17\,000$), which is the oxygen-carrying protein in muscle. At roughly the same time, a broad picture of the tertiary structure of the much more complex haemoglobin ($M_r = 64\,500$), the oxygen-carrying protein in the blood, emerged from over twenty years' study of the protein by Max Perutz. In 1962, Kendrew and Perutz received the Nobel Prize for Chemistry in recognition of their work. Since that time, fairly detailed pictures of the molecular structures of other proteins (including enzymes) have been arrived at. The synthesis of proteins is now one of the greatest challenges to the organic chemist.

17.10 Enzymes

17.10.1 Properties of enzymes

Enzymes are proteins which act as biological catalysts for reactions occurring in all living organisms. Like other catalysts, they speed up the rates of reactions enormously. The enzyme *catalase*, for example, which catalyses the decomposition of hydrogen peroxide and hence prevents the accumulation of this toxic material in living systems, speeds up this reaction by a factor of the order of 10^7 compared with the rate of the uncatalysed reaction. Thus, enzymes are generally extremely efficient catalysts. A compound whose reaction is catalysed by an enzyme is known as that enzyme's **substrate**. In general, enzymes display a very high degree of **specificity** in their catalytic action, i.e. they will usually act only on one specific substrate, although there are some enzymes which will catalyse one particular *type* of reaction. Thus, the enzyme *urease* will act on *urea* (carbamide) but not on *methylurea* (methylcarbamide). Some are even more specific. The enzyme *lactic dehydrogenase*, which catalyses the oxidation of 2-hydroxypropanoic acid (*lactic acid*) $CH_3CH(OH)COOH$ to 2-oxopropanoic acid (*pyruvic acid*) $CH_3COCOOH$, will act only on the L-isomer and not at all on the D-isomer.

Enzymes are generally named by prefixing *-ase* with the name of the enzyme's substrate. Some examples are given in *Table 17.3*.

Table 17.3

Enzyme	Source	M_r	Reaction catalysed
lysozyme	egg-white	14 000	hydrolysis of polysaccharide
catalase	liver	250 000	$2H_2O_2 \rightarrow 2H_2O + O_2$
urease	soya bean	480 000	$H_2NCONH_2 + H_2O \rightarrow CO_2 + 2NH_3$
amylase	saliva	—	starch $\rightarrow$ maltose
sucrase	yeast	—	sucrose $\rightarrow$ glucose + fructose

In some cases, a non-protein compound (with M_r of the order of hundreds only) must be combined with the enzyme in order for it to display catalytic activity. Such a compound, which has no catalytic properties of its own, is known as a **co-enzyme**, or co-factor. The most outstanding general features of enzymic catalysis are the specificity and efficiency of the catalysis, and the very mild conditions under which it takes place. For example, various proteolytic enzymes will catalyse the cleavage of peptide chains in proteins (often at specific points) at room temperature in aqueous solution at pH close to 7. In contrast, the chemist has to heat a peptide with 6 molar HCl under reflux for several hours, in order to bring about cleavage of the peptide bonds and, even then, the cleavage is entirely unspecific.

17.10.2 Mode of action

In general terms, the high activity and specificity of an enzyme are thought to be due to the conformation, or three-dimensional shape, of a region of the enzyme molecule known as the **active site**. According to Emil Fischer's 'lock and key' hypothesis (1894) of enzyme action, there is a particular three-dimensional fit for the enzyme and substrate molecules, analogous to a lock-and-key situation, whereby the enzyme holds the substrate molecule in a position which facilitates reaction (*see Figure 17.2*).

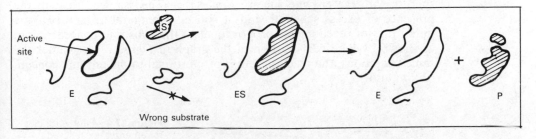

Figure 17.2 The lock-and-key mechanism of enzyme action

Although the active site is the region of chemical reaction, the tertiary structure of the enzyme can, nevertheless, influence its activity. More recent modifications of the theory consider an 'induced fit' of substrate and enzyme also to be possible, in which the enzyme is made to adopt a particular conformation about the substrate. Whatever the specific mode of action, an enzyme lowers the activation energy of the reaction it is catalysing (*see Figure 17.3*). The enzyme and substrate interact via the active site to form an enzyme–substrate complex (the formation of which will require a very low activation energy), in which the two are sometimes covalently bound. This complex may then break down into either reactants

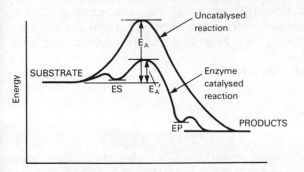

Substrate	Enzyme	Activation energies	
S	E	E_A	E_A'
		$(kJ \ mol^{-1})$	$(kJ \ mol^{-1})$
H_2NCONH_2	urease	137	37
H_2O_2	catalase	79	23

(*See also Table* 17.3)

Figure 17.3 Energy profile for an enzyme-catalysed reaction

or products. In general, it is assumed that little, if any, recombination of product and enzyme takes place:

$$E \; + \; S \; \underset{k_{-1}}{\overset{k_1}{\rightleftharpoons}} \; ES \; \overset{k_2}{\rightarrow} E + \; product(s)$$

enzyme   substrate   complex

Evidence for these general views on the mode of action of enzymes comes from a variety of sources, including kinetic studies. In 1913, L. Michaelis and M. L. Menten studied the kinetics of the hydrolysis of sucrose with the enzyme sucrase (*invertase*). They showed that, in the presence of excess substrate and at constant temperature and pH, the initial rate of reaction was proportional to the substrate concentration, as shown by the linear portion of the graph in *Figure 17.4*. (Under the same conditions, the reaction rate is proportional to the enzyme concentration also.)

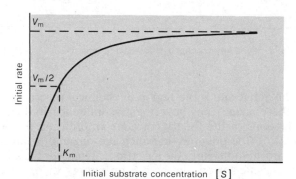

Figure 17.4   Rate v. [S] plot for an enzyme-catalysed reaction

Initial substrate concentration [S]

However, with increasing substrate concentration there comes a point where the linear relationship breaks down and the rate of reaction becomes independent of the substrate concentration. This is explained by assuming that a stage is reached when all the active sites of the enzyme molecules are temporarily occupied, so that increasing the concentration of the substrate in solution does not increase the reaction rate, as the enzyme and substrate remain linked for a finite time before decomposing into product.

The Michaelis–Menten equation

$$Rate = \frac{V_m[S]}{K_m + [S]}$$

summarizes these experimental results and can also be derived from simple kinetic arguments ($K_m = (k_{-1} + k_2)/k_1$). V_m represents the maximum reaction rate, under the given conditions of temperature and pH, and [S] the initial substrate concentration. K_m, the Michaelis constant, represents the substrate concentration at which the reaction is occurring at exactly half its maximum rate, V_m.

17.10.3 Enzyme inhibition

More information on the mode of action of enzymes is derived from studies of enzyme inhibition. The activity of an enzyme may be inhibited by molecules which can occupy the active site and so prevent the enzyme from acting on its substrate. An important example of such inhibition is the mode of action of the 'sulpha drug' 4-aminobenzenesulphonamide (*sulphanilamide*, *see* Section 14.4), which is very effective against certain disease-causing bacteria. These bacteria require the compound 4-aminobenzenecarboxylic acid in the growth medium for normal growth. If the corresponding sulphonamide is present in excess, however, it can compete with the acid for the active site of an enzyme involved in the metabolism of the bacteria, because the molecules are so similar:

Since the bacteria cannot utilize the sulphonamide in their metabolic growth, they are deprived of the essential acid and hence die. This is an example of **competitive inhibition** and it is also **reversible**, i.e. the sulphonamide could be displaced from the active sites by a large excess of the carboxylic acid.

The mode of action of certain organophosphorus compounds, originally studied as nerve gases and later as insecticides, illustrates **irreversible inhibition**. The compound di-(1-methylethyl)fluorophosphate reacts chemically with the hydroxyl side-chain of the amino-acid serine, present in an enzyme like *cholinesterase*, to form a strong covalent bond in an irreversible step. Since this particular enzyme is involved in the transmission of nerve impulses, inhibition of its action leads to breakdown of the normal pattern of transmission:

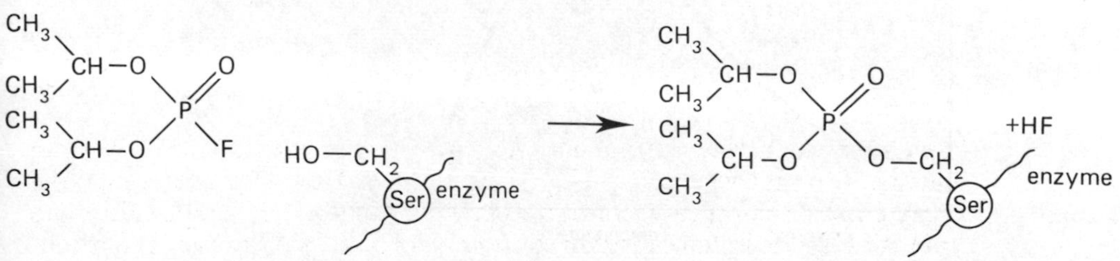

Another practical use of enzyme inhibition has been the treatment of chronic alcoholism by the drug *disulfiram*. In the body, ethanol is normally oxidized first to ethanal, and then to ethanoic acid, in two enzymically controlled reactions. The oxidation of ethanal to ethanoic acid is thought to require the enzyme *alcohol dehydrogenase* and a co-enzyme. It seems that the drug competes with the co-enzyme for the active site and hence inhibits the enzyme's action. Consequently, ethanal builds up in the body and symptoms of extreme nausea, etc., soon develop. These are exploited in the aversion therapy of alcoholics.

17.10.4   Denaturation

As implied in the section on enzyme kinetics, temperature and pH also have considerable influence on the activities of enzymes. Most enzymes function satisfactorily only over a relatively narrow pH range of the order

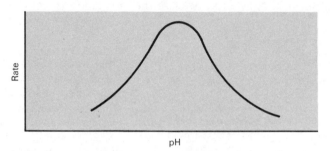

Figure 17.5   Rate dependence on pH for an enzyme-catalysed reaction

3–9, although pepsin, which aids the breakdown of protein in the digestive juices of the stomach, is active at pH ~ 2. There is usually an optimum value at a fixed temperature at which an enzyme's activity is greatest (*see Figure 17.5*). At low pH values, basic sites in the enzyme molecule become protonated, e.g. $-NH_2 \rightarrow \overset{+}{N}H_3$ and $-COO^- \rightarrow -COOH$, while at high pH values deprotonation of ionizable groups takes place. The balance of electronic interactions between groups and side-chains is thus disturbed and both the active site and the secondary and tertiary structures can be affected.

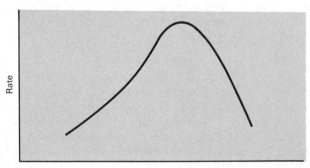

Figure 17.6   Rate dependence on temperature for an enzyme-catalysed reaction

Although an increase in temperature generally increases the rate of a reaction, enzyme-controlled reactions show a more complex dependence on temperature (*see Figure 17.6*). Initially, an increase in temperature increases the reaction rate in the normal, non-linear way, but gradually this trend is offset by the denaturation of the enzyme as the heat begins to break down intermolecular bonds and the tertiary structure begins to collapse. There is thus an optimum temperature for enzyme activity of about 37°C. However, not all enzymes are so sensitive to temperature and some can function efficiently at temperatures well in excess of this figure. Thus, the enzyme subtilisin, extracted from the bacterium *bacillus subtilis* and employed in the 'enzyme detergents' (introduced in the 1960s) for the breakdown of proteinaceous 'biological stains', retains its activity even at 60°C.

Enzymes can also be deactivated by certain metal ions, such as Hg^{2+} and Pb^{2+}, which can react with —SH groups.

17.10.5 Uses of enzymes

An enzyme (*urease*) was first isolated in a pure state in 1926, but it was not until 1965 that the complete structure of an enzyme was determined. The enzyme was lysozyme, isolated from egg-white, and David Phillips and his workers showed that it consisted of a single chain of 129 amino-acid units (20 different kinds). This enzyme catalyses the splitting of polysaccharide material in bacterial cell walls, and exhaustive studies (including X-ray diffraction work) have now revealed the whole tertiary structure and much detail of the actual active site, which appears to be a deep cleft within the molecule. Other enzyme structures have since been determined but, although ribonuclease was synthesized in the laboratory in 1969, the first *total* synthesis of an enzyme has yet to be accomplished.

Much of this exploratory work has suggested that it may be possible to synthesize much smaller molecules which, if they still contain the correct active site, should display catalytic activity, and this is one approach to making use of enzymes. Some enzyme preparations have been used in industry for years, of course, as in brewing, the food industry and the manufacture of antibiotics. Although the remarkable efficiency and specificity of enzymes have always been attractive to organic chemists, their use has been tempered by their instability to certain conditions of pH and temperature. Recent technological developments, however, mean that the use of enzymes, particularly from bacterial sources, will assume greater significance in the future. Enzymes can now be stabilized, either by bonding them to a solid polymer support, or else by incorporating them into gels which are permeable to substrates. These *immobilized* enzymes can be used in chemical processes with the advantages that no difficult separation of products and enzyme is necessary and that the process can be operated as a continuous, rather than batch, process. Possible future uses include the manufacture of glucose and fructose from cellulosic material like sawdust and waste paper, the specific oxidation of alkanes into alcohols, and the preparation of various foods. A comparison of the laboratory/industrial conditions (500°C and 20 MN m^{-2} pressure) and biological conditions (the enzyme nitrogenase in the nitrogen-fixing bacteria *azotobacter*) for the production of ammonia from nitrogen indicates that an enzyme-catalysed manufacturing process could be of immense economic importance.

17.11   Nucleic acids and protein biosynthesis

The **nucleic acids** are giant condensation polymers containing a series of **nucleotide** units. A nucleotide contains the sequence phosphate–sugar–base (*see Figure 17.7*). The bases to be found in nucleic acids are of the *purine* or *pyrimidine* classes and are generally limited to *adenine, cytosine, thymine, guanine* and *uracil*. The sugars are *ribose* or *deoxyribose* (which lacks an oxygen atom at the C_2 position of ribose).

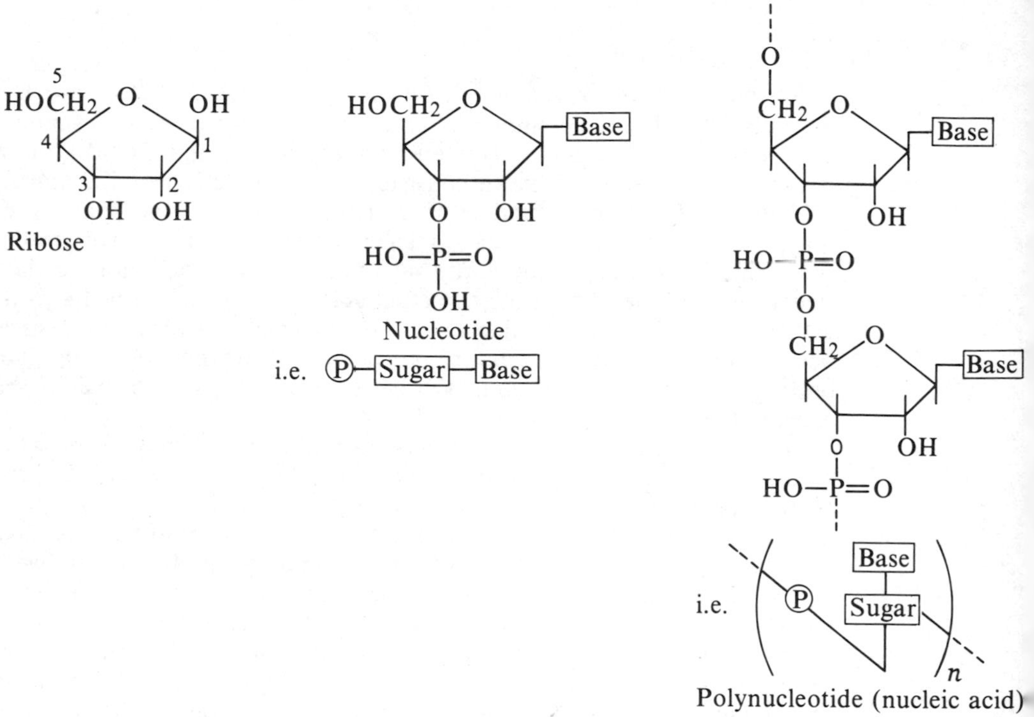

Figure 17.7   Molecular sequences of nucleotides and polynucleotides

Nucleic acids containing ribose are called **ribonucleic acids (RNA)**; those which contain deoxyribose are called **deoxyribonucleic acids (DNA)**. DNA is the most familiar nucleic acid and has an M_r of about 10 million, corresponding to about 300 000 nucleotide units. As long ago as the 1940s, it was recognized that DNA, found in the chromosomes within the cell

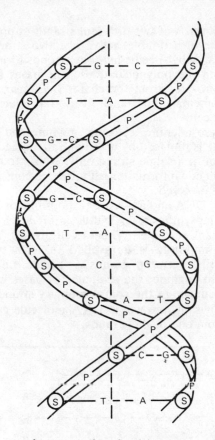

Figure 17.8 The double helix of DNA (P, phosphate; S, sugar; A, adenine—pairs with T, thymine; C, cytosine—pairs with G, guanine)

nucleus, was involved in the storage and transmission of genetic information (*inheritance*). The full, immensely complex structure of DNA was finally worked out following the suggestion, by the molecular biologists James Watson and Francis Crick, that the molecule has a double-helix structure (*see Figure 17.8*); Watson, Crick and Maurice Wilkins (who carried out X-ray work on DNA) were awarded the 1962 Nobel Prize for Medicine for this work.

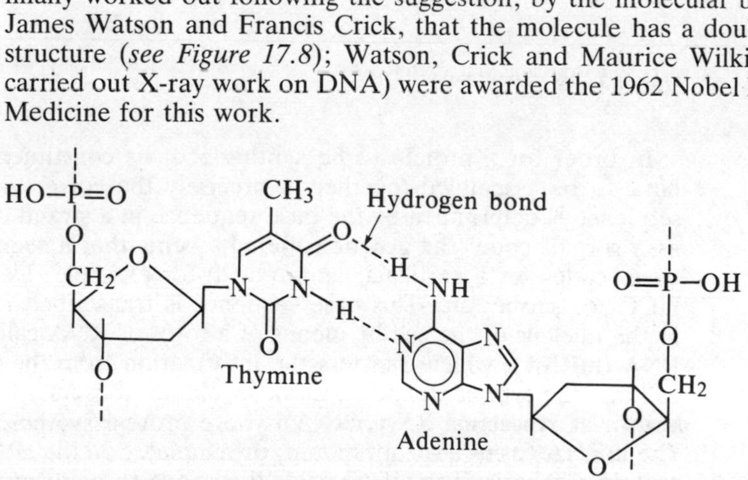

Figure 17.9 Example of base-pairing by hydrogen bonds

The DNA molecule consists of two different strands of polynucleotide intertwined in the shape of a double helix. The two strands are held together in a particular way by specific hydrogen bonds formed between specific pairs of bases in the polynucleotides. The bases face inwards while the sugars face outwards from the chains (*see Figure 17.9*). This base-pairing is of fundamental importance in the reproduction, or **replication**, of DNA in the cell.

In DNA, the bases are adenine, cytosine, guanine and thymine. In RNA, however, thymine is replaced by uracil and the molecule consists, *not* of a double helix but of a single strand only. RNA, too, is involved in the transmission of genetic information although, in contrast to DNA, three different sorts are involved.

In their various forms, DNA and RNA are responsible for the synthesis in biological systems (biosynthesis, or synthesis *in vivo*) of proteins, including the all-important enzymes. When a cell divides (**mitosis**), the double-stranded DNA molecule is unravelled to form two separate strands. Since base-pairing between strands is specific, i.e. adenine to thymine, and cytosine to guanine, the sequence of bases in one strand determines the base sequence in the complementary strand (*see Figure 17.10*). Thus, on cell division, the whole DNA molecule can be reproduced, or **replicated**, from one of the strands,

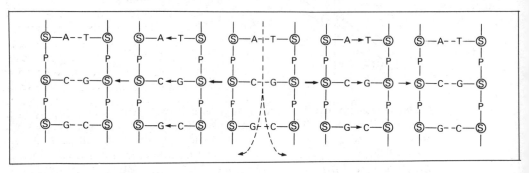

Figure 17.10   Replication of DNA

In order for a protein to be synthesized, its constituent amino-acids have to be assembled together in precisely the correct sequence. This sequence is determined by the base sequence in a strand of DNA acting as a sort of code (**the genetic code**). It seems that a sequence of three bases codes for a particular amino-acid; for example, UGU for valine, UCG for serine, etc. This base sequence is transcribed from the DNA in the nucleus of the cell by means of a form of RNA called **messenger-RNA (mRNA)**, which transfers the information from the nucleus of the cell into the cytoplasm and on to the insoluble ribosomes (made of RNA, known as **ribosomal RNA, rRNA**) where protein synthesis takes place. The mRNA forms a sort of mould, or template, on the rRNA, ready for protein synthesis. The amino-acids then have to be assembled from the cytoplasm. This is achieved with the aid of yet another kind of RNA,

soluble-RNA (sRNA, also sometimes referred to as *transfer*-RNA), which picks up an activated amino-acid in the cell (there must therefore be at least twenty different sRNAs) and transports it to the template or mRNA for it to be incorporated in the correct sequence. When the protein has been assembled, the unstable mRNA breaks down into nucleotides, while the stable sRNA remains in solution in the cytoplasm (*see Figure 17.11*).

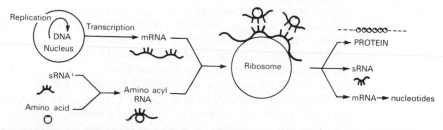

Figure 17.11   Protein synthesis in the cell

Thus, it is the DNA in the nucleus of cells which carries the essential genetic code for producing the proteins and enzymes required by a living organism. If a chemical change in the base sequence of DNA is brought about, i.e. if a **mutation** takes place, then the precise amino-acid sequence of a protein or enzyme will be altered. This can have serious consequences. For example, the substitution of one valine unit for one particular glutamic acid unit in haemoglobin's 574 amino-acid units causes the genetically inherited blood disorder called 'sickle-cell' anaemia. The inactivation of an enzyme by the same sort of process causes the condition known as phenylketonuria. The amino-acid phenylalanine is normally oxidized to tyrosine with the aid of a particular enzyme. If this enzyme is inactive in an individual, then the phenylalanine is partially converted into 'phenylketone' which is excreted in the urine. The accumulation of phenylalanine in the blood of young children can lead to mental retardation, unless this amino-acid is rigorously excluded from the diet.

17.12 Origin of life

With so much known about the chemistry of life processes, scientists have been tempted to speculate about ways in which life on earth could have evolved from a collection of simple chemicals, on the premise that life appeared as a result of processes controlled only by the established laws of physics and chemistry. Although theories of life's *abiological* origins are unlikely to be proved in the normal sense, i.e. demonstrated to have taken place, many of their postulates can be tested experimentally.

The starting point for chemical evolution is taken to be a primitive earth atmosphere of a strongly *reducing* kind, subject to a supply of energy from such sources as heat, ultraviolet light, radioactivity and lightning (electrical discharge). In 1953, the American chemist Stanley

Miller subjected a gaseous mixture of CH_4, NH_3, H_2 and H_2O to a continuous electrical discharge for many hours. Initially, compounds such as HCN and HCHO were formed, but gradually these disappeared, and in the final products Miller detected several amino-acids, which are found in proteins, together with other simple organic acids. Other experiments by the American Nobel Prize-winner Melvin Calvin and his workers, using electron bombardment as the energy source, have produced a wider range of products including the base adenine. Thus, the plausibility of some of the ideas of chemical evolution have been reinforced by experiment. Other work has shown how carbohydrates could be formed by alkaline condensations of methanal HCHO and how condensations of sugars and amino-acids into polysaccharides and peptides, respectively, could be brought about by HCN. The more difficult problems of the origin of the asymmetry of biological molecules, and of protein synthesis, etc., are under active investigation at the present time.

Questions

1* (a) (i) Write down the molecular formula of glucose. Hence write down the molecular formula of maltose, a disaccharide containing only glucose.
    (ii) To what class of macromolecules do glycogen and cellulose belong?
        State one chemical and one physical difference between glycogen and cellulose.
    (iii) Give the physiological function of glycogen and cellulose.
    (b) (i) Write an equation for the complete oxidation of glucose. The enthalpy of combustion of glucose as measured by bomb calorimetry is 2826 kJ mol^{-1}. If 1 g of glucose is oxidized completely to carbon dioxide and water by muscle, the measured heat output is 10 kJ. Account for this value.
    (ii) State one major difference between the mechanism of oxidation in a bomb calorimeter and in a living cell.
        How is this difference achieved?
    (iii) In what location in muscle cells does the production of carbon dioxide mainly take place?   JMB

2   The hydrolysis of starch may be accomplished in a test-tube containing a 1% starch solution and the salivary enzyme amylase, placed in a water bath.
    (a) By what simple method could the progress of the hydrolysis be followed?
    (b) At approximately what temperature would you expect the hydrolysis to proceed most efficiently? Explain your answer.
    (c) In investigating the enzymic hydrolysis what control experiment could also be set up?
    (d) The hydrolysis of starch produces the disaccharide maltose. Describe a chemical test which could be carried out to demonstrate the presence of such a sugar in the solution.

(e) In what ways would the acid hydrolysis of the starch solution differ, both in terms of reaction conditions and product, from the enzymic hydrolysis?

3* (a) Give the molecular formula of each of the following sugars: (i) glucose, (ii) maltose, (iii) maltotriose (a trisaccharide formed from glucose).
    (b) A mixture of the sugars listed in (a) constitutes corn syrup. Would the boiling point of a solution of corn syrup, for which the average relative molecular mass of the sugar mixture is 410, be higher or lower than that of a solution of sucrose of the same concentration by mass? (Relative molecular mass of sucrose = 342.)
    (c) Give a brief explanation for your answer in (b).
    (d) The first stage in the manufacture of corn syrup involves heating a starch suspension with acid. Account for the changes in (i) viscosity and (ii) reducing power which occur during this process.   JMB

4* (a) (i) Name the naturally occurring material used in the manufacture of Viscose rayon.
        (ii) What is the chemical difference between Viscose rayon and Tricel?
       (iii) Describe the spinning process used to produce Tricel fibres.
    (b) What is the difference between the water absorption properties of Viscose rayon and those of Tricel? Explain the difference.
    (c) Water absorption is an important property of textiles and textile fibres.
        (i) Give **two** advantages of high water absorption.
       (ii) Give **one** disadvantage of high water absorption.
    (d) How may the water absorption of cotton fibres be reduced?
                                                                              JMB

5 **Either**
   Distinguish between a monosaccharide, a disaccharide and a polysaccharide, giving a specific example of each.
   Compare and contrast the reactions of glucose and of starch with
   (a) iodine,
   (b) alkaline $Cu^{2+}(aq)$ ('Fehling's solution'),
   (c) dilute aqueous sulphuric acid, and
   (d) ammoniacal silver nitrate ('Tollens's reagent').
   Or
   Distinguish between an amino-acid, a polypeptide and a protein, giving a specific example of each.
   By what chemical reactions is it possible to obtain
   (a) an amino-acid from a protein,
   (b) 2-aminopropanoic acid $(CH_3CH(NH_2)CO_2H)$ from propanoic acid $(CH_3CH_2CO_2H)$,
   (c) a polyamide such as nylon 66 from compounds containing only 6 carbon atoms per molecule?
   Ethanoic (*acetic*) acid and ethylamine are both liquids at ordinary temperature, but aminoethanoic acid (glycine) is a crystalline solid. What structural property of glycine explains this difference?   London

6 (a) The amino acid alanine, $CH_3CH(NH_2)COOH$, can be prepared from
       2-bromopropanoic acid by reaction with a large excess of ammonia.
       Write an equation for this reaction and indicate the mechanism.
       Why is it advisable to use a large excess of ammonia?
   (b) If an aqueous solution of alanine is electrolysed at pH 1.0, the amino
       acid migrates to the cathode, whereas at pH 13.0 it migrates to the
       anode. Write structures for the migrating species in each case
   (c) Samples of natural alanine and the synthetic compound behave dif-
       ferently when a solution of each is examined in plane-polarized light.
       State the difference in behaviour and give the reason for this
       difference.

                                                                      JMB

7* This question is about a substance whose trivial name is glutamic acid
    (relative molecular mass 147) and which has the structure

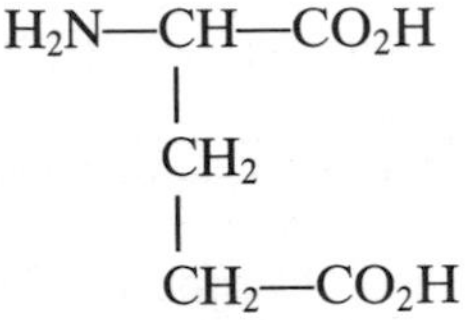

    H_2N—CH—CO_2H
            |
           CH_2
            |
           CH_2—CO_2H

   (a) (i) On the structural formula above, circle the asymmetric carbon
            atom.
       (ii) What physical property of the glutamic acid is due to the presence
            of an asymmetric carbon atom?
   (b) Draw the structural formula of the principal organic ions formed
       when glutamic acid is reacted with a slight excess of (i) hydrochloric
       acid, (ii) sodium hydroxide.
   (c) Using reactions chosen from the following list, devise a reaction
       scheme for converting glutamic acid into a trihydric alcohol (i.e. three
       OH groups per molecule):

       $RCl \xrightarrow{KCN} RCN$

       $RCN \xrightarrow{2H_2} RCH_2NH_2$

       $RCO_2H \xrightarrow{PCl_5} RCOCl$

       $RCO_2H \xrightarrow{LiAlH_4} RCH_2OH$

       $RCOCl \xrightarrow{NH_3} RCONH_2$

       $RCONH_2 \xrightarrow[\text{acid}]{\text{Dilute}} RCO_2H$

       $RNH_2 \xrightarrow{HNO_2} ROH$

(d) What volume of 1 M sodium hydroxide solution would be required
exactly to neutralize a solution containing 0.294 g of glutamic acid?

Nuffield

8** The table below gives data about a number of amino acids which occur
in proteins:

Name	Abbreviation	Relative molecular mass	R_f value in phenol	R_f value in butan-1-ol/ethanoic (acetic) acid
alanine	ala	89	0.43	0.38
aspartic acid	asp	133	0.13	0.24
glycine	gly	75	0.33	0.26
leucine	leu	131	0.66	0.73
lysine	lys	146	0.62	0.14
phenylalanine	phe	165	0.64	0.68
serine	ser	105	0.30	0.27
valine	val	117	0.58	0.60

A small polypeptide was hydrolysed with 6 M acid and the resulting amino
acids were separated by two-way chromatography. The chromatogram
is reproduced below:

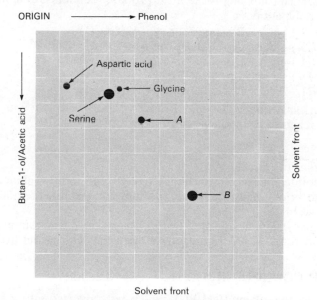

ORIGIN ⟶ Phenol

Butan-1-ol/Acetic acid

Aspartic acid
Glycine
Serine
A
B
Solvent front
Solvent front

Chromatogram of amino acids from the polypeptide

(a) Determine the R_f values in both solvents of the amino acids labelled
A and B on the chromatogram and identify them.

(b) The polypeptide was reacted with 1-fluoro-2,4-dinitrobenzene (FDNB). After hydrolysis, DNP glycine was identified by chromatography. What information does this give about the sequence of amino acids in the polypeptide?

(c) When 1.00 g of the polypeptide, whose molar mass was found to be 780 g mol^{-1}, was subjected to the action of carboxypeptidase for 12 h, the amino acids released were:

serine 2.5×10^{-3} mol
aspartic acid 1.2×10^{-3} mol
amino acid A 1.1×10^{-3} mol

What information does this result give about the sequence of amino acids in the polypeptide?

(d) Quantitative estimation of the amino acids in the polypeptide showed that they were present in the following molar ratio:

serine   3
aspartic acid   2
amino acid A   1
amino acid B   1
glycine   1

Using this information and information from the previous sections, which of the following is the most probable sequence of amino acids in the polypeptide?

(i)   (NH$_2$)-gly-ser-B-asp-ser-ser-asp-A-(CO$_2$H)
(ii) (NH$_2$)-gly-asp-A-ser-B-asp-ser-ser-(CO$_2$H)
(iii) (NH$_2$)-gly-ser-B-asp-A-asp-ser-ser-(CO$_2$H)
(iv) (NH$_2$)-ser-ser-asp-A-ser-B-asp-gly-(CO$_2$H)
(v)   (NH$_2$)-A-asp-ser-ser-ser-B-asp-gly-(CO$_2$H)

The correct sequence is .... Explain your reasoning for each of the sequences (i) to (v).

                                                                    Nuffield

9*   Answer *either* (a) or (b).
(a) A tripeptide X is known to contain one glycine and two alanine units. Indicate briefly how you would attempt to establish the amino-acid sequence in X.
    When glycylalanine is heated to 175°C a product Y is obtained which does not react with 2,4-dinitrofluorobenzene. On partial hydrolysis, Y gives two compounds. Explain.

    NH$_2$CH$_2$CONHCH(CH$_3$)CO$_2$H   (glycylalanine)

    How may the alkene (CH$_3$)$_2$CH=CH$_2$ be converted into leucine?

    (CH$_3$)$_2$CHCH$_2$CH(NH$_2$)CO$_2$H   (leucine)

(b) ....
                                                                    Cambridge

10* Answer **one** part of this question.
Either (a) Treatment of the dipeptide alanylglutamic acid with 2,4-dini-
trofluorobenzene gives yellow derivative (A). What is the structure of
this derivative, and what products would you expect after boiling the
derivative (A) with 6 M hydrochloric acid?

$$NH_2CH(CH_3)CONHCHCO_2H$$
$$|$$
$$CH_2CH_2CO_2H$$
    (alanylglutamic acid)

Explain why boiling of the tripeptide alanylglycylglutamine with 6 M
hydrochloric acid gives the amino acids glycine, alanine, and glutamic
acid in equimolar amounts.

$$NH_2CH(CH_3)CONHCH_2CONHCHCO_2H$$
$$|$$
$$CH_2CH_2CONH_2$$

    (alanylglycylglutamine)

Describe how the three amino acids might have been separated from the
above peptide hydrolysate.
Or ....

                                                                    Cambridge

11 (a) Show how amino-acids can link together to produce a polymer and
        draw a section of a typical protein using the formula $H_2NCHRCOOH$
        for a typical amino-acid.
    (b) Explain the meanings of *primary*, *secondary* and *tertiary* structures
        as applied to proteins. What bonds maintain the secondary structure?
    (c) Explain what happens to the structure of a protein when a protein
        solution is boiled and what observations would therefore be made.
    (d) What is the general name given to proteins which affect the rates of
        chemical reactions in biological systems? Give one example of such
        a protein and state the reaction which it controls.
    (e) Give two other functions of proteins in biological systems and name
        a specific protein in each case.

12** Egg white consists essentially of a mixture of proteins and glucose in
    aqueous solution.
    (a) (i)   What chemical test would show that a reducing sugar is present?
        (ii) Why is it not possible to determine the glucose content
              polarimetrically?
        (iii) Describe briefly how to obtain a solution of the glucose free from
              proteins, without denaturing the proteins.
        (iv) State **one** chemical test to show that the sugar in solution is an
              aldose and not a ketose.
        (v)   Why is the glucose removed before egg white is preserved by
              drying? How is the removal effected?
    (b) (i)   Name **one** method to show that egg white contains several
              proteins.

(ii) Describe briefly the conformational changes that occur when proteins are heated.

(iii) The pH of egg white is 8.70. What is the approximate hydrogen ion concentration?

(iv) When pepsin and lemon juice are added to egg white which has been boiled and cooled, the egg white slowly goes clear and liquefies. Explain why no such changes are observed when the lemon juice is omitted.

JMB

13* Many enzymes require the presence of another substance, called a co-enzyme or co-factor, before they can show any catalytic activity.

(a) (i) Show the linkage which is characteristic of the polymer chain in an enzyme, by drawing its structural formula.

(ii) State the difference in molecular mass between enzymes and co-enzymes.

Starch is hydrolysed to glucose by salivary amylase. A reasonable rate of hydrolysis is observed if $1\,cm^3$ of saliva diluted 1:100, $5\,cm^3$ of 1% starch solution, $2\,cm^3$ of 1% sodium chloride solution and $2\,cm^3$ of a buffer solution of pH 6.6 are mixed and incubated at 35°C.

(b) (i) Describe briefly, giving procedure and expected observation, a simple method for following the progress of the reaction.

(ii) Describe briefly how you would vary the mixture in order to demonstrate that sodium chloride is promoting the activity of the enzyme.

(iii) Describe briefly how you would vary the mixture in order to demonstrate that the chloride ion, not the sodium ion, is responsible for this effect.

(iv) If varying concentrations of sodium chloride solution, up to 30% mass/volume, were used, sketch the graph you would expect to obtain by plotting rate of hydrolysis of starch against concentration of sodium chloride.

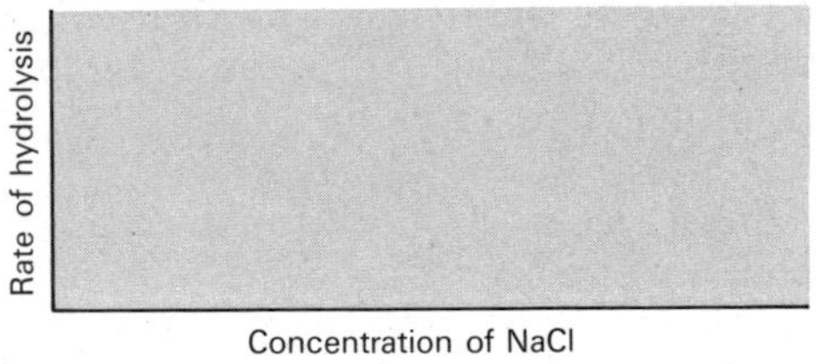

Concentration of NaCl   JMB

14* The enzyme butanedioic (*succinic*) dehydrogenase catalyses the following reaction:

$HOOCCH_2CH_2COOH$ + FAD
butanedioic   flavine adenine
acid   dinucleotide

$\longrightarrow$ $HOOCCH{=}CHCOOH$ + $FADH_2$
trans-butenedioic   reduced flavin
(*fumaric*) acid   adenine dinucleotide

(a) For this reaction, write down the following:
  (i)   the substrate,
  (ii) the co-enzyme,
  (iii) the type of reaction catalysed.

(b) Butanedioic dehydrogenase has a Michaelis constant (K_m) of 5×10^{-4} mol l^{-1}. Sketch a curve to illustrate the relationship between the amount of *trans*-butanedioic acid produced per unit time and the substrate concentration at a fixed enzyme concentration.

(c) The pK_a values for the ionization of the two carboxyl groups in butanedioic acid are 4.2 and 5.6. Write down the structural formula of the species that predominates at physiological pH (that is approximately pH 7).

(d) Explain why propanedioic (*malonic*) acid ($HOOCCH_2COOH$), which has similar pK_a values, is a competitive inhibitor.

(e) Ethanoic (*acetic*) acid (pK_a 4.8) is not affected by the enzyme nor is it a competitive inhibitor. What does this suggest about the nature of the interaction between butanedioic dehydrogenase and its substrate?

JMB

15 (a) Write down the three types of component molecules found in all nucleotides (names rather than formulae will suffice).
Show the arrangement of the above components in a typical nucleotide.

(b) State the relationship between nucleic acids and nucleotides and give a generalized formula for a nucleic acid.

(c) Name the two major classes of nucleic acid and give **one** physical and **one** chemical difference between them.

(d) What are the main differences in biological function of the two classes of nucleic acid?

(e) When mutation occurs, which nucleic acid is involved and what chemical alteration takes place?

JMB

16* (a) (i)   State the function of DNA.
       (ii) State the function of m-RNA.
       (iii) How do DNA and m-RNA compare in stability with respect to hydrolysis?
       (iv) Relate this stability to the functions of DNA and m-RNA.

(b) Consider fats and sugars.
       (i)   Which class contains the greater percentage by mass of oxygen?
       (ii) Hence predict which has the greater enthalpy of combustion per mole of carbon, and explain your answer.
       (iii) Suggest why it is advantageous for animals which hibernate to accumulate fatty tissue as a food store before hibernation.

JMB

17* (a) The amino-acid valine has the following structural formula:

$$CH_3CH(CH_3)CH(NH_2)COOH$$

(i)   Write down the structural formula of the dipeptide valylvaline.
(ii) Write down the structural formula of polyvaline.
(iii) Comment on the solubility of polyvaline in water and give a reason for your answer.

(b) Polyvaline may be made using the synthetic polynucleotide

$$(GUC\ GUC\ GUC)_n$$

where G, U and C represent the nucleotides containing the bases guanine, uracil and cytosine, respectively.
(i) What is the genetic code for the amino-acid valine?
(ii) Name two other major components which must be added to the polynucleotide and the amino-acid to produce polyvaline in a test-tube and state their function.

(c) In a polynucleotide where adenine (A) replaces the base uracil, the polypeptide produced is polyaspartic acid.
(i) What is the genetic code for aspartic acid?
(ii) Comment on the solubility of polyaspartic acid in water and give a reason for your answer.

JMB

Chapter 18
Chemicals from oil and coal

18.1 Oil and gas

Oil, or petroleum, is one of the most valuable natural resources at present available to man and its intelligent use is of fundamental importance to his material well-being, both now and in the future. The chief oil-bearing regions of the world are the Americas, Africa, the Middle East, Russia and the North Sea, and hence the availability and use of oil may be subject to political and economic factors entirely beyond the control of the chemist. Oil has long been exploited as a fuel, but the combustion processes involved are irreversible and much thought needs to be given to this permanent depletion of a valuable resource. More recently, since the end of the Second World War, the organic chemicals industry has changed from being coal-based to oil-based, and oil is now an essential source of a vast range of chemicals from detergents, plastics, pharmaceuticals and synthetic fibres to agrochemicals and even synthetic protein. At the present time, it has been estimated that about 16% of all the recoverable oil known to be available in the world has been used and, since the future is likely to see increasing demands on oil as a raw material for chemicals manufacture, energy in the future is likely to be increasingly derived from another fossil fuel, coal, and from nuclear reactions.

Oil is thought to have derived from the decomposition, by the action of anaerobic bacteria, of micro-organisms found in seas and lakes millions of years ago. Shallow seas and lakes, rich in micro-organisms, are thought to have provided animal and vegetable matter which, under layers of mud and silt, underwent decomposition followed by complex chemical reactions under the conditions of high pressures and temperatures. Oil was slowly formed as a thick, black liquid which, under pressure, permeated sedimentary rock strata, where it is now to be found, not in pools but trapped rather like water held in a sponge.

Wherever oil is found, natural gas is also found, although in varying quantities. Until relatively recently, this natural gas (essentially methane together with some ethane, propane and butane) was burned off at the well-head but, where there is a substantial quantity, it is now utilized as a fuel and may be piped great distances underground. During the 1960s, large oil deposits, along with associated natural gas, were discovered in the North Sea. Gas can also be found in the absence of oil ('non-associated gas'). The Leman field, developed by Shell and Esso, is one of the world's largest offshore gas fields and by the early 1970s natural gas ('North Sea Gas') had replaced coal-gas as fuel throughout the UK.

18.2 Refining of oil

Oil is an extremely complex mixture of compounds whose composition varies enormously with the place of origin (*see* Section 18.3). The constituents of the mixture are predominantly hydrocarbons of various sorts, although there may also be varying amounts of compounds containing nitrogen, sulphur and oxygen. In the treatment of crude oil at the refinery, the oil is first subjected to **primary distillation** which separates the oil into *fractions* which boil over definite temperature ranges. (It is quite impossible to separate even the fractions into their *individual* components.) The oil is heated to about 500°C and then introduced into the bottom of a tall *fractionating tower* (up to about 75 m in height), in which the temperature decreases towards the top (*see Figure 18.1*). The column is divided into compartments fitted with devices called 'bubble-caps' which allow the lighter, more volatile components to pass up the column while the heavier, less volatile ones condense and flow into trays below (*see Figure 18.2*). Several fractions are collected directly, while the residue is distilled in a vacuum distillation tower.

The main fractions are then usually fractionally distilled again to obtain a more efficient separation, or else carefully processed. The following compositions are quite typical:

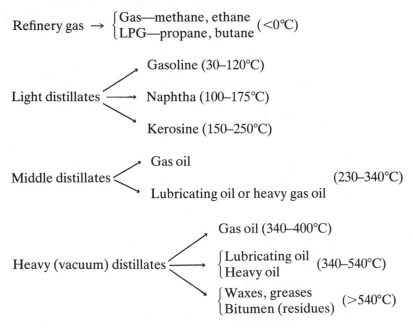

$$\text{Refinery gas } \rightarrow \begin{cases} \text{Gas—methane, ethane} \\ \text{LPG—propane, butane} \end{cases} (<0°\text{C})$$

Gasoline (30–120°C)

Light distillates ⟶ Naphtha (100–175°C)

Kerosine (150–250°C)

Gas oil

Middle distillates   (230–340°C)

Lubricating oil or heavy gas oil

Gas oil (340–400°C)

$$\text{Heavy (vacuum) distillates} \longrightarrow \begin{cases} \text{Lubricating oil} \\ \text{Heavy oil} \end{cases} (340–540°\text{C})$$

$$\begin{cases} \text{Waxes, greases} \\ \text{Bitumen (residues)} \end{cases} (>540°\text{C})$$

The resulting fractions are described in more detail below.

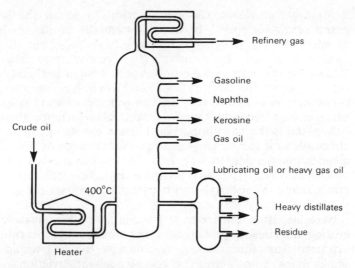

Figure 18.1 Fractional distillation of crude oil

(i) *Refinery gas*. This contains the simpler alkanes such as methane, ethane, etc., which were formerly dissolved under pressure in the crude oil. Methane and ethane can be used as fuel for the distillation processes. The alkanes propane and butane are liquefied under pressure and stored in the very characteristic huge, round 'Hortonspheres' to be found on any refinery site. They are sold as **liquefied petroleum gas (LPG)**, (*see* Section 2.5(i)).

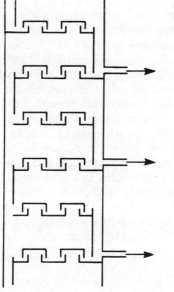

Figure 18.2 Cross-section of interior of fractionating tower

(ii) *Light distillates*. **Gasoline** is mainly used for the preparation of petrol or 'motor spirit'. It consists essentially of alkanes but has to be blended with alkenes (from *cracking*), branched-chain alkanes, cycloalkanes and aromatic hydrocarbons (from *catalytic reforming—see* Section 18.3.2) for this purpose. An unblended 'straight-run' fraction has a low octane number (*see* Section 2.4.2) and would have very poor anti-knock properties. Gasoline is also used to provide aviation spirit; a blend of 'straight-run' fractions is used to make an important one known as JP4.

Naphtha is the most important fraction for the manufacture of petrochemicals as it can be *cracked* to provide a range of simple alkanes and alkenes (*see* Section 18.3.1). It is also used in the ICI steam-reforming process (*see* Section 2.5(ii)) for the manufacture of hydrogen, and in the manufacture of carboxylic acids by the BP oxidation route (*see* Section 9.7).

Kerosine, often referred to as 'paraffin', is used extensively for domestic heating purposes. For this reason, the aromatic hydrocarbons have to be extracted from this fraction, as otherwise the fuel would burn with a smoky flame. Some kerosine is also blended into aviation spirit and some *cracked* or *reformed* to provide petrochemicals feedstock.

(iii) *Middle distillates*. **Gas oil** is used as diesel fuel (DERV fuel) and for oil-fired heating installations. Some is also *cracked* to provide feedstock for petrochemicals. **Lubricating oil** is often blended with additives for wear protection and finds extensive application.

(iv) *Heavy distillates*. The very high-boiling residue from primary distillation is subjected to a similar distillation under reduced pressure. The fractions obtained are heavier ones than those from the middle distillates and are used for oil-fired furnaces, lubrication and *cracking*. The non-volatile residue yields hydrocarbon **greases** and **waxes** and **bitumen**, which is used in the preparation of road surfaces, etc.

Since all the fractions described above are still complex mixtures, there is in fact a great deal of overlap between them, and the descriptions are necessarily a simplified guide to the sources of the various materials. The fractions themselves often have their compositions altered by solvent extraction (as in the cases of kerosine, when aromatic compounds are removed, and of heavy lubricating oil, when waxes are removed). In addition, a great deal of sulphur is removed from the oil fractions, both because sulphur can poison catalysts used in subsequent processing of the oil, and because combustion, in car engine or furnace, would provide extensive atmospheric pollution by sulphur dioxide. The removal is brought about by a process called **hydrodesulphurization**, in which the oil is heated with hydrogen at about 360°C and a pressure of 4–6 MN m^{-2}. Sulphur-containing compounds are broken down and the sulphur is combined with hydrogen to produce hydrogen sulphide. This is then burned in a limited supply of air to produce sulphur. Any sulphur dioxide which is produced is reacted with more hydrogen sulphide to form sulphur:

$$2H_2S + O_2 \rightarrow 2S + 2H_2O$$
$$2H_2S + SO_2 \rightarrow 3S + 2H_2O$$

The sulphur, so obtained, makes a very substantial contribution to the amounts which are available for sulphuric acid manufacture.

18.3 Processing of oil fractions

In the primary and vacuum distillation stages, no chemical changes in the nature of the compounds present are brought about; a physical separation of various groups of hydrocarbons is all that is really accomplished. The relative amounts of the different fractions obtained from the crude oil clearly depend on its source. The figures below illustrate this:

	Gasoline/naphtha	Kerosine	Gas oil	Fuel oil
North Sea oil	23%	15%	24%	38%
Arabian heavy oil	18%	11.5%	18%	52.5%

North Sea oil, for example, has a high cycloalkane and aromatic hydro-carbons content but a low alkene content. Such variations are unsatis-factory when a consistent demand for various fractions must be met, so a way has to be found for maintaining supplies. In addition, most of the 'straight-run' fractions from distillation are of too poor a quality to permit their immediate use for their intended purpose and need to be improved by blending with higher quality components. For these reasons, a con-siderable amount of chemical processing of the oil fractions becomes necessary, in order to provide the right quality products in the right amounts. A variety of techniques are used in the processing of the fractions.

18.3.1 Cracking

Cracking involves the chemical breakdown of higher, less volatile hydro-carbons into smaller molecules. It provides reactive alkenes for petro-chemicals manufacture and volatile alkanes for blending with gasoline, etc., e.g.

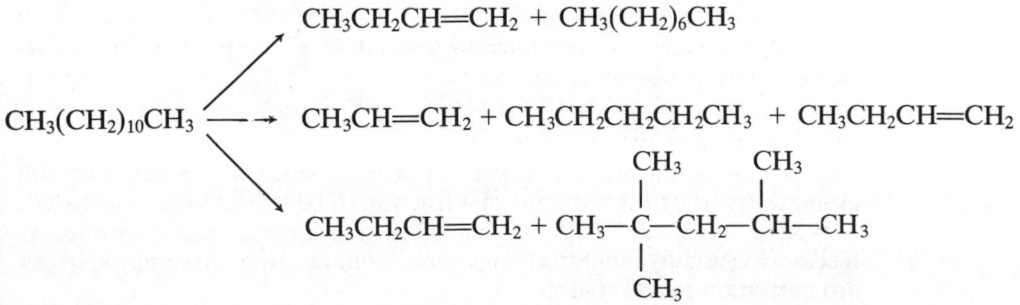

$$CH_3CH_2CH{=}CH_2 + CH_3(CH_2)_6CH_3$$

$$CH_3(CH_2)_{10}CH_3 \longrightarrow CH_3CH{=}CH_2 + CH_3CH_2CH_2CH_2CH_3 + CH_3CH_2CH{=}CH_2$$

$$CH_3CH_2CH{=}CH_2 + CH_3{-}\underset{\underset{CH_3}{|}}{\overset{\overset{CH_3}{|}}{C}}{-}CH_2{-}\underset{}{\overset{\overset{CH_3}{|}}{CH}}{-}CH_3$$

The breakdown of the hydrocarbon chains is essentially a random process so that many different products can be formed by cracking. In the past, *thermal* cracking has been widely used, involving the heating of heavy oils with steam to a high temperature (~900°C). However, much better control over the products can be achieved by means of *catalytic* cracking. A hydrocarbon fraction is fed into a reactor (catalytic cracking unit, or 'cat cracker') containing a synthetic silica/alumina catalyst (similar to a molecular sieve, *see* Section 2.3), in very fine powder form which flows like a liquid. Using such a fluid-bed catalyst, temperatures of about 490°C can be used to bring about cracking at almost atmospheric pressure. Whereas thermal cracking involves a free-radical mechanism, catalytic cracking involves a carbocation (i.e. ionic) mechanism. Catalytic cracking is more efficient and leads to higher quality products for blending with gasoline, etc., following fractional distillation.

Most alkenes are produced by catalytic cracking and are extremely important feedstocks for a wide range of manufacturing processes.

18.3.2   Reforming

Reforming involves molecular rearrangements, rather than breakdown, of hydrocarbons and produces a wider range of important products. The feedstock for reforming is heated to about 540°C in the presence of a catalyst and at a pressure of about 1–5 $MN\,m^{-2}$. The catalyst is usually platinum and hence this catalytic reforming is sometimes referred to, colloquially, as 'platforming'. In contrast to cracking, which is exothermic, reforming is an endothermic process. A wide variety of molecular rearrangements can take place, some of which are summarized below.

(i) Dehydrocyclization
This produces cycloalkanes from straight-chain ones. Cycloalkanes are valuable for improving the octane rating of gasoline, e.g.

$$CH_3CH_2CH_2CH_2CH_2CH_2CH_3 \longrightarrow \qquad + \quad H_2$$

Cycloalkanes can also be dehydrogenated to valuable aromatic hydrocarbons (*see* (ii) below).

(ii) Aromatization (dehydrogenation)
This is one of the most important reactions since it enables the vital aromatic hydrocarbon content of a fraction to be substantially increased. A typical 'reformate' may contain as much as 50% aromatics, while the feedstock (usually naphtha) may only contain 10%. Examples of an aromatization reaction are:

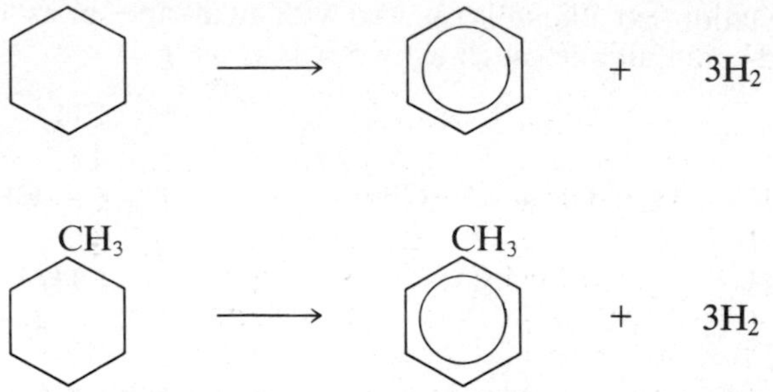

The addition of aromatics to 'straight-run' gasoline improves its octane rating considerably.

(iii) Isomerization

Using special platinum catalysts active at relatively low temperatures (150°C), branched-chain alkanes can be made from straight-chain isomers. Once again, these products boost the octane rating of gasoline considerably, e.g.

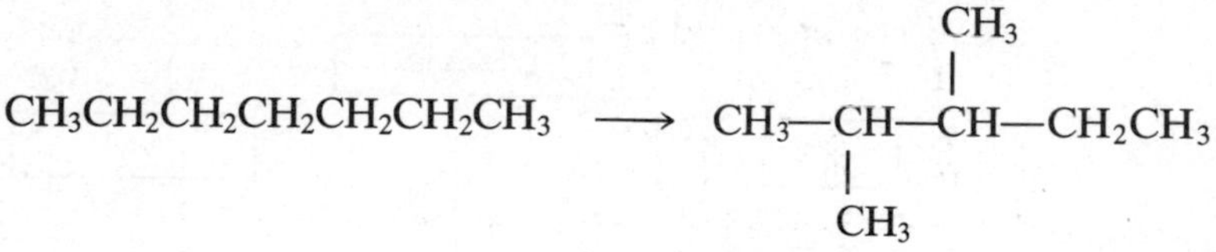

(iv) Hydrocracking

This cracking is carried out in the presence of a controlled amount of hydrogen and yields smaller, saturated molecules, e.g.

$$CH_3CH_2CH_2CH_2CH_2CH_2CH_2CH_3 + H_2 \rightarrow 2CH_3CH_2CH_2CH_3$$

18.3.3 Polymerization and alkylation

These reactions afford a method of utilizing C_3 and C_4 hydrocarbons. The degree of polymerization is usually limited to simple dimerization and trimerization, e.g.

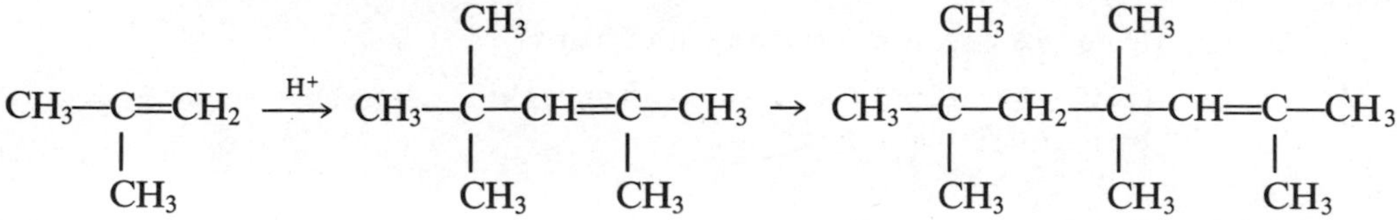

The dimer can be hydrogenated to 'iso-octane'. 2-methylpropene is also, separately, polymerized to poly(butenes), which are widely used as engine lubricant additives, using a phosphoric acid catalyst at 200°C and about 7 MN m^{-2}.

In alkylation, an alkene is reacted with an alkane, in excess, to produce branched-chain alkanes such as 'iso-octane', e.g.

$$CH_3-CH-CH_3 + CH_2=C-CH_3 \xrightarrow{H_2SO_4} CH_3-C-CH_2-CH-CH_3$$

These branched-chain alkanes can be blended with gasoline to increase its octane rating.

18.4 Chemicals from oil

Some of the important materials which have their origins in feedstock derived from oil are summarized in *Figure 18.3*. Details of some of the individual manufacturing processes are to be found at various places in the text.

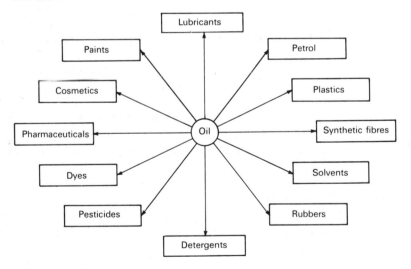

Figure 18.3 Industrial chemicals derived from oil

18.5 Coal

In the past, a high proportion of industrial organic chemicals was extracted from **coal-tar,** obtained from coal. Since the Second World War, however, oil has supplanted coal as the major source. Thus, in 1948 oil was responsible for about 11% of the UK chemical industry's feedstock, but by 1970 the figure had risen to over 90%, with a corresponding reduction in the use of coal. Since the replacement of town gas by natural gas, the main source of coal chemicals has been the steel industry which makes metallurgical coke for the smelting of iron ore.

Nevertheless, coal is the only indigenous fossil fuel available in massive reserves and it seems likely that by the beginning of the twenty-first century it will increasingly replace oil as a fuel. Furthermore, if oil supplies become scarce or prohibitively expensive, coal-based feedstocks can be expected to play a greater role in supplying the organic chemicals industry. Consequently, research into coal and its products continues and several major projects are being investigated by the National Coal Board, sometimes in co-operation with other companies such as BP and British Gas.

18.6 Processing of coal

Coal resulted from the slow decomposition of wood from forests, under pressure in the absence of air and with the aid of bacterial action over many millions of years. The extremely complex chemical structures present in the wood were gradually broken down into simpler ones which in turn, in very old coal deposits, produced carbon. This process is known as **carbonization**. Like oil, coal does not have a constant composition and the variation in carbon content is reflected in the various grades of coal. The carbon content increases from *lignite* (60–75% carbon), through *bituminous coal* (84–91% carbon) to *anthracite* (93% carbon). The average hydrogen content of coal is much lower than that of oil (about 5% compared with 13%) so that chemical processing of coal is designed to increase this hydrogen content. The products are used to produce Substitute Natural Gas (SNG), gasoline and chemical feedstocks. There are three main processes, which are described in the following paragraphs.

18.6.1 Liquefaction or solvent extraction

Modern methods have evolved from the work of the German chemist Friedrich Bergius, who was awarded the 1931 Nobel Prize for Chemistry. Coal is heated and extracted with a solvent to produce a thick, tarry liquid. The solvent may be a liquid such as *anthracene oil* (*see* Section 18.6.3(4)), at about 400°C and 0.7 MN m^{-2} pressure, or else it may be a gas at high pressure and above its critical temperature (for example, methylbenzene at 400°C and 20 MN m^{-2} pressure), when a coal-in-gas solution is formed. The 'coal solution' is filtered from ash and carbon residues and subjected to *hydrocracking*. At high temperatures and pressures, in the presence of hydrogen and a hydrogenation catalyst, the extract is broken down into a variety of simpler compounds. The resulting mixture is fractionally distilled to give a light fraction, rich in aromatic hydrocarbons (often referred to as BTX after benzene, toluene and xylene) and a heavy oil fraction. The solvent can be recycled. The extraction process removes up to 40% of the coal feed. After processing, up to 220 l of gasoline and 360 l of distillate oils can be recovered from

1 tonne (t) of coal. Because of the high octane rating of aromatic hydrocarbons, this method has great potential for production of high-quality
gasoline from coal, should this become necessary.

18.6.2 Gasification

Coal is strongly heated until the volatile materials are driven off. The
residue, or *char*, is then reacted with superheated steam/oxygen (or air)
to form synthesis gas (*see* Section 2.5(ii)), a process developed by
Lurgi Ruhrchemie and known as the Lurgi process:

$$C \ + \tfrac{1}{2}O_2 \rightleftharpoons CO \qquad\qquad\qquad \Delta H^{\ominus} = -121 \text{ kJ}$$
$$C \ + H_2O \rightleftharpoons CO \ + H_2 \qquad\qquad \Delta H^{\ominus} = +120 \text{ kJ}$$
$$CO + H_2O \rightleftharpoons CO_2 + H_2 \qquad\quad \Delta H^{\ominus} = -43 \text{ kJ}$$
$$C \ + 2H_2 \rightleftharpoons CH_4 \qquad\qquad\qquad \Delta H^{\ominus} = -89 \text{ kJ}$$

High pressure (2–7 MN m^{-2}) and temperatures (>1000°C) favour the
formation of methane, both by the reaction shown above and by reduction
of carbon monoxide. The synthesis gas may be treated in three ways.

(i) If the molar H_2/CO ratio is adjusted to about 3:1 then, in the
presence of a suitable catalyst, the mixture will undergo **catalytic meth-
anation**, a reaction which is the reverse of the steam-reforming of natural
gas, namely

$$CO + 3H_2 \rightleftharpoons CH_4 + H_2O \qquad\qquad \Delta H^{\ominus} = -205 \text{ kJ}$$
and also
$$CO_2 + 4H_2 \rightleftharpoons CH_4 + 2H_2O \qquad\qquad \Delta H^{\ominus} = -163 \text{ kJ}$$

(ii) Back in 1925, the German chemists Franz Fischer and Hans Tropsch
demonstrated that mixtures of hydrogen and carbon monoxide could be
converted into hydrocarbons, by carefully controlling the composition of
the mixture and its reaction, under elevated temperatures and pressures
in the presence of suitable catalysts. The reaction became known as the
Fischer–Tropsch synthesis:

$$nCO + (2n + 1) H_2 \xrightarrow{\text{Fe catalyst}} C_nH_{2n+2} + nH_2O$$

$$nCO + 2nH_2 \rightarrow C_nH_{2n} + nH_2O$$

The greater the proportion of hydrogen present, the greater the proportion of alkanes among the products. This proportion can be increased
by passing synthesis gas, together with steam, over an iron catalyst:

$$CO + H_2 + H_2O \rightleftharpoons CO_2 + 2H_2$$

Thus, as well as obtaining Substitute Natural Gas (SNG) from synthesis
gas, a mixture of hydrocarbons may be obtained by the Fischer–Tropsch
process. In the largest plant of its kind, operating at Sasolburg in South
Africa, about 1 t of synthetic oil is obtained from 4–5 t of coal. The
mixture can be fractionally distilled to give a range of useful products,
including a high-quality gasoline. Alkenes, so produced, can be converted

into alcohols (*see* Section 3.3.1(v)), aldehydes and ketones (OXO reaction, *see* Section 3.3.1(vi)) and carboxylic acids and their esters (via oxidation of the aldehydes).

(iii) Finally, in a third reaction, carbon monoxide, in a 1:2 molar ratio with hydrogen, can be reduced to methanol:

$$CO + 2H_2 \xrightleftharpoons[250°C, \, 5\,MN\,m^{-2}]{Catalyst} CH_3OH \qquad \Delta H^{\ominus} = -109 \, kJ$$

This is particularly useful because the Mobil Research Corporation have developed a method of making gasoline from methanol (*see* Section 9.7). In addition, Monsanto have a route to ethanoic acid, which starts from methanol:

$$CH_3OH + CO \rightarrow CH_3COOH$$

The utility of the gasification of coal is summarized in *Figure 18.4*.

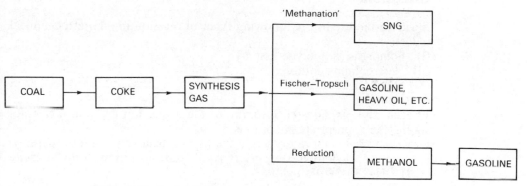

Figure 18.4 Coal gasification and associated processes

18.6.3 Carbonization

When coal is heated to a temperature of the order of 1000°C, coal-gas, ammonia and coal-tar are driven off and coke is left as residue. The **coal-tar**, which may contain over 500 distinct compounds, can be fractionally distilled and yields the following fractions:

(1) *Light oil,* up to 170°C, containing mainly benzene and alkylbenzenes, such as methyl- and the dimethyl-benzenes. This oil accounts for between 1% and 6% of the coal-tar.
(2) *Middle oil,* collected between 170°C and 230°C, containing mainly naphthalene $C_{10}H_8$ and phenols, and accounting for between 6% and 8% of the coal-tar.
(3) *Heavy oil* (creosote oil), collected between 230°C and 270°C, containing naphthalene and naphthalenol, and accounting for 8–16% of the coal-tar.
(4) *Anthracene oil,* collected between 270 and 400°C, containing mainly anthracene $C_{14}H_{10}$, and accounting for 50–60% of the coal-tar.

The remaining material is *pitch* which is used as road-binding material, etc. It is the light oil which yields most of the aromatic hydrocarbons (BTX), only about 10% of the total coal-tar being used to supply organic chemicals.

In the *low-temperature* carbonization process, the coal is heated in a fluidized bed, under reduced pressure, at about 400–600°C. The coal-tar produced is of a different composition to that from *high-temperature* carbonization, consisting of alkanes and cycloalkanes.

Finally, the coke obtained from coal can be used to provide ethyne (*see* Section 4.5) which is a very versatile starting material for organic synthesis. All of these coal-based processes have lost ground during the upsurge in the oil economy but may yet play a more important role in contributing organic chemicals for industrial use.

Questions

1* Illustrate the use of the following types of reaction in the petrochemical industry:
(i) homolysis (homolytic fission),
(ii) electrophilic addition,
(iii) polymerization,
(iv) oxidation.
In each case, explain the meaning of the terms and give one example, stating the appropriate conditions.
Describe briefly laboratory experiments by which you could prepare:
(a) poly(phenylethene) (*polystyrene*) starting from ethylbenzene ($C_6H_5CH_2CH_3$), (b) a nylon.

JMB

2 Petroleum is the major industrial source of alkanes, alkenes and aromatic hydrocarbons which, in turn, are the starting materials for the production of many important chemical compounds and materials of use and benefit to man.
(a) Outline the methods by which hydrocarbons are obtained from crude petroleum on an industrial scale and indicate what particular hydrocarbons are isolated.
(b) Review the major synthetic pathways by which the hydrocarbons are converted into other compounds and materials.

London

3 The major process used in the petrochemical industry today is steam cracking of hydrocarbon feedstocks at 900°C. A typical analysis of the product from a cracking furnace is shown in the table opposite:
(a) How are these components of the gas mixture separated on a commercial scale?
(b) Show by means of equations and brief notes on reaction conditions how ethene is converted into (i) ethane-1,2-diol, (ii) ethenyl ethanoate (*vinyl acetate*), and (iii) phenylethene (*styrene*). For each of these compounds, name a commercial outlet.

Products of steam cracking of naphtha at 900°C

Product	Yield (% by mass)	Product	Yield (% by mass)
ethene	30	fuel oil	5
propene	15	methane	14
butane/butenes	5	ethane	4
buta-1,3-diene	4	propane	1
gasoline	21	hydrogen	1

(c) Using an example of your own choice, show, by means of equations and brief notes on reaction conditions, how propene is converted into a material in extensive everyday use.

JMB

4* Ethene is made on an industrial scale by cracking naphtha in the presence of steam at 900°C at just above atmospheric pressure.
(a) (i) State what you understand by the term *cracking*.
    (ii) What is the function of the steam in this process?
    (iii) Why is the process operated at this comparatively low pressure?
    (iv) Show by means of an equation the first chemical step in the purely thermal cracking of octane.
    (v) Thence show how ethene is formed.
Ethene is frequently called the *building block chemical*.
(b) (i) Show by means of equations and conditions how ethene can be converted into either chloroethene (*vinyl chloride*) or phenylethene (*styrene*).
    (ii) Name a material in everyday use made from chloroethene or from phenylethene.
    (iii) Give the structural formula for the material in (ii).

JMB

5* (a) Ethene is often referred to as the *building block chemical* of the petrochemical industry.
    (i)   What is the name of the chemical process which is used to form ethene from petroleum naphtha?
    (ii) What is the basis of this process in terms of the relative stabilities of alkanes and alkenes?
    (iii) Name five compounds derived from ethene which are in everyday use.
    (iv) Indicate schematically by equations how any **one** of the compounds you have named in answer to (iii) is prepared from ethene.
    (b) In order to be commercially viable, an ethene plant must have a production capacity of some 500 000 tonnes per annum.
    (i)   Give two reasons why the production cost per tonne decreases with increasing plant capacity.
    (ii) Describe two risks (economic or otherwise) which are attendant on operating a large-scale plant.

JMB

Chapter 19
Fats, oils and waxes

19.1 Introduction

These compounds are widely distributed throughout the plant and animal kingdom. Those which are liquid at room temperature are referred to as **oils**, those which are solid at room temperature are **fats** and those with a melting point above body temperature are usually called **waxes**. This classification is thus based on a physical property. Chemically, these compounds are generally high relative molecular mass esters of long-chain (usually C_{16} or C_{18}) alkanoic or alkenoic acids, commonly referred to as *fatty acids*. The most abundant of all *saturated* fatty acids is hexadecanoic acid, also known as *palmitic acid*. The most abundant of *all* fatty acids is the unsaturated *cis*-octadec-9-enoic acid, also known as *oleic acid*. Nearly all such naturally occurring acids possess an even number of carbon atoms in their molecules; some common ones are listed in *Table 19.1*.

Most of the fats and oils are esters of fatty acids and the trihydric alcohol propane-1,2,3-triol, or *glycerol*, and are known as **triglycerides**. There are exceptions; beeswax, from which the honey-bee makes its comb, is a mixture of esters of C_{26}–C_{28} acids and saturated monohydric

Table 19.1

Systematic name	Trivial name	Structure	Melting point (°C)
tetradecanoic acid	myristic acid	$CH_3(CH_2)_{12}CO_2H$	54
hexadecanoic acid	palmitic acid	$CH_3(CH_2)_{14}CO_2H$	63
octadecanoic acid	stearic acid	$CH_3(CH_2)_{16}CO_2H$	70
cis-octadec-9-anoic acid	oleic acid	$CH_3(CH_2)_7CH\overset{cis}{=\!=}CH(CH_2)_7CO_2H$	16
cis,cis-octadeca-9,12-dienoic acid	linoleic acid	$CH_3(CH_2)_4CH\overset{cis}{=\!=}CHCH_2CH\overset{cis}{=\!=}CH(CH_2)_7CO_2H$	−5
cis,cis,cis-octadeca-9,12,15-trienoic acid	linolenic acid	$CH_3CH_2CH\overset{cis}{=\!=}CHCH_2CH\overset{cis}{=\!=}CHCH_2CH\overset{cis}{=\!=}CH(CH_2)_7CO_2H$	−11

alcohols, while spermaceti, which is obtained from the head of the sperm whale, is an ester of palmitic acid and the alcohol hexadecanol $CH_3(CH_2)_{14}CH_2OH$ (also called *cetyl alcohol*). Carnauba wax (m.p. = 85–90°C) contains esters of C_{27}, C_{29} and C_{30} alcohols.

Oils contain a higher proportion of esters of unsaturated acids than do fats, and are thus liquids at room temperature. They are more common in the vegetable kingdom than in the animal kingdom (hence the common description, vegetable oils and animal fats), although fish oils actually contain a greater proportion of unsaturated acids than the vegetable oils. A brief comparison is made below:

animal fat	e.g. Beef tallow (m.p. = 68°C)	28% palmitic acid 19% stearic acid 38% oleic acid	(% of total acid content)
vegetable oil	e.g. Linseed oil (m.p. = 8°C)	57% linolenic acid 14% linoleic acid 18% oleic acid	

In nature, fatty material is a reserve food. It is stored in various parts of the animal body, often as a subcutaneous layer of the skin where its insulating power, as well as its high calorific value (about 36 kJ g^{-1}, which is higher than for protein or carbohydrate), is particularly valuable. For a healthy diet, man requires a daily intake of about 100 g of fat; oxidation of this material takes place mainly in the liver. In the plant kingdom, fatty material is a common constituent of the food store of the seed.

The ever-increasing demand for animal fats has gradually led to their replacement, for many purposes, by vegetable oils. Animal fats and oils are usually separated from fatty tissue by heating it with steam, a process known as *rendering*. Vegetable oils are extracted from seed either by solvent extraction, or by milling using considerable pressure. A wide range of oils such as groundnut, cottonseed, sunflower, rapeseed, palm and olive oils are used to make margarine (*see* Section 19.3), cooking oils and salad dressings. Other oils such as linseed, castor and tung oils are used in the manufacture of paints and varnishes.

19.2   Chemistry of glycerides

In most oils and fats the fatty acids are combined with propane-1,2,3-triol (*glycerol*) in the form of esters called triglycerides:

$$
\begin{array}{c}
\quad\quad\quad CH_2OCOR' \\
\quad\quad\quad \diagup \\
RCOO\!-\!CH \\
\quad\quad\quad \diagdown \\
\quad\quad\quad CH_2OCOR''
\end{array}
$$

The three acid components of the triglyceride are not often identical (i.e. R– = R'– = R''–) and there is almost invariably at least one unsaturated acid. The higher the proportion of saturated acid components of glycerides, the more likely it is that the material will become solid at room temperature, e.g.

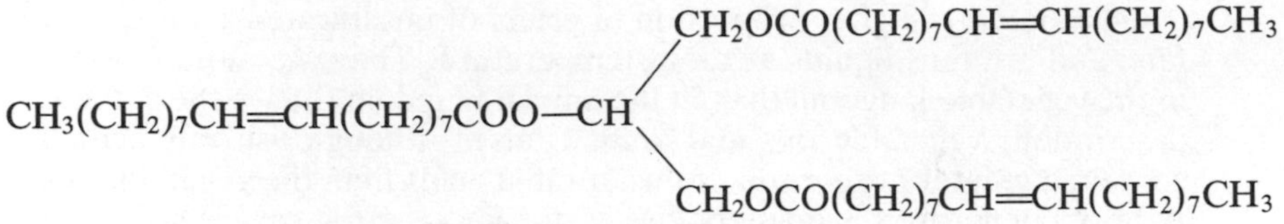

$$CH_2OCO(CH_2)_7CH=CH(CH_2)_7CH_3$$

$$CH_3(CH_2)_7CH=CH(CH_2)_7COO-CH$$

$$CH_2OCO(CH_2)_7CH=CH(CH_2)_7CH_3$$

triolein (from olive oil)
liquid

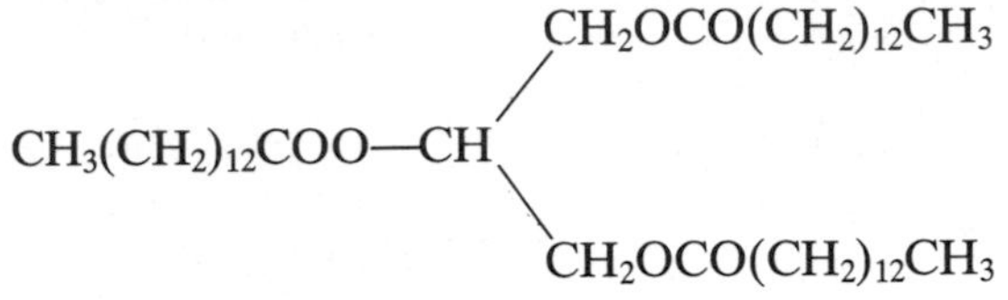

$$CH_2OCO(CH_2)_{12}CH_3$$

$$CH_3(CH_2)_{12}COO-CH$$

$$CH_2OCO(CH_2)_{12}CH_3$$

trimyristin (from nutmeg)
m.p. = 59°C

The oils are non-volatile liquids, which are insoluble in water and, usually, in ethanol. The fats are also insoluble in water and can generally exist in more than one crystalline form.

In general, the chemical properties of the triglycerides are those shown by either esters or alkenes (*see* Sections 10.1 and 3.3, respectively).

19.2.1 Hydrolysis

Hydrolysis is achieved by heating a triglyceride with aqueous alkali for some time. The products of the hydrolysis are *glycerol* and the salts of the component fatty acids. These salts are **soaps** (*see* Section 19.4) and the alkaline hydrolysis is therefore often referred to as **saponification**:

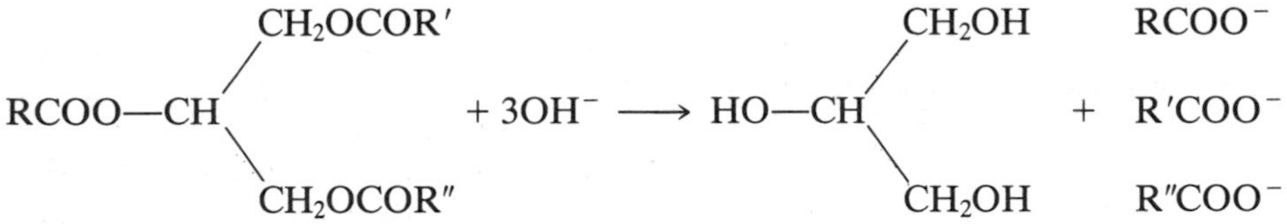

$$CH_2OCOR'$$
$$RCOO-CH$$
$$CH_2OCOR''$$
$$+ 3OH^- \longrightarrow$$
$$CH_2OH$$
$$HO-CH$$
$$CH_2OH$$
$$+ \begin{matrix} RCOO^- \\ R'COO^- \\ R''COO^- \end{matrix}$$

The *saponification number* of a fat or oil is defined as the number of milligrams of potassium hydroxide which is required to saponify 1 g of fat or oil. A typical value would lie between 180 and 200. This number gives a rough indication of the average relative molecular mass of the glycerides.

In the body, hydrolysis of glycerides is brought about with the aid of the *lipase* enzymes which catalyse the hydrolysis of triglycerides to fatty acids and the 2-monoglyceride:

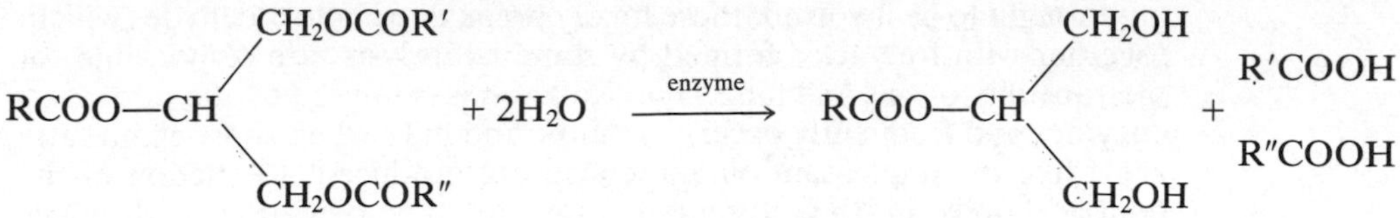

$$\begin{array}{c} \text{CH}_2\text{OCOR}' \\ | \\ \text{RCOO—CH} \\ | \\ \text{CH}_2\text{OCOR}'' \end{array} + 2\text{H}_2\text{O} \xrightarrow{\text{enzyme}} \begin{array}{c} \text{CH}_2\text{OH} \\ | \\ \text{RCOO—CH} \\ | \\ \text{CH}_2\text{OH} \end{array} + \begin{array}{c} \text{R}'\text{COOH} \\ \\ \text{R}''\text{COOH} \end{array}$$

19.2.2 Hydrogenation

The double bonds in the unsaturated fatty acid components of glycerides can be reduced by catalytic hydrogenation. The reduction increases the degree of saturation of oils and fats and raises their melting points ('*hardening*').

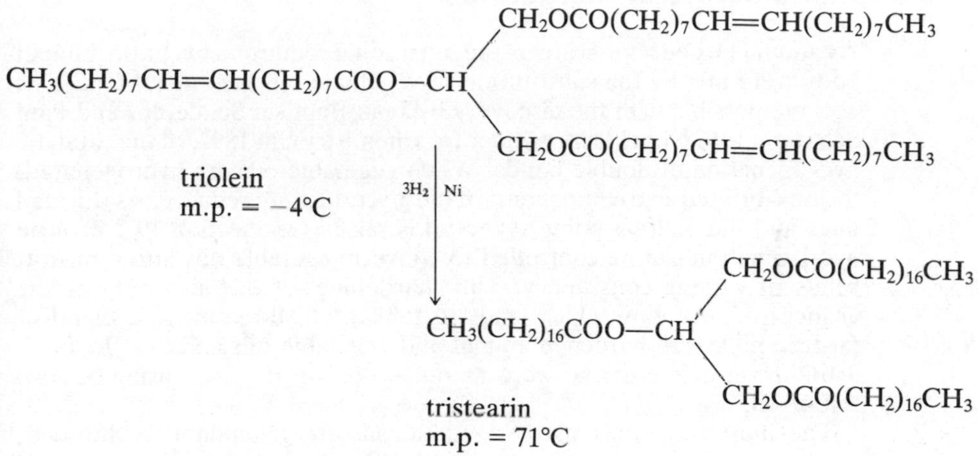

$$\begin{array}{c} \text{CH}_2\text{OCO(CH}_2)_7\text{CH}=\text{CH(CH}_2)_7\text{CH}_3 \\ | \\ \text{CH}_3(\text{CH}_2)_7\text{CH}=\text{CH(CH}_2)_7\text{COO—CH} \\ | \\ \text{CH}_2\text{OCO(CH}_2)_7\text{CH}=\text{CH(CH}_2)_7\text{CH}_3 \end{array}$$

triolein
m.p. = −4°C

3H_2 | Ni

$$\begin{array}{c} \text{CH}_2\text{OCO(CH}_2)_{16}\text{CH}_3 \\ | \\ \text{CH}_3(\text{CH}_2)_{16}\text{COO—CH} \\ | \\ \text{CH}_2\text{OCO(CH}_2)_{16}\text{CH}_3 \end{array}$$

tristearin
m.p. = 71°C

This type of reaction is of great importance in the manufacture of margarine (*see* Section 19.3).

19.2.3 Iodination

A measure of the degree of unsaturation of a fat or oil can be obtained by adding iodine across the double bonds. In the Wijs method, a known excess of a solution of iodine(I) chloride is added to a solution of the fat or oil in tetrachloromethane. After allowing the mixture to stand in the dark for about 1 h, the solution is then back-titrated with standard sodium thiosulphate solution to determine the amount of iodine absorbed. The *iodine number* is the number of grams of iodine absorbed by 100 g of fat or oil. A typical value, for corn oil, would be 120; for a very unsaturated oil, like linseed oil, it would be nearer 180.

19.2.4 Oxidation

The double bonds of unsaturated fatty acid components of glycerides make them susceptible to atmospheric oxidation. Peroxide intermediates are thought to be involved; these finally break down into aldehydes which, together with free acids formed by some hydrolysis, are responsible for the rancidity of oils and fats. This oxidation is catalysed by the *lipoxidase* enzymes and frequently occurs in butter and milk when short-chain fatty acids like the unpleasant butanoic acid are produced. Reduction of the double bonds in the glycerides prevents this oxidation and hence 'hardened' vegetable oils (*see* Section 19.2.2) are frequently used for frying.

Unsaturated oils such as linseed and tung oils are oxidized in complex reactions to form thin, flexible films, on drying, and are used in paints and varnishes.

19.3 Manufacture of margarine

As animal fats became scarcer, the nutritional requirements of the human body were met by the substitution of vegetable oils for animal fats. This became possible with the discovery by Jean Baptiste Senderens and Paul Sabatier (1912 Nobel Prize-winner for Chemistry), in 1897, of the catalytic hydrogenation of double bonds. When vegetable oils are hydrogenated, the unsaturated acid components of the glycerides are reduced to saturated ones and the melting point of the oil is raised (*see* Section 19.2.2). The hydrogenation can be controlled to convert vegetable oils into semi-soft solids of varying consistency. This 'hardening' of oils also reduces the chances of oxidation which leads to rancidity. Margarine is a blend of fat-free milk with hardened animal and vegetable oils and fats. In 1977, 380 000 ton of margarine were manufactured in the UK, using 324 000 ton of oil.

The most commonly used vegetable oils are groundnut, cottonseed, sunflower, palm kernel and coconut oils. The crude oils are first extracted with alkali to remove any free acids, bleached by heating with Fuller's Earth to give a light-coloured oil, and finally deodorized by heating with steam to drive off volatile impurities. Hydrogen, under 0.1–0.5 MN m^{-2} pressure, is then bubbled through the oil at 100–180°C in the presence of a suspension of finely divided nickel as catalyst. The hydrogenation can be stopped at any point when the oil has reached the required consistency. The more unsaturated acid components tend to be reduced more quickly than others and some of the *cis* double bonds are isomerized to the *trans* form, which also raises the melting point.

The semi-soft product is churned with milk which has been inoculated with certain micro-organisms to reproduce a butter flavour, and the law requires that Vitamins A and D also be added. Other additives include colouring, flavouring, anti-oxidants and emulsifying agents. The final product, margarine, is a water-in-oil emulsion, the maximum permitted water content (to which salt is often added) being 16%. The emulsion is stabilized and contains water droplets dispersed throughout the continuous phase of oil. Milk is an emulsion of fats and oils in water, stabilized by proteins; when the cream, which contains a higher proportion of fats and oils, is churned the emulsion is inverted and butter is formed.

Hardened oils are also widely used for cooking and for making salad dressing, which is also an emulsion. This latter is usually stabilized with an agent such as *1-glycerylstearate* and thickened with carbohydrate polymers called *alginates*.

Very large intakes of saturated oils and fats may be implicated in cardiovascular and coronary heart diseases, when blood vessels become blocked by the accumulation of fatty deposits (particularly those containing esters of the steroid alcohol *cholesterol*). The substitution of the more soluble *unsaturated* fats and oils, particularly those containing linoleic acid which cannot be synthesized by the body but which is an essential ingredient of the human diet, is generally recommended to sufferers of these diseases. For this purpose, certain commercial oils and fats are advertised as being 'high in polyunsaturates'.

19.4 Soaps and detergents

Ordinary soaps are sodium or potassium salts of fatty acids present in natural fats and oils. They are manufactured by the alkaline hydrolysis or **saponification** of the glycerides in fats and oils and the products are the salts of the acids and propane-1,2,3-triol (*glycerol*) (*see* Section 19.2.1).

Although olive oil was first used for soap manufacture, other vegetable oils can be used, by hardening them first to a suitable consistency. Usually, beef and mutton tallows are blended together with vegetable oils such as coconut and palm kernel oils. The blended oils and fats are then bleached with sulphuric acid/Fuller's Earth at 90°C under reduced pressure, and impurities removed by presses. Subsequent heating with steam at 160°C under reduced pressure removes volatile impurities and deodorizes the blend. Saponification to yield the soap base is generally carried out by heating the blend with alkali under pressure in a vertical column; hydrolysis occurs very rapidly at elevated temperatures. The process is a continuous one and the quantity of alkali added is carefully controlled in a fully automated process. As the soap base is removed from the column, after cooling, it is washed with salt solution to dissolve the glycerol and the two products then separated in a centrifugal extractor. The glycerol solution is concentrated and the glycerol then distilled under reduced pressure to give the pure liquid. Any free acids in the crude soap base are carefully neutralized and the soap then dried and processed into household and toilet soaps, soap powders, soap flakes, etc.

Both soaps and soapless detergents are examples of **detergents**, or **surfactants** (surface-active agents) as they are sometimes called. Their function, in cleansing or 'deterging', is to increase the wetting power of the water by decreasing interfacial tension and to dislodge and disperse dirt particles. A typical soap molecule, for example sodium octadecanoate (*sodium stearate*), consists of a long hydrocarbon chain with an ionized carboxyl group at one end. Since the hydrocarbon chain is incapable of hydrogen-bonding within the loose water structure, it is described as the **hydrophobic** (*water-hating*) part of the molecule. The carboxyl group, on the other hand, is readily solvated by water and is described as the

hydrophilic (*water-loving*) portion of the molecule. Consequently, soap molecules tend to orient themselves at the surface of water, with the hydrophilic portion penetrating the water and the hydrophobic portion trying to leave the water. In doing this, they reduce the surface tension

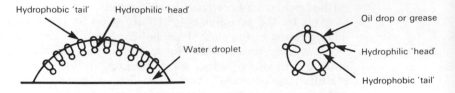

Figure 19.1 The action of detergent molecules

of the water, as shown in *Figure 19.1*, and droplets tend to break up and spread across a surface. In addition, the hydrophobic chains of the molecules tend to enclose dirt particles and 'float' them off soiled surfaces. The major disadvantage of soap, however, is that it is largely ineffective, or at least is very uneconomic to use, in hard water. This is because the magnesium and calcium ions, responsible for the hardness of water, combine with soap molecules to form insoluble salts. The soap is therefore precipitated out of solution as a 'scum' and is thus prevented from functioning properly.

This problem was overcome by the introduction of soapless detergents, which now have the larger share of the market. The soapless detergents are chemically quite different from soaps. The most commonly used ones are sulphonated alkylbenzenes. The long alkyl chain of such a molecule represents the hydrophobic portion, while the sulphonate group is the hydrophilic portion. Unlike the calcium and magnesium salts of long-chain fatty acids, the corresponding salts of sulphonic acids are freely soluble in water so that such soapless detergents can be used in hard water.

The alkylbenzenes are manufactured by Friedel–Crafts reactions of benzene with either C_{10}–C_{15} alkenes (obtained from the cracking of hydrocarbon waxes) or secondary chloroalkanes:

$$CH_3(CH_2)_9CH{=}CH_2$$

$$CH_3(CH_2)_9CHCH_3$$
$$\underset{Cl}{|}$$

$$\xrightarrow{AlCl_3} \quad CH_3(CH_2)_9CH\text{—}\bigcirc$$
$$\underset{CH_3}{|}$$

The alkylbenzenes are sulphonated, either with oleum or gaseous sulphur trioxide, the mixture diluted with water, and the acid separated from the detergent. The sulphonated alkylbenzene is then neutralized with alkali:

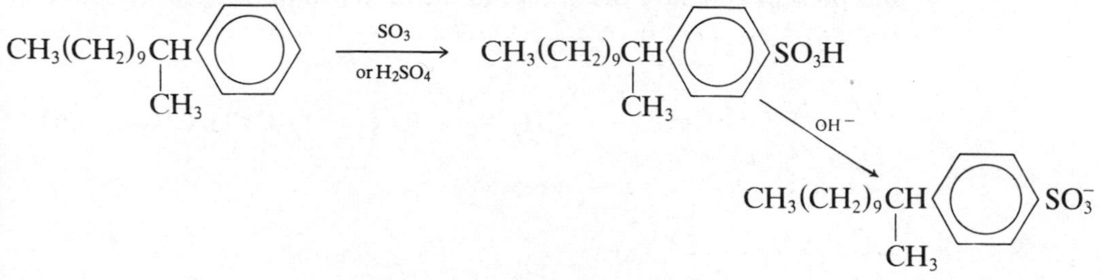

$$CH_3(CH_2)_9CH\langle\bigcirc\rangle \xrightarrow[\text{or } H_2SO_4]{SO_3} CH_3(CH_2)_9CH\langle\bigcirc\rangle SO_3H$$
$$\quad\quad\overset{|}{CH_3} \quad\quad\quad\quad\quad\quad\quad\quad\quad\overset{|}{CH_3}$$

$$\xrightarrow{OH^-} CH_3(CH_2)_9CH\langle\bigcirc\rangle SO_3^-$$
$$\quad\quad\quad\quad\quad\quad\overset{|}{CH_3}$$

The alkyl side-chain must be essentially linear, i.e. unbranched. This is so that the detergent can be broken down, or degraded, in the environment by the action of bacteria. Such detergents are said to be **biodegradable** and are sometimes described as 'soft' detergents. In the past, non-bio-degradable or 'hard' detergents caused both considerable problems in sewage plants and unpleasant foaming in rivers, but their use is now banned in many countries and they have been replaced by the bio-degradable ones.

Another class of soapless detergents consists of the *sulphated* alcohol type, a well-known example being Teepol (R = C_{10}–C_{18} below), e.g.

$$RCH{=}CH_2 \xrightarrow[\text{or } SO_3]{H_2SO_4} \underset{OSO_3H}{R{-}\overset{|}{C}H{-}CH_3} \xrightarrow{OH} \underset{OSO_3^-}{R{-}\overset{|}{C}H{-}CH_3}$$

or

$$CH_3(CH_2)_{16}CH_2OH \xrightarrow[\substack{\text{or}\\ H_2SO_4}]{SO_3} CH_3(CH_2)_{16}CH_2OSO_3H \xrightarrow{OH^-} CH_3(CH_2)_{16}CH_2OSO_3^-$$

Both the soaps and the soapless detergents of the sulphonated and sulphated kind are described as *anionic* detergents, or surfactants, because of the nature of their hydrophilic groups. *Cationic* detergents include quaternary ammonium compounds (*see* Section 11.1) which have useful germicidal properties as well:

$$CH_3(CH_2)_{16}COOH \xrightarrow[200{-}300°C]{NH_3} CH_3(CH_2)_{16}CN \xrightarrow[\substack{100{-}150°C\\ 1.5{-}7 \text{ MN m}^{-2}}]{H_2/Ni}$$

a fatty acid

$$CH_3(CH_2)_{16}CH_2NH_2 \xrightarrow{CH_3Cl} CH_3(CH_2)_{16}CH_2\overset{+}{N}(CH_3)_3 \ Cl^-$$

Non-ionic detergents, too, have been developed and have proved useful because they have a low foaming power. They are also unaffected by any ions present and have been used in the dispersal of oil slicks at sea, e.g.

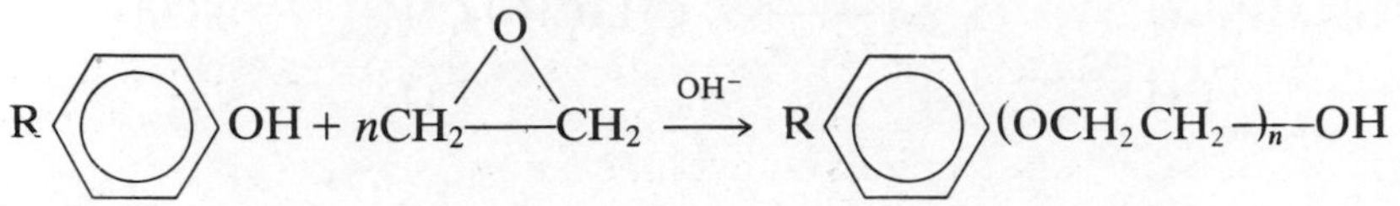

$$R\langle\bigcirc\rangle OH + nCH_2{-\!\!-}CH_2 \xrightarrow{OH^-} R\langle\bigcirc\rangle(OCH_2CH_2{-\!)_n}{-}OH$$

or

$$CH_3(CH_2)_mCH_2OH + nCH_2{-\!\!-}CH_2 \rightarrow CH_3(CH_2)_mCH_2(OCH_2CH_2)_nOH$$

Various additives are employed with detergents such as phosphates for water-softening, etc., bleaches, fluorescers* and perfumes. In addition to their washing functions, detergents also find uses in leather-softening oils, paints, lubricating greases and cosmetics.

Questions

1 (a) Explain whether you would expect the following triglyceride to be a hard solid, a soft solid or a liquid at room temperature.

$$CH_2OCOC_{15}H_{31}$$
$$|$$
$$CHOCOC_3H_7$$
$$|$$
$$CH_2OCOC_{17}H_{33}$$

   (b) Write an equation for the alkaline hydrolysis of the above compound. To what use may the products be put?
   (c) After refluxing of 0.5 g castor oil with 10 cm³ of 0.5M ethanolic potassium hydroxide, the resulting solution required 6.5 cm³ (corrected) of 0.5M hydrochloric acid for neutralization. What is the saponification number of the castor oil?
   (d) Write equations for the complete oxidation of the fat *tristearin* $C_{57}H_{110}O_6$ and of the sugar glucose $C_6H_{12}O_6$. The respective standard enthalpies of combustion are about 34 600 and 2850 kJ mol⁻¹. Which is the source of greater energy? Explain your answer, making reference to the chemical compositions of the two classes of compound.

* These are compounds which absorb in the ultraviolet region of the spectrum and then emit light of longer wavelength in the visible, usually blue, region of the spectrum to give a 'blue whiteness'.

2*   (a) The principal fat in olive oil is propane-1,2,3-triyl tri-*cis*-octadec-9-enoate (*glyceryl trioleate*).
        (i) Give the structural formula of this compound.
        (ii) Give the structural formula of a product obtained on hydrogenation.
     (b) Describe briefly how the process of hydrogenation is carried out.
     (c) Name (i) a chemical method and (ii) a physical method for following the progress of hydrogenation.
     (d) For use in foods, fats are never fully hydrogenated. Give two reasons for this.

                                                                          JMB

3*   (a) (i)   The melting point of a margarine is the minimum temperature at which no solid remains in equilibrium with the liquid. The table below shows the percentage by volume of solid fat for a soft margarine S and a hard margarine H at various temperatures. Plot this information and deduce the approximate melting points of the soft margarine and the hard margarine.

| | Solid fat (percentage by volume) | |
Temperature (°C)	S	H
10	18	66
21	7	59
27	2	52
33	0	43
38	0	27

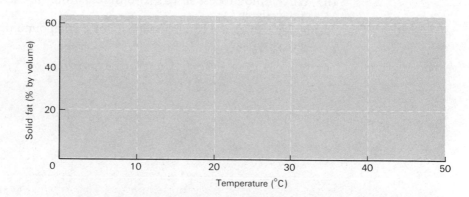

        (ii) The iodine values of these margarines are 55 and 86. Which value refers to margarine S and how do you account for the difference in values in terms of the molecular constitution of the fats?
        (iii) Name **one** of the two important components, other than fat and emulsifier, which must be present in margarine to meet legal requirements.

(b) Oils and fats are liable to spoilage (rancidity) through enzymic hydrolysis.

    (i)   Name the main fatty acid produced in this way from contaminated milk.

    (ii) Why do vegetable oils *seem* less susceptible to this kind of deterioration than milk fats?

    (iii) Explain why free fatty acids need to be removed before an oil is suitable for making an edible emulsion, such as salad cream.

    (iv) Name **one** thickening agent used in edible emulsions and explain why it is needed.

JMB

4   What is understood by the word 'detergent'? Explain in detail the differences in molecular structure between the various types of detergent, both soapy and soapless, and mention the advantages and disadvantages of each type.

Nuffield

5*   (a) (i)   Define the term *alkylation*.

           (ii) Show, by means of an equation and a named catalyst, how 1-chlorododecane ($C_{12}H_{25}Cl$) reacts with benzene.

           (iii) By means of an equation and brief notes on reaction conditions, show how the product formed in (ii) may be converted into a compound suitable for use as a detergent.

           (iv) To which class of detergent does the compound formed in (iii) belong?

    (b) (i)   By means of equations, show how epoxypropane is made on a large scale.

           (ii) Why is the process in (b) (i) different from that used to make epoxyethane?

           (iii) Name and give a commercial use for a compound derived from epoxypropane.

JMB

Chapter 20
Synthetic polymers

20.1 Introduction

Polymers are macromolecules which are made up of innumerable repetitions of a small molecular unit, or units. They may have relative molecular masses of the order of thousands. There are naturally occurring polymers, such as cellulose and rubber, but this chapter deals with those important polymers which have been synthesized by the organic chemist.

There are two simple ways in which the formation of a synthetic polymer—**polymerization**—may be brought about. In **addition polymerization**, a simple molecule—the **monomer**—is caused to *add on* to itself to produce huge molecular chains containing many thousands of units, e.g.

$$n\text{A} \quad \rightarrow \quad \text{—A—A—A—A—A—A—A—, etc} \quad \rightarrow \quad \text{(A)}_n$$

monomer   polymer

where n = a very large number.

Examples of such addition polymers are the poly(alkene) plastics, such as poly(ethene) (polythene) and PVC. The polymerization of the monomer is usually initiated by free-radicals, although ionic initiators are used in certain cases.

In **condensation polymerization**, two molecular units or monomers, in general, are employed. Each monomer possesses two identical functional groups which can react with those of the other monomer by a condensation reaction (namely, one involving the elimination of small molecules such as H_2O or HCl), so that chain growth proceeds to give huge molecules, e.g.

$$n\text{HO—A—OH} + n\text{H—B—H} \rightarrow \text{—A—B—A—B—A—B—} \rightarrow \text{(A—B)}_n \text{H} + n\,H_2O$$

Examples of condensation polymers are the polyester and polyamide synthetic fibres. Both types of polymerization are exothermic processes.

A polymer which contains two alternating monomer units is called an **alternating copolymer**, e.g.

—A—B—A—B—A—B—A—B—A—B—A—B—

but it is possible to produce copolymers in which the order of monomer units is random and these are known as **random copolymers**, e.g.

—A—B—A—A—A—B—B—A—A—A—A—A—B—B—

Special techniques also allow polymers to be made in which a sequence of one kind of monomer unit is followed by a sequence of another, and these are termed **block copolymers**, e.g.

—A—A—A—A—A—A—A—B—B—B—B—B—A—A—A—A—

A particular sequence of monomer units may also be *grafted* on to a polymer chain to produce a **graft copolymer**, e.g.

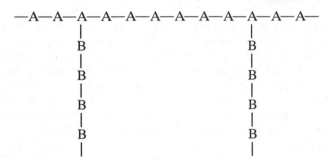

The termination of chain growth in either type of polymerization is a randomly occurring event, so that the final polymer consists of macro-molecules of varying sizes, i.e. the individual molecules have identical chemical compositions (excepting some of the copolymers) but different relative molecular masses. A typical distribution is shown in Figure 20.1;

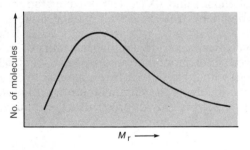

Figure 20.1 Molecular mass distribution of a polymer

a polymer's physical properties, such as viscosity and softening point, often depend on whether this spread is broad or narrow. Thus, it is the bulk properties of the polymer, rather than those of individual polymer molecules, which are studied and which are important.

In a simple polymer like poly(ethene), the chain of CH_2 units of which such a molecule is comprised, although unbranched, is not really a *straight* chain but rather a zig-zag shape:

$$-CH_2-CH_2-CH_2-CH_2-CH_2-CH_2-$$

Rotation about the 'backbone' chain of single covalent bonds can occur relatively freely and the polymer chain is therefore flexible. The molecular forces between such chains are of the weak van der Waals type but regions of very close alignment of the unbranched chains are to be found in the polymer. These regions are called **crystallites** and they confer some degree of crystallinity on the polymer as a whole. The more symmetrical and regular the molecular structure, the more crystalline the polymer is likely to be. If, on the other hand, the monomer from which the polymer chain is constructed is rather bulky, then the free rotation may be restricted and the polymer chain will be stiffer and more rigid. If, as in **cross-linked** polymers, the molecular chains are actually chemically linked by covalent bonds, then the polymer will be even tougher. Crystallinity is influenced by the size of side-groups in the chain, the flexibility of the chain, and the nature of the intermolecular forces operating between chains. Any factor which enhances the close alignment of chains will also enhance the polymer's crystallinity.

Since polymers are seldom perfectly crystalline, they do not have very sharp melting points. At lower temperatures, when there is little molecular motion, they are in an amorphous but rigid, glass-like state, but at the **glass transition temperature** T_g the motion of the chains is such as to cause the polymer to assume a more flexible, rubber-like condition. At the **melting point** T_m, the crystallites melt, crystallinity eventually disappears and the polymer becomes a viscous liquid.

Thus, by careful attention to the molecular composition of the polymer chain, a wide range of physical properties such as density, solubility, crystallinity and elasticity may be controlled, at least to some degree. Consequently, an enormous range of synthetic polymers exists and an understanding of the relations between physical characteristics and molecular structure has enabled chemists to tailor the properties of polymers to suit specific technical demands. The chemists' ability to do this is demonstrated by the polymers discussed in the following sections.

20.2 Plastics

Plastics have become so familiar as replacements for more traditional materials like wood, glass or metal that several have become household names. By 1976, the total quantity of plastics consumed in the UK exceeded 2 million tonnes per annum and the problem of plastic-waste disposal is now the subject of research. Plastics are synthetic polymers which, when heated, can be made to flow and hence have the property of *mouldability*. Some can be softened and hardened many times over by heating and cooling; these are called **thermoplastics**. Others soften on heating but then undergo a chemical change and become permanently hard; these are called **thermosetting** plastics. About 70% of all plastics produced are thermoplastics.

The monomers used in plastics manufacture are all derived from oil; about 2% of all oil products are used for this purpose, but coal may play a more important part in the future in the event of oil shortages.

20.2.1   Thermoplastics

Thermoplastics are generally manufactured in the form of powders or granules and, because they can be heated and cooled many times without undergoing any chemical change, they can be moulded and fabricated by various techniques into an enormous range of products and consumer goods. The absence of any pronounced colour to the polymers is a great advantage in this respect. Plastics generally have higher T_g values than elastomers or synthetic fibres, since their easy deformation is not a desirable property.

(i) Poly(ethene)

Structurally, this is the simplest thermoplastic and is made by the polymerization of ethene. It was first made, accidentally, by Reginald Gibson and Eric Fawcett at ICI's Northwich plant in 1933. The accidental introduction of oxygen into a reactor containing ethene caused it to polymerize at high temperature and pressure instead of combining with another reactant. Its development was hastened by the onset of the Second World War since its unique electrical insulating properties made it ideal for use with radar equipment.

The polymerization of alkenes is initiated by a source of free-radicals, such as an organic peroxide like di(benzenecarbonyl) peroxide; when traces of oxygen are used as initiator it is believed that a peroxide is first formed. On heating, peroxides readily yield free-radicals:

$$RO\text{---}OR \xrightarrow{\text{heat}} 2RO^{\cdot} \qquad \text{initiation}$$

e.g.

$$(C_6H_5COO)_2 \rightarrow 2C_6H_5COO^{\cdot}$$

Attack by radicals on ethene molecules forms new radicals and a chain reaction ensues:

$$RO^{\cdot} + CH_2{=}CH_2 \rightarrow ROCH_2\text{---}CH_2^{\cdot} \xrightarrow{CH_2{=}CH_2} ROCH_2CH_2CH_2CH_2^{\cdot} \rightarrow, \text{etc.,}$$
$$\text{propagation}$$

Eventually, chain growth is terminated by combination of radicals and disproportionation of chains:

$$RO(CH_2)_xCH_2CH_2^{\cdot} + {}^{\cdot}CH_2CH_2(CH_2)_yOR \begin{cases} RO(CH_2)_xCH_2CH_2CH_2CH_2(CH_2)_yOR \\ \qquad\qquad \text{combination} \\ RO(CH_2)_xCH{=}CH_2 + CH_3CH_2(CH_2)_yOR \\ \qquad\qquad \text{disproportionation} \end{cases}$$

Termination

The polymer chains produced are so long (x and y are very large numbers) that two RO— groups at the end of a chain have no effect on the physical and chemical properties of the polymer and it remains essentially a hydrocarbon.

However, as a result of radicals abstracting hydrogen atoms from points along the growing chain, these chains are not entirely linear but have short side-chains branching out at various points. Consequently, there are few regions of close alignment of molecular chains and there is a fairly low degree of crystallinity in the polymer (~50%). This form of poly(ethene), with a density of 0.92 g cm^{-3}, is known as the low density (LD) form:

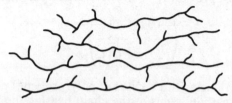

Since the hydrogen atoms are so small, there is free rotation about the C—C bonds and the polymer is extremely flexible. Since the molecular forces between chains are of the weak van der Waals type, the polymer also has a low softening point (about 80°C).

LD poly(ethene) is manufactured by two routes. One involves the addition of carefully controlled amounts of oxygen to ethene under very high pressure (200 MN m^{-2}) and at 170°C, as developed by ICI; the other process was developed by Phillips Petroleum Company and the Standard Oil Company and uses a chromium oxide catalyst at much lower pressures. This method is used in the UK by BP:

$$n\text{CH}_2{=}\text{CH}_2 \xrightarrow[170°C]{200 \text{ MN m}^{-2}} ({-}\text{CH}_2{-}\text{CH}_2{-})_n \xleftarrow[\substack{3\text{--}5 \text{ MN m}^{-2} \\ 60\text{--}200°C}]{\text{Cr}_2\text{O}_3} n\text{CH}_2{=}\text{CH}_2$$

$$T_g = {<}0°C$$
$$T_m = 110°C$$

In 1953, the German chemist Karl Ziegler, working at the Max Planck Institute in Mülheim, discovered that ethene could be polymerized at room temperature and pressure, with the aid of a catalyst consisting of a combination of a trialkyl aluminium R$_3$Al and titanium(III) chloride TiCl$_3$. This discovery, too, stemmed from a chance observation made as a result of the accidental contamination of a reaction vessel. The mechanism of this reaction is an ionic one and does not involve free-radicals. As a result, completely linear, unbranched polymer chains are produced and these pack more closely together, giving the polymer more crystalline regions (~90% crystalline) and hence greater rigidity:

$$T_g = -122°C$$
$$T_m \approx 130°C$$

This high density (HD) form of poly(ethene) has a higher density (0.95 g cm^{-3}) and a higher softening point $(>100°C)$ than the LD form.

Both forms are manufactured in the UK and poly(ethene) is the largest tonnage plastic produced. Its strength, transparency, flexibility and durability ideally fit it for use as packaging film and containers of all sorts (e.g. washing-up liquid bottles).

(ii) Poly(propene)

In 1953–54, the Italian chemist Giulio Natta, working at Milan Polytechnic, applied Ziegler's catalysts to the polymerization of the alkene propene, which had not previously been successfully polymerized.

$$n\text{CH}_2\text{==CHCH}_3 \longrightarrow \left(\text{-CH}_2\text{-CH-}\underset{\text{CH}_3}{|}\right)_n$$

He succeeded in producing three kinds of poly(propene) polymer. In the **isotactic** form, the carbon atoms of the chain bearing the methyl groups all have an identical configuration in space. In the **syndiotactic** form, these configurations alternate, while in the **atactic** form the configurations are randomly arranged:

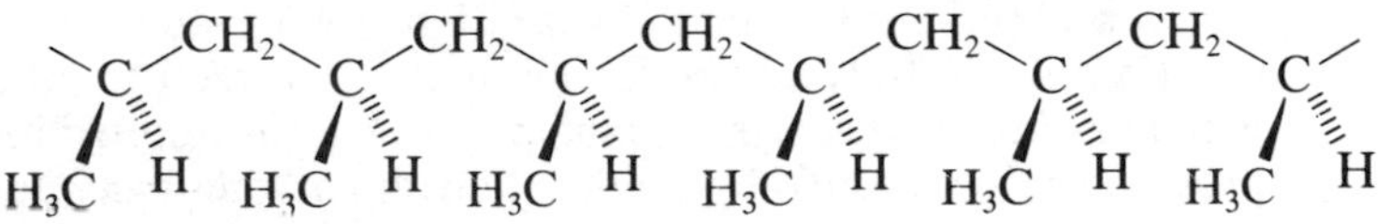

isotactic form
$T_g = -20°C$
$T_m = 170°C$

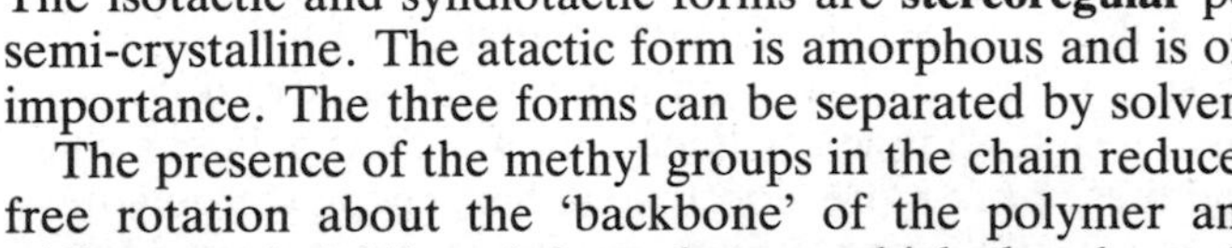

syndiotactic form

The isotactic and syndiotactic forms are **stereoregular** polymers and are semi-crystalline. The atactic form is amorphous and is of no commercial importance. The three forms can be separated by solvent extraction.

The presence of the methyl groups in the chain reduces the amount of free rotation about the 'backbone' of the polymer and increases the stiffness and rigidity of the polymer, which thus has a higher softening point (about 140°C) than poly(ethene). The rather bulky methyl groups also reduce the alignment of the molecular chains, so that poly(propene), with a density of 0.91 g cm^{-3}, is lighter than any other thermoplastic. The isotactic form, which is the commercially important form, was first manufactured in the UK by ICI in 1958. The rigidity and lightness of poly(propene) has made it an ideal plastic for the manufacture of chairs and other articles and it has also been produced in fibre form (Ulstron), mainly for industrial use. The propene feedstock is generally obtained from the cracking of a C_5–C_8 oil distillate.

One interesting aspect of propene polymerization is that an asymmetric centre is established each time a monomer unit reacts with the growing polymer chain. However, even in the stereoregular forms, the polymer is not optically active because there exist planes of symmetry—through each carbon atom in the isotactic form and through each second carbon atom of the syndiotactic form:

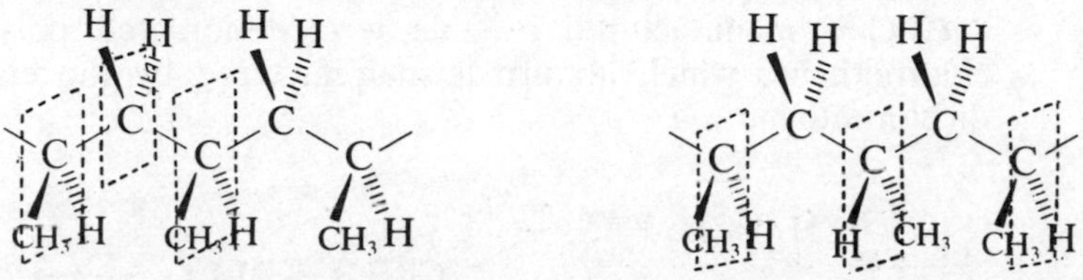

In 1963, just 10 years after their momentous discoveries, Ziegler and Natta were jointly awarded the Nobel Prize for Chemistry in recognition of the tremendous impact that their work had on the plastics industry.

(iii) Poly(phenylethene)

This polymer, also known as *polystyrene*, is one of the most familiar plastics and is the third most important thermoplastic, after 'polythene' and PVC. It is made by the free-radical polymerization of phenylethene, which is made by the dehydrogenation of ethylbenzene (*see* Section 12.4.2(iv)):

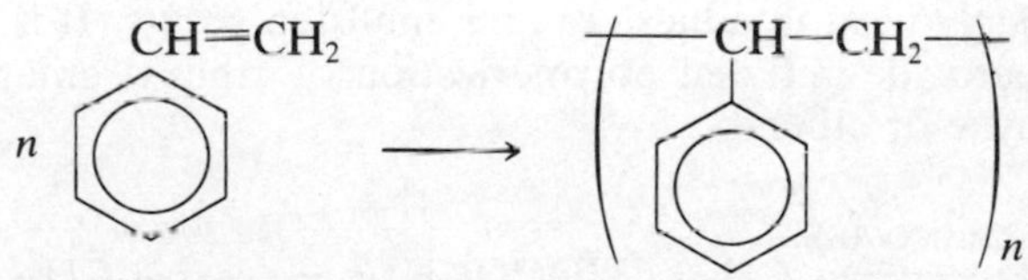

$$T_g \approx 100°C$$
$$T_m \approx 230°C$$

The benzene rings stiffen the polymer chain and the polymer forms a hard, transparent resin which is resistant to chemical attack and which softens at about 100°C. It is widely used in the form of 'expanded' poly(phenylethene) which has outstanding thermal insulation properties (e.g. as in vending cups).

(iv) Poly(chloroethene)

This widely used polymer, also known as *polyvinylchloride* or **PVC**, differs from most of the other important thermoplastics in that its useful properties are controlled primarily by additives. Unlike the stereoregular forms of poly(propene), the configurations of the carbon atoms bearing the chlorine atoms in the polymer chain are randomly arranged, so that the polymer possesses very little crystallinity. However, the presence of the polar C—Cl bonds increases the strength of the intermolecular forces and raises the softening point of the polymer ($\sim$180°C). At this temperature, the polymer tends to eliminate hydrogen chloride but the softening point may be lowered, either by copolymerization or by incorporating into the polymer a polar, high boiling-point liquid (e.g. a dialkyl benzene-1,2-dicarboxylate) called a *plasticizer*. The plasticizer acts as a sort of internal lubricant and eases the molecular movement of the polymer chains, so improving the flexibility of PVC. In this form, PVC is widely used in the manufacture of clothing and electrical insulation. The rigidity of the untreated form is utilized in piping and plumbing accessories.

PVC is manufactured by the peroxide-initiated polymerization of chloroethene, which in turn is manufactured by the cracking of 1,2-dichloroethane:

$$CH_2{=}CH_2 + Cl_2$$
$$CH_2{=}CH_2 + 2HCl + \tfrac{1}{2}O_2$$
$$\searrow \nearrow$$
$$ClCH_2{-}CH_2Cl \xrightarrow[-HCl]{500°C} CH_2{=}CHCl$$
$$b.p. = -14°C$$

$$\xrightarrow[\text{peroxide}]{20\text{-}25°C} \left(CH_2{-}CH \atop \underset{Cl}{\big|} \right)_n \qquad \begin{array}{l} T_g = 80°C \\ T_m = 180°C \end{array}$$

Chloroethene is also copolymerized with ethenyl ethanoate to produce a softer, less brittle, polymer ideal for the manufacture of gramophone records. Poly(ethenyl ethanoate) is an important polymer in its own right, being widely used in adhesives and emulsion paints. It is manufactured by the peroxide-initiated polymerization of ethenyl ethanoate, derived from ethyne or ethene:

$$HC{\equiv}CH + CH_3COOH \xrightarrow[200°C]{Zn(OCOCH_3)_2} CH_2{=}CHOCOCH_3 \xrightarrow{\text{peroxide}} \left(CH_2{-}CH \atop \underset{OCOCH_3}{\big|} \right)_n$$
$$b.p. = 72°C$$
$$(+H_2O)$$

$$\xrightarrow{200°C}$$

$$CH_2{=}CH_2 + \tfrac{1}{2}O_2 + CH_3COOH$$

$$T_g = 28°C$$

(v) Poly(methyl 2-methylpropenoate)
This polymer was first manufactured in 1936 (by ICI) and is made by the

peroxide-initiated polymerization of methyl 2-methylpropenoate (*see* Section 8.4):

$$n\text{CH}_2{=}\text{C}\begin{smallmatrix}\text{CH}_3\\\text{CO}_2\text{CH}_3\end{smallmatrix} \longrightarrow \left(\text{CH}_2{-}\underset{\underset{\text{CO}_2\text{CH}_3}{|}}{\overset{\overset{\text{CH}_3}{|}}{\text{C}}}\right)_n \qquad T_g \approx 100°\text{C}$$

The polar and bulky ester groups make the polymer relatively rigid and the atactic, amorphous form is a hard, transparent resin which absorbs very little light. It is sometimes known as an *acrylic* plastic and is sold under trade names such as Perspex.

(vi) Poly(tetrafluoroethene)

This polymer is formed by the free-radical polymerization of tetrafluoroethene (*see* Section 5.4(v)) and was discovered owing to the accidental polymerization of a cylinder of the gaseous monomer, initiated by oxygen:

$$n\text{CF}_2{=}\text{CF}_2 \quad \xrightarrow[\text{7 MN m}^{-2}]{\text{O}_2,\ 50\text{--}200°\text{C}} \quad (\text{CF}_2{-}\text{CF}_2)_n \qquad \begin{matrix}T_g = <-100°\text{C}\\ T_m = 327°\text{C}\end{matrix}$$

The bulky fluorine substituents prevent free rotation about the C—C bonds and the chain assumes a twisted, zig-zag conformation in space. Consequently, the polymer does not melt but becomes a sort of viscous gel at about 327°C. Thus, the polymer does not really flow and is difficult to fabricate, since it is also insoluble in most solvents. Some of the polymer's outstanding properties include unusual resistance to chemical attack, non-flammability and low coefficient of friction. This latter property is responsible for its use as an important surface coating, most widely encountered in 'non-stick' surfaces such as are found in kitchen utensils. The polymer is marketed under names such as Fluon, Teflon and PTFE.

20.2.2 Thermosetting plastics

In contrast to thermoplastics, the polymer chains of a **thermosetting plastic** become chemically cross-linked during the heating process, so that a hard, rigid structure is produced which does not soften on reheating. Probably the oldest plastic of this type is Bakelite, introduced by the Belgian chemist Leo Baekeland in 1907.

(i) Phenolic resins

Baekeland discovered that when phenol, in slight excess, and methanal are heated in the presence of *acid*, condensation reactions occur and —CH$_2$OH groups are substituted in the ring. Subsequently, these phenolic alcohols condense together and a short-chain linear polymer, known as a 'novolak', is produced. This is soluble, fusible material, without any free —CH$_2$OH groups, but by further treatment with alkaline methanal, or certain amines, an insoluble cross-linked polymer can be made, on heating.

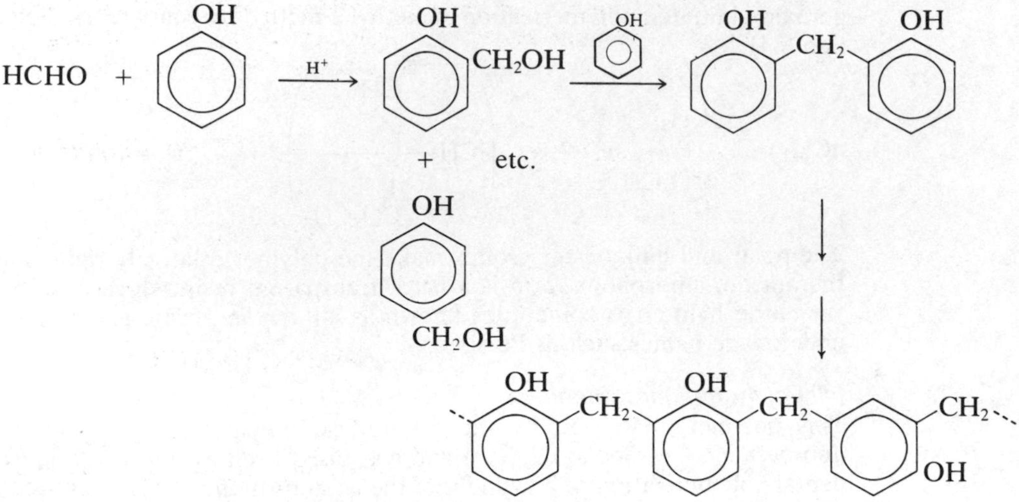

HCHO + (phenol, OH) $\xrightarrow{H^+}$ (phenol) CH$_2$OH (+ ortho isomer) $\longrightarrow$ (phenol)-CH$_2$-(phenol)

+ (phenol) CH$_2$OH etc.

... (phenol)-CH$_2$-(phenol)-CH$_2$-(phenol)-CH$_2$ ... with OH

no cross-linking   a 'novolak'

However, in the presence of an excess of methanal and *alkali* as catalyst, short-chain polymers called 'resols' are first obtained. On heating, these chains increase in length to form 'resitols' which finally cross-link to form a hard, insoluble thermosetting plastic:

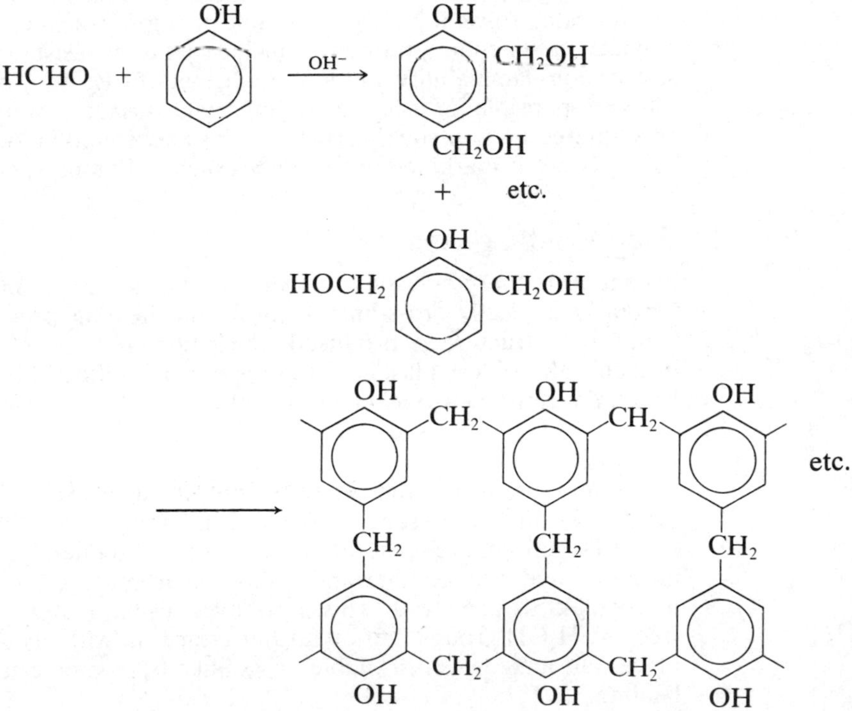

HCHO + (phenol, OH) $\xrightarrow{OH^-}$ (phenol) CH$_2$OH with CH$_2$OH

+ etc.

HOCH$_2$-(phenol, OH)-CH$_2$OH

$\longrightarrow$ cross-linked network of (phenol) rings linked by CH$_2$ bridges (OH groups on rings)   etc.

Similar condensations occur when carbamide is treated with methanal and, as with the phenolic plastics or resins, cross-linking or 'curing' is brought about by heating, usually under slightly acidic conditions:

$$H_2NCONH_2 + HCHO \xrightarrow{OH^-} H_2NCONHCH_2OH \longrightarrow HOCH_2NHCONHCH_2OH \text{ etc.}$$

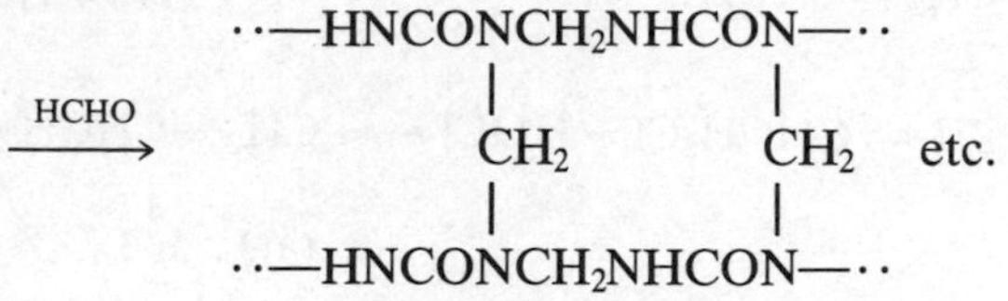

$$\xrightarrow{HCHO}$$

$$\begin{array}{ccc} \cdots-HNCONCH_2NHCON-\cdots \\ | \qquad\qquad | \\ CH_2 \qquad\quad CH_2 \quad \text{etc.} \\ | \qquad\qquad | \\ \cdots-HNCONCH_2NHCON-\cdots \end{array}$$

Amines such as *melamine* have also been condensed with methanal to produce important plastics. These thermosetting plastics are used in adhesives and in mouldings and laminates (e.g. Formica).

(ii) Epoxy resins

These polymers, which are notable for their bonding properties, are utilized in adhesives such as Araldite and are made by condensation of epoxy compounds with dihydric phenols or alcohols. A typical example is the polymer made from 1-chloro-2,3-epoxypropane and a dihydric phenol known by the trivial name diphenylol propane;

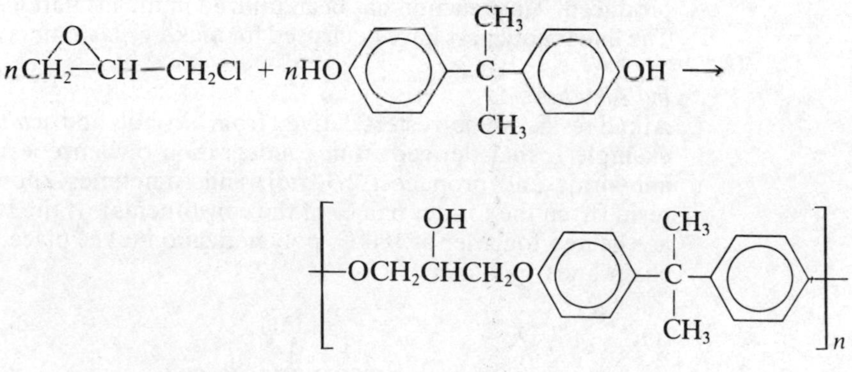

$$n CH_2\overset{O}{\frown}CH-CH_2Cl + nHO\langle \bigcirc \rangle - \underset{CH_3}{\overset{CH_3}{\underset{|}{\overset{|}{C}}}}-\langle \bigcirc \rangle OH \longrightarrow$$

$$\left[-OCH_2\underset{}{\overset{OH}{\underset{|}{CH}}}CH_2O\langle \bigcirc \rangle - \underset{CH_3}{\overset{CH_3}{\underset{|}{\overset{|}{C}}}}-\langle \bigcirc \rangle - \right]_n$$

The phenolic component is manufactured by the condensation of phenol with propanone:

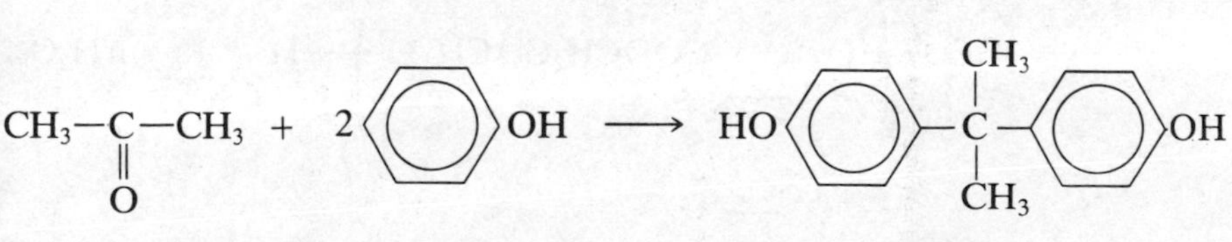

$$CH_3-\underset{\underset{O}{\|}}{C}-CH_3 + 2\langle \bigcirc \rangle OH \longrightarrow HO\langle \bigcirc \rangle - \underset{CH_3}{\overset{CH_3}{\underset{|}{\overset{|}{C}}}}-\langle \bigcirc \rangle OH$$

The polymer produced is 'cured', i.e. cross-linked, by heating with certain amines, when terminal epoxy groups of the chains react with the amine, cross-linking the chains.

The epoxy compound is manufactured by chlorinating propene and reacting the resulting 3-chloropropene with chloric(I) acid followed by a suspension of calcium hydroxide:

$$CH_2{=}CHCH_3 + Cl_2 \xrightarrow[\text{0.2 MN m}^{-2}]{\text{500°C}} CH_2{=}CHCH_2Cl$$

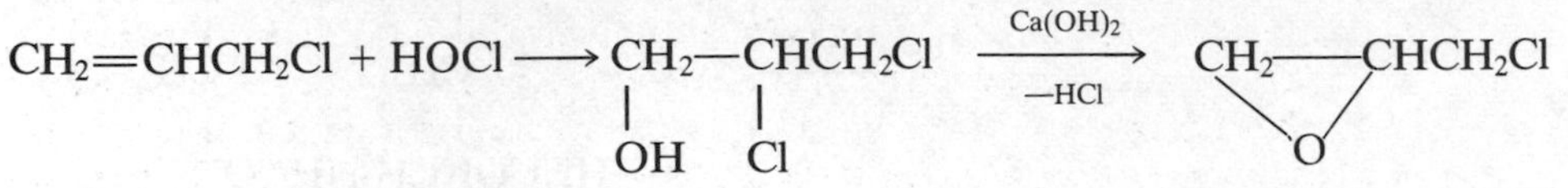

$$CH_2{=}CHCH_2Cl + HOCl \longrightarrow \underset{\underset{OH}{|}}{CH_2}{-}\underset{\underset{Cl}{|}}{CH}CH_2Cl \xrightarrow[\text{—HCl}]{\text{Ca(OH)}_2} CH_2{-}CHCH_2Cl \text{ (epoxide)}$$

(iii) Polyurethanes
These polymers are obtained from reaction of isocyanates with alcohols, e.g.

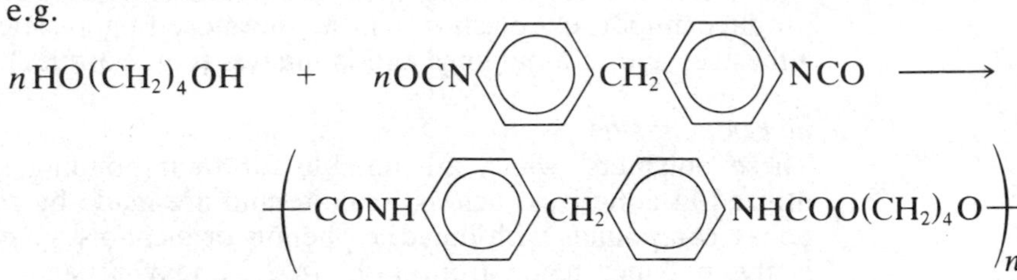

If triols and di-, or tri-, isocyanates are employed, cross-linking can take place and when a blowing agent, such as CCl_3F, is incorporated, the heat of polymerization causes it to vaporize and a polyurethane foam is produced. This reaction has been utilized in the insulation of cavity walls. The linear polymers have been used for making elastomers (e.g. Spandex).

(iv) Alkyd resins
Alkyd resins are polyesters derived from *alc*ohols and *ac*i*d*s. An important example is that derived from condensation of benzene-1,2-dicarboxylic anhydride and propane-1,2,3-triol, and sometimes known as a *glyptal* resin (from the former names of the constituents). If the two components are heated together at 180°C, polymerization takes place to form a long-chain linear polymer:

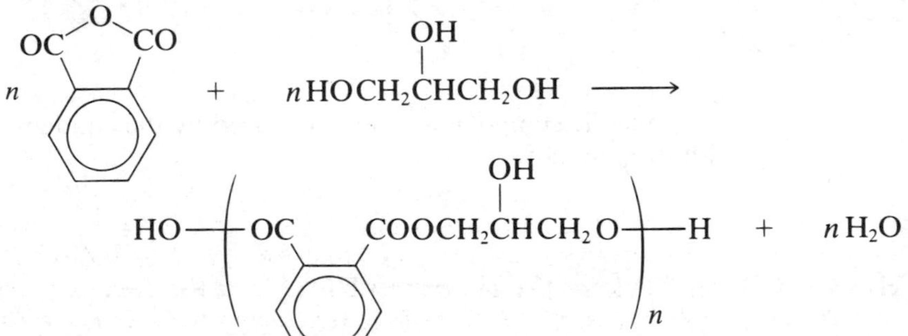

This can be dissolved in a suitable solvent and applied as a surface coating; on heating, the solvent evaporates and a hard, smooth, protective surface results, owing to cross-linking of the chains.

20.3 Synthetic fibres

Man-made fibres have replaced the traditional cotton and wool for many purposes, since the chemist has been able to provide textiles with specially desired properties. Man-made fibres may be based on natural materials, like cellulose, or on wholly synthetic compounds. It is the latter on which synthetic fibres are properly based. Some of the most important and well-known ones are polyesters and polyamides, as exemplified by the familiar names Terylene and Nylon. Both of these polymers are in fact thermoplastics and were extensively studied by the American chemist Wallace Carothers when working for the Du Pont company during the 1920s and 1930s. Only his work on polyamides proved immediately fruitful and the result was the first totally synthetic fibre, known as **Nylon** and first produced in the USA in 1938.

Apart from the necessity for synthetic fibres to be chemically unreactive, their strength and abrasion resistance are two highly important features. The strength can be expressed in terms of *tenacity* and *extensibility*, as indicated by a typical stress/strain graph for a polymer (*Figure 20.2*).

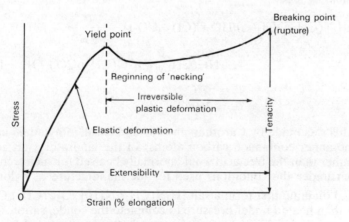

Figure 20.2 Plot of stress v. strain for a synthetic fibre

The initial, linear portion of the graph represents rapid *elastic* deformation of the polymer, as bonds in the molecular chains are stretched and twisted; the gradient, which reflects the degree of crystallinity of the polymer, is a reflection of the *rigidity modulus*, or stiffness, of the polymer and is high for a hard material and low for a soft one. Above the yield point, plastic deformation of the polymer begins as the polymer chains begin to disentangle and reorient themselves, and very considerable elongation can occur, although there is irreversible deformation (*see creep*, Section 20.4). When the randomly arranged crystallites begin to align themselves along the directional axis of strain, the process occurs *gradually* along the length of the filament, rather than *uniformly* throughout the sample, the diameters of the two parts of the filament remaining unchanged and the length of one increasing at the expense of the other. This phenomenon is known as **necking**. The plastic deformation is known as **drawing** and when it occurs at ordinary temperatures it is referred to as **cold-drawing**. It is of great importance in producing textile fibres (e.g. Terylene is readily cold-drawn), as the crystallinity and hence strength are usually increased appreciably. Eventually, at high strains, irreversible viscoelastic flow of the polymer occurs as the molecular chains slip over each other and the filament finally breaks.

As with other polymers, the crystallinity of the fibre, which will determine its tenacity and stiffness, is influenced by side-groups in the polymer chain, by its flexibility and by the intermolecular forces. The latter, naturally, also affect the melting point and power of water retention.

(i) Nylon

Nylon is actually a generic name covering the polyamide class of synthetic fibres. The first successful nylon to be made was derived from the condensation of two difunctional monomers, 1,6-diaminohexane and hexanedioic acid:

$$nH_2N(CH_2)_6NH_2 + nHO_2C(CH_2)_4CO_2H \xrightarrow[2\ MN\ m^{-2}]{270°C}$$

$$H\text{-}(NH(CH_2)_6NHCO(CH_2)_4CO)_n\text{-}OH + (2n-1)H_2O$$

$$T_g \approx 57°C$$
$$T_m \approx 260°C$$

This was made by Carothers in 1935 and is called **nylon 66** because each monomer contains 6 carbon atoms. In the laboratory, an acid chloride rather than the free acid would normally be used for such a condensation, but under the conditions used for the manufacture of nylon the diacid and diamine first form a salt $[H_3\overset{+}{N}(CH_2)_6\overset{+}{N}H_3][^-O_2C(CH_2)_4CO_2^-]$, which is then heated under pressure to complete the condensation. The starting materials for this polymerization are both manufactured from cyclohexane, although some hexanedinitrile has been manufactured by the electrolytic reduction of propenenitrile:

$$2CH_2\!\!=\!\!CHCN + 2H_2O + 2e^-$$

$$\searrow \text{(+ 2OH}^-\text{)}$$

$$HO_2C(CH_2)_4CO_2H \xrightarrow[\text{heat}]{NH_3} H_2NOC(CH_2)_4CONH_2 \xrightarrow[]{-H_2O} NC(CH_2)_4CN$$

$$\xrightarrow[Ni]{H_2} H_2NCH_2(CH_2)_4CH_2NH_2$$

Another nylon, **nylon 6**, is made by the polymerization of a difunctional monomer obtained from the ring-opening of 6-aminohexanoic acid lactam (a *lactam* is a cyclic amide):

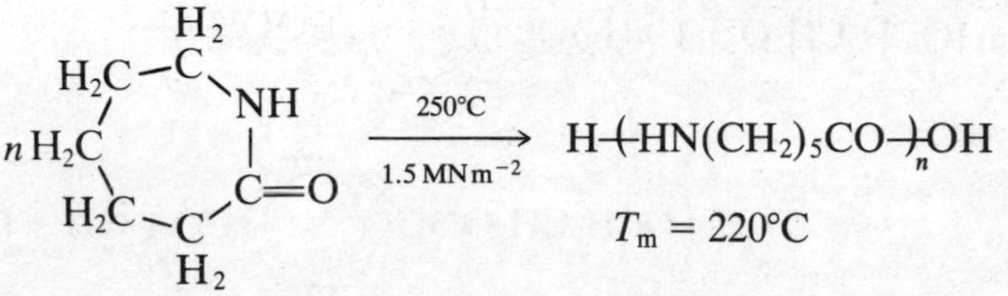

$$n \quad \xrightarrow[1.5\,\text{MN m}^{-2}]{250°C} \quad H\!\!-\!\!(HN(CH_2)_5CO\!\!-\!\!)_n OH$$

$$T_m = 220°C$$

The lactam is manufactured by a rearrangement reaction of cyclohexanone oxime which is derived from cyclohexane, again, as starting material:

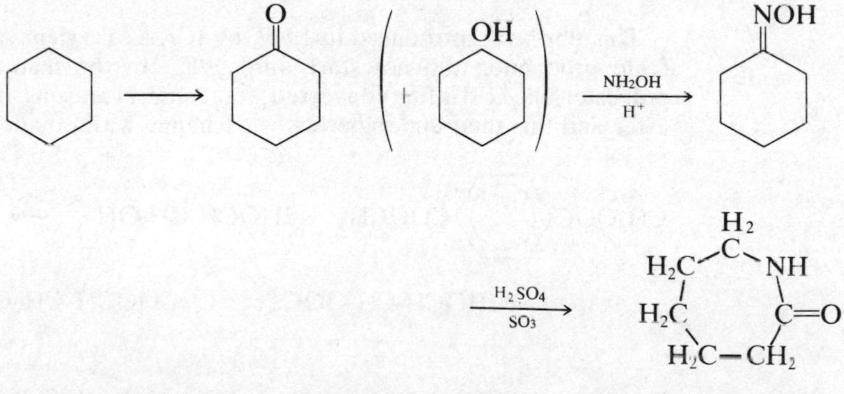

$$\xrightarrow[H^+]{NH_2OH}$$

$$\xrightarrow[SO_3]{H_2SO_4}$$

The hydrogen bonding in the nylons leads to very low water retention:

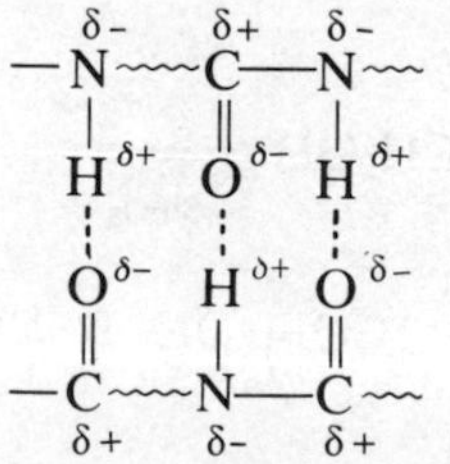

The excellent properties of strength, elasticity, abrasion resistance, non-shrinkage and quick-drying make them very versatile fibres. They are generally unaffected by solvents but, being polyamides, they are slowly attacked by strong acids.

(ii) Polyesters

Carothers' fundamental exploratory work on polyesters was taken up by Rex Whinfield and James Dickson, working in the research laboratories of the Calico Printers Association. In 1941, they succeeded in producing a polyester fibre based on the condensation of ethane-1,2-diol with 1,4-benzenedicarboxylic acid:

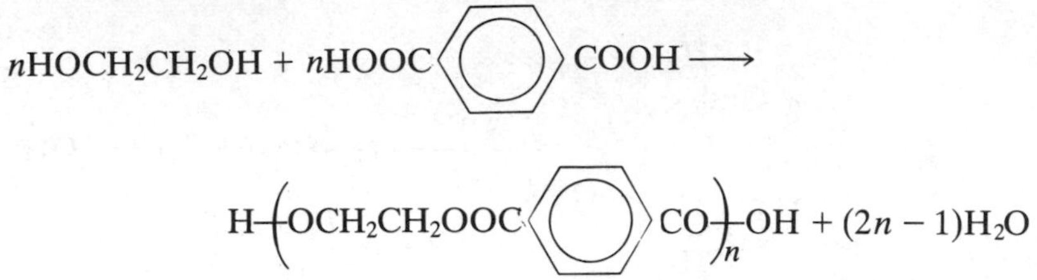

$$n\text{HOCH}_2\text{CH}_2\text{OH} + n\text{HOOC}\langle\bigcirc\rangle\text{COOH} \longrightarrow$$

$$\text{H}\!\!-\!\!\left[\text{OCH}_2\text{CH}_2\text{OOC}\langle\bigcirc\rangle\text{CO}\right]\!\!-\!\!\text{OH} + (2n-1)\text{H}_2\text{O}$$

$$T_\text{g} = 67°\text{C}$$
$$T_\text{m} = 265°\text{C}$$

This fibre was introduced in 1949, by ICI, as **Terylene**, although large-scale production did not start until 1955. In the manufacture of the polyester, the acid is first converted, for technical reasons, into its dimethyl ester and this then undergoes ester exchange with ethane-1,2-diol:

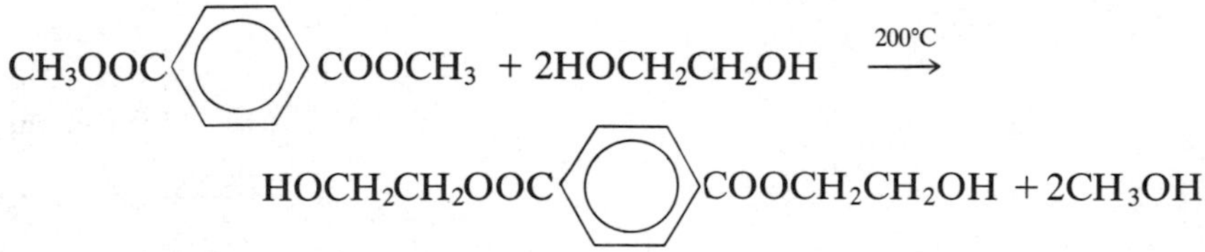

$$\text{CH}_3\text{OOC}\langle\bigcirc\rangle\text{COOCH}_3 + 2\text{HOCH}_2\text{CH}_2\text{OH} \xrightarrow{200°\text{C}}$$

$$\text{HOCH}_2\text{CH}_2\text{OOC}\langle\bigcirc\rangle\text{COOCH}_2\text{CH}_2\text{OH} + 2\text{CH}_3\text{OH}$$

The product is then polymerized by heating to about 270°C under reduced pressure in the presence of a catalyst, with continuous removal of the diol which is produced:

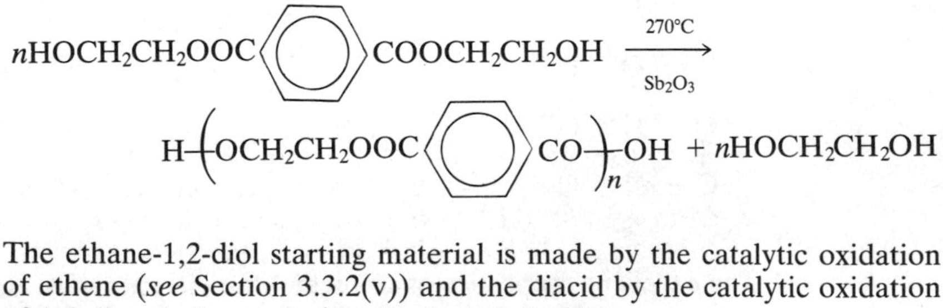

$$n\text{HOCH}_2\text{CH}_2\text{OOC}\langle\bigcirc\rangle\text{COOCH}_2\text{CH}_2\text{OH} \xrightarrow[\text{Sb}_2\text{O}_3]{270°\text{C}}$$

$$\text{H}\!\!-\!\!\left[\text{OCH}_2\text{CH}_2\text{OOC}\langle\bigcirc\rangle\text{CO}\right]\!\!-\!\!\text{OH} + n\text{HOCH}_2\text{CH}_2\text{OH}$$

The ethane-1,2-diol starting material is made by the catalytic oxidation of ethene *(see* Section 3.3.2(v)) and the diacid by the catalytic oxidation of 1,4-dimethylbenzene *(see* Section 12.5.2(ii)).

The benzene rings increase the stiffness of the chains and raise the T_m of the polymer. The 1,4-diacid is symmetrical and permits a close alignment of chains, particularly on cold-drawing; the 1,2-diacid results in bulky groups protruding from the chains, thus lowering the crystallinity and T_m of the polymer, which becomes a hard, glass-like resin. Terylene possesses many of the advantages of nylon, although it is more sensitive to strong acid hydrolysis. Its T_g is sufficiently low for pleats to be ironed into the fabric, but sufficiently high to prevent them being washed out in hot water.

(iii) Polyalkenes

The most widely used polyalkene synthetic fibre is poly(propenonitrile), first produced by Du Pont in 1945. Propenonitrile is polymerized with the aid of a peroxo salt in solution:

$$nCH_2{=}CHCN \longrightarrow \left(CH_2{-}\underset{\underset{CN}{|}}{CH}\right)_n$$

$$T_g \approx 100°C$$
$$T_m = 317°C$$

The propenonitrile starting material is normally manufactured by the 'ammoxidation' of propene (cf. Section 3.3.2(v)):

$$2CH_2{=}CHCH_3 + 2NH_3 + 3O_2 \xrightarrow[\text{400–470°C}]{\text{catalyst}} 2CH_2{=}CHCN + 6H_2O$$

The small cyano group permits fairly close packing of molecular chains, which are held together by quite strong hydrogen bonds:

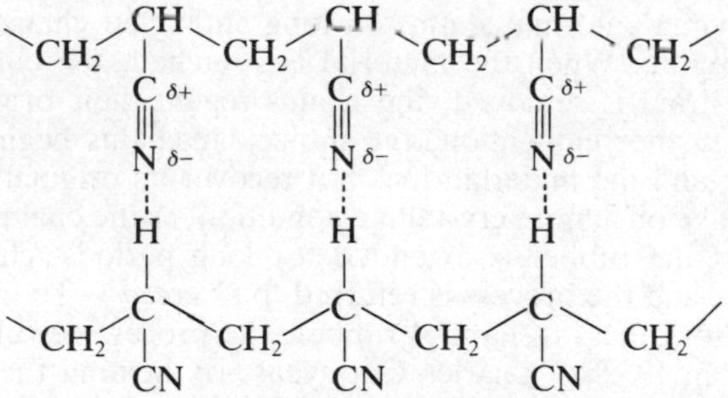

Consequently, the polymer is quite crystalline and hard with a high softening point (>300°C). Sometimes known as an 'acrylic' fibre, poly(propenonitrile) is extensively used in clothing (e.g. Courtelle, Orlon, Acrilan, etc.), although it is quite an expensive fibre.

Poly(propenonitrile) has recently become extremely important in the preparation of **carbon fibres**. The polymer, in the form of filaments, is wound on a rigid frame and oxidized at about 200°C until a complex 'ladder' polymer is produced. This is then released from the tension of the frame and carbonized in an inert atmosphere, hydrogen cyanide, ammonia and water being eliminated. Finally, the material is heated to 1800–3000°C, when all elements other than carbon are eliminated and a graphite type of crystal structure is produced.

20.4   Elastomers

Elastomers are polymers which possess the property most characteristic of natural rubber, i.e. of being easily deformed under stress but of recovering immediately the tension is removed. This elasticity is shown only by flexible polymers with a low glass transition temperature, i.e. ones which at room temperature have already passed from the totally amorphous state to a rubber-like condition.

Natural rubber is a hydrocarbon, a 1,4-addition polymer of methylbuta-1,3-diene (*isoprene*), in which the carbon–carbon double bonds are in the *cis* arrangement, i.e. the methyl and hydrogen atoms of the double bond lie on the same side of it:

$$T_g = -75°C$$
$$T_m = 28°C$$

In the polymer's normal state, the long entangled chains are coiled up to some degree. When the material is stretched, the chains uncoil but when the strain is removed, the chains regain their original entangled state. On further elongation, the molecular chains begin to slide over each other and the material does not recover its original condition, but begins to take on a more crystalline condition, as the chains become more ordered. If the rubber is extended for long periods, chain slip occurs irreversibly and the process is referred to as *creep*. To impart a greater degree of toughness to natural rubber, the process of **vulcanization** was developed in 1839 by Charles Goodyear. By heating the rubber with a measured amount of sulphur, occasional cross-linking of the chains is brought about. This cross-linking still permits some parts of the polymer chains to extend themselves and stretch out, but it prevents chains slipping over each other and hence reduces creep. It also prevents the polymer molecules from dissolving in organic solvents. However, there is still space for solvent molecules to penetrate the rubber, and hence vulcanized rubber will swell in such a liquid:

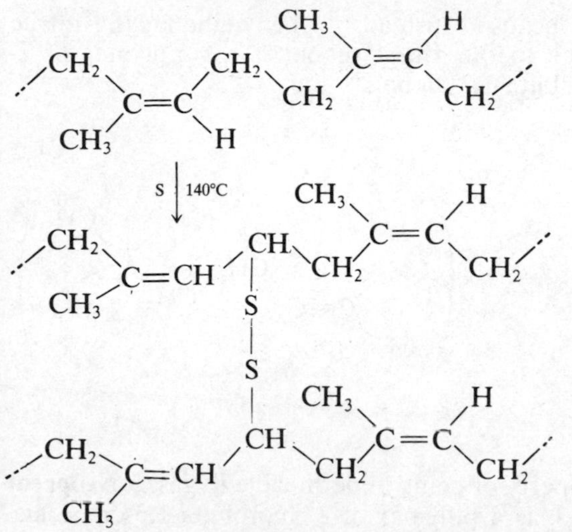

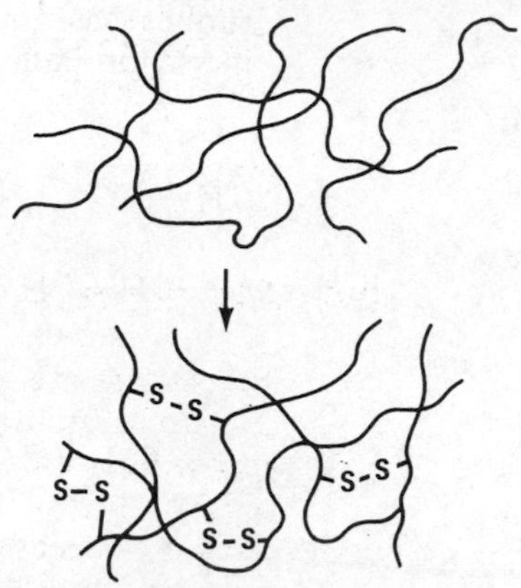

If vulcanization is carried to extremes, the polymer will become hard and brittle like ebonite and will not absorb any solvent.

Another kind of rubber is **gutta-percha**, obtained from trees of the *Palaquium* family. Owing to the more rigid *trans* arrangement of the double bonds, this material is hard and non-resilient, quite unlike natural rubber:

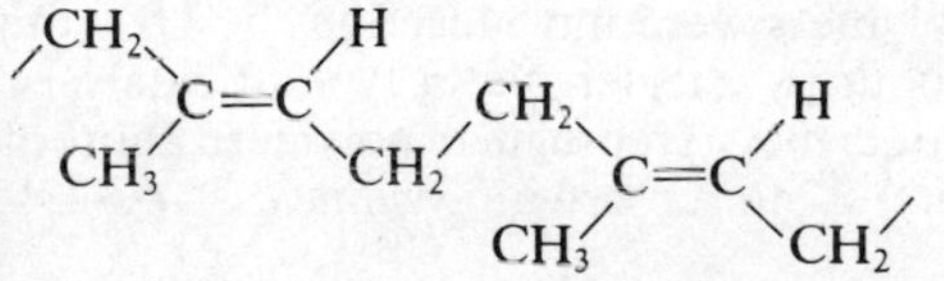

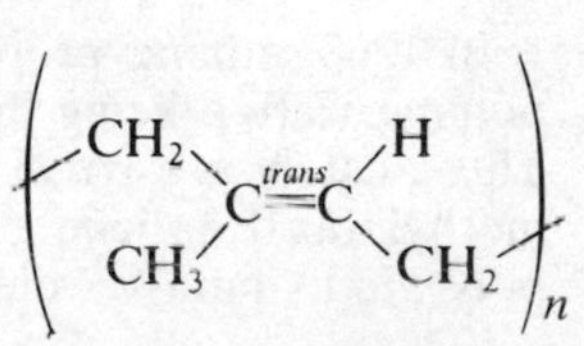

Various synthetic elastomers have been produced with a view to improving properties such as oil resistance and resistance to oxidation. The most important one is a copolymer of buta-1,3-diene and phenyl-ethene, known as **SBR rubber** (*s*tyrene *b*utadiene *r*ubber):

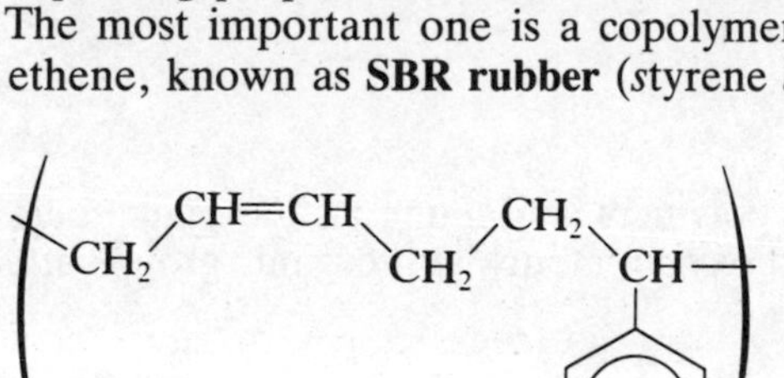

The monomers are polymerized randomly in the approximate molar ratio of 3:1, respectively. The buta-1,3-diene is obtained from the C_4 fraction of naphtha cracking.

Propenonitrile can also be used instead of phenylethene; the nitrile rubber has good resistance to oil. Butyl rubber is a copolymer of 2-methylpropene and methylbuta-1,3-diene:

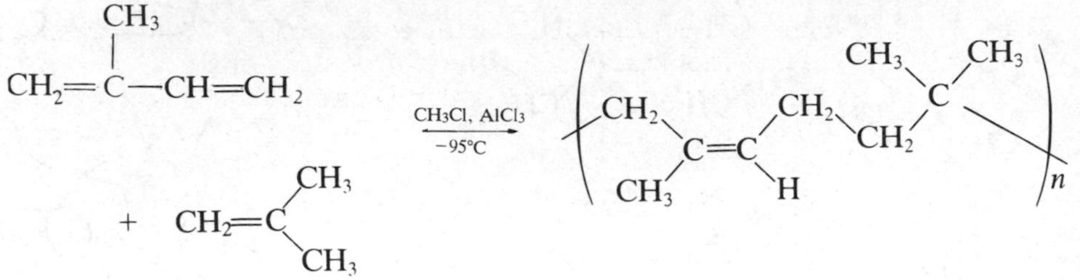

and has the important property of being impermeable to gases. **Neoprene rubber**, made first in 1933, is a polymer of 2-chlorobuta-1,3-diene and shows excellent resistance to solvents:

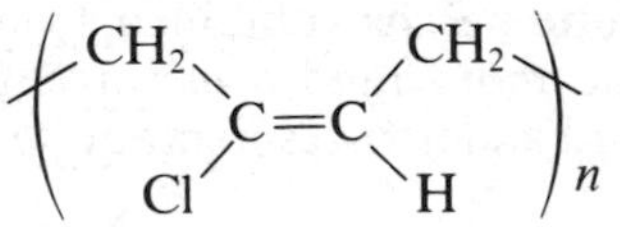

$$T_g = -50°C$$

In 1968, ethene–propene copolymers were introduced in the UK, the polymerization being brought about by Ziegler–Natta type of catalysts. These catalysts have also permitted the stereoregular polymerization of methylbuta-1,3-diene (*isoprene*) to form *synthetic cis*-1,4-poly(methylbuta-1,3-diene).

All rubbers containing $\diagdown$C=C$\diagup$ bonds are sensitive, to some extent, to atmospheric oxidation and have to be mixed with anti-oxidants to inhibit this (*but see silicone rubbers*, Section 20.5).

20.5 Silicones

The silicones are polymers containing an inorganic 'backbone' of alternating silicon and oxygen atoms with organic groups attached to silicon:

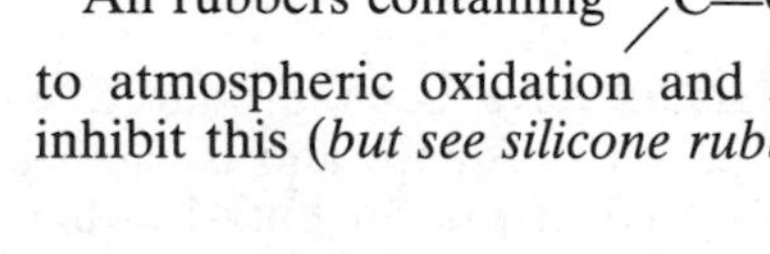

where R is commonly CH_3— or C_6H_5—

Although they were unwittingly discovered by Professor Frederic Kipping at the turn of the century, they were only characterized and properly developed as a result of commercial research carried out by Corning Glass Works and the General Electric Company in the USA during the 1930s (although some work was carried out, quite independently, in Russia at about the same time). Both these companies were looking for new and novel materials which might rival glass and the acrylic plastics for certain applications.

Silicone polymers are made by condensation polymerization of silicon diols and triols obtained by the hydrolysis of organosilicon chlorides. In 1940–41, the American chemist Eugene Rochow discovered his 'Direct Synthesis' of methylchlorosilanes. Chloromethane is passed over a finely divided mixture of silicon with copper, as catalyst, at about 300°C. The products are separated by very careful fractional distillation:

$$CH_3Cl + Si \xrightarrow[Cu]{300°C} CH_3SiCl_3 \ + \ (CH_3)_2SiCl_2 \ + \ (CH_3)_3SiCl$$

$$\begin{array}{ccc} \text{b.p.} = 66°C & \text{b.p.} = 70°C & \text{b.p.} = 57°C \\ 10\% & 75\% & 4\% \end{array}$$

These chloro compounds all undergo hydrolysis quite readily, but they have different roles to play in the polymerization process. When dimethyl dichlorosilane is added to water, it hydrolyses to the corresponding diol which spontaneously condenses with other molecules to form polymers, some of which are cyclic. If, however, the mixture is stirred in dilute acid for some time, an equilibrium mixture of long-chain silicone polymers is produced:

$$\underset{\substack{|\\CH_3}}{\overset{\substack{CH_3\\|}}{Cl-Si-Cl}} + 2H_2O \longrightarrow \underset{\substack{|\\CH_3}}{\overset{\substack{CH_3\\|}}{HO-Si-OH}} + 2HCl$$

$$\ldots + \underset{\substack{|\\CH_3}}{\overset{\substack{CH_3\\|}}{HO-Si-OH}} + \underset{\substack{|\\CH_3}}{\overset{\substack{CH_3\\|}}{HO-Si-OH}} + \underset{\substack{|\\CH_3}}{\overset{\substack{CH_3\\|}}{HO-Si-OH}} + \ldots \longrightarrow$$

$$\underset{\substack{|\\CH_3}}{\overset{\substack{CH_3\\|}}{-Si-O}}-\underset{\substack{|\\CH_3}}{\overset{\substack{CH_3\\|}}{Si-O}}-\underset{\substack{|\\CH_3}}{\overset{\substack{CH_3\\|}}{Si-O}}- \ \text{i.e.} \ HO{\left(\underset{\substack{|\\CH_3}}{\overset{\substack{CH_3\\|}}{Si-O}}\right)}_n H \ + (n-1)H_2O$$

The average length of the silicone polymer chains may be controlled by varying the proportions of trimethylchlorosilane which is added to the material before hydrolysis. Since this compound is only monofunctional, once it has been hydrolysed and condensed with a polymer chain it prevents any further chain growth at that point and is thus referred to as a *chain-stopper*. The higher the proportion of $(CH_3)_3SiCl$ added, the shorter is the average length of silicone chains produced:

$$
\begin{array}{c}
CH_3 \\
| \\
CH_3\!-\!Si\!-\!Cl + H_2O \\
| \\
CH_3
\end{array}
\longrightarrow
\begin{array}{c}
CH_3 \\
| \\
CH_3\!-\!Si\!-\!OH + HCl \\
| \\
CH_3
\end{array}
\underrightarrow{\quad
\begin{array}{cc}
CH_3 & CH_3 \\
| & | \\
HO\!-\!Si\!-\!O\!-\!Si\!-\!\cdots \\
| & | \\
CH_3 & CH_3
\end{array}
\quad}
$$

$$
\begin{array}{ccc}
CH_3 & CH_3 & CH_3 \\
| & | & | \\
CH_3\!-\!Si\!-\!O\!-\!Si\!-\!O\!-\!Si\!-\!O\!-\!\cdot\!-\!\cdot\!-\!\cdot \\
| & | & | \\
CH_3 & CH_3 & CH_3
\end{array}
$$

In contrast, the CH_3SiCl_3 produces a trifunctional compound on hydrolysis which permits cross-linking of silicone chains to occur:

$$
\begin{array}{cc}
CH_3 & CH_3 \\
| & | \\
\cdots\!-\!O\!-\!Si\!-\!O\!-\!Si\!-\!OH \\
| & | \\
CH_3 & CH_3
\end{array}
+
\begin{array}{c}
CH_3 \\
| \\
HO\!-\!Si\!-\!OH \\
| \\
OH
\end{array}
+
\begin{array}{c}
CH_3 \\
| \\
HO\!-\!Si\!-\!O\!-\!\cdots \\
| \\
CH_3
\end{array}
\qquad
\begin{array}{cccc}
CH_3 & CH_3 & CH_3 & CH_3 \\
| & | & | & | \\
\cdots\!-\!O\!-\!Si\!-\!O\!-\!Si\!-\!O\!-\!Si\!-\!O\!-\!Si\!-\!O\!-\!\cdot \\
| & | & | & | \\
CH_3 & CH_3 & & CH_3 \\
& & | & \\
& & O & +\,H_2O \\
& & | &
\end{array}
$$

$$
\begin{array}{cc}
CH_3 & CH_3 \\
| & | \\
\cdots\!-\!O\!-\!Si\!-\!O\!-\!Si\!-\!OH \\
| & | \\
CH_3 & CH_3
\end{array}
+
\begin{array}{c}
OH \\
| \\
HO\!-\!Si\!-\!OH \\
| \\
CH_3
\end{array}
+
\begin{array}{c}
CH_3 \\
| \\
HO\!-\!Si\!-\!O\!-\!\cdots \\
| \\
CH_3
\end{array}
\qquad
\begin{array}{cccc}
CH_3 & CH_3 & & CH_3 \\
| & | & | & | \\
\cdots\!-\!O\!-\!Si\!-\!O\!-\!Si\!-\!O\!-\!Si\!-\!O\!-\!Si\!-\!O\!-\!\cdot \\
| & | & | & | \\
CH_3 & CH_3 & CH_3 & CH_3
\end{array}
$$

The silicones illustrate, particularly well, the way the chemist can tailor the properties of a polymer to suit particular purposes. In devising a new polymer, Rochow specified that it should contain the minimum amount of carbon and hydrogen, in order to differentiate it from other organic polymers, and no carbon–carbon bonds, in order to avoid charring at high temperatures. The Si—O link provided the required, flexible 'backbone' which was to be thermally stable.

The simple, linear silicone polymers are generally colourless, unreactive fluids whose viscosities are controlled by the proportions of chain-stopper incorporated during polymerization. Cross-linking of the polymer chains reduces the flexibility of the chains and the resulting polymers may be very viscous liquids or low-melting greases. When extensive cross-linking takes place, *silicone resins* are produced. Cross-linking by way of methyl groups ($-CH_2\!-\!CH_2-$) produces *silicone elastomers*:

$$
\begin{array}{ccc}
\text{CH}_3 & \text{CH}_3 & \text{CH}_3 \\
| & | & | \\
-\text{O}-\text{Si}-\text{O}-\text{Si}-\text{O}-\text{Si}-\text{O}- \\
| & | & | \\
\text{CH}_3 & \text{CH}_3 & \text{CH}_3
\end{array}
\qquad
\begin{array}{ccc}
\text{CH}_3 & \text{CH}_3 & \text{CH}_3 \\
| & | & | \\
-\text{O}-\text{Si}-\text{O}-\text{Si}-\text{O}-\text{Si}-\text{O}- \\
| & | & | \\
\text{CH}_3 & \text{CH}_2 & \text{CH}_3
\end{array}
$$

$$
\xrightarrow[\text{heat}]{\text{ROOR}}
$$

$$
\begin{array}{ccc}
\text{CH}_3 & \text{CH}_3 & \text{CH}_3 \\
| & | & | \\
-\text{O}-\text{Si}-\text{O}-\text{Si}-\text{O}-\text{Si}-\text{O}- \\
| & | & | \\
\text{CH}_3 & \text{CH}_3 & \text{CH}_3
\end{array}
\qquad
\begin{array}{ccc}
\text{CH}_3 & \text{CH}_2 & \text{CH}_3 \\
| & | & | \\
-\text{O}-\text{Si}-\text{O}-\text{Si}-\text{O}-\text{Si}-\text{O}- \\
| & | & | \\
\text{CH}_3 & \text{CH}_3 & \text{CH}_3
\end{array}
$$

These silicone elastomers are of particular importance because they can withstand very considerable extremes of temperature (e.g. $-100°C$ to $250°C$) while still performing satisfactorily, and they are resistant to attack by oxygen, ozone and ultraviolet light, since they contain no double bonds.

The silicones have been used in an enormous range of applications, depending on their outstanding properties of chemical inertness, water-repellency, electrical insulation and thermal stability at high temperatures.

The excellent thermal stability and electrical insulation properties have made silicones ideal for insulation at high temperatures, and the room temperature vulcanizing (RTV) *silicone rubbers* are frequently used for encapsulating small electronic items. The water-repellent properties have been utilized in the shower-proofing of mackintoshes and in the treatment of brickwork and masonry, where they increase water 'run off' and also help prevent erosion. Small amounts of *silicone emulsions* are extremely effective in combating 'foaming' which often occurs in dyebaths, textile processing and fermentations, and the 'anti-stick' properties of silicones have caused them to be used for mould-release agents and on backing paper of wallpapers and adhesives. Significantly, the lack of chemical reactivity has been taken advantage of in the manufacture of a variety of prosthetic devices and artificial organs, for use in the human body.

Questions

1   (a) Define what is meant by each of the following terms:
       (i)   a random copolymer,
       (ii) an alternating copolymer,
       (iii) a block copolymer.
    (b) (i)   Define what is meant by the *glass transition temperature*, T_g.
       (ii) State three structural factors which determine this temperature.
       (iii) Is T_g for poly(ethanediyl benzene-1,4-dicarboxylate) (Terylene) higher or lower than that for nylon 6?
       (iv) Give a reason for your answer in (iii).
       (v)   How do you account for the fact that pleats in a polyester (Terylene) garment are not removed during washing of the garment?   JMB

2 (a) An ethenyl (*vinyl*) monomer A, on polymerization, gives long-chain
       molecules which may be represented as:

   —A—A—A—A—A—A—A—A—A—

   Using a similar representation,
   (i)   draw a single chain of a random copolymer of A and a second
         monomer B,
   (ii) draw a single chain of the block copolymer of A and B,
   (iii) and draw a single chain of a graft copolymer where B is grafted
         on to the polymer A.
   (b) (i) Name the monomer which is copolymerized with buta-1,3-diene
            to produce a synthetic rubber similar to natural rubber.
       (ii) What type of copolymer is this?
   (c) The polymerization of an ethenyl (*vinyl*) monomer may be represented
       in its simplest form by the change

   $$nCH_2{=}CHX \rightarrow +CH_2{-}CHX{+}_n$$

   (i) What type of organic compound can be used to initiate a free
        radical polymerization?
   (ii) Name the three essential stages in the mechanism of a free radical
        polymerization.
   (d) Consider the following two polymers:

   $$+CH_2{-}CH+_n \qquad +CH_2{-}CH+_n$$
   $$\qquad | \qquad\qquad\qquad |$$
   $$\qquad CN \qquad\qquad\qquad C_6H_5$$
   P   Q

   (i) Name the polymers P and Q.
   (ii) Explain why P and Q have different physical properties.

                                                                      JMB

3 (a) Describe what is meant by the terms *homolysis* and *heterolysis*.
   (b) Under the action of heat a molecule A_2 breaks down to form two
       radicals:

   $$A_2 \xrightarrow{\text{heat}} 2A^{\cdot}$$

   These radicals react with phenylethene (*styrene*) to give a new radical
   species:

   $$A^{\cdot} + C_6H_5CH{=}CH_2 \rightarrow A{-}CH_2CH^{\cdot}$$
   $$\qquad\qquad\qquad\qquad\qquad |$$
   $$\qquad\qquad\qquad\qquad\qquad C_6H_5$$

   (i)   Explain how this reaction may be regarded as a chain initiation
         process.
   (ii) Write the equation for a typical chain propagation step involving
         the new radical and a phenylethene (*styrene*) molecule.
   (iii) What polymer results from the propagation of such a chain?

(iv) Show a possible method of chain termination.
(v) Could such a polymer exhibit optical activity?

St. Andrews

4   (a) Give the structural formula of hexanedioic (*adipic*) acid.
    (b) Give the structural formula of 1,6-diaminohexane.
    (c) What feature of the molecules in (a) and (b) enable them to be used
        to make polymeric material?
    (d) Indicate by means of equations how the compounds in (a) and (b)
        interact to form a polymer.
    (e) What is the name of the chemical group formed in (d)?
    (f) Give the structural formula of the repeating unit of this polymer.
    (g) What is the common name of this polymer?
    (h) What type of polymerization process did you show in (d)?
    (j) How is the material made in (d) converted into a fibre?
    (k) What determines the strength of a fibre?
    (l) Which particular physical property of the fibre in (j) makes it especially
        suitable for use in ropes and ladies' stockings?

JMB

5* (a) A dicarboxylic acid (HOOC—X—COOH) and a diol (HO—Y—OH)
        undergo a chemical reaction to produce a polymeric product.
        (i)   Give the equation for the first step in the polymerization reaction.
        (ii) Give the formula of the product formed by the complete reaction
              of *n* molecules of the dicarboxylic acid with *n* molecules of the
              diol.
        (iii) How can the chain length of the polymer be controlled in this
              polymerization?
        (iv) Name **two** physical properties of the polymer which are deter-
              mined by the average chain length of the polymer.
    (b) (i)   Give the equation for the reaction between epoxypropane and
              ethane-1,2-diol.
        (ii) Why is a triol sometimes used in this type of reaction to produce
              a polyurethane?
        (iii) How is an expanded polyurethane produced?
    (c) State **two** uses of the product formed by the reaction between an
        epoxyalkane and a long-chain alcohol.

JMB

6   (a) Consider the polymerization of phenol with methanal (*formaldehyde*).
        (i)   Give the equation for the ring substitution reaction between
              phenol and methanal.
        (ii) Give the equation for the linking together of these ring substituted
              compounds.
        (iii) Give the formula of the final product.
        (iv) To what class of polymeric materials does the product of this
              polymerization belong?
        (v)   Why is the final polymer insoluble in all known solvents?

(b) The curve below shows the relationship when the tensile force is plotted against the extension for a sample of Terylene.

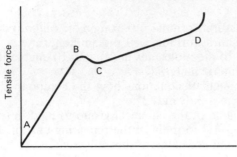

Extension

(i) Account for the shape of the curve, paying particular attention to the sections A to B and C to D.
(ii) Describe the changes in the polymer structure which occur in the region C to D.

JMB

7 (a) What type of polymerization is involved in the following process?

$$n CF_2 {=} CF_2 \rightarrow +CF_2{-}CF_2+_n$$

(b) How is the monomer made and what is the name of the polymer?
(c) What special property of the polymer has led to its widespread use in kitchen utensils, for example?
(d) Why is this polymer more difficult to fabricate than other similar ones?
(e) To what feature of the polymer's structure are these special properties due?

8 (a) Natural rubber is a polymer of 2-methylbuta-1,3-diene (*isoprene*) with a *cis* arrangement about the double bonds. Illustrate a section of the polymer's structure.
(b) Gutta percha consists of the above polymer molecules in the *trans* form. Explain why it is hard and non-resilient compared with natural rubber.
(c) Explain why vulcanization reduces the tendency of rubber to become sticky, particularly when heated.
(d) Describe the composition of any two synthetic rubbers and name one advantage each possesses over natural rubber.

9 (a) Write equations for the reactions between water and (i) methyltrichlorosilane, (ii) dimethyldichlorosilane, and (iii) trimethylchlorosilane.
(b) Show how polymers can be made from the methylchlorosilanes and explain the functions of each of the three types in (a).
(c) Outline the manufacturing route to the methylchlorosilanes.

(d) Give one example in each case of the use of silicones in (i) the building industry, (ii) the textile industry, and (iii) the electrical industry.

(e) Illustrate the structural difference between a silicone polymer and a silicone elastomer and state two advantages which silicone elastomers have over other synthetic elastomers.

10*   Answer all parts of this question.

(i)   'Terylene' (Dacron) is a synthetic fibre with the following structure, where x is a large number:

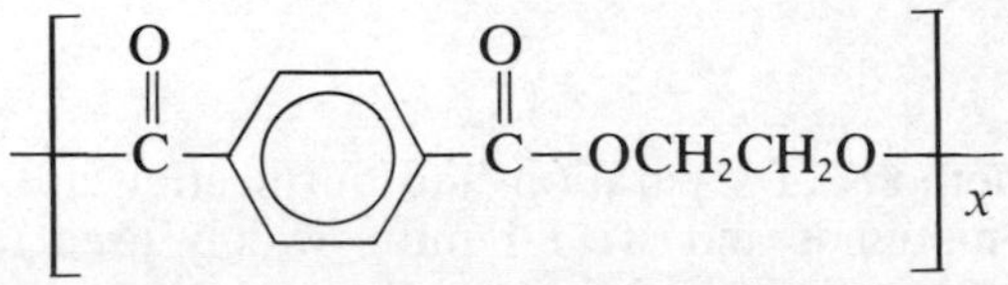

How could it be synthesized from simple organic starting materials?

(ii) What would be the effect on the structure and properties of the Terylene if 5% by weight of benzoic acid ($C_6H_5CO_2H$) were added to the reaction mixture during synthesis?

(iii) Discuss whether the following isomer of Terylene would be a good fibre:

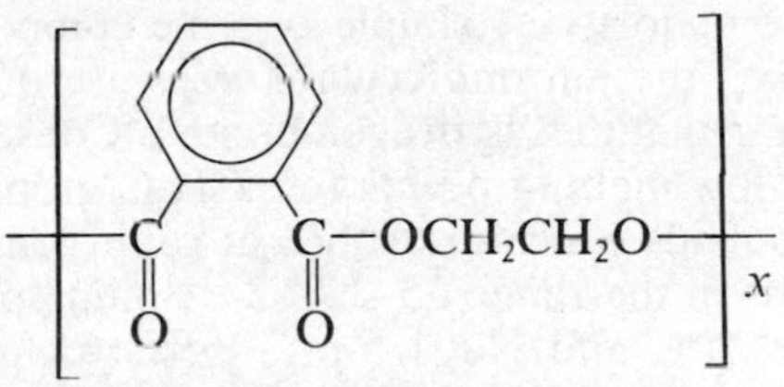

(iv) Why does a Terylene shirt slowly dissolve in dilute sulphuric acid?

Cambridge

11   Describe with examples the types of chemical reaction that lead to the formation of synthetic polymers.

How are the properties of synthetic polymers affected by their composition and structure?

Nuffield

Chapter 21
Physical methods of separation and purification

Physical methods of separation and purification are vital to a study of organic compounds and are far more widely used than with inorganic compounds. Many reactions between organic compounds do not proceed, under normal conditions, to completion, but instead reach a state of dynamic equilibrium. The desired product has therefore to be separated from the equilibrium mixture. Even when no equilibrium is involved, many organic reactions are accompanied, to some extent, by competing reactions (side-reactions) and, therefore, a separation of reaction products is necessary. Very few organic reactions produce just a single organic product.

Fortunately, the majority of simple organic compounds are covalent compounds in which the intermolecular forces are of the weak van der Waals type (or are sometimes hydrogen bonds). Consequently, they tend to have relatively low melting points (cf. CH_3Cl m.p. $-98°C$ and NaCl m.p. $808°C$; compounds which are solids at room temperature normally have melting points in the range $25-250°C$). Boiling points, too, are low (cf. CH_3Cl b.p. $-24°C$ and NaCl b.p. $1465°C$; compounds which are liquids at room temperature generally boil in the range $25-200°C$), as are molar enthalpies of vaporization. These properties permit the use of such techniques as sublimation and distillation. Organic compounds are also frequently soluble in a range of solvents of varying polarity and thus techniques such as crystallization, chromatography and solvent extraction can be widely practised.

21.1 Criteria of purity

The success of a separation or purification is usually judged by the criteria of melting point (for solids) and boiling point (for liquids), although these physical values may occasionally be unreliable.

21.1.1 Solids

Most organic solids which are simple compounds melt sharply at a relatively low temperature, which is characteristic of the individual compound. The melting point of a solid may be determined by a variety of methods.

In the school laboratory a quick, convenient and generally reliable method is to place a few dry crystals of the solid on to the bulb of a suitable thermometer, which is then enclosed in the apparatus shown in *Figure 21.1*. The outer tube is heated slowly with a burner, and the temperature carefully noted when the crystals begin to melt and when melting is complete. The two values should not differ by more than 2 degrees for a pure compound; the melting point is normally quoted as this melting range.

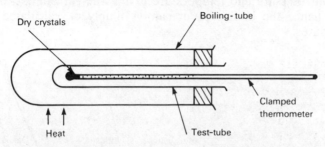

Figure 21.1   A simple method of melting point determination

If the compound is not pure, then the melting point will not be sharp and melting will occur over a temperature range. Further purification will therefore be necessary. Furthermore, the presence of impurity in a compound always *depresses* the melting point, so that if a melting point turns out to be higher than the accepted value for a particular compound, the substance cannot be that compound. The depression of melting point allows one to distinguish between two solids with identical, or very closely similar, melting points by the **mixed melting point** method. One solid is mixed with the other and the melting point determined. If it remains unchanged, then the two solids are identical; a depression in melting point indicates that the two solids are not identical.

The melting points of an enormous range of organic compounds are recorded in the chemical literature (textbooks, learned journals, etc.), so that compounds may be identified by their melting points. Carbonyl compounds, for example, are often identified in the laboratory by converting them into their 2,4-dinitrophenylhydrazones (*see* Section 8.3.3). These are sharp-melting, crystalline derivatives which are easily purified by crystallization (*see* Section 21.2.1). The melting points of the 2,4-DNP derivatives below show how some aldehydes and ketones could thus be recognized:

methanal	166°C
ethanal	168°C
propanal	155°C
propanone	128°C
butan-2-one	115°C

21.1.2   Liquids

The boiling of a liquid involves more complex behaviour than the melting of a solid. The normal boiling point of a liquid is the temperature at which the saturated vapour pressure of the liquid becomes equal to the external pressure (normally atmospheric pressure; standard atmospheric pressure is 101.32 kN m^{-2}, or approximately 0.1 MN m^{-2}). The boiling point is therefore not a constant value and must be quoted together with the pressure at which it has been recorded. The relationship between saturated vapour pressure and temperature is not a linear one (*see Figure 21.2*), and hence the boiling point cannot simply be calculated at different pressures.

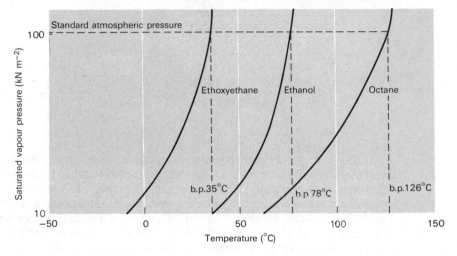

Figure 21.2 Plot of saturated vapour pressure v. temperature for various liquids

The boiling point of a pure liquid can be measured in the school laboratory by the following method (*see Figure 21.3*). A small volume of liquid is placed in the glass tube, together with some anti-bumping granules, and the thermometer positioned so that the bulb lies in the constriction. The tube is gently heated with a micro-burner until drops of liquid appear on the thermometer bulb and the liquid gently refluxes at the constriction. Under these conditions, equilibrium is reached between liquid and vapour and the temperature recorded is the boiling point of the liquid (care must be taken if the liquid is flammable).

The presence of a non-volatile solute in a liquid raises the boiling point, but the presence of other liquid impurities can cause variable effects. According to Raoult's law, the vapour pressure exerted by a component of an *ideal** solution is proportional to the mole fraction of that component

---

* In an *ideal* solution of liquids A and B, the intermolecular forces A—A are considered to be indistinguishable from B—B and A—B ones.

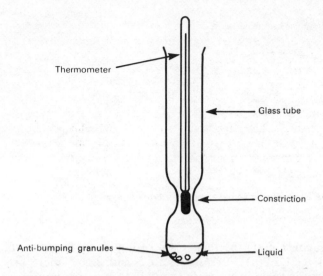

Figure 21.3 A simple method of boiling point determination

present at constant temperature. Propan-1-ol (b.p. = 97°C) and
propan-2-ol (b.p. = 83°C), for example, form a virtually ideal solution
so that the boiling point of an equimolar mixture of the two (i.e. mole
fraction of each component is 0.5) is about 90°C, i.e. midway between
the boiling points of the two pure components. However, components
have to be very closely similar in structure before a mixture of them will
approximate to an ideal solution, so that such mixtures are generally the
exception rather than the rule. *Ideal behaviour* is shown in *Figure 21.4*.

Mixtures more commonly show either positive or negative deviation
from Raoult's law, with the result that a mixture may have a boiling point
either higher or lower than that anticipated from ideal behaviour. Certain
mixtures (*constant-boiling mixtures* or *azeotropes*) also boil at a constant
temperature without change in composition. These facts are important

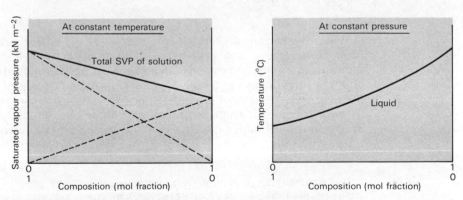

Figure 21.4 Ideal behaviour of a two-component solution of liquids

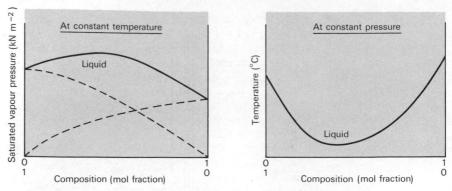

Figure 21.5 Non-ideal behaviour of a two-component solution of liquids showing positive deviation

when considering the separation of liquids by fractional distillation (*see* Section 21.2.2(ii)). In *Figures 21.5* and *21.6*,* there are maxima and minima in the curves, corresponding to constant-boiling mixtures.

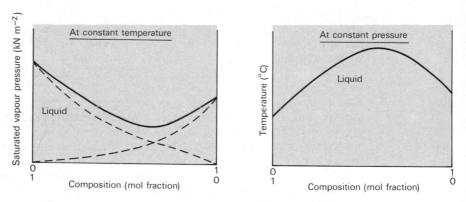

Figure 21.6 Non-ideal behaviour of a two-component solution of liquids showing negative deviation

When cyclohexane and ethanol are mixed, for example, there is a slight increase in volume and heat is absorbed from the surroundings. This indicates that some intermolecular bonds have been broken. In fact, the addition of cyclohexane to the alcohol reduces the degree of hydrogen bonding between alcohol molecules and increases the escaping tendency of each component from the mixture. This is an example of positive deviation from Raoult's law as illustrated by the saturated vapour pressure/composition curve (*Figure 21.5*). The corresponding boiling

---

* The curves relating saturated vapour pressure or temperature to the composition of the *vapour phase* in equilibrium with the liquid phase have been omitted from these diagrams (but *see* Section 21.2.2(ii)).

point/composition curve shows that solutions of cyclohexane and ethanol may boil at temperatures below the boiling point of either component.

In contrast, when trichloromethane is mixed with methyl ethanoate, there is a slight contraction in volume and heat is given out. Hydrogen bonds are formed between the two liquids (although the H in $CHCl_3$ is not bonded directly to a very electronegative atom, there is a substantial inductive effect exerted by the three chlorine atoms so that the H atom certainly carries a $\delta+$ charge), and hence the escaping tendency of the two liquids from the solution is diminished. This is an example of negative deviation from Raoult's law and the corresponding boiling point/composition curve shows that mixtures may boil at temperatures greater than the boiling point of either component (*Figure 21.6*).

The separations of mixtures which have just been discussed are dealt with in Section 21.2.2(ii).

The purity of a liquid can be most accurately assessed, not by its boiling point, but by use of gas–liquid chromatography (*see* Section 21.2.4(iv)).

21.2 Methods of separation

21.2.1 Crystallization

Crystallization is the most general method for the purification of solids. Its use depends on the fact that, in general, the solubility of a solid solute in a particular solvent increases with increasing temperature (*see Figure 21.7*). Thus, when a hot, virtually saturated solution is allowed to cool,

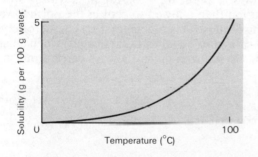

Figure 21.7 Solubility curve of benzenecarboxylic acid

the solubility of the solute decreases and hence the solute begins to crystallize from the solution. If cooling is slow, then large crystals are usually obtained. Any small amount of *soluble* impurity will generally remain dissolved in the solution (sometimes called the 'mother liquor') because the solution will not be saturated with respect to the impurity, even at lower temperatures. Any *insoluble* impurity must be removed from the solution by rapid filtration, before the solution is allowed to cool.

The choice of solvent is important. Obviously, the solvent chosen must be chemically inert, i.e. it must not react with the solute. The solute should be reasonably soluble in the hot solvent but should have a very low solubility in the cold solvent, otherwise recovery of solute by crystallization will be poor. The solvent should not be too volatile, otherwise a lot will be lost by evaporation on heating but, on the other hand, its boiling point should not be too high because the solute may then melt before entering the solution; this often means that the solute separates out as an oil, rather than crystals, on cooling. As a general rule-of-thumb, *'like dissolves like'* so that a non-polar solvent, such as *'light petroleum'* or benzene, might be chosen for a non-polar solute, while more polar solvents, such as ethyl ethanoate or ethanol, would be appropriate for polar compounds. Sometimes mixed solvent systems are used and crystallization is achieved by adding to the hot solution another miscible solvent, in which the solute is less soluble. For example, in crystallizing 2,4-DNP derivatives of carbonyl compounds (*see* Section 8.3.3), the derivative is often dissolved in the minimum volume of hot ethanol, and water is then added dropwise until the solution becomes turbid. The solution is then warmed again, or one drop more of ethanol added, until the solution clears and it is then allowed to cool, when crystallization occurs.

Very often, organic crystals are discoloured by traces of impurity; these may usually be removed by heating the solution, immediately prior to crystallization, with a very small amount (about 0.1 g) of 'activated' or 'decolorizing' charcoal. This has very considerable powers of adsorption so only a very small amount must be used if adsorption, and hence loss, of solute is to be avoided. After boiling with the charcoal, the solution is readily filtered (but not by suction filtration as this would cause evaporation of solvent) into a preheated conical flask (to prevent rapid cooling), and the solution then allowed to cool in the normal way.

The following example illustrates the general technique of crystallization, although special care must be taken (as indicated) when a flammable solvent is being used. Crystallizations are always carried out in conical

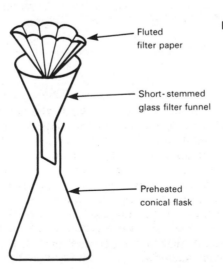

Fluted
filter paper

Short-stemmed
glass filter funnel

Preheated
conical flask

Figure 21.8 Simple filtration apparatus

flasks (to reduce loss of solvent by evaporation). Filtration of a hot solution to remove insoluble impurities or 'decolorizing' charcoal is always performed with fluted filter paper and short-stemmed funnel (*Figure 21.8*). The final crystals are isolated by filtration under reduced pressure (suction filtration), using a Buchner flask and funnel (*Figure 21.9*). The crystals are usually washed with a small volume of cold solvent and dried in air on filter paper.

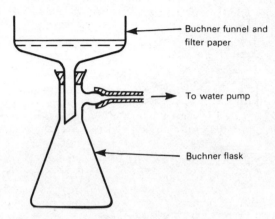

Buchner funnel and
filter paper

Figure 21.9 Apparatus for suction filtration

To water pump

Buchner flask

Example About 40 cm³ of water are poured into a 100 cm³ conical flask and heated to just below the boiling point (*note:* if an organic solvent was being used the heating would have to be carried out with the aid of a hot-water bath). About 1.5 g (the mass should be known accurately) of a sample of benzenecarboxylic acid, contaminated with a very small amount of a dye (such as methylene blue), is taken and added in successive spatula measures to the hot solvent until no more will dissolve. A little more water is then added and the solution stirred with a glass rod until all the solid has just dissolved. This procedure is repeated until all the sample has gone into solution. About half a spatula measure of 'decolorizing' charcoal is then added to the hot solution (if the temperature has fallen considerably it should be reheated) which, after being stirred and allowed to stand for about 2 min, is then carefully but rapidly filtered through a preheated glass funnel, bearing fluted filter paper, and into a preheated conical flask (the glassware can be heated in steam or hot water). The contents of the flask are then slowly allowed to cool to room temperature and the deposited crystals are then filtered off under suction. The crystals are carefully pressed down on to the filter paper with a glass rod and washed with a few cm³ of cold water. Suction is continued until nearly all the solvent has been removed and the crystals are then carefully transferred to a piece of filter paper and allowed to dry in a current of warm air (alternatively, they can be placed in a desiccator). Finally, the dry crystals are weighed to determine the efficiency of the recovery. A melting point determination (benzenecarboxylic acid m.p. 122°C) will show whether further recrystallization is necessary or not.

21.2.2   Distillation

(i) Simple distillation

Simple distillation is seldom effective unless an organic liquid is being separated from solids. However, once a preliminary separation has been effected an organic liquid may finally be distilled and a fraction (*see* below) with the correct boiling point collected. The apparatus is shown in *Figure 21.10*.

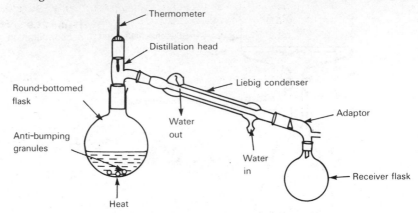

Figure 21.10   Apparatus for simple distillation

The thermometer is positioned so that it reads the temperature of the vapour as it passes into the side-arm and then into the condenser. The condenser is fitted with an adaptor so that the condensed liquid can be collected in a receiver, without loss by evaporation. The distillation flask may be heated by an electric mantle, or else with the aid of a hot-water bath (for temperatures up to 100°C) or oil bath (a silicone oil is particularly suitable, since it may be safely heated to 250°C or more).

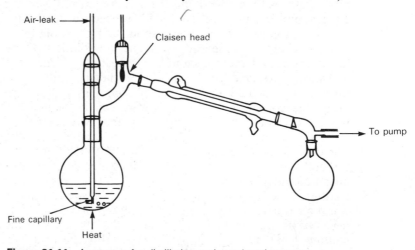

Figure 21.11   Apparatus for distillation under reduced pressure

For high-boiling liquids, or liquids which begin to decompose near their boiling points, the distillation may be carried out under reduced pressure so that the boiling point is reduced (*see* Section 21.1.2). In this case, the normal distillation head is replaced by a Claisen head bearing an 'air-leak', which is a glass inlet tube drawn out into a fine capillary (*Figure 21.11*). The adaptor is connected to a water pump (or oil pump if a very low pressure is required) and, as the pressure is reduced, a fine stream of air bubbles is drawn through the capillary and prevents 'bumping' during distillation. In any distillation, the distillation flask should never be more than half full of liquid to be distilled.

(ii) Fractional distillation
Fractional distillation is usually far more effective in purifying a liquid than simple distillation.

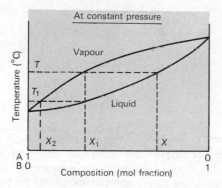

Figure 21.12 Temperature v. composition plot for vapour and liquid phases of a liquid mixture

Suppose a mixture contains two volatile liquids A and B. At the boiling point of the mixture (i.e. the temperature at which its saturated vapour pressure is equal to the external pressure), the compositions of the liquid and vapour phases are not identical, the vapour containing a greater proportion of the more volatile component than the liquid. Thus, *Figure 21.12* shows two curves, one indicating the change in composition of the liquid and the other the change in composition of the vapour at various temperatures, the total vapour pressure being constant.

When a liquid mixture of composition X is at its boiling point T, then the composition of the vapour in equilibrium with it is given by X_1. If this vapour was collected as distillate, it would still be a mixture of the two liquids. If it was then itself distilled, the composition of the vapour at the boiling point T_1 would be approximately X_2. If this was collected as distillate and then distilled again, it can be seen that, after a few more operations of this sort, pure A could be obtained from the mixture. In fractional distillation, a similar result is obtained in *one* distillation by fitting the distillation flask with a fractionating column. There are many different designs of such columns but, commonly, a tall glass column, loosely filled with small pieces of capillary tube, is employed (*Figure 21.13*).

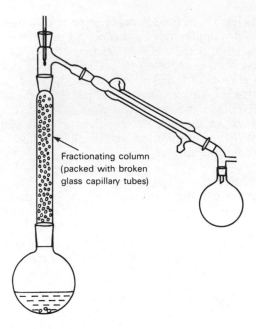

Figure 21.13 Apparatus for fractional distillation

Fractionating column
(packed with broken
glass capillary tubes)

When the liquid mixture is heated in the distillation flask, the first vapour to come off from the liquid condenses at the bottom of the fractionating column. As this liquid runs back it is vaporized, as it meets more hot vapour coming from the flask, and passes a little further up the column this time, before condensing again on the cold surface provided by the column packing. As this process is repeated over and over again, it amounts to a series of distillations in which the vapour approaching the top of the column is getting progressively richer in the more volatile component A. Eventually, if the distillation is performed carefully and slowly, pure A will pass out of the top of the column and will be collected as distillate. In time, pure B will be left in the distillation flask. Many pairs of isomers may be separated in this way and partial separations into fractions are obtained in the fractional distillation of crude oil (*see* Section 18.2).

However, those liquids which form constant-boiling mixtures (*azeotropes*) cannot be separated into both pure components by fractional distillation. An important example is the ethanol–water mixture, which shows positive deviation from Raoult's law and which exhibits a minimum in its temperature/composition curves at constant pressure (*Figure 21.14*).

Suppose a mixture of composition X is distilled. Then, as distillation proceeds, the liquid in the distillation flask will become richer in water, until finally pure water is left in the flask. The distillate will never be pure ethanol but will finally become a constant-boiling mixture of 95.6% by mass ethanol and 4.4% water (*see also* Section 6.2). If, however, benzene is added to this constant-boiling mixture, a three-component system is

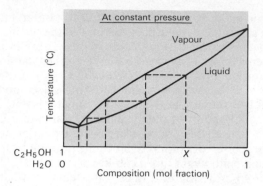

Figure 21.14 Temperature v. composition plot for a two-component system which forms a constant-boiling mixture

produced and fractional distillation of it first yields a mixture of the three liquids, then a mixture of benzene and ethanol, and finally pure ethanol. This is how pure ethanol is obtained industrially.

(iii) Steam distillation

This technique is particularly useful in the case of liquids which tend to undergo decomposition, or oxidation, near their boiling point, or for conveniently isolating an organic liquid from an aqueous suspension or mixture. To be successfully steam-distilled, the liquid must be immiscible with water (or have an extremely low solubility in water), should obviously not react in any way with water and should have a reasonably high saturated vapour pressure at the temperature of distillation (which will be just less than 100°C at standard atmospheric pressure).

Consider the steam distillation of a liquid X. If a mixture of X and water, which are immiscible, is vigorously agitated so that the two layers do not separate out, then each liquid can establish an equilibrium with the vapour phase and exert its saturated vapour pressure at constant temperature. It therefore follows that the total saturated vapour pressure exerted by such a mixture does not depend on the relative proportions of the two liquids present (so long as each is present in the liquid phase at equilibrium) but is simply the sum of the saturated vapour pressures of the two liquids at constant temperature (Dalton's Law of Partial Pressures, total pressure $P = p_{H_2O} + p_X$).

The number of moles n of each compound present in the vapour phase at equilibrium is proportional to its saturated vapour pressure at a fixed temperature:

$$\frac{n_X}{n_{H_2O}} = \frac{p_X}{p_{H_2O}} = \frac{m_X \times M_{H_2O}}{M_X \times m_{H_2O}}$$

where m represents the mass of a component in the vapour state and M is its molar mass. Therefore

$$\frac{m_X}{m_{H_2O}} = \frac{p_X \times M_X}{p_{H_2O} \times M_{H_2O}}$$

This expression shows how the ratio, by mass, of X to water in the distillate depends on the saturated vapour pressure of X at the distillation temperature, and on the molar mass. For chlorobenzene, this ratio is about 2:1, while for phenylamine it is about 5:1. When phenylamine is prepared by the reduction of nitrobenzene (*see* Section 14.2), the free amine is not obtained directly but via a complex salt. This is decomposed by the addition of alkali and the amine is then conveniently isolated by steam distillation. The boiling point of phenylamine at standard atmospheric pressure is 184°C, and at this temperature some oxidation tends to occur and the distillate is often discoloured. Since the saturated vapour pressures of phenylamine and water are 3.60 kN m^{-2} and 97.72 kN m^{-2}, respectively, at 99°C, it follows that, on steam distillation, phenylamine and water will distil over at this temperature, if the pressure is standard atmospheric pressure (101.32 kN m^{-2}).

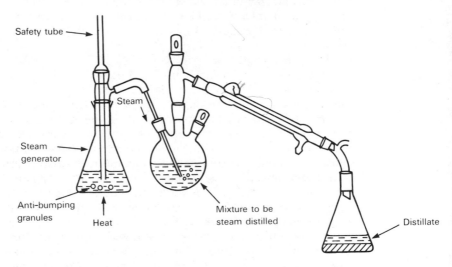

Safety tube

Steam

Steam generator

Anti–bumping granules

Heat

Mixture to be steam distilled

Distillate

Figure 21.15 Apparatus for steam distillation

An interesting separation of isomers can be achieved by steam distillation. When phenol is treated with dilute nitric acid, a mixture of 2-nitro- and 4-nitrophenols is obtained. If the black, oily mixture is steam distilled, the 2-nitrophenol is collected in the distillate and solidifies on cooling. If the residue in the distillation flask is cooled in ice, crystals of 4-nitrophenol appear. The 4-nitro isomer hydrogen bonds with water but the 2-nitro isomer exhibits *intramolecular* hydrogen bonding and is volatile in steam (*see* Section 13.3.1(v)).

In carrying out steam distillation, steam is normally passed continuously into the distillation flask (which should not be more than half full), which need not be heated directly (*Figure 21.15*). The organic liquid is separated from the water in the distillate using a separating funnel and then dried over a suitable drying agent.

21.2.3   Solvent extraction

In preparative organic chemistry, the desired product is often produced
in an aqueous suspension or solution. In these circumstances, solvent
extraction is widely used in the isolation of the product. The mixture is
shaken with an immiscible solvent, in which the desired product is soluble,
in a separating funnel and the two layers then allowed to settle out
(*Figure 21.16*). The bottom layer is then carefully run off, via the tap,

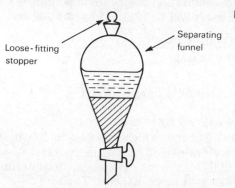

Figure 21.16   Separating funnel

Loose-fitting
stopper

Separating
funnel

into a clean flask. The aqueous solution is extracted several times (most
of the solvents commonly used are less dense than water and form the
top layer) and the combined organic extracts dried by standing in contact
with a suitable drying agent, such as anhydrous calcium chloride (but this
cannot be used for drying alcohols or amines since they react; anhydrous
magnesium sulphate may be used). The organic solvent is then removed
from the dried extract by distillation.

In general, non-polar solvents which are immiscible with water, like
'*light petroleum*' or benzene, are used to extract non-polar compounds,
and more polar solvents, like trichloromethane, for more polar compounds. In practice, ethoxyethane is the most widely used solvent for
extraction purposes since it has very good solvent properties and a low
boiling point (35°C), but great care must be taken when using it because
of its high flammability.

If a solute is added to two immiscible solvents A and B, which are in
contact with each other, then it distributes itself between the two solvents
roughly according to its solubility in each. At constant temperature,
assuming there is no dissociation or association in one of the solvents,
then:

$$\frac{\text{Concentration of solute in A}}{\text{Concentration of solute in B}} = \text{a constant } (K)$$

K is termed the **distribution**, or **partition**, **constant** and the relationship
above is called the **Distribution Law** or **Partition Law**. Using this, it can
be shown that, with a given volume of solvent for extraction purposes,
three or four extractions with smaller volumes of solvent are more efficient, in terms of recovery of solute, than a single extraction with the
whole volume of solvent.

Suppose that 100 cm^3 of an aqueous solution of phenol (2.5 g) is to be extracted with pentan-1-ol in order to recover the phenol, and that 100 cm^3 of pentan-1-ol is made available. The distribution, or partition, constant for phenol distributed between pentan-1-ol and water is 16.

(i) Extraction with 100 cm^3 solvent
Let x g be the mass of phenol which dissolves in the 100 cm^3 pentan-1-ol. Then the mass concentration of phenol in pentan-1-ol will be x g per 100 cm^3 and in the water will be $(2.5 - x)$ g per 100 cm^3. Therefore

$$K = \frac{\text{Concentration in pentan-1-ol}}{\text{Concentration in water}} = 16 = \frac{x}{(2.5 - x)}$$

$$\Rightarrow \quad x = 2.35$$

mass of phenol obtained by a single extraction = 2.35 g

(ii) Two extractions with 50 cm^3 solvent each
Let y g be the mass of phenol which dissolves in 50 cm^3 of pentan-1-ol. Then the mass concentration of phenol in pentan-1-ol will be y g per 50 cm^3 or $2y$ g per 100 cm^3 solvent. The mass concentration of phenol in the water will be $(2.5 - y)$ g per 100 cm^3. Then

$$K = 16 = \frac{2y}{(2.5 - y)}$$

$$\Rightarrow \quad y = 2.22$$

mass of phenol obtained by first extraction – 2.22 g

After this extraction, the mass of phenol remaining in the water is 0.28 g. Let z g now be the mass of phenol extracted by another 50 cm^3 portion of pentan-1-ol. Then

$$K = 16 = \frac{2z}{(0.28 - z)}$$

$$\Rightarrow \quad z = 0.25$$

mass of phenol extracted by second extraction = 0.25 g

Therefore, the total mass of phenol extracted by the second method is 2.47 g (i.e. 2.22 g + 0.25 g), compared to 2.35 g in the first method. Where larger quantities of solute are concerned, three or four extractions will be even more efficient.

21.2.4 Chromatography

This method embraces a variety of techniques in which components of a mixture are distributed between two phases, one moving and the other fixed, so that they eventually become concentrated in separate regions or zones. The oldest technique is **adsorption chromatography**, in which the components of a mixture in solution are differentially adsorbed on a solid stationary phase. In the more recent **partition chromatography**, developed by the chemists Richard Synge and Archer Martin (for which they were awarded the 1952 Nobel Prize for Chemistry), the components of the mixture are distributed, or partitioned, between two phases, the stationary one being a liquid on a solid support and the moving phase being either another liquid, or else a gas. Organic compounds which ionize to some extent can often be separated by **ion-exchange chromatography**, in which certain ions displace others from ion-exchange resins.

(i) Column adsorption chromatography
Usually, a solution of the mixture to be separated (the *moving phase*) is allowed to percolate through a tall glass column, loosely packed with a suitable adsorbent, commonly *alumina* or *silica* (the *stationary phase*).

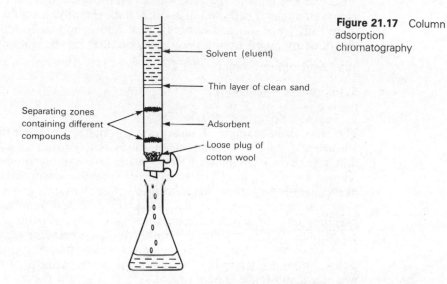

Solvent (eluent)

Thin layer of clean sand

Separating zones containing different compounds

Adsorbent

Loose plug of cotton wool

Figure 21.17   Column adsorption chromatography

When the mixture has been adsorbed on the column, the column is then slowly washed, or *eluted*, with a suitable solvent (the *eluent*). The components of the mixture begin to move down the column at different rates, depending on how strongly adsorbed they are, and if they are coloured a series of coloured zones or bands begin to separate out (hence the name **chromatography**, from the Greek *chroma* meaning colour, and *graphe* meaning a writing). In general, a non-polar solvent such as '*light petroleum*' (a hydrocarbon mixture) will elute non-polar materials, which are only poorly adsorbed on the column, and more polar solvents will have to be used to remove more strongly adsorbed materials from the column. As the solvent passes down the column, it is collected at the bottom in a conical flask (the liquid is called the *eluate*) and the flasks are changed as different materials come off the column (*Figure 21.17*). If the components of the mixture are colourless, and hence cannot be seen on the column, then the eluate must be tested for each component at regular intervals, or else collected as a series of small volumes (*fractions*).

Example: separation of 2- and 4-nitrophenylamines A column is made up
by pouring into a column (an old burette is suitable) a slurry of chromatographic alumina (about 70 g) in benzene (*care!*). The column is
tapped gently so as to allow the alumina to settle evenly as the solvent
passes down the column. The tap is closed when the solvent level is just
above the alumina (on no account must a column be allowed to run dry
as the adsorbent will then 'channel' and become uneven). A very thin
layer of clean sand is placed on top of the alumina and then there is added
carefully, so as not to disturb the top of the column, a mixture of the
nitrophenylamines (~0.5 g) dissolved in benzene. The concentrated solution is allowed to run on to the column so that adsorption occurs. The
column is then eluted with this solvent and eluate collected as soon as
orange drops appear. The yellow band on the column is then eluted with
ethoxyethane and collected in another flask. Finally, the solvent is evaporated off (*care:* on a water-bath in a fume cupboard) and the solids
identified by melting point determination after crystallization:

m.p. 2-nitrophenylamine = 71°C

m.p. 4-nitrophenylamine = 149°C

(ii) Thin layer chromatography (TLC)
This, too, is an example of adsorption chromatography but this time the
adsorbent (again often alumina or silica) is employed as a coating, or
thin layer, on glass plates. These plates are usually prepared in the
laboratory, using a special spreader which spreads a slurry of the adsorbent
over a number of glass plates to give, after drying, a layer of suitable
thickness. Alternatively, special TLC plates (the adsorbent is on a plastic
backing) may be purchased, but these are more expensive. Although
preparative TLC is often used in research laboratories for separating very
small amounts of material, TLC is most widely used for analytical purposes, either to determine the nature of an impurity, or to determine
whether a substance is pure or not.

The mixture to be analysed is placed as a very small spot (with the aid
of a capillary tube) on the plate, a short distance above the bottom of the
plate. The plate is then stood vertically in a glass tank or jar, containing
a suitable solvent whose level lies just below the spot on the plate (*Figure
21.18*). As the solvent rises up the plate, the components of the mixture
travel different distances up the plate, according to their degree of adsorption. Since there is only a thin layer of adsorbent, this process takes place
much more rapidly than in column chromatography. If the components
of the mixture are coloured, then coloured spots will begin to separate
out but if they are colourless, they have to be 'visualized' later. When the
solvent front has moved almost to the top of the plate, the plate is
removed from the tank and the position of the front marked on the plate.
The plate is then examined when all the solvent has evaporated (*Figure
21.19*). The positions of colourless components may be located by standing
the plate in a tank containing a little solid iodine, which fills the tank with
iodine vapour. This causes the components to show up as brown spots
after about 5 min. The components of a mixture may be identified by

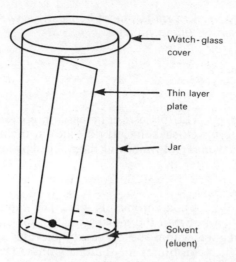

Figure 21.18 Thin layer chromatography (TLC)

recording their R_f, or *reference*, values. The R_f value is the ratio of the distance travelled by the substance in question up the plate to the distance travelled by the solvent front, and is therefore $\leqslant 1$. Known substances can also be 'spotted' on the plate during analysis, in order to aid identification. The R_f value is a constant for a particular substance only under the precise conditions of the experiment, so the adsorbent and solvent must always be quoted.

Example: identification of products of the nitration of phenol A spot of the product mixture is placed on a silica plate and benzene is

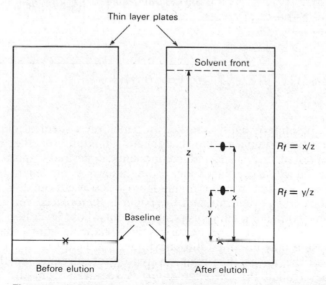

Figure 21.19 Thin layer chromatography (TLC)

used as eluent. If the product spots cannot be clearly seen, they should be located with the aid of iodine vapour:

R_f 2-nitrophenol = 0.9
R_f 4-nitrophenol = 0.4
R_f 2,4-dinitrophenol = 0.2
R_f 2,4,6-trinitrophenol = 0.05

In the school laboratory, TLC plates can be easily prepared by dipping microscope slides into a well-shaken slurry of silica in trichloromethane (*care:* poisonous) and then rapidly removing them and allowing the solvent to evaporate.

(iii) Paper chromatography

This is an example of partition chromatography. The components of a mixture are partitioned between the moving organic solvent phase and water, which is contained within the cellulose molecules in chromatographic paper and which constitutes the stationary phase (*Figure 21.20*). The technique is similar to that used in TLC.

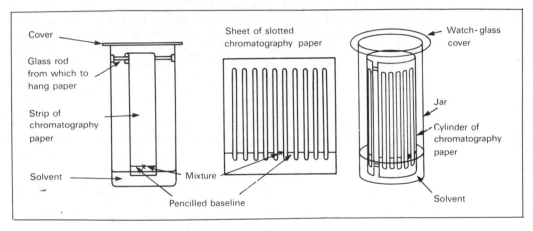

Figure 21.20   Paper chromatography

A baseline is lightly marked in pencil on a sheet of chromatography paper (very similar to filter paper) and a mixture, together with authentic samples of any suspected components if desired, 'spotted' on to the baseline with a capillary tube. (Sometimes it is convenient to use 'slotted' paper, i.e. paper which is divided into strips by cut-out slots.) Any solvent is allowed to evaporate and the paper is then placed vertically in a tank or jar, containing an appropriate solvent whose level lies just below the baseline. The paper can be supported by hanging from a suitable support, or else it may be rolled into a cylinder and held together by paper clips. When the solvent has risen an appropriate distance up the paper, the paper (called the *chromatogram*) is removed from the tank and the position of the solvent front marked, before the solvent is allowed to evaporate. If the compounds are coloured, then the spots are easily located and the R_f values determined. If they are colourless, then the chromatogram must be sprayed with a suitable reagent which will react with the compounds to form coloured spots.

Example: identification of amino-acids in a mixture A source of amino-acids (e.g. natural lemon juice or a protein hydrolysate) is taken and one drop of the mixture placed on the baseline of a sheet of chromatography paper. Along the baseline are also placed drops of authentic amino-acids, such as valine, glycine and leucine. The paper is then supported in a tank

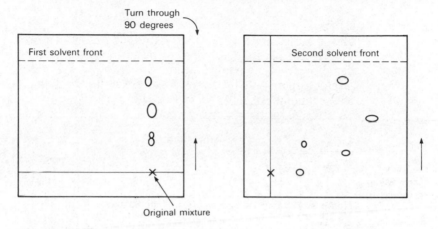

Figure 21.21 Two-way paper chromatography

containing the solvent (a suitable solvent mixture is the upper organic layer obtained from shaking together butan-1-ol, water and pure ethanoic acid in the volume ratio 5:4:1). When the solvent front has advanced a suitable distance, the paper is removed from the tank, the position of the solvent front marked, and the solvent allowed to evaporate. The amino-acids are detected by spraying the paper with a solution of *ninhydrin* in propanone (*care:* avoid skin contact, as it will stain the skin) and then heating the chromatogram, either with a hair-dryer, or else in an oven at about 110°C, for 10 min. Most amino-acids form a purple spot with this reagent, although one or two (e.g. proline) give a yellow colour. The positions of the spots should be indicated lightly in pencil, as these colours fade fairly quickly.

In some cases, the amino-acids may not be fully separated by this process in which case, *before* spraying with ninhydrin, the chromatogram may be turned through 90 degrees and then placed in a different solvent (e.g. a water–ethanol–ammonia mixture), and the chromatogram 'run' as before. This is called *two-way chromatography* (*Figure 21.21*).

Paper discs can also be used instead of sheets of chromatographic paper, as illustrated in *Figure 21.22*.

Example: chromatographic identification of the N-terminal amino-acid of a dipeptide In order to determine the N-terminal acid of a simple peptide, the peptide is reacted with 1-fluoro-2,4-dinitrobenzene (*see* Section 17.9). The FDNB-peptide is then hydrolysed to give a mixture of the constituent amino-acids, with the N-terminal acid present as its FDNB derivative (this is yellow). The mixture is extracted with ethoxyethane to obtain the FDNB derivative, and a sample of this extract is then chromatographed on a paper disc, using the top layer from a mixture of equal volumes of butan-1-ol and 2 molar ammonia solution as solvent. A small impregnated wick is used to feed solvent to the disc, as shown (*Figure 21.22*).

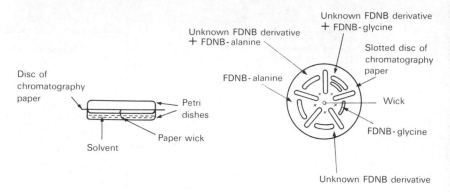

Figure 21.22 Circular disc paper chromatography

Suppose the peptide was simply a dipeptide containing glycine and alanine. To identify the N-terminal acid, the behaviour of the FDNB derivative would be compared with that of derivatives of glycine and alanine as standards during chromatography. The diagram shows how alanine could be correctly identified.

(iv) Gas–liquid chromatography (GLC)
This is a form of partition chromatography in which the stationary phase is a non-volatile liquid adsorbed on a solid support, and the moving phase is a chemically inert gas, usually nitrogen. The method was introduced by Archer Martin and Anthony James in 1952.

A long, narrow, glass or stainless-steel tube, or coil, contains the liquid coating a solid support (e.g. powdered firebrick) and is contained in an oven, so that the temperature can be precisely controlled (*Figure 21.23*). The moving phase, or 'carrier' gas, passes through this column and, when a mixture is injected into the heated column by means of a self-sealing rubber cap, the components of the mixture partition themselves between the liquid and the gas. They are gradually carried through the column at a rate which depends on the degree to which they are distributed in the liquid or the gas, those mainly in the gas phase coming off the end of the column first. As a component comes off the column it is detected, usually either by a flame ionization detector (the conductivity of a small hydrogen

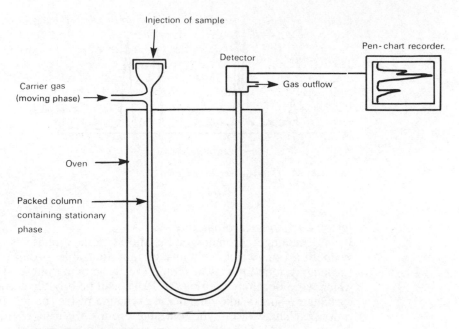

Figure 21.23 Gas–liquid chromatography

flame suddenly increases when an organic compound enters it owing to ionization of the compound), or by a katharometer (in which the thermal conductivity of a strip of platinum wire is altered by a sudden change in the composition of the gas surrounding it). In either case, an electrical signal from the detector is amplified and transmitted to a pen-chart recorder, which plots peaks corresponding to eluted components against a time base. The area under a peak is proportional to the amount of material eluted and it is desirable that the peaks should be well separated (*resolved*) from one another. Known compounds can be separately chromatographed under identical conditions (gas pressure, temperature, liquid used in column). The time taken from the moment they are injected on to the column to the moment each is detected at the end of the column is known as the *retention time* and may be used to identify the components of a mixture.

The injection inlet is heated, in order to vaporize less volatile substances, but essentially non-volatile substances like ionic solids cannot be investigated by GLC. This form of chromatography is extremely useful in analysis and in determining the purity of liquids; for example, it is widely used in the oil industry for examining the components of oil fractions, and for quality control in many industries. It is employed by the police for the routine quantitative analysis of ethanol in blood and urine samples of drunken drivers, and is of fundamental importance in forensic science in the analysis of materials ranging from drugs to explosives. A typical GLC chart is shown in *Figure 21.24*.

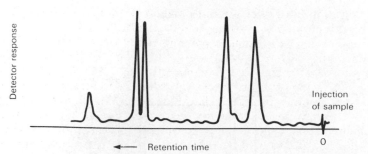

Figure 21.24   Gas–liquid chromatogram

(v) Ion-exchange chromatography

In ion-exchange chromatography, the stationary phase is an insoluble synthetic resin which contains certain ionizable groups, capable of exchanging ions for ions in solution in the moving phase. These resins, which were developed during the 1930s, can be broadly classed as *cation* exchange resins (acidic) and *anion* exchange resins (basic). In the *strong* cation exchange resins, the ionizable groups are usually sulphonic acid groups —SO_3H, linked to benzene rings which are part of a resin matrix of cross-linked poly(phenylethene); in the *weak* ones, the ionizable groups are usually carboxyl groups —CO_2H. Cation exchange resins are capable of exchanging the hydrogen ions of these groups for other cations in solution. The adsorption of cations is naturally accompanied by the release of an equivalent amount of hydrogen ions into the solution. (The degree to which the weak groups are ionized depends, of course, on the pH of the solution.)

$$\overset{\text{resin}}{\sim\!\sim\!\sim} SO_3^-H^+ + M^+ \;\rightleftharpoons\; \sim\!\sim SO_3^-M^+ + H^+$$

$$\sim\!\sim\!\sim CO_2^-H^+ + M^+ \;\rightleftharpoons\; \sim\!\sim CO_2^-M^+ + H^+$$

In the case of anion exchange resins, the *strong* ones generally contain quaternary ammonium groups —$NR_3^+OH^-$, attached to a similar resin matrix, while the *weak* ones contain amine groups such as —NH_2 or —NR_2:

$$\overset{\text{resin}}{\sim\!\sim\!\sim} NR_3^+OH^- + X^- \;\rightleftharpoons\; \sim\!\sim NR_3^+X^- + OH^-$$

$$\sim\!\sim\!\sim NH_3^+OH^- + X^- \;\rightleftharpoons\; \sim\!\sim NH_3^+X^- + OH^-$$

(after treatment with H_2O)

In the practice of ion-exchange chromatography, a column is packed with the wet resin, in a manner similar to the preparation of an adsorption chromatographic column, and 'back-washed' with de-ionized water to remove any debris and air-pockets. A solution of the ions to be exchanged is placed on the column and, when they have been adsorbed by the resin, they can be differentially displaced, if necessary by elution of the column with different buffer solutions. When elution is complete, the resin can be regenerated with strong acid or alkali.

Although ion-exchange columns are probably most widely used for removing ions from solution, as in the de-ionization of water or the removal of metal ion contaminants, they are of great importance in organic chemistry for the separation of amino-acid mixtures, nucleotides and sometimes proteins. The following outline illustrates the principles involved. Suppose that a mixture of the three amino-acids aspartic acid, alanine and lysine are to be separated (for structures *see* Section 17.8). They are first of all adsorbed on to a strong cation exchange resin:

$$\sim\!\!\sim\!\!\sim \overset{\text{resin}}{SO_3^- H^+} + H_2N\!\!-\!\! \rightleftharpoons \sim\!\!\sim\!\!\sim SO_3^- \overset{+}{H_3N}\!\!-$$

The amino-acids are then differentially displaced from the column by elution with various buffer solutions. A comparison of isoelectric points shows that, at a pH of about 3, aspartic acid will be essentially in zwitterion or dipolar form and will not be strongly adsorbed by the resin. It can, therefore, be eluted, or displaced, from the resin with a buffer solution of this approximate pH. Alanine and lysine, on the other hand, have higher isoelectric points and at this pH will remain adsorbed on the resin, as they are more basic than aspartic acid. As the pH of the buffer solution eluent is raised, so alanine will be displaced. Finally, an alkaline buffer solution will elute the strongly basic lysine.

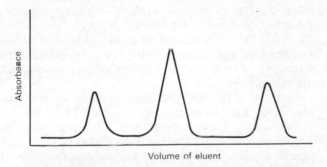

Volume of eluent

Figure 21.25   Ion-exchange chromatogram

In fully automatic amino-acid analysers, mixtures are separated on ion-exchange columns in this way and the eluted fractions are quantitatively analysed colorimetrically, following the addition of ninhydrin reagent. A 'printout' then records the absorbance of the fractions against the volume of eluate collected, as shown in *Figure 21.25*; the peaks correspond to different amino-acids.

Ion-exchange resins are also used in organic chemistry as catalysts for reactions such as esterification and hydrolysis.

To summarize, adsorption column chromatography and ion-exchange chromatography are frequently used for separating mixtures, while TLC, paper chromatography and GLC are generally used for analysing mixtures.

Questions

1* You are required to purify a non-polar organic substance S which has a melting point of 45–50°C in its impure state.
(a) State two criteria which are important in choosing a suitable solvent for recrystallizing the substance S.
(b) If the substance S contains sand as an impurity, describe how you would obtain a pure sample of S.
(c) How would you obtain a pure sample of S, if it is contaminated with benzenecarboxylic acid? (Assume that S is neutral.)
(d) If the substance S contains water as an impurity, describe how you would remove the water and obtain pure S.
(e) The melting point of the pure substance S will differ from the original melting point in two important respects. What are they?

JMB

2 (a) Organic compounds can frequently be removed from aqueous solution by extraction with ethoxyethane (*diethyl ether*). This operation is normally performed using a separating funnel.
(i)   Would you expect the ether layer to be the upper or lower layer in the funnel? How could you check this experimentally?
(ii) Name a reagent which may be used to dry the ether layer.
(iii) How would you recover a solid from the ether solution? State clearly the type of process involved, outline a suitable apparatus and describe briefly any safety precautions which should be observed. You may answer by means of a labelled sketch.
(b) The resulting solid would normally then be purified by recrystallization from a suitable solvent, and its identity and purity checked by melting point determination.
(i) In what **two** ways does the melting point of the purified solid differ from that of the impure compound?
(ii) Given a supply of the genuine pure compound, describe, briefly, what additional melting point determination you would make in order to confirm the identity of your substance. State what results you would expect.

JMB

3 (a) In two comparative experiments, a student mixed (i) trichloro-methane, and (ii) tetrachloromethane with equal volumes of propan-one (*acetone*). In the first experiment he reported a rise in temperature of 10°C and in the second a fall of 2°C. State whether these changes are exothermic or endothermic, and explain the observations.
(b) Sketch the curves showing the variation of (i) vapour pressure at constant temperature, and (ii) boiling point at constant pressure, with changing concentration of mixtures of trichloromethane and propanone.
It is customary to express the concentration of these mixtures in terms of the mole fraction of the components. Explain the meaning of the term mole fraction.

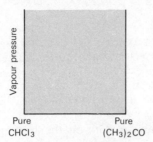

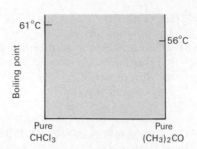

(c) (i) Two liquids A and B form an ideal system. [In the space below,] sketch a vapour pressure/composition diagram for such a system.
(ii) Give an example of two liquids which would behave in this way.
(d) Two isomers of molecular formula C_5H_{12}, pentane and 2,2-di-methylpropane, have boiling points of 36°C and 9°C, respectively. Explain this difference in boiling point.

London

4*   This question is about five liquids of which the formulae, relative molecular masses and boiling points are given in the table:

Name	Formula	Relative molecular mass	Boiling point (°C)
cyclopentane	⬠	70	49
butanone	$CH_3COCH_2CH_3$	72	80
1-aminobutane	$CH_3CH_2CH_2CH_2NH_2$	73	78
1,2-dibromoethane	CH_2BrCH_2Br	188	132
pentane	$CH_3CH_2CH_2CH_2CH_3$	72	36

(a) (i) In which liquid would hydrogen bonding occur?
(ii) Draw the structural formulae of two molecules of this liquid, showing a hydrogen bond between them.
(b) On the following formula of butanone, show the position and nature of the partial charges contributing to the dipole moment of the molecule.

```
   H      H  H
   |      |  |
H—C—C—C—C—H
   |   ‖  |  |
   H   O  H  H
```

(c) Which pair of liquids from the list would you expect to exhibit least deviation from Raoult's law when mixed together?
(d) Which of the following temperatures is most reasonable for the boiling point of an equimolar mixture of 1-aminobutane and pentane? Put a ring around your choice.

(i) 44°C, (ii) 57°C, (iii) 70°C.
Justify your answer.
(e) When an equimolar mixture of butanone and pentane at room temperature is made, would you expect a temperature change and if so, in what direction?
Justify your answer.

Nuffield

5* Discuss the influence of the following factors on the boiling points of organic substances:
(a) the length of the carbon chain,
(b) the shape of the molecule,
(c) bond polarity (the inductive effect),
(d) hydrogen bonding.
The data given below may be used to illustrate your answer:

	Relative molecular mass	b.p. (°C)
CH_3COCH_3	58	56.5
$CH_3CH_2CH_2OH$	60	97
CH_3CH_2Cl	64.5	12.5
CH_3COOH	60	118
$CH_3CH_2CH_2CH_3$	58	0
$(CH_3)_2CHCH_3$	58	−12
$CH_3CH_2CH_2CH_2CH_3$	72	36
$C(CH_3)_4$	72	9.5

JMB

6 Among the techniques used by chemists for the isolation and separation of substances from complex systems or reaction mixtures are:
(a) simple distillation,
(b) fractional distillation,
(c) solvent extraction,
(d) chromatography, and
(e) crystallization.
Outline the physico-chemical principles underlying *four* of the above techniques. Illustrate the usefulness of each technique you have chosen by means of one example of an actual application. The application may relate either to an industrial or to a laboratory situation.

London

7* Describe, giving brief experimental details and the principles of the methods you describe, how the italicized compound could be separated pure and in good yield in the laboratory from each of the following mixtures. Physical, chemical or a combination of both types of method may be used.

(a) *methanol* and butan-2-ol,
(b) *butan-2-ol* and water,
(c) *methanol* and propanone.

Describe a **chemical** test in each case which would allow you to show the absence of substantial contamination in the purified sample. (Methanol has a boiling point of 65°C and is completely miscible with water and with many non-polar solvents. Butan-2-ol has a boiling point of 99°C, is only partially miscible with water but completely miscible with many non-polar solvents. Propanone has a boiling point of 56°C and is completely miscible with water and with many non-polar solvents.)

JMB

8* How would you obtain a pure sample of each component from **four** of the following mixtures of compounds? State clearly the physical and/or chemical principles on which your reasoning is based.

(i)

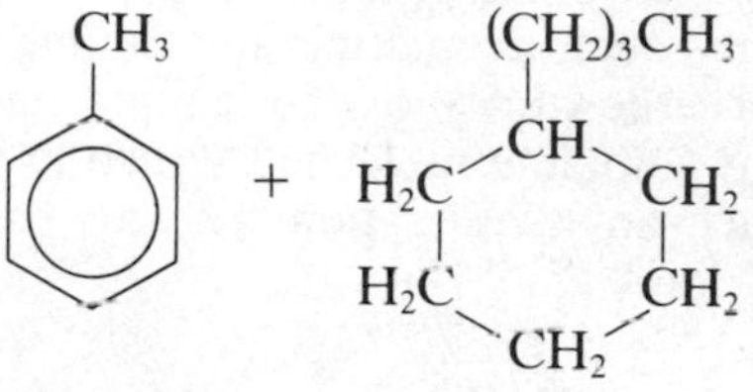
CH3
+
(CH2)3CH3
CH
H2C
CH2
H2C
CH2
CH2

b.p. 110°C b.p. 181°C

(iv)

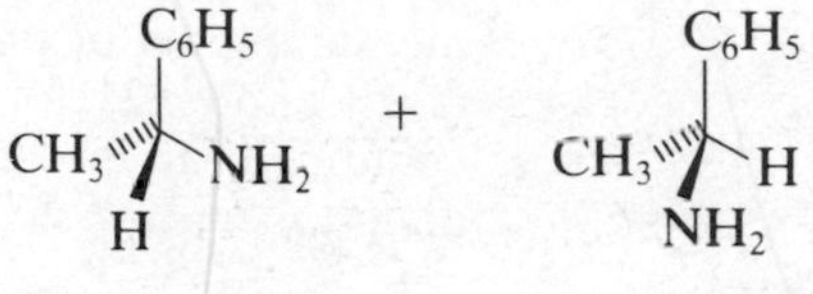
C6H5
CH3
NH2
H
+
C6H5
CH3
H
NH2

b.p. 187°C b.p. 187°C

(ii)

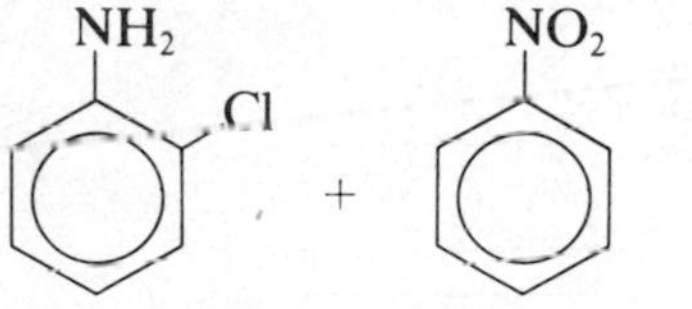
NH2
Cl
+
NO2

b.p. 209°C b.p. 211°C

(v)

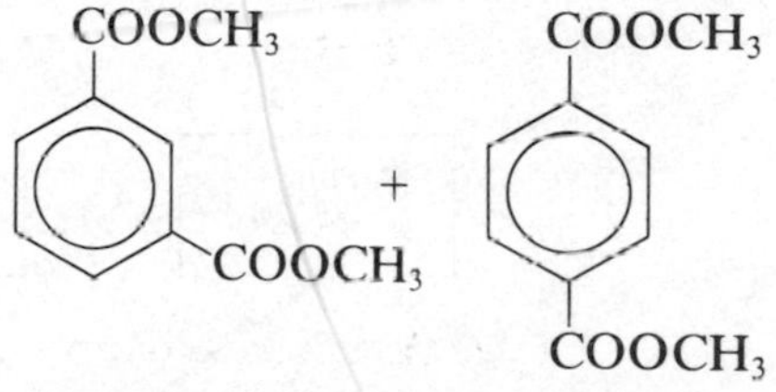
COOCH3
COOCH3
+
COOCH3
COOCH3

m.p. 131°C m.p. 141°C

(iii)

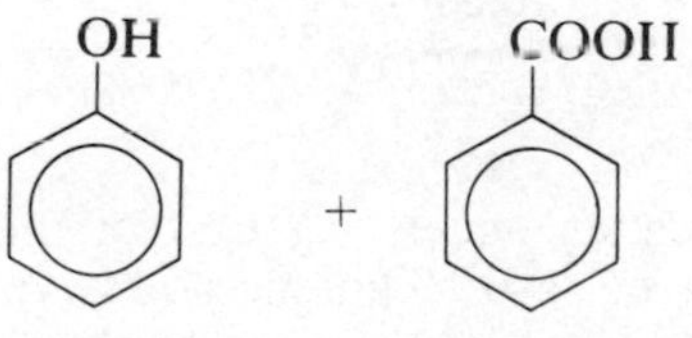
OH
+
COOH

m.p. 43°C m.p. 122°C

(vi)

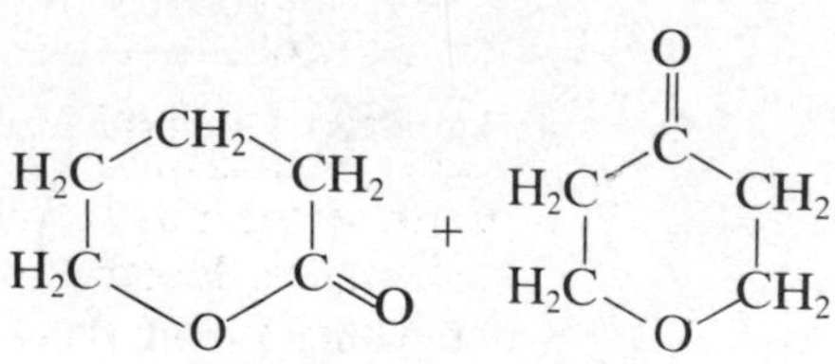
CH2
H2C
CH2
H2C
C
O
O
+
O
C
H2C
CH2
H2C
CH2
O

b.p. 218°C b.p. 166°C

Cambridge

Chapter 22
Determination of molecular structures

22.1 Introduction

When a new organic compound is prepared or isolated for the first time, its purity will first be checked by chromatographic methods, and by measurements of physical constants such as melting and boiling points (*see* Chapter 21 for details). Once the purity of the sample has been confirmed, an investigation into its complete molecular structure will be undertaken (*see Figure 22.1*). Such an investigation will involve (1) a

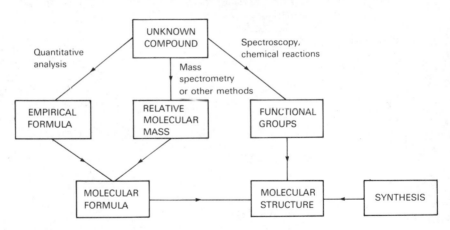

Figure 22.1 Routes to molecular structure

determination of the empirical and/or molecular formula of the compound, (2) a determination of the relative molecular mass of the compound, and (3) the identification of functional groups present in the compound by spectroscopic investigation and by study of its physical and chemical properties. In the case of very complex molecules, other physical aids, such as the technique of X-ray diffraction, may also have to be called upon (*see* proteins, for example, Section 17.9). Finally, it is conventional to confirm the proposed structure by a total synthesis of the compound, using established reactions, where this is possible.

22.2   Determination of empirical formula

The empirical formula is deduced from a quantitative analysis of the compound. In principle, a known mass of the *dry* compound is heated to a high temperature in excess, pure, *dry* oxygen. Carbon is oxidized to carbon dioxide, hydrogen to water and nitrogen to nitrogen oxides. The gas stream is passed over various materials, designed to remove elements such as halogens, phosphorus and sulphur (these elements are analysed for by other specialized methods), and then over very hot copper which reduces the nitrogen oxides to nitrogen. The water vapour is then absorbed by magnesium chlorate(VII) and the carbon dioxide by *soda-lime*, before the amount of nitrogen remaining is measured. The masses of water and carbon dioxide absorbed are then determined. This technique of micro-combustion analysis was first introduced, in 1911, by the Austrian chemist Fritz Pregl, for which he was awarded the 1923 Nobel Prize for Chemistry, and it enables samples consisting of only a few milligrammes to be analysed. If it is known that the compound does not contain elements other than C,H,O and (possibly) N, then any difference between the sum of the elemental percentages and 100% can be attributed to the oxygen content of the sample, which is not usually directly ascertained. The following example shows the method of calculation of elemental percentages from experimental results.

Example On combustion analysis 0.360 g of a pure organic compound was found to yield 0.528 g of carbon dioxide and 0.216 g of water. No nitrogen was obtained. What is the empirical formula of the compound?
   From these figures, it is first necessary to work out the actual masses of carbon and hydrogen present in the original sample:

Mass of carbon in 44 g CO_2 $(M_r = 44)$ $= 12$ g

Mass of carbon in 0.528 g CO_2 $\qquad = 12 \text{ g} \times \dfrac{0.528 \text{ g}}{44 \text{ g}} = 0.144 \text{ g}$

Mass of hydrogen in 18 g H_2O $(M_r = 18) = 2$ g

Mass of hydrogen in 0.216 g H_2O $\qquad = 2 \text{ g} \times \dfrac{0.216 \text{ g}}{18 \text{ g}} = 0.024 \text{ g}$

These masses were present in 0.360 g of sample:

% carbon $\quad = \dfrac{0.144 \text{ g}}{0.360 \text{ g}} \times 100 = 40.00$

% hydrogen $= \dfrac{0.024 \text{ g}}{0.360 \text{ g}} \times 100 = 6.66$

Since there was no nitrogen (or other elements such as Cl,P,S) present, oxygen must account for the remaining 53.34% by mass.

From these analytical figures, the empirical formula can be calculated:

	C	H	O
% by mass	40.00	6.66	53.34
molar ratio	$\dfrac{40.00}{12}$	$\dfrac{6.66}{1}$	$\dfrac{53.34}{16}$
of atoms			

$$3.33 \quad : \quad 6.66 \quad : \quad 3.33 \quad \text{or } 1:2:1$$

empirical formula is CH_2O

In the school laboratory, some quantitative analysis of organic compounds can be carried out, using the Schöniger oxygen flask technique. An organic solid, supported on a piece of aluminium foil connected to two copper leads by nichrome wire, is heated electrically in an atmosphere of pure oxygen in a conical flask (*Figure 22.2*). On combustion of the solid, the carbon dioxide produced is absorbed in standard alkali, which is subsequently titrated against standard acid (*note*: a safety screen should be used during the combustion). The carbon content of the solid can thus be determined.

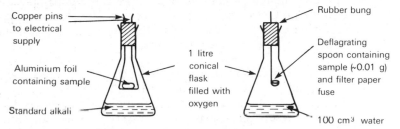

Copper pins to electrical supply

Aluminium foil containing sample

Standard alkali

1 litre conical flask filled with oxygen

Rubber bung

Deflagrating spoon containing sample (~0.01 g) and filter paper fuse

100 cm³ water

Figure 22.2   Schöniger oxygen flask technique

To determine the halogen content of a compound, the sample can be placed in a deflagrating spoon, together with a filter paper fuse; this is ignited and the spoon then placed firmly in the conical flask of oxygen. The hydrogen halide produced is absorbed in water and the solution neutralized before titration against standard silver nitrate solution, with potassium chromate(VI) as indicator. Liquids may be placed in a gelatin medicine capsule before being ignited on the spoon.

22.3   Determination of molecular formula

22.3.1   Gas syringe method

If the empirical formula of a compound is known, then a determination of its relative molecular mass allows the molecular formula to be found.

The relative molecular mass of a volatile liquid may be determined by a gas-syringe method, in which the volume of a known mass of the liquid is measured at a temperature at which the sample is completely vaporized and at constant pressure. This volume is then corrected to a figure for standard conditions (0°C, 101.3 kN m^{-2}) and, since the molar volume of

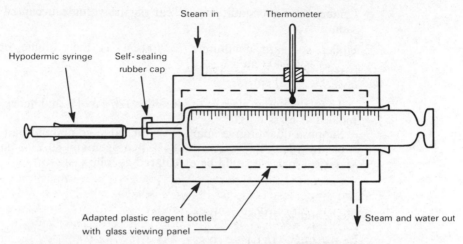

Figure 22.3 Apparatus for gas syringe method of M_r determination

an ideal gas is $22\,400$ cm^3 mol^{-1} under these conditions, the relative molecular mass may be calculated. A known mass of the liquid is usually introduced into a gas syringe, using a hypodermic syringe, the hypodermic syringe being weighed before and after the injection of the sample to give the mass of sample. The gas syringe is maintained at constant temperature, either by partial immersion in a water-bath, or else by enclosure in a steam jacket (*Figure 22.3*). The steady volume occupied by the vaporized sample in the gas syringe is read and the temperature and pressure noted.

The necessary calculation is outlined below.

Example 0.359 g of a volatile organic liquid was injected into a gas syringe and found to occupy a volume of 94 cm^3 at 98°C and 98.6 kN m^{-2} pressure. What is the relative molecular mass of the liquid?

The volume occupied by the vapour should first be corrected to a volume for standard conditions (0°C, 101.3 kN m^{-2}), using the gas law $PV/T = $ const. for a fixed mass of gas.

Let $P_1 = 98.6$ kN m^{-2}, $V_1 = 94$ cm^3, $T_1 = 371$ K (temperature *must* be in Kelvin) and $P_2 = 101.3$ kN m^{-2}, $T_2 = 273$ K. Then

$$\frac{P_1 V_1}{T_1} = \frac{P_2 V_2}{T_2} \quad \text{for a fixed mass of gas}$$

$$\frac{98.6\ \text{kN m}^{-2} \times 94\ \text{cm}^3}{371\ \text{K}} = \frac{101.3\ \text{kN m}^{-2} \times V_2}{273\ \text{K}}$$

$$V_2 = \frac{98.6\ \text{kN m}^{-2} \times 94\ \text{cm}^3 \times 273\ \text{K}}{101.3\ \text{kN m}^{-2} \times 371\ \text{K}}$$

$$V_2 = 67\ \text{cm}^3$$

Under standard conditions, $67 \, cm^3$ is the volume occupied by $0.359 \, g$ sample

Under standard conditions, $22\,400 \, cm^3$ is the volume occupied by

$$0.359 \, g \times \frac{22\,400 \, cm^3}{67 \, cm^3} = 120 \, g$$

Thus, the molar mass of the liquid is $120 \, g \, mol^{-1}$ and hence the relative molecular mass is 120.

Suppose quantitative analysis of the same compound had shown that it contained 10.04% carbon, 0.84% hydrogen and 89.12% chlorine. The empirical formula could be calculated as follows:

	C	H	Cl
elemental %	10.04	0.84	89.12
molar ratio of atoms	$\dfrac{10.04}{12.00}$:	$\dfrac{0.84}{1.00}$:	$\dfrac{89.12}{35.45}$
	0.84 :	0.84 :	2.51
	1 :	1 :	3

Thus, the empirical formula is $CHCl_3$ and, since $M_r = 120$, the molecular formula must be $CHCl_3$ ($12 + 1 + (3 \times 35.5) = 119.5$), and not $C_2H_2Cl_6$ or some other multiple. This example is designed to illustrate the calculations involved; the actual experimental method is unlikely to give results quite so accurately.

In the case of non-volatile organic compounds, M_r values can be derived from colligative property measurements, such as the elevation of the boiling point or the depression of the freezing point of a suitable solvent by the solute. However, the values so obtained are not always very accurate. Specialized techniques generally have to be employed in the determination of very high M_r values; for example, those of polymers (*see also* proteins, Section 17.9).

In practice, however, in research and industrial laboratories relative molecular masses of organic compounds are likely to be determined by **mass spectrometry**.

22.3.2   Mass spectrometry

A mass spectrometer is an instrument designed to separate a stream of ionized atoms or molecules, according to their different masses, and to record these masses. In successive operations, the mass spectrometer (1) vaporizes the sample, (2) ionizes the particles of the sample by bombardment with electrons (or sometimes by other means), (3) accelerates the resulting unipositive (usually) ions and then separates them, according to their different masses, by passage through a powerful magnetic field, (4) records the relative abundances of the ions of different mass, either electronically or photographically. The basic features of a mass spectrometer are outlined in *Figure 22.4*.

As the charged particles pass through the magnetic field, they are deflected to an extent which depends on their mass, the lighter particles being deflected the most. The particles travel along a curved path which may be taken as part of a circle of radius r. It can be shown that

$$\frac{m}{e} = \frac{B^2 r^2}{2V}$$

where m and e are the mass and charge, respectively, of a particle, B is the magnetic field strength, and V is the accelerating voltage. Thus, to focus individual beams of ions at the collector plate (the pathway r is then fixed), the magnetic field strength can be gradually increased, when ion beams of increasing mass strike the collector, or else the accelerating voltage can be reduced.

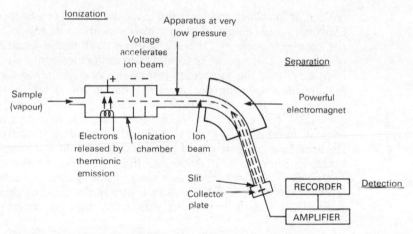

Figure 22.4   Mass spectrometer

If the magnetic field is 'swept' in this way, then the *mass spectrum* of the sample is traversed and the relative abundances of ions of different mass recorded. The results are usually printed out as a mass spectrum in which a series of vertical lines appear, corresponding to different masses. The spectrum is 'normalized' by giving the most abundant ion mass a peak of relative intensity 100 and adjusting the other peaks, or lines, in the spectrum accordingly.

The mass spectrum of methane is shown in *Figure 22.5*.

When methane molecules are introduced into the mass spectrometer, they are bombarded by electrons which cause some of them to lose an electron and form a unipositive ion CH_4^+. The CH_4^+ ion is called the *parent*, or *molecular, ion* and it appears in the mass spectrum at a mass value of 16, corresponding to the relative molecular mass of methane. Other lines appear in the spectrum, as a result of fragmentations of the

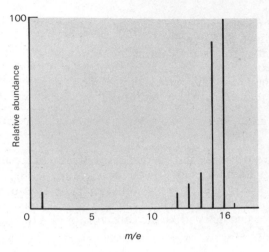

Figure 22.5 Mass spectrum of methane

molecular ion which are discussed in a later section (*see* Section 22.4.2(ii)). However, since natural carbon consists of about 98.9% ^{12}C and 1.1% of the heavier isotope ^{13}C, there is another peak in the spectrum at a value of $(m + 1)$, corresponding to methane molecules containing this heavier isotope. Thus, the molecular ion peak does not always appear at the highest mass value in the spectrum but it will nearly always be more intense than these other isotopic peaks. For a hydrocarbon molecule containing one carbon atom, like methane, the ratio of the $(m + 1)$ to m peak heights will be about 1:99, corresponding to the probability of the molecule containing a ^{13}C rather than ^{12}C. As the number of carbon atoms in a molecule increases, so the probability of the molecule containing one ^{13}C atom increases. Thus, if a molecule contains two carbon atoms, then there is a 2.2% chance of a molecule containing a ^{13}C. The ratio of $(m + 1)$ to m peak heights in the mass spectrum will therefore

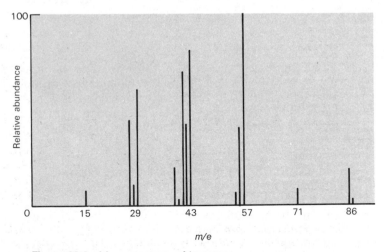

Figure 22.6 Mass spectrum of hexane

be 2.2:97.8. If this measurement was taken from the mass spectrum and converted into a percentage, the figure of 2.2% would be obtained. When this is divided by the natural isotopic abundance of ^{13}C, which is 1.1%, the answer is 2, i.e. the molecule whose mass spectrum has just been examined contains two carbon atoms. This useful way of estimating the number of carbon atoms present in a molecule from its mass spectrum is illustrated by the example in *Figure 22.6.** The spectrum clearly shows the molecular ion at 86, so the relative molecular mass of the compound is immediately established as 86. There is a small peak at 87, corresponding to molecules containing one atom of ^{13}C. The ratio of $(m + 1)$ to m peak heights is 1:15, giving the chances of the molecule containing one ^{13}C as $1/15 \times 100 = 6.6\%$. Thus, $6.6\% \div 1.1\%$ indicates that the molecule contains six carbon atoms. Since $M_r = 86$, the mass spectrum indicates that the molecular formula for the compound is C_6H_{14}. Thus, it would not be necessary to know the empirical formula of a compound to determine its molecular formula, if mass spectrometry was being used.

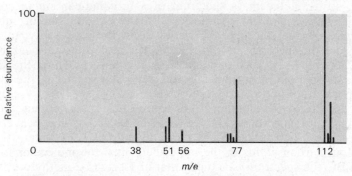

Figure 22.7 Mass spectrum of chlorobenzene

Chlorine and bromine also produce very characteristic isotopic peaks, as a result of their natural isotopic composition. Natural chlorine consists of 75.5% ^{35}Cl and 24.5% of the heavier isotope ^{37}Cl. Thus, the mass spectrum of a molecule containing one chlorine atom will show a peak at two units higher than the molecular ion peak, and the ratio of $(m + 2)$ to m peak heights will be 1:3.

Consider the spectrum in *Figure 22.7*. Chlorine is clearly indicated by the peaks at m and $(m + 2)$, which are in the ratio 3:1, indicating that there is one chlorine atom present in the molecule. There is also a $(m + 1)$ peak, as a result of ^{13}C. The $(m + 1)/m$ ratio is 7/100, and hence the chance of the molecule containing one ^{13}C is 7%. This indicates that there are six carbon atoms in the molecule ($7 \div 1.1 = 6.4$). The molecular

---

* These calculations have been applied to compounds containing C and H only; other elements may introduce complications (*see*, for example, the paragraphs dealing with halogeno compounds).

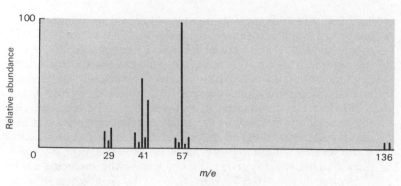

Figure 22.8   Mass spectrum of 1-bromo-2-methylpropane

ion peak indicates a M_r value of 112, and thus it becomes apparent that the molecular formula of the compound is C_6H_5Cl (i.e. chlorobenzene).

The natural isotopic composition of bromine is 50.5% ^{79}Br and 49.5% ^{81}Br, so again the mass spectrum of a bromine-containing compound would show m and $(m + 2)$ peaks. If there was only one bromine atom in the molecule, these peaks would be in the ratio of, virtually, 1:1. Consider the spectrum in *Figure 22.8*. The molecular ion peak is at 136, with an $(m + 2)$ peak of almost equal intensity. Thus, the molecule contains an atom of bromine. In this particular spectrum, the molecular ion peak is not particularly intense (for reasons, see later) so it is impossible, in this actual diagram, to measure the $(m + 1)/m$ peak ratio. The original spectrum shows this value to be 0.045, so that converting this figure into a percentage (4.5%) and dividing by 1.1% shows that there are four carbon atoms in the molecule. Thus, the molecular formula is C_4H_9Br (the spectrum is actually that of 1-bromo-2-methylpropane).

When there is more than one halogen atom in the molecule, then peaks can also arise at $(m + 4)$, etc. (*see Figure 22.9*), and application of statistical theory enables the ratio of peak heights to be calculated.*

With a very high resolution mass spectrometer, the relative molecular mass of a compound can be determined with such accuracy (i.e. to several decimal places) that a molecular formula may be assigned even though several formulae have the same integral relative molecular masses. Thus, the following three substances could be distinguished, despite the fact that the integral relative molecular mass is 28 in each case:

CO	N_2	C_2H_4
27.994 91	28.006 15	28.031 30

Mass spectrometry provides an extremely versatile technique for investigating the molecular structures of organic compounds, and many laboratories now employ a GLC–mass spectrometer link so that mixtures

---

* The number of peaks and their relative abundances are given by the expansion of the polynomial $(a + b)^n$, where a and b represent the percentage natural abundance of each isotope and n is the number of atoms of the element in the molecule.

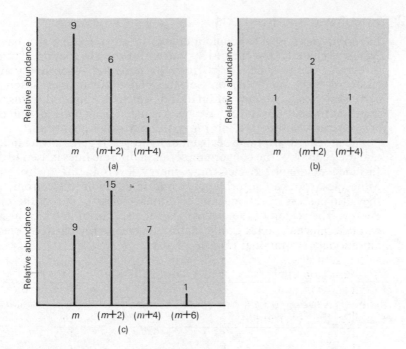

Figure 22.9 (a) Molecule containing 2 Cl; (b) molecule containing 2 Br; (c) molecule containing 2 Cl and 2 Br

may be separated and the individual components identified. Compounds which are thermally unstable cannot, of course, be investigated satisfactorily by mass spectrometry, as no molecular ion would be produced and there are also compounds which give very weak molecular ion peaks, owing to the instability of the molecular ion.

22.4 Identification of functional groups

22.4.1 Chemical methods

When sufficient sample is available for work on a semi-micro scale, then much information may be derived from investigations of the sample's chemical properties, and by carrying out specific tests for various functional groups. Previous chapters have described reactions characteristic

of the main functional groups such as $\Large \rangle C{=}C\langle$, $\Large {\rightarrow}C{-}Cl$, $\Large \rangle C{=}O$,

—COOH etc., including those which are the basis of specific tests. A sound knowledge of these reactions is clearly necessary before the identification of functional groups can be attempted. Measurements of physical properties such as boiling point, density, pK_a, etc., can also be useful.

22.4.2 Spectroscopic methods

In recent years, spectroscopic methods in organic chemistry have completely dominated the chemist's strategy for the elucidation of structures of organic compounds. Apart from the remarkable sensitivity and precision of these methods, one of the major advantages has been the extremely small amount of material which is required. This is very important when substances have been isolated from natural sources (such as hormones, antibiotics, etc.) in extremely small quantities.

The atomic absorption spectrum of hydrogen is generally a familiar one and the analysis of the converging series of discrete lines is used to develop the idea of quantized electronic energy levels in the hydrogen atom. Molecules, too, can absorb energy but, in addition to electronic energy, they also possess vibrational and rotational energy as part of their internal energy. The region of the electromagnetic spectrum in which absorption will take place depends on the nature of the energy change—electronic, vibrational or rotational (*Figure 22.10*).

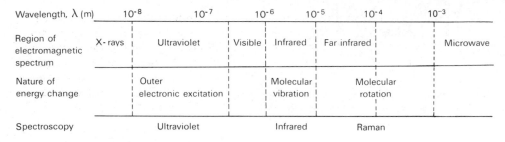

Wavelength, λ (m)	10^{-8}	10^{-7}	10^{-6}	10^{-5}	10^{-4}	10^{-3}
Region of electromagnetic spectrum	X-rays	Ultraviolet	Visible	Infrared	Far infrared	Microwave
Nature of energy change		Outer electronic excitation		Molecular vibration	Molecular rotation	
Spectroscopy		Ultraviolet		Infrared	Raman	

Figure 22.10 Electromagnetic spectrum

(i) One of the regions of the electromagnetic spectrum most useful to the organic chemist is the infrared region, because it is here that organic molecules can be excited to higher vibrational energy levels. These levels are quantized, just like electronic energy levels, so that discrete wavelengths of energy are absorbed when molecules are promoted to higher levels. Furthermore, these wavelengths are characteristic of vibrational energy changes associated with particular chemical bonds, and may therefore be used to detect the presence of these bonds. For example, below are shown some of the possible bending and stretching vibrations associated with the hydroxyl group:

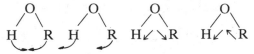

bending   bending   stretching   stretching

Each vibrational state can have its energy increased by discrete amounts, resulting in the absorption of energy of specific wavelengths. The O—H

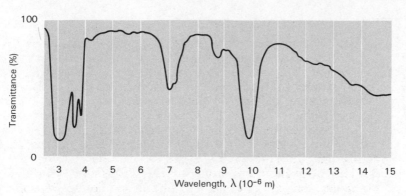

Figure 22.11   IR spectrum of methanol

stretching vibrations lead to absorption at wavelengths in the region of 2.75×10^{-6} m and O—H bending and C—O stretching vibrations in the region of 9.5×10^{-6} m and 7.5×10^{-6} m.

Infrared spectra are recorded by an infrared spectrophotometer which can utilize the sample as a liquid film, as a solution, or as a suspension, or 'mull', in a hydrocarbon (Nujol). The spectrum records the percentage transmittance (or conversely, the absorbance) against the wavelength of the infrared radiation. The infrared spectrum of methanol is shown in *Figure 22.11*. The peak at 3×10^{-6} m (broad because of intermolecular hydrogen bonding) is very characteristic of the hydroxyl group.

Table 22.1

Functional group	*Wavelength, λ $(10^{-6}$ m)*
—NH$_2$	~2.8
—OH	~3.0
$-\overset{\mid}{\underset{\mid}{C}}$—H (alkane)	~3.4
$\overset{}{\underset{H}{\diagup}}$C=O	~3.5 and 3.6
—COOR	~5.7
$>$C=O	~5.8
$>$C=C$<$	~6.0
$-\overset{\diagdown}{\underset{\diagup}{C}}$—O—$\overset{\diagup}{\underset{\diagdown}{C}}$—	~9.0
⬡	~13–15

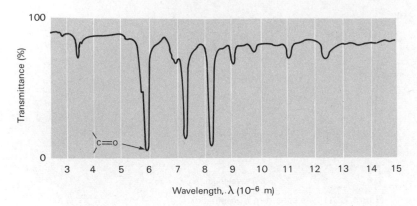

Figure 22.12   IR spectrum of propanone CH_3COCH_3

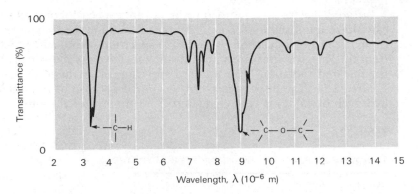

Figure 22.13   IR spectrum of ethoxyethane $C_4H_{10}O$

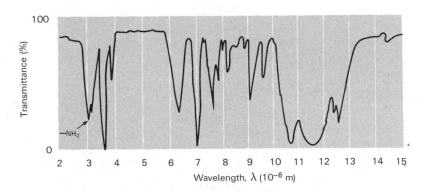

Figure 22.14   IR spectrum of cyclohexylamine $C_6H_{11}NH_2$

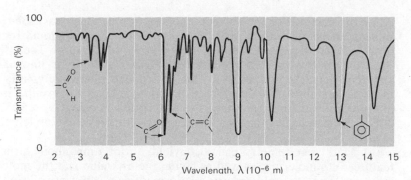

Figure 22.15 IR spectrum of 3-phenylpropenal $C_6H_5CH{=}CH{-}CHO$

Characteristic wavelengths, at which common functional groups absorb, are summarized in *Table 22.1*. The infrared spectra of some simple organic compounds are shown in *Figures 22.12–22.15* to illustrate these absorptions.

Careful analysis of an infrared spectrum can provide more detailed information than just the functional group(s) present, but this is beyond the scope of this section. As well as indicating the presence of functional groups, the infrared spectrum acts as a sort of chemical, or molecular, 'fingerprint' of a molecule. Comparison of the infrared spectra of an unknown compound and an authentic compound will immediately serve to confirm or reject a hypothesis that they are identical, without the 'unknown' compound having to be analysed. Such a rapid method as the recording of a compound's infrared spectrum is also of great use in following the course of column adsorption chromatography (*see* Section 21.2.4(i)), for example, when different compounds are being eluted.

The following example gives an interesting illustration of the importance and convenience of infrared spectroscopy. (Chloromethyl)benzene was heated under reflux with a methanolic solution of potassium hydroxide.

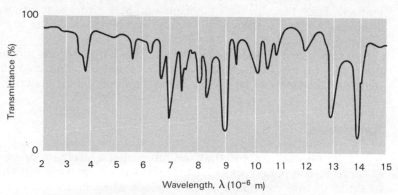

Figure 22.16 IR spectrum of $C_8H_{10}O$

After the removal of the solvent, the reaction product was isolated by solvent extraction with ethoxyethane and, after the drying and removal of the ethoxyethane, was distilled to give a colourless liquid. The infrared spectrum of this product is shown in *Figure 22.16*. It is immediately apparent that the product is not the anticipated phenylmethanol, since there is no absorption characteristic of the hydroxyl group. In the interpretation of this spectrum, the absorption at about 9×10^{-6} m is critical as it indicates the functional group—this task is now left to the reader.

(ii) Mass spectrometry can also yield information about structural features of a molecule, as well as giving the M_r value and the molecular formula. When the molecular ion is formed, as a result of collision with a high energy electron, it is in an excited state and usually breaks down, to a lesser or greater extent, to form smaller ions and neutral fragments. Different types of compounds tend to show characteristic modes of breakdown, or fragmentation, and hence the identification of other ions in the mass spectrum can suggest structural features of the molecule. For example, in the mass spectrum of C_6H_{14} shown in *Figure 22.6* there are characteristic peaks at m/e values of 71,57,43,29 and 15. These can be explained by the following fragmentation process:

$$[CH_3CH_2CH_2CH_2CH_2CH_3]^+ \rightarrow [CH_3CH_2CH_2CH_2CH_2]^+ + CH_3 \rightarrow$$

$m/e = 86$ $\qquad\qquad\qquad\qquad\qquad$ $m/e = 71$

$$[CH_3CH_2CH_2CH_2]^+ + CH_2, \text{etc.}$$

$$m/e = 57$$

These fragmentation ions all fit the formula C_nH_{2n+1}, and correspond to successive losses of neutral CH_2 fragments. The molecule is clearly that of hexane.

In the mass spectrum of C_6H_5Cl in *Figure 22.7* there is a characteristic peak at $m/e = 77$, corresponding to the aromatic $C_6H_5^+$ ion.

Ultraviolet absorption spectroscopy is also widely used for unsaturated molecules, although its scope is more restricted than that of infrared. Nuclear magnetic resonance (NMR) spectroscopy gives valuable information about the chemical environment of hydrogen atoms within a molecule, but the analysis of ultraviolet and NMR spectra is beyond the scope of this book.

Spectroscopic methods then have exerted a tremendous influence on the work of the organic chemist and have enabled the most complicated structural problems to be solved, using the minimum of material.

22.5 Synthesis

It is conventional, and certainly desirable, that a structure proposed for an organic compound should be confirmed by carrying out the total synthesis of that compound from simple starting materials, using well-established reactions. If the synthetic material proves to be identical in properties to the compound (except perhaps in properties such as optical activity, *see* Section 16.3) then the proposed structure is usually accepted as being established. Some of the many fundamental reactions used in organic synthesis are summarized in Chapter 15.

Questions

1*   A substance, **X**, has the following composition by mass:

carbon   52.2%,
hydrogen   13.0%,
oxygen   34.8%

(a) What is the empirical formula of **X**?
(b) When completely vaporized, 0.023 g of **X** is found to occupy a volume of 11.2 cm^3 (corrected to s.t.p.). What is the relative molecular mass of **X**?
(c) What is the molecular formula of **X**?
(d) Write down two different structural formulae which are consistent with your answer to (c).
(e) **X** reacts with iodine in aqueous, alkaline solution to produce a yellow precipitate of triiodomethane (*iodoform*). Write the structural formula of substance **X** and show by means of equations how the triiodomethane is formed.

JMB

2*   The percentage composition by mass of an organic compound **X** was found to be C 85.7%, H 14.3%. 0.21 g of **X** were found to occupy a volume of 84 cm^3 at s.t.p. Ozonolysis of **X** produced two compounds **Y** and **Z**. Both **Y** and **Z** formed a crystalline derivative with 2,4-dinitrophenylhydrazine, but only **Y** gave a positive silver mirror test.
(a) Deduce the empirical formula of **X**.
(b) Deduce the molecular formula of **X**.
(c) Write down the structural formulae of (i) **X**, (ii) **Y**, (iii) **Z**, (iv) the compound formed when **X** reacts with hydrogen bromide gas.

JMB

3*   A substance **X** contains C = 62.07%, H = 10.34% and O = 27.59% (by mass). In an experiment it was found that 0.1704 g of the vapour of **X** occupied a volume of 94.8 cm^3 at 100°C and atmospheric pressure.
(a) Determine the empirical formula of **X**.
(b) Determine the molecular formula of **X**.
(c) On oxidation, **X** yields a monobasic acid **Y**, containing C = 48.65%, H = 8.11% and O = 43.24% (by mass). What is the empirical formula of **Y**?
(d) Give a possible structural formula for **Y**.

JMB

4*   (a) Carbon consists of 99% of the isotope ^{12}C and 1% of the isotope ^{13}C. One way of estimating the number of carbon atoms in the molecule of a compound is to examine the mass spectrometer peaks which the molecule gives. The peak corresponding to the second highest mass is caused by the molecule ion having only ^{12}C in it but the highest mass peak is due to a molecule ion with one ^{13}C atom in it.

A hydrocarbon X gives a peak at mass M and a smaller peak at mass $M + 1$. There are no significant peaks at higher mass than this. The peak at mass M is 12.5 times as intense as the peak at mass $M + 1$.

When a quantity of X was completely oxidized, 0.11 g of CO_2 and 0.023 g of water were formed. X decolorizes bromine water and 1 mol of X reacts with 1 mol of bromine molecules.

What is the molecular formula and probable structure of X?

(b) Bromine consists of a mixture of isotopes ^{79}Br and ^{81}Br. Assuming no bonds are broken in the mass spectrometer and that hydrogen has only one significant isotope, how many peaks will be given by the compound CH_2Br_2 and what will each peak be due to?

Nuffield

5**   Two halogen derivatives of ethane, A and B, have (within ± 1%) equal gas densities. Their mass spectra show peaks ascribed to A^+ and to B^+, as follows:

	Mass number	Relative intensity
A^+	134	100
	136	66
	138	11
B^+	132	100
	134	100
	136	33
	138	4

Explain these observations and derive formulae for the compounds A and B. How many isomers do they each have? Could you differentiate between the isomers?

C = 12, F = 19; Cl = 35 (75%), 37 (25%).   Oxford

6*   (a) Describe a simple mass spectrometer and briefly outline the principles on which it is based.

(b) The figure opposite represents the mass spectrometer trace in the molecular ion region, of a compound containing x chlorine atoms and y bromine atoms.

Chlorine consists of isotopes ^{35}Cl and ^{37}Cl in the ratio 3:1, and bromine of ^{79}Br and ^{81}Br in the ratio 1:1, all other atomic species in the molecular ion being mono-isotopic. Determine x and y.

(c) The bond dissociation energy of a diatomic molecule X_2 is 490 kJ. The ionization potential of X_2 is 1160 kJ, and the energy required to bring about the reaction $X_2^+ \rightarrow X^+ + X^{\cdot}$ in the mass spectrometer is 820 kJ.

Determine the ionization potential of the radical $X^{\cdot}$.

Cambridge

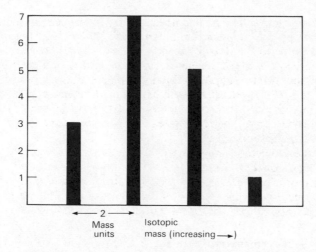

7*   An organic bromide **A** contains C, 36.4% and H, 3.0%: the highest peaks
      in its mass spectrum occur at m/e values of 262, 264 and 266 in the
      approximate intensity ratio 1:2:1. **A** reacts with potassium cyanide to give
      B which on hydrolysis yields an acid **C**, $C_{10}H_{10}O_4$: oxidation of **C** yields
      another acid **D**, $C_8H_6O_4$, heating of which yields **E**, $C_8H_4O_3$.
      Deduce the identities of compounds **A–E** and answer the following
      questions:
      (i)   **F** is an isomer of **A** which also yields **E** in the above reaction sequence:
            identify **F** and explain how **A** and **F** could be distinguished (a)
            chemically, and (b) spectroscopically.
      (ii) Why is **D** a stronger acid than benzenecarboxylic (*benzoic*) acid?
      (iii) What commercial importance has one of the positional isomers of
            E?
      Suggest, on mechanistic grounds, what product would result from the
      reaction of **E** with sodium hydride, NaH.
      (Relative atomic masses H,1; C,12; Br, 79.9)   St Andrews

8*   This question is about the identification of two organic substances, **A** and
      B.
      Substance **A** contains carbon, hydrogen and oxygen only and its mass
      spectrum is shown in *Figure 1*.
      Substance **A** melts at 178 K and boils at 329 K, gives a crystalline derivative
      with 2,4-dinitrophenylhydrazine but does not reduce Fehling's solution.
      Substance **B** was prepared directly from **A** by reduction and its infrared
      absorption spectrum is shown in *Figure 2*.
      B was converted into its monobromo derivative by reaction with sodium
      bromide and concentrated sulphuric acid. The mixture was refluxed and
      then steam distilled. The crude product was purified to give a monobromo
      derivative of melting point 184 K and boiling point 332 K.

      (a) What is the relative molecular mass of **A**?
      (b) Write down the probable formulae of the two main fragments of **A**
          produced in the mass spectrometer.

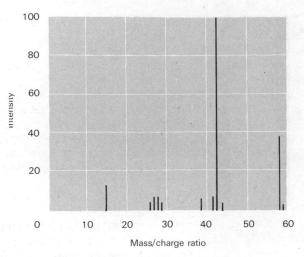

Figure 1. *Substance A—mass spectrum*

Some characteristic infrared absorptions:

Group	Wavelength (10^{-6} m)
$\diagdown$C$=$C$\diagup$	6.1
⬡	6.3, 6.7
—CH$_3$	6.8–7.0, 7.3
—CHO	5.8
$\diagdown$CHOH	9.0
—CH$_2$OH	9.5

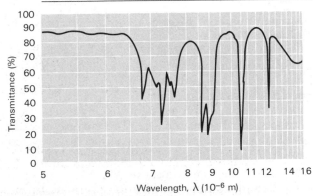

Figure 2.   *Substance B—infrared absorption*

(c) What can be deduced from the facts that A reacts with 2,4-dinitro-phenylhydrazine but not with Fehling's solution?

(d) Give the name of A and its formula.

(e) By interpreting the infrared absorption spectrum, what groups in B can be identified?

(f) Give the name of B and its formula.

(g) This part of the question is about the preparation of the monobromo derivative of B.

    (i) What would be the *two* main constituents of the steam distillate?

    (ii) What apparatus would you use to separate these constituents?

    (iii) How would you obtain a fully purified product from the crude material?

Nuffield

9   Chemists make extensive use of 'physical methods' to elucidate molecular structures. Briefly review the application of *three* of the following techniques for this purpose: (a) Dipole moment measurements; (b) Infrared spectroscopy; (c) Mass spectroscopy; (d) X-ray analysis.
(No details about instrumentation or measurement techniques are required. Instead, emphasis should be given to the kind of information that can be obtained and to the way in which this helps towards the elucidation of molecular structures.)

London

10**   From lemon grass, an oil can be extracted of relative molecular mass 152 and analysis gives C 79.0%, H 10.6%, O 10.4% by mass.
The oil reacts with 2,4-dinitrophenylhydrazine, forming a red crystalline derivative, and reaction with hydrogen at normal pressure using a nickel catalyst produces a primary alcohol of relative molecular mass 158. Write equations for these reactions and deduce as much as you can about the structural formula of the oil.
On vigorous oxidation of the lemon grass oil, three products are obtained in equimolar quantities: ethanedioic acid, $(CO_2H)_2$, propanone and 4-oxopentanoic acid (oxo refers to a carbonyl group). Write down the structural formulae of the three products and, by means of diagrams, indicate the alternative sequences in which they might have been linked before oxidation.
Dehydration of the lemon grass oil, using potassium hydrogensulphate, causes a cyclic structure to form which is 2-(4-methylphenyl)propane. Write down the structural formula of this compound. Decide which of your alternative sequences could have given rise to this compound and hence deduce a structural formula for lemon grass oil.

Nuffield

11**   Outline *very briefly* simple tests you would use to determine the functional groups present in an unknown compound containing carbon, hydrogen and oxygen only.
The toxic principle in buttercups is an unstable oil called protoanemonin, $C_5H_4O_2$, which is readily hydrolysed by acid or base to A, $C_5H_6O_3$. Unlike

protoanemonin, A dissolves with effervescence in sodium bicarbonate solution and can be converted to a phenylhydrazone.

On treatment with hydrogen in the presence of platinum, protoanemonin gives B, $C_5H_8O_2$. B does not dissolve in sodium bicarbonate solution but does so in sodium hydroxide solution; acidification of this solution gives an acid C, $C_5H_{10}O_3$, which on standing reverts to B. When C is treated with sodium hydroxide and iodine it gives triiodomethane (*iodoform*) and an acid D, $C_4H_6O_4$, which is readily converted to an anhydride on heating. Explain the above reactions and deduce a probable structure for protoanemonin.

Cambridge

Revision questions

1   Show, giving suitable examples to illustrate your answer, how the physical
    properties of organic compounds are determined by their molecular
    structures.
    Explain why pentane melts at $-130°C$ and boils at $36°C$, while 2,2-
    dimethylpropane melts at $-16°C$ and boils at $10°C$.

                                                                    JMB

2   Explain the meaning of the terms
    (a) nucleophilic addition,
    (b) electrophilic addition,
    (c) addition–elimination.
    Illustrate your answer by at least **two** examples of each type of reaction
    chosen to show the similarities and differences in these reactions.

                                                                    JMB

3* An organic compound **A**, with a relative molecular mass of 178, contains
    74.2% C, 7.9% H, and 17.9% O. Boiling **A** with aqueous alkali gave a
    volatile compound **B** which did not give a positive haloform test. Acidi-
    fication of the alkaline solution gave **C** which was soluble in aqueous
    sodium carbonate solution. A 0.100 g sample of **C** neutralized 7.35 cm^3
    of aqueous sodium hydroxide solution (0.100 mol dm^{-3}). Reduction of
    either **A** or **C** with lithium tetrahydridoaluminate (LiAlH$_4$) yielded **D**
    which reacted with ethanoic (*acetic*) anhydride to form **E**, C$_{10}$H$_{12}$O$_2$.
    Oxidation of **D** yields benzene-1,4-dicarboxylic acid.
    Identify compounds **A** to **E**, giving your reasoning.

                                                        Oxford & Cambridge

4* For **five** of the following pairs of compounds, describe a chemical test
    which would enable you to distinguish one member of the pair from the
    other. Describe the effect of the reagent(s) you choose on both
    compounds.
    (a) CH$_3$OCH$_3$ and CH$_3$CH$_2$OH
    (b) CH$_3$CH$_2$CH(OH)CH$_3$ and CH$_3$CH$_2$CH$_2$CH$_2$OH
    (c) CH$_3$CH$_2$CH$_2$CH$_2$CHO and CH$_3$CH$_2$COCH$_2$CH$_3$
    (d) C$_6$H$_5$CH$_2$OH and HOC$_6$H$_4$CH$_3$
    (e) CH$_3$CH$_2$NH$_2$ and CH$_3$CN
    (f) C$_6$H$_5$OH and C$_6$H$_5$COOH

                                                        Oxford & Cambridge

5   Addition and addition–elimination reactions are a common feature of the
    chemistry of carbonyl compounds. Give an account of the general mech-
    anisms of these reactions.
    Discuss the similarities and differences which exist between these two
    types of reaction by considering the reactions between
    (a) ethanal and lithium tetrahydridoaluminate,

(b) propanone and hydrazine,

(c) ethanoyl (*acetyl*) chloride and ammonia,

(d) ethanoic (*acetic*) acid and ethanol.

Show how the principles of these reactions find application in the production of (i) Terylene, (ii) nylon 66.

JMB

6   Summarize the reactions, if any, of simple alkanes, alkenes, alkynes and of benzene with the following reagents:

(a) bromine,

(b) potassium permanganate, and

(c) sulphuric acid.

Clear structural formulae of reaction products should be given and approximate experimental conditions stated.

For each of the types of hydrocarbon, select one reaction and briefly discuss its mechanism.

London

7*   By reference to the mechanism of the reactions explain the following observations:

(a) The rate of reaction of ethanal with phenylhydrazine varies with pH.

(b) When chlorobenzene is mixed with silver nitrate solution and dilute nitric acid, no precipitate is observed, whereas chloroethane gives a white precipitate under similar conditions.

(c) Bromoethane is hydrolysed by aqueous sodium hydroxide more easily than 2-bromopropane.

(d) Bromoethene reacts less readily with hydrogen bromide than ethene does.

(e) Phenol has a pK_a of 9.9, whereas methanol has a pK_a of 16.

JMB

8*   Write down, giving your reasons, a structural formula for **four** of the following compounds:

(i)   A carbonyl compound C_8H_8O which gives a product $C_7H_6O_2$ upon treatment with aqueous NaOI (I_2 in aqueous NaOH solution).

(ii) A compound $C_6H_5C_3H_6NO_2$ which is soluble in water at both pH 2 and pH 10 but which can be precipitated from aqueous solution at pH 6.

(iii) An alcohol $C_4H_{10}O$ which is not changed by acidified aqueous potassium dichromate(VI) ($K_2Cr_2O_7$) solution.

(iv) A compound C_6H_7N which after treatment with $NaNO_2$ in dilute hydrochloric acid solution at 273 K followed by phenol (hydroxybenzene) gives a bright orange colour.

(v)   An optically active compound C_5H_8O which gives a yellow precipitate with 2,4-dinitrophenylhydrazine and which does not absorb hydrogen in the presence of platinum catalyst.

Cambridge

9* Draw all the stereoisomers of the following compounds indicating which
   are optically active:
   (a) $CH_3CH(OH)CH(OH)CH_3$
   (b) $CH_3CH\!=\!CHCH\!=\!CHCH_3$
   (c) $CH_3CH\!=\!C\!=\!CHCH_3$
   (d) ClHC—CHCl
       | |
       ClHC—CHCl

                                                                    Oxford

10 (a) State the Markovnikov rule for predicting the direction of electrophilic
       addition of hydrogen bromide to alkenes. Explain in detail the rule
       in terms of the relative stabilities of primary, secondary, and tertiary
       carbocations.
   (b) Propene reacts with hydrogen bromide to give a substance A, C_3H_7Br.
       Substance A, when heated with aqueous potassium hydroxide, gives
       an alcohol B.
       (i) Derive structures for A and B.
       (ii) Explain and illustrate the meaning of the terms *base, nucleophile,*
            and *inductive effect* by referring to the reactions of substance A
            with potassium hydroxide under various conditions.

                                                                      JMB

11 Outline **one** series of reactions by which **each** of the following macromolecules is synthesized on an industrial scale from petroleum or coal
   tar:
   (a) poly(ethene)
   (b) nylon 66.
   Indicate reagents and the conditions used and name the substances formed
   at the different stages.
   What is the essential difference in the polymerization process by which
   these two macromolecules are made from monomeric materials?
   Compare the properties of poly(ethene) and nylon 66 and indicate why
   one of these materials is more suitable than the other for making fibres.

                                                                      JMB

12 Describe the reactions between each of the following pairs of compounds:
   (a) ethanal and sodium hydrogensulphite,
   (b) 2-bromopropane and potassium cyanide.
   In each case give
       (i)   the type of reaction taking place,
       (ii) a balanced equation,
       (iii) the name of the organic product of the reaction,
       (iv) the mechanism of the change.
   (c) Compare and contrast the reactions between (i) hydrogen cyanide
       and propanone, and (ii) hydrogen bromide and ethene.

                                                                      JMB

13* Comment critically on **three** of the following statements. Elaborate upon
     them where you agree with them and offer improvements where you do
     not.
     (a) Halogenoalkanes (*alkyl halides*) are susceptible to nucleophilic sub-
         stitution because of the polarity of the carbon–halogen bond—the
         more polar the bond, the more readily will halogenoalkanes undergo
         this type of reaction.
     (b) The polarity of the carbonyl group induces a dipole in the adjacent

         $\overset{\gamma}{} \overset{\beta}{} \overset{\alpha}{}$

         (α–) CH bonds of carbonyl compounds $[CH_3CH_2CH_2CO\sim]$. The α-
         hydrogens of these compounds are therefore more acidic than are
         the hydrogens of alkanes.
     (c) Ethanoyl (*acetyl*) chloride is more reactive towards water than is
         chloroethane (*ethyl chloride*) because the carbonyl group in the former
         is more polar than the CH_2 group in the latter.
     (d) Alkenes react readily with bromine water, whereas alkanes do not,
         because the carbon–carbon double bond is weaker than the carbon–
         carbon single bond.
     (e) Water and petrol do not mix because like will only dissolve like.
                                                            Oxford & Cambridge

14* 'One of the principal methods for the synthesis of substituted aromatic
     compounds involves the nitration of the parent aromatic hydrocarbon
     followed by further transformations of the products.'
     Illustrate this statement by discussing the synthesis of the following com-
     pounds from either benzene or methylbenzene (*toluene*):
     (a) iodobenzene

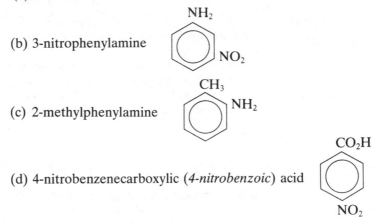

     (b) 3-nitrophenylamine

     (c) 2-methylphenylamine

     (d) 4-nitrobenzenecarboxylic (*4-nitrobenzoic*) acid

     Your answer should contain the names or formulae of all reagents, the
     structures of intermediate compounds and, where appropriate, the struc-
     tures of principal by-products. Important reaction conditions should be
     specified, e.g. temperatures, if important in obtaining high yields of
     correct products. Details of isolation, separation and purification are **not**
     required.
                                                            JMB

15 Show, by the use of suitable labelled diagrams, the type and geometry
    of the bonds in
    (a) alkanes,
    (b) alkenes,
    (c) alkynes.
    Discuss those properties of (i) tetrachloromethane (*carbon tetrachloride*),
    (ii) ethene and (iii) benzene, which are dependent upon the type and
    geometry of the bonding in the molecule.

                                                                    JMB

16* Answer **one** part of this question:
     Either (a) You are given a supply of radioactive sodium cyanide Na*CN
     in which the radioactivity is located at the carbon atom as the isotope
     ^{14}C. Show how you would use this reagent to carry out the following
     conversions in which the radioactive carbon atom is to be located at the
     position marked with an asterisk.
     (i) $CH_3CH_2Br \rightarrow CH_3CH_2{}^*CN$
     (ii) $CH_3I \rightarrow CH_3{}^*COOCH_3$
     (iii)

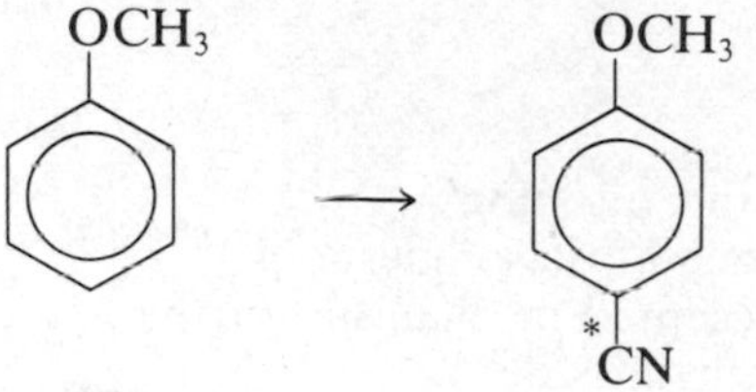

     (iv)

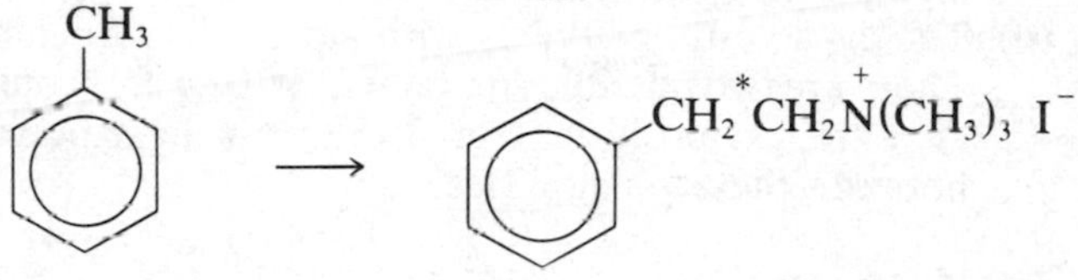

     (v)

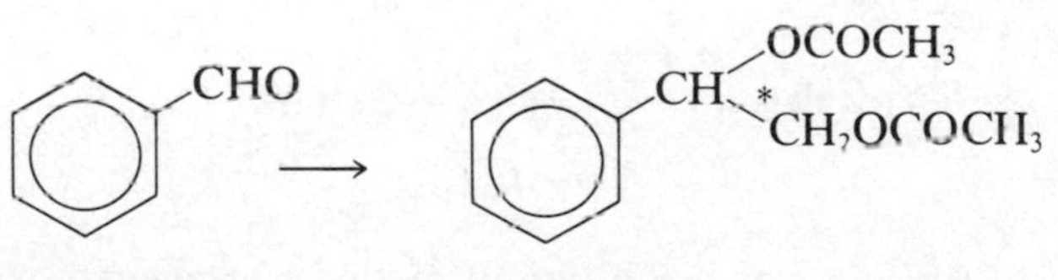

     Or (b) A possible way in which glycine, a component of protein in all
     living things, was first produced in the solar system is this sequence of
     reactions:

$$CH_2O + NH_3 \rightarrow CH_2{=}NH \xrightarrow{HCN} H_2NCH_2CN \xrightarrow{H_2O} H_2NCH_2COOH \text{ (glycine)}$$

     Write detailed mechanisms for these reactions. Glycine is soluble in water
     and not in organic solvents. How might its formula be better drawn?

                                                                    Cambridge

17* Discuss, with examples, the role of catalysis in organic chemical reactions. Write out reasonable structures which explain **three** of the following reactions:

(a) $C_6H_{14}O_2 \xrightarrow[H_2O]{H_3O^+} C_2H_4O + C_2H_6O$

(b) $C_3H_6O \xrightarrow[D_2O]{NaOD} C_3D_6O$

(c) $C_5H_8 \xrightarrow{H_2/Pt} C_5H_{10}$

(d) $C_7H_8 \xrightarrow{Br_2/Fe} C_7H_7Br$

Oxford

18 Arrange the following compounds in order of increasing base strength. Give a clear statement of your reasoning:

CH_3NH_2, $(CH_3)_2NH$, $(CH_3)_4N^+OH^-$, $C_6H_5NH_2$, $C_6H_5CONH_2$, $4\text{-}NO_2C_6H_4NH_2$

Oxford

19* Answer **one** part of this question:
Either (a) Answer **each** of these questions:
(i) Write out the accepted mechanism for the hydrolysis of an ester in aqueous sodium hydroxide.
(ii) The cyclic ester A reacts with aqueous hydrochloric acid to give B. There are two possible sites at which the water could attack compound A, either at position 1 or 3. Suggest an experiment to distinguish between these possibilities.

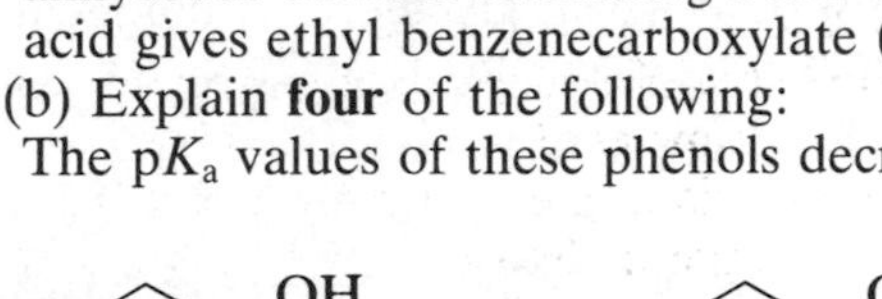

$$^3CH_2 - O \qquad\qquad CH_2OH$$
$$\qquad\qquad \xrightarrow[H_2O]{H^+}$$
$$^2CH_2 - {}^1C{=}O \qquad\quad CH_2 - COOH$$

A   B

(iii) Explain why a solution of methyl benzenecarboxylate (*benzoate*) in anhydrous ethanol containing a few drops of concentrated sulphuric acid gives ethyl benzenecarboxylate (*benzoate*).
Or (b) Explain **four** of the following:
(i) The pK_a values of these phenols decrease in the order shown:

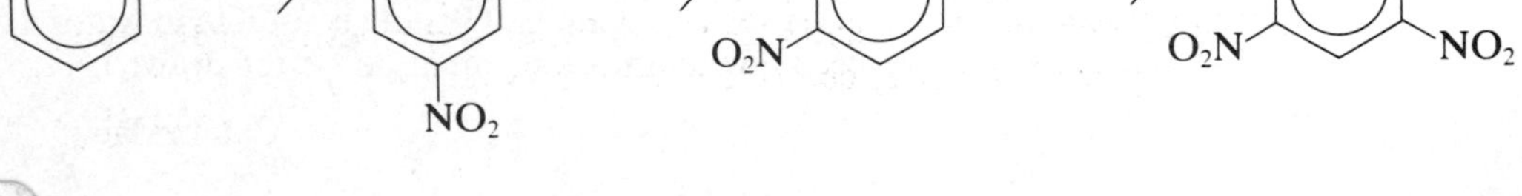

(ii) Only one of the following is a suitable solvent for the Friedel–Crafts reaction between benzene and ethanoyl (*acetyl*) chloride (CH_3COCl) to give phenylethanone (*acetophenone*) ($C_6H_5COCH_3$):

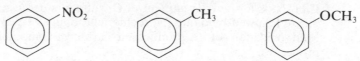

(iii) Different carbohydrates can be fermented by various bacteria. A simple test for the occurrence of fermentation is the addition of an acid/base indicator to the medium.

(iv) Benzene has a melting point of 5°C, methylbenzene (*toluene*) ($C_6H_5CH_3$) has a melting point of −95°C.

(v) C_6H_5COCl and $C_6H_5COOCH_3$ both give the same product ($C_6H_5CONH_2$) on treatment with ammonia, but C_6H_5COCl reacts much faster than $C_6H_5COOCH_3$.

Cambridge

20* An optically active compound **A**, $C_9H_{10}O$ forms a condensation product when treated with hydroxylamine.

Treatment of compound **A** with warm ammoniacal silver nitrate results in the production of metallic silver.

Reduction of compound **A** with lithium tetrahydridoaluminate in ethoxyethane followed by acidification gives an optically active compound **B**, $C_9H_{12}O$ which, on treatment with concentrated sulphuric acid, is converted to a product **C**, C_9H_{10}.

Ozonolysis of **C** followed by decomposition of the ozonide with water gives methanal and a compound **D**, C_8H_8O, which reacts with iodine in aqueous sodium hydroxide on warming to give triiodomethane (*iodoform*). Subsequent acidification of the reaction mixture gives a precipitate of benzenecarboxylic (*benzoic*) acid.

Deduce the structures of the compounds **A**, **B**, **C**, and **D**.

Explain why **A** and **B** are optically active and suggest why compound **A** loses its optical activity on treatment with cold dilute alcoholic potassium hydroxide.

JMB

21   (a) Describe, giving detailed diagrams of the mechanisms, the reaction of
       (i)   benzene with chlorine in the presence of iron(III) chloride,
       (ii) 1-chloropropane with aqueous sodium hydroxide.
     (b) Using these mechanisms explain the following:
       (i)   2-chloropropane is less easily hydrolysed than 1-chloropropane by aqueous sodium hydroxide,
       (ii) 1-chloro-2-nitrobenzene reacts with aqueous sodium hydroxide at a lower temperature than does chlorobenzene,
       (iii) benzenecarbaldehyde (*benzaldehyde*) and chlorine in the presence of iron(III) chloride react only with difficulty.

JMB

22*   A halogenoalkane, **A**, of formula C_3H_7Cl, on heating with an alcoholic
       solution of potassium cyanide yields **B**, of formula C_4H_7N. When **B** is
       reduced with lithium tetrahydridoaluminate a substance **C**, of formula
       $C_4H_{11}N$, is formed. Treatment of **C** with an aqueous solution of sodium
       nitrite followed by dilute hydrochloric acid gives **D**, of formula $C_4H_{10}O$.
       Mild oxidation of **D** produces a carbonyl compound with reducing
       properties.
       From the above reactions deduce possible structures for **A**, **B**, **C** and **D**.
       When **A** is treated with dilute sodium hydroxide and the product gently
       oxidized, the resultant compound is a non-reducing carbonyl compound.
       Using this additional information choose between the possible structures
       of **A**.

                                                                              JMB

23*   **Explain why**
       (a) The order of reactivity of the following halogen derivatives towards
           dilute aqueous alkali is

           bromomethane > bromoethane > bromobenzene

       (b) In the reaction of ethanal with hydrazine hydrochloride, sodium
           ethanoate (*acetate*) is added.
       (c) The order of reactivity of the following carbonyl compounds towards
           hydrogen cyanide is

           ethanal > propanone > benzenecarbaldehyde (*benzaldehyde*).

       (d) Ethanal is unchanged by water whereas ethanoyl (*acetyl*) chloride
           reacts very rapidly.

                                                                              JMB

24*   Study this reaction scheme:

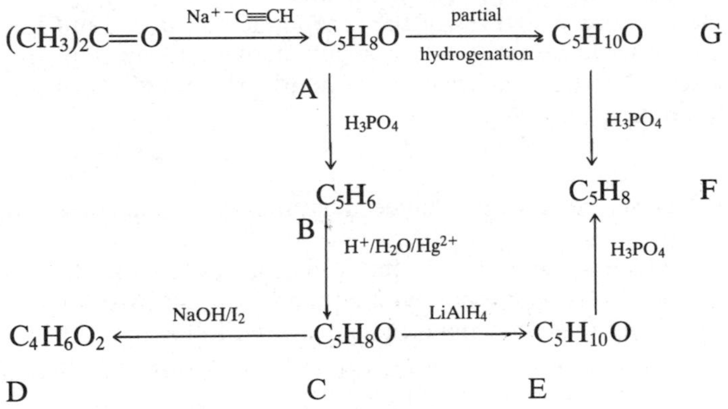

$$(CH_3)_2C{=}O \xrightarrow{\text{Na}^+{}^-C{\equiv}CH} C_5H_8O \xrightarrow[\text{hydrogenation}]{\text{partial}} C_5H_{10}O \quad G$$

$$A \downarrow {}_{H_3PO_4} \qquad\qquad \downarrow {}_{H_3PO_4}$$

$$C_5H_6 \qquad\qquad C_5H_8 \quad F$$

$$B \downarrow {}_{H^+/H_2O/Hg^{2+}} \qquad\qquad \uparrow {}_{H_3PO_4}$$

$$C_4H_6O_2 \xleftarrow{\text{NaOH/I}_2} C_5H_8O \xrightarrow{\text{LiAlH}_4} C_5H_{10}O$$

$$D \qquad\qquad\qquad C \qquad\qquad E$$

       (i)   Write one structure for **each** of compounds A to G.
       (ii) Name compounds F **and** G.
       (iii) What products would be obtained from ozonolysis of compound F?
                                                                        Cambridge

25*   Compound A, C_7H_{14}, reacts with hydrogen in the presence of a platinum
      catalyst to give B, C_7H_{16}. Compound B contains an asymmetric carbon
      atom, but the sample produced is racemic.
      When A is treated with ozone and then water, two compounds C and D
      are produced. Compound C liberates carbon dioxide from sodium car-
      bonate solution, but compound D does not. Treatment of D with sodium
      iodate(I) (*sodium hypoiodite*) solution gives a yellow precipitate, and
      acidification of the filtrate gives C.
      Suggest possible structures for compounds A, B, C and D, and explain
      the reactions involved.

                                                                    Oxford

26   What is a polymer?
      (a) Outline the synthesis of a nylon and give the structure of the final
          product.
      (b) Show how benzene-1,4-dicarboxylic acid (*terephthalic acid*) and
          ethane-1,2-diol react to form a polymer. In what way would the
          products differ if a small quantity of ethanol were added to the
          reactants?
      (c) A natural polymer is made up from molecules of the general formula
          $RCHNH_2COOH$. Suggest a basic structure for this polymer.
          When R is CH_3CHOH the molecule is *threonine*.
          Predict what reaction might take place when *threonine* is treated
          with
          (i) nitrous acid,
          (ii) phosphorus pentachloride.

                                                                      JMB

27*   (a) An organic compound with the molecular formula $C_4H_4O_4$ undergoes
          the following reactions:
          (i)   The compound decolorizes bromine water and reduces a purple
                aqueous solution of potassium manganate(VII).
          (ii) 1 mol of the compound reacts completely with 1 mol of bromine
                Br_2.
          (iii) 1 mol of the compound reacts completely with 2 mol of a mono-
                acidic base.
      Suggest a structure for the compound.
      (b) Outline, giving equations and essential reaction conditions, the indus-
          trial production of
          (i)   ethane-1,2-diol from ethene,
          (ii) chloroethene ($CH_2{=}CHCl$) from ethene, and
          (iii) propan-2-ol from propene.
          Give **one** major use for each of the products of (ii) and (iii).

                                                                      JMB

28*   (a) When 4-chloro(chloromethyl)benzene (*para-chlorobenzylchloride*),
          I, is boiled with alkali, only one chlorine atom is lost. The product
          of the hydrolysis is therefore either II or III.

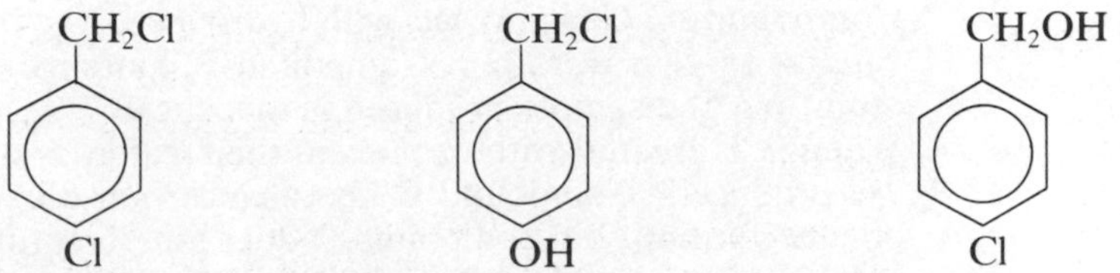

CH$_2$Cl   CH$_2$Cl   CH$_2$OH

Cl   OH   Cl

I   II   III

(i) Devise **one** chemical test for **each** of the possible products which could be used to determine which one was actually formed.

(ii) Which of the products, II or III, would you expect to observe for this reaction, and why?

(b) The carbon–oxygen bonds in cyclic ethers containing small rings (epoxides) are quite easily broken:

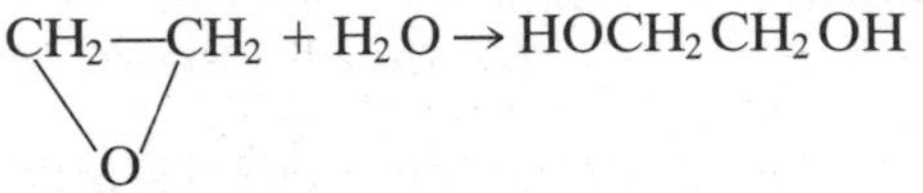

$$CH_2\!-\!CH_2 + H_2O \rightarrow HOCH_2CH_2OH$$
$$\diagdown O \diagup$$

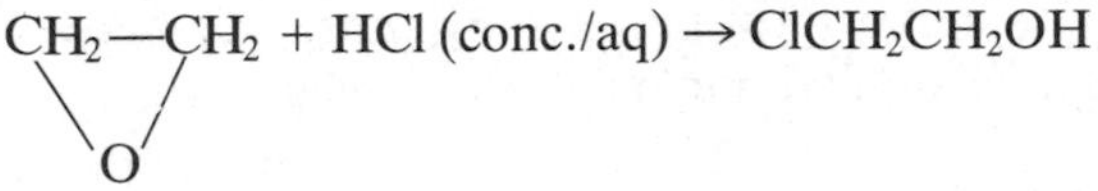

$$CH_2\!-\!CH_2 + HCl\ (conc./aq) \rightarrow ClCH_2CH_2OH$$
$$\diagdown O \diagup$$

What are the possible products from the reaction between $CH_3\!-\!CH\!-\!CH_2$ and concentrated aqueous hydrochloric acid, and
$$\diagdown O \diagup$$

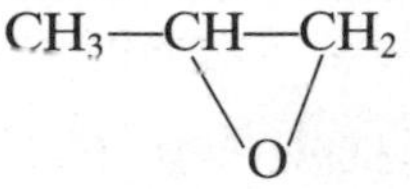

how could you distinguish between them?

(c) 3-Chloropropene (*allyl chloride*) labelled with carbon-14 as shown, hydrolyses to give **two** products:

$$^{14}CH_2\!=\!CH\!-\!CH_2\!-\!Cl + H_2O \rightarrow HO\!-\!{}^{14}CH_2\!-\!CH\!=\!CH_2 +$$
$$^{14}CH_2\!=\!CH\!-\!CH_2\!-\!OH$$

What products would you expect to obtain from the methyl-substituted equivalent, $CH_3CH\!=\!CH\!-\!CH_2\!-\!Cl$? Assuming that you could easily separate the products, how could you recognize which was which? (Give one chemical test for each.)

Oxford & Cambridge

29*   Three different compounds, each of molecular formula C_8H_8O, give yellow precipitates with 2,4-dinitrophenylhydrazine and are reduced to compounds of formula $C_8H_{10}O$ by lithium tetrahydridoaluminate (LiAlH$_4$).

Given that these compounds do not differ merely by substituent orientation around a ring, suggest structures for each one.

How, using chemical tests, would you decide which compound is which? Suggest how the following compounds could be synthesized, assuming all three of the above compounds are available as the only organic starting materials:

(a) benzenecarboxylic (*benzoic*) acid,
(b) one of the isomeric methylphenylmethanols

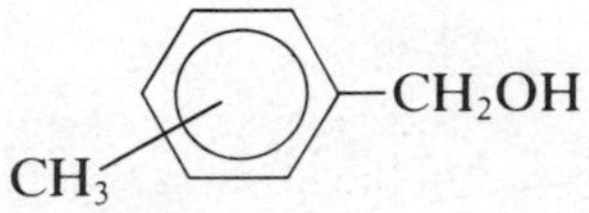

(c) $C_6H_5CH_2CH(OH)CH(C_6H_5)CHO$.

Oxford & Cambridge

30*   Account for **four** of the following observations:
(a) CH_3COCCl_3 is hydrolysed by aqueous alkali with carbon–carbon bond cleavage, whereas CH_3COCH_3 is not.
(b) The hydrogen atoms of benzene will exchange with the deuterium of deuterium oxide in strongly acidic media but not in basic media.
(c) Benzene reacts with bromine by substitution, whereas under similar conditions ethene reacts by addition.

(d) A carbon–oxygen bond in epoxyethane, $CH_2\!-\!CH_2$, is more readily cleaved by nucleophiles than is a carbon–oxygen bond in ethoxyethane (*diethyl ether*).
(e) Unsealed bottles of benzenecarbaldehyde (*benzaldehyde*) (b.p. 179°C) tend to develop a solid white crust round the neck. The crust is sparingly soluble in cold water but is soluble in aqueous sodium hydroxide solution.
(f) Ethanonitrile (*methyl cyanide*) is much less toxic than is sodium cyanide.

Oxford & Cambridge

31*   Show, by giving essential reagents, conditions and equations, how the following conversions can be carried out.
(a) propan-1-ol to propan-2-ol,
(b) 1-chloropropane to propanoyl chloride.
(c) ethanoic (*acetic*) acid to propanoic acid.

JMB

32   Describe, giving examples, how **four** of the following may be used in organic chemistry: (a) chlorine; (b) sulphuric acid; (c) anhydrous aluminium chloride; (d) sodium hydroxide; (e) magnesium; (f) HNO_2; (g) ozone.

Oxford

33* Write down the possible structures for a compound of molecular formula
     C_4H_6. One of these, **A**, undergoes the following reactions:

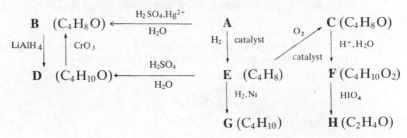

     1 mol of **F** gives 2 mol of **H** in oxidation by HIO_4. Of compounds **A–H**
     only **H** reacts with ammoniacal silver nitrate. Suggest structures for
     compounds **A–H**, give your reasons, and explain the chemistry involved.
                                                                    Oxford

34   Suggest reaction mechanisms to account for each of the following, explain-
     ing any terms that you use:
     (a) The addition of hydrogen chloride to propene results in the formation
         of 2-chloropropane rather than 1-chloropropane.
     (b) Substances such as ammonia and hydrogen cyanide add more easily
         to the C=O bond than do bromine or hydrogen iodide. The latter
         compounds add more easily to the C=C bond than do the former.
     (c) The polymerization of chloroethene (*vinyl chloride*), CH_2=CHCl,
         is catalysed by a peroxide such as di(benzenecarbonyl) peroxide
         (*dibenzoyl peroxide*) $C_6H_5COOOCOC_6H_5$.
                                                                    London

35   What is the diazotization reaction? Describe the usual experimental
     conditions and explain the reasons for their adoption.
     What types of compound can be coupled with the products of the dia-
     zotization reaction? Discuss with examples the applications of the
     reaction.
     8.0 cm³ of a saturated solution (at room temperature) of phenylamine
     (*aniline*) was diazotized. When the reaction mixture was heated, 76.4 cm³
     of nitrogen gas was produced, measured at room temperature. Assuming
     that no side reactions occurred, calculate the solubility of phenylamine
     in water at room temperature.
                                                                    Nuffield

36* The compound **A**, of elemental composition by mass C 56.0%: H 7.0%:
     O 37.0%, is readily oxidized by acidified potassium dichromate(VI) to
     an acid, **B**. To neutralize 118 mg of **B**, 20.0 cm³ of 0.1M sodium hydroxide
     is required.
     A reacts by addition with sodium hydrogensulphite. **A** does not undergo
     an addition reaction with bromine in tetrabromomethane but forms a
     mixture containing five different halogenated substitution products
     (excluding optical isomers) with this reagent.

Deduce the structures of **A** and **B** giving your reasons and the equations for the reactions you discuss.
Suggest a synthesis of **A** from a suitable alkene.

JMB

37 Discuss the use and importance of each of the following reagents in organic syntheses:
(a) lithium tetrahydridoaluminate,
(b) anhydrous aluminium chloride,
(c) a mixture of concentrated nitric and sulphuric acids,
(d) peroxoethanoic acid (*peroxoacetic acid*).

JMB

38* (a) Name the compounds whose formulae are shown below:

A $CH_3CH_2CONH_2$

B $CH_3CH_2CHICH_3$

C

D CH_3COOCH_3

E    $CH{=}CH_2$

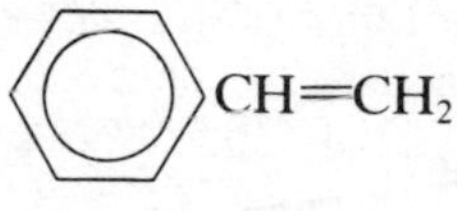

(b) For *each* of the compounds A to E above, select *one* of the reagents listed below. State clearly the formulae of the products of the reaction of the compound with that reagent, including inorganic ones where appropriate, and indicate the conditions for a successful reaction. The reagents are

Dilute aqueous sodium hydroxide   Bromine in tetrachloromethane
A mixture of concentrated nitric   Bromine in aqueous sodium
and sulphuric acids   hydroxide

London

39 Illustrate the essential features of the reactions of saturated and unsaturated aliphatic hydrocarbons and of aromatic hydrocarbons by considering the reaction of
(a) butane,
(b) but-2-ene,
(c) benzene,
(d) methylbenzene (*toluene*),
with (i) halogens, (ii) oxidizing agents and (iii) nitrating agents.

JMB

40*   Outline a synthesis of 4-methylphenylamine CH_3—⬡—NH_2 from

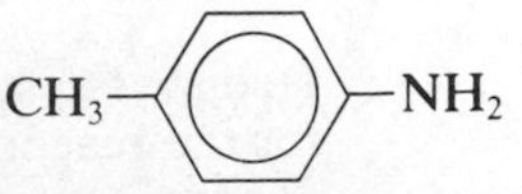

benzene, indicating the reagents and conditions for each reaction you
mention.
By means of equations and brief notes on reaction conditions, show how
the following compounds could be prepared from 4-methylphenylamine:

(a) 4-methylphenol CH_3—⬡—OH

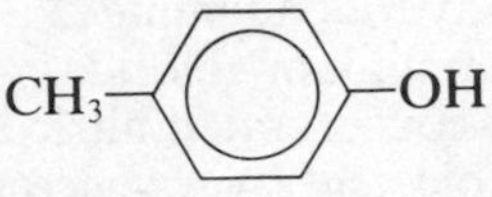

(b) 4-methylphenylhydrazine,
(c) methylbenzene (*toluene*).

Given that a pure substance is 4-methylphenylamine **or** N-methylphenylamine **or** ethylamine, describe briefly **two** tests you would carry
out to identify the substance.

                                                                    JMB

41*   The following scheme shows the inter-relationships of a number of compounds which can be prepared from ethanol:

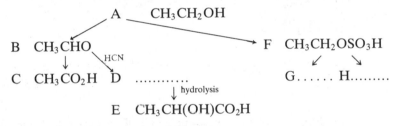

(a) Give the formulae and names of compound D, and of two possible
    compounds, G and H.
(b) Describe the approximate experimental conditions needed to convert
    ethyl hydrogensulphate (compound F) into the compounds G and H
    you have chosen.
(c) (i) What reagents may be used for the conversion of ethanol into
       ethanal (compound B) and thence into ethanoic acid (compound
       C)?
   (ii) What experimental conditions are required if ethanal is to be
       obtained as the main reaction product?
(d) (i) What feature of the structure of compound E suggests that it
       should be possible to resolve it into optical isomers?
   (ii) In practice, if compound E is synthesized by the procedure given,
       it would not be optically active. Briefly explain why.
(e) Assume you are given a mixture of ethanal (compound B) and ethanoic
    acid (compound C) and are asked to determine the composition of
    this mixture, i.e. find the molar ratio of the two compounds.
    Suggest an analytical procedure which may be used for this purpose.
                                                                    London

42**   An optically inactive base, **A**, $C_{10}H_{15}NO$ isolated from barley, forms a
       crystalline compound **B**, $C_{11}H_{18}INO$ on treatment with an excess of
       iodomethane (*methyl iodide*). When compound **B** is heated with moist
       silver oxide, a product **C**, C_8H_8O can be obtained by distillation.
       C gives a purple colour with iron(III) chloride solution and can be
       converted to **D**, $C_7H_6O_3$ by oxidation with potassium permanganate.
       Nitration of the methyl ether of **D** gives a single mononitro derivative **E**.
       Deduce the structures of **A**, **B**, **C**, **D** and **E**.

                                                                          Oxford

43*   In five of the following pairs indicate which compound is the stronger
       acid, giving your reasons:
       (a) $ClCH_2CO_2H$ and CH_3CO_2H

       (b) O_2N—⟨benzene ring with OH⟩   and   O_2N—⟨benzene ring⟩—OH

       (c) $CH_3NH_3^+ \ Cl^-$ and $C_6H_5NH_3^+ \ Cl^-$
       (d) HO_2CCO_2H and $HO_2CCH_2CH_2CO_2H$
       (e) $C_6H_5SO_3H$ and $C_6H_5CO_2H$
       (f) HCN and H_2S

                                                                          Oxford

44*   Describe and explain how you would distinguish by chemical experi-
       ment(s) between the members of each of the following pairs of isomers:
       (a) CH_3COCH_3 and C_2H_5CHO

       (b) H_3C—⟨benzene ring⟩—Cl   and   ⟨benzene ring⟩—CH_2Cl

       (c) Cyclohexane and hexene
       (d) $ClCH_2CO_2H$ and $HOCH_2COCl$
       (e) $(CH_3)_3COH$ and $(CH_3)_2CHCH_2OH$
       Give the results of the experiment(s) you describe on each compound.

                                                                          London

45**   When dry hydrogen chloride is passed through an alcohol, **A**, of molecular
       formula, $C_3H_8O_3$, another alcohol, **B**, of molecular formula, $C_3H_6OCl_2$
       is produced. If **B** is oxidized with chromic acid a neutral compound, **C**,
       is produced which does not reduce warm ammoniacal silver nitrate. **C**
       has the formula $C_3H_4OCl_2$ and reacts with aqueous sodium hydrogen-
       sulphite to give a white precipitate. When an aqueous solution of potas-
       sium cyanide is added to a cold aqueous suspension of this precipitate
       and the mixture is subsequently heated under reflux with a concentrated
       alcoholic solution of potassium cyanide, the product, **D**, has the molecular
       formula $C_6H_5ON_3$. **D** on hydrolysis with hot aqueous sodium hydroxide
       and acidification gives a tribasic acid, **E**, of molecular formula $C_6H_8O_7$.

Deduce the structure of **A** and **E** and explain, by means of equations, the formation of the substances **B**, **C** and **D**. Indicate in each case the type of reaction involved.

Outline a synthesis of **A** from propene. What is the industrial importance of this synthesis?

JMB

46* An organic compound gives the following reactions.
(a) With warm water an acidic solution is formed.
(b) With phenol in the presence of a weak base, an ester is formed.
(c) With diethylamine, an N,N-disubstituted amide is formed.
(d) With benzene in the presence of anhydrous aluminium chloride, diphenylmethanone is formed.

Deduce a possible structure for the compound and explain fully its reactions.

Describe **one** simple chemical test which can be carried out to confirm your deductions.

JMB

47* Suggest experiments which would distinguish the compounds in each of the following pairs:
(a) Ethanamide (*acetamide*) and ammonium ethanoate.

(b) Cyclohexene [hexagon structure] and benzene.

(c) Benzenecarboxylic (*benzoic*) acid and phenol.
(d) Ethanol and propan-2-ol.
Comment on points of interest in **two** of these pairs.

Oxford

48* A compound of molecular formula $C_{10}H_7O_4Cl$ is believed to have the structural formula

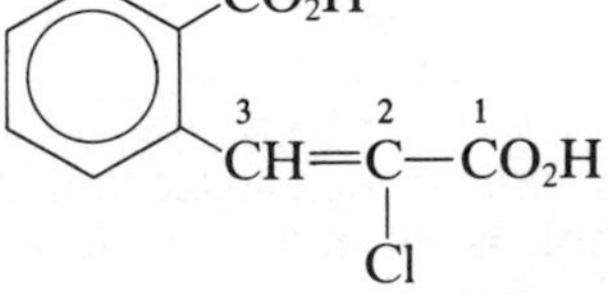

[structure: benzene ring with CO_2H group and side chain $\overset{3}{CH}=\overset{2}{C}-\overset{1}{CO_2H}$ with Cl on carbon 2]

Suggest experiments which may be used to
(a) establish the presence in the molecule of two acidic hydrogen atoms;
(b) confirm the presence of the carbon–carbon double bond in the side chain;
(c) show that the chlorine is positioned
(i) in the side chain of the molecule,
(ii) on the 2-carbon atom relative to the carboxyl group and not on the 3-carbon atom.

How may the above compound be converted into benzene-1,2-dicar-boxylic acid, $C_6H_4(CO_2H)_2$?

London

49*   Account for **four** of the following facts:
  (a) The hydrolysis of halogenoalkanes (*alkyl halides*) is catalysed by bases but not by acids.
  (b) The reaction between alcohols and hydrogen bromide (to give bromoalkanes) is catalysed by acids but not by bases.
  (c) The hydrolysis of esters is catalysed by both acids and bases.
  (d) The product $C_6H_6Cl_6$, obtained by adding 3 mol of chlorine to benzene, consists of a number of isomers.
  (e) Iodoethane (*ethyl iodide*) turns brown when left in a bottle on the bench, whereas bromoethane (*ethyl bromide*) does not.

Oxford & Cambridge

50   Explain fully the principles underlying the following techniques and their use in organic chemistry:
  (a) steam distillation,
  (b) distillation under reduced pressure,
  (c) solvent extraction,
  (d) paper chromatography.

JMB

Index

Information on individual compounds should be found by consulting the contents list for the chapter dealing with the relevant functional group; there is extensive cross-referencing between chapters. Generally only tradenames and popular names of individual chemicals are listed here.